Springer Series in Biophysics 13

Nils G. Walter • Sarah A. Woodson • Robert T. Batey
Editors

Non-Protein Coding RNAs

 Springer

Editors

Dr. Nils G. Walter
Associate Professor of Chemistry
Department of Chemistry
University of Michigan
930 N. University
Ann Arbor, MI 48109-1055
USA

Dr. Sarah A. Woodson
Professor of Biophysics
Department of Biophysics
Johns Hopkins University
3400 North Charles Street
Baltimore, MD 21218
USA

Dr. Robert T. Batey
Associate Professor of Chemistry and
Biochemistry
Department of Chemistry and Biochemistry
University of Colorado at Boulder
Box 215, Boulder, CO 80309-0215
USA

ISSN 0932-2353
ISBN 978-3-540-70833-9   e-ISBN 978-3-540-70840-7

Library of Congress Control Number: 2008931054

Cover design: WMXDesign GmbH, Heidelberg, Germany

Printed on acid-free paper

9 8 7 6 5 4 3 2 1

springer.com

Preface

The 2006 Nobel Prize in Physiology or Medicine was awarded to the discoverers of RNA interference, Andrew Fire and Craig Mello. This prize, which follows "RNA" Nobels for splicing and RNA catalysis, highlights just one class of recently discovered non-protein coding RNAs. Remarkably, non-coding RNAs are thought to outnumber protein coding genes in mammals by perhaps as much as four-fold. In fact, it appears that the complexity of an organism correlates with the fraction of its genome devoted to non-protein coding RNAs. Essential biological processes as diverse as cell differentiation, suppression of infecting viruses and parasitic transposons, higher-level organization of eukaryotic chromosomes, and gene expression are found to be largely directed by non-protein coding RNAs.

Currently, bioinformatic, high-throughput sequencing, and biochemical approaches are identifying an increasing number of these RNAs. Unfortunately, our ability to characterize the molecular details of these RNAs is significantly lacking. The biophysical study of these RNAs is an emergent field that is unraveling the molecular underpinnings of how RNA fulfills its multitude of roles in sustaining cellular life. The resulting understanding of the physical and chemical processes at the molecular level is critical to our ability to harness RNA for use in biotechnology and human therapy, a prospect that has recently spawned a multi-billion dollar industry.

This book assembles chapters from some of the experts in Biophysics of RNA to provide a snapshot of the current status of this dynamic field. While by necessity incomplete, this book aims to survey a number of the better characterized non-protein coding RNAs and the biophysical techniques used to study them. It is written for students and researchers at all levels of accomplishment interested in understanding how non-protein coding RNAs work and how biophysical and computational approaches can be used to delineate the molecular underpinnings of RNA function. Many topics are approached with the goal of describing how biophysical tools and techniques have been used to address fundamental questions in the biology of non-protein coding RNAs, rather than a description of RNAs themselves. In this light, we hope that the book will be of particular use to junior scientists seeking to tackle new problems in RNA biology from the vantage of biophysics.

Following a foreword featuring a general overview of the lessons from the biophysical study of RNA, the first three chapters aim to describe how theory, simulation, and experimental probing can be used to unveil the thermodynamics and kinetics governing RNA folding and dynamics. Chapters 4–6 are devoted to small self-cleaving ribozymes, as understood through the lens of X-ray crystallography, ensemble and single molecule fluorescence, and chemical probing. Subsequent chapters tackle increasingly complex RNAs and their protein complexes. In particular, Chaps. 7–9 focus upon large ribozymes that use more sophisticated mechanisms of catalysis and even recruit proteins to facilitate function in the cellular environment. As genetic regulation appears to be an increasingly important role for non-coding RNAs, Chaps. 10 and 11 concentrate on how X-ray crystallography, NMR spectroscopy, and fluorescence techniques have revealed how riboswitches specifically recognize small molecule metabolites to affect gene expression. Many modern non-protein coding RNAs are assembled into large ribonucleoprotein complexes (RNPs) and Chaps. 12–14 yield insights into how these particles are assembled to form a functional complex. These large RNP machines are by necessity highly dynamic entities that must adopt a number of conformations, as revealed in studies of the ribosome by cryo-electron microscopy in Chap. 15. Finally, non-coding RNAs often interact with other cellular machineries to enable their function, as discussed in Chaps. 16 and 17. We hope that our selection of topics is both timely and stimulating for the rapidly growing RNA community and beyond.

USA   Nils G. Walter
September 2008   Sarah A. Woodson
                                                                                             Robert T. Batey

Contents

Contributors

Robert Batey
Department of Chemistry and Biochemistry, Campus Box 215, University
of Colorado-Boulder, Boulder, CO 80309, USA, Robert.Batey@colorado.edu

Janina Buck
Institut für Organische Chemie und Chemische Biologie, Zentrum für
Biomolekulare Magnetische Resonanz, Johann Wolfgang Goethe-Universität,
Max-von-Laue-Strasse 7, N160-314, 60438 Frankfurt am Main, Germany,
buck@nmr.uni-frankfurt.de

Kathleen Collins
Department of Molecular and Cell Biology, University of California at Berkeley,
Berkeley, CA 94720-3200, USA, kcollins@berkeley.edu

Andrew Feig
Department of Chemistry, Wayne State University, 5101 Cass Ave., Detroit, MI
48202, USA, afeig@chem.wayne.edu

Joachim Frank
Howard Hughes Medical Institute, Department of Biochemistry and Molecular
Biophysics and Department of Biological Sciences, Columbia University, 650
West 168th Street, NY 10032, USA, Joachim@wadsworth.org

Haixiao Gao
Wadsworth Center, Empire State Plaza, Albany, NY 12201-0509, USA,
hxgao@hotmail.com

Michael Harris
Center for RNA Molecular Biology, Department of Biochemistry, CWRU -
School of Medicine, Cleveland, OH 44106, USA, meh2@cwru.edu

Jeffrey Kief
Department of Biochemistry and Molecular Genetics, Denver School of
Medicine, University of Colorado, 12801 East 17th Ave, Rm L18-9110, Aurora,
CO 80045, USA, Jeffrey.Kief@ucdenver.edu

Jamie LeBarron
Wadsworth Center, Empire State Plaza, Albany, NY 12201-0509, USA,
jslebarron@yahoo.com

Neocles Leontis
Department of Chemistry, Bowling Green State University, 141 Overman Hall,
Bowling Green, OH 43403, USA, leontis@bgsu.edu

Kyoshi Nagai
Structural Studies Division, MRC Laboratory of Molecular Biology, Hills Road,
Cambridge CB2 2QH, UK, kn@mrc-lmb.cam.ac.uk

Jonas Noeske
Institut für Organische Chemie und Chemische Biologie, Zentrum für
Biomolekulare Magnetische Resonanz, Johann Wolfgang Goethe-Universität,
Max-von-Laue-Strasse 7, N160-314, 60438 Frankfurt am Main, Germany,
noeske@nmr.uni-frankfurt.de

Anna Marie Pyle
266 Whitney Avenue, Room 334A Bass Building, Yale University, New Haven,
CT 06511, USA, anna.pyle@yale.edu

Tariq Rana
Department of Biochemistry and Molecular Pharmacology, University of
Massachusetts Medical School, Worcester, MA 01605, USA,
tariq.rana@umassmed.edu

Harald Schwalbe
Institut für Organische Chemie und Chemische Biologie, Zentrum für
Biomolekulare Magnetische Resonanz, Johann Wolfgang Goethe-Universität,
Max-von-Laue-Strasse 7, N160-314, 60438 Frankfurt am Main, Germany,
schwalbe@nmr.uni-frankfurt.de

William Scott
Department of Chemistry, University of California, 1156 High Street, Santa Cruz,
CA 95064, USA, wgscott@chemistry.ucsc.edu

Garrett Soukup
Department of Biomedical Sciences, Creighton University School of Medicine,
2500 California Plaza, Omaha, NE 68178, USA, garrettsoukup@creighton.edu

Devarajan (Dave) Thirumalai
Department of Chemistry and Biochemistry, University of Maryland, College
Park, MD 20742, USA, thirum@umd.edu

Ignacio Tinoco
Department of Chemistry, University of California, Berkeley, CA 94720-1460,
USA, intinoco@lbl.gov

Olke Uhlenbeck
Department of Biochemistry, Molecular Biology and Cell Biology, Hogan 2-100,
2205 Tech Drive, Evanston, IL 60208, USA,
o-uhlenbeck@northwestern.edu

Nils Walter
Department of Chemistry, University of Michigan, Ann Arbor, 930 N. University,
MI 48109-1055, USA, nwalter@umich.edu

Jens Wöhnert
Institut für Molekulare Biowissenschaften, Zentrum für Biomolekulare
Magnetische Resonanz, Johann Wolfgang Goethe-Universität,
Max-von-Laue-Strasse 9, N200-2.04, 60438 Frankfurt am Main, Germany,
woehnert@bio.uni-frankfurt.de

Sarah Woodson
T.C. Jenkins Department of Biophysics, Johns Hopkins University, 3400 N.
Charles St., Baltimore, MD 21218, USA, swoodson@jhu.edu

Chapter 1
RNA 3D Structural Motifs: Definition, Identification, Annotation, and Database Searching

Lorena Nasalean, Jesse Stombaugh, Craig L. Zirbel, and Neocles B. Leontis(⊠)

Abstract Structured RNA molecules resemble proteins in the hierarchical organization of their global structures, folding and broad range of functions. Structured RNAs are composed of recurrent modular motifs that play specific functional roles. Some motifs direct the folding of the RNA or stabilize the folded structure through tertiary interactions. Others bind ligands or proteins or catalyze chemical reactions. Therefore, it is desirable, starting from the RNA sequence, to be able to predict the locations of recurrent motifs in RNA molecules. Conversely, the potential occurrence of one or more known 3D RNA motifs may indicate that a genomic sequence codes for a structured RNA molecule. To identify known RNA structural motifs in new RNA sequences, precise structure-based definitions are needed that specify the core nucleotides of each motif and their conserved interactions. By comparing instances of each recurrent motif and applying base pair isostericity relations, one can identify neutral mutations that preserve its structure and function in the contexts in which it occurs.

1.1 Defining Motifs at Different Levels of Structure

Defining and identifying recurrent modular motifs in 3D structures and developing bioinformatic methods to find them in sequences will improve RNA gene finding and RNA 3D structure prediction. In 2005, the RNA Ontology Consortium (http://roc.bgsu.edu/) was created as an umbrella organization to convene and coordinate working groups to reach scientific consensus on the best ways to define, classify and annotate RNA structural motifs for bioinformatics applications, including (1) identifying RNA genes in genomic sequences; (2) predicting their secondary structures from sequence and readily obtainable experimental data; (3) inferring their function(s); and (4) modeling their three-dimensional structures (Leontis et al. 2006).

N.B. Leontis
Department of Chemistry, Bowling Green State University, 141 Overman Hall, Bowling Green, OH 43403, USA
e-mail: leontis@bgsu.edu

N.G. Walter et al. (eds.) *Non-Protein Coding RNAs*
doi: 10.1007/978-3-540-70840-7_1, © Springer-Verlag Berlin Heidelberg 2009

This is an area of active research in which a variety of approaches are being investigated (Leontis and Westhof 2003; Leontis et al. 2006). For comprehensive discussions of new RNA 3D structures and motifs and their functional roles the reader is referred to recent reviews (Hendrix et al. 2005; Holbrook 2005).

1.1.1 Hierarchical Architectures and Folding of Structured RNA Molecules

Like proteins, RNA molecules fold hierarchically in time and space to form specific 3D structures necessary for molecular function. Local secondary structure elements – primarily short helices capped by hairpin (terminal) loops – form in the first stages of folding. In subsequent folding stages, these elements coalesce into local domains composed of helical elements organized by multi-stem junctions. Some of these helices are formed by complementary sequences distant in the RNA sequence. In the final, slowest stages of folding, the native, compactly folded tertiary structure is produced, as the correct tertiary interactions are established between structural domains (Thirumalai and Woodson 1996; Zhuang et al. 2000; Thirumalai et al. 2001; Rangan et al. 2003). While RNA 3D structures can be very large and complex, they are hierarchical and modular. As is the case for proteins, the global structures of RNA molecules change more slowly than their sequences or secondary structures. These features help us to analyze and understand them.

1.1.2 Defining the Modular Units of RNA Structure

We gain a better understanding of RNA structures by identifying modular subunits of structure and their interactions at each hierarchical level of organization.

Primary Sequence. At the level of the sequence, the modular subunits are individual nucleotides, covalently linked 5'-to-3' by phospho-diester bonds. Each nucleotide consists of three chemical moieties – the base, the sugar and the phosphate. When RNA molecules fold, the nucleotides interact with each other in characteristic ways. The most specific and best understood interactions involve the bases – base–base, base–sugar, and base–phosphate interactions. Base–base interactions include edge-to-edge pairing interactions mediated by hydrogen-bonding, face-to-face stacking interactions and (rare) edge-to-face perpendicular interactions. Sugar–sugar, sugar–phosphate, and metal- or solvent-mediated phosphate–phosphate interactions also occur and contribute to the stability of complex RNA structures, but are harder to classify and to relate to sequence information. Although the sugar-phosphate backbone of RNA is very flexible, it is possible to classify the observed conformations of nucleotides and dinucleotides in discrete, recurrent patterns that can be associated with certain motifs or sub-motifs (Sykes and Levitt 2005; Richardson et al. 2008).

Watson–Crick Helices and Secondary Structure. Single-stranded RNA molecules fold back on themselves to juxtapose Watson–Crick complementary sequence in an anti-parallel fashion. This produces Watson–Crick helices, the fundamental modular units of secondary structure. The helices are composed of the canonical Watson–Crick (WC) basepairs, AU, UA, CG, and GC, as well as "wobble" GU and UG pairs; the Watson–Crick basepairs are the modular subunits of secondary structure and they stack on each other in a regular, recurrent way. The helices are generally short (no longer than 10–15 WC pairs) because they are interrupted or terminated at their ends by nominally unpaired stretches of sequence that are called, depending on where they occur in the secondary structure, hairpin, internal, or multi-helix junction "loops." Bases in loops are usually depicted in secondary structures as not forming basepairs. In general, about 60% of the nucleotides of a structured RNA form Watson–Crick basepairs.

RNA 3D Structures. The 3D structures of a relatively small number of RNA molecules are determined to atomic resolution by X-ray crystallography or NMR spectroscopy each year. Although small in size compared to the 3D protein database and all the RNAs known from genomes, the RNA 3D structure database has expanded rapidly in the last few years. This data shows that most "loops" in 2D representations in fact form specific *3D motifs*, characterized by non-Watson–Crick base-pairing, base-stacking and base-phosphate interactions between loop nucleotides. For example, in a survey of the 3D structures of rRNA in the 70S ribosomes of *E. coli* and *T. thermophilus* and the 50S subunit of *H. marismortui*, only ~59% of bases form standard WC basepairs, and ~7% make, in addition, at least one non-WC base pair (Stombaugh et al. submitted). Of the rRNA bases, ~20% form one or more non-WC basepairs but no WC pair while ~21% do not base pair at all. However, most of the unpaired bases participate in base-stacking, base–phosphate, or RNA–protein interactions. Thus, the loops comprise a significant fraction of the nucleotides of structured RNA molecules and most of these nucleotides interact with other nucleotides, proteins or ligands.

1.1.3 Modular and Recurrent 3D Motifs

Modular 3D Motifs. Most 3D motifs are flanked by WC basepairs and they are modular in the sense that they can be attached to or inserted within any double helix and still form the same 3D structure. These observations suggest the following general definition: "Modular RNA 3D motifs are autonomous sets of interacting nucleotides that form a defined 3D structure." This definition distinguishes structural motifs from sequence motifs and full motifs from sub-motifs and emphasizes the physical interactions of the nucleotides rather than the sequence identity of each nucleotide. While the Watson–Crick helix is the most important RNA 3D motif, here we will focus on motifs that comprise non-Watson–Crick basepairs. When one or more flanking Watson–Crick pairs form tertiary interactions with the "loop" nucleotides of the motif, they are best considered part of the motif. For example, in the C-loop motif, both flanking WC pairs form base triples with the nucleotides of the C-loop (Lescoute et al. 2005). Even when the flanking basepairs do not form base-triples, they usually interact with nucleotides of the 3D motif by stacking. Therefore it is not surprising that the flanking basepairs are often conserved or show a strong statistical preference. Thus the flanking base pair for UNCG hairpin loops is usually a *cis* Watson–Crick CG base pair (abbreviated "cWW" – see below) and the flanking base pair in the eleven nucleotide GAAA loop-receptor is *cis* Watson–Crick (cWW) GU (Cate et al. 1996a). Therefore, we generally include the flanking cWW pairs in the 3D motif.

Recurrent 3D Motifs. Many 3D motifs are recurrent. Homologous RNA molecules usually contain the same motifs at corresponding positions in their structures as a result of evolutionary conservation. Recurrent motifs also occur in unrelated RNA molecules (or at non-equivalent positions of homologous molecules) as a result of convergent evolution. There are instances of the same recurrent motif sharing a set of core nucleotides that can be superposed in 3D space; each core nucleotide bears the same relationship to neighboring nucleotides as do the equivalent nucleotides in the other instances of the motif. Thus two helices of the same length are instances of the same motif because they can be superposed base-by-base, with equivalent bases in each helix base pair and stacked in geometrically similar ways. Instances also exist of a recurrent motif having common base pairing and base-stacking interactions but differing significantly in sequence or in strand topology. The (generally unknown) set of all sequences that form a particular 3D motif is its "*sequence signature*." When we speak of a recurrent RNA 3D motif we are actually talking about all the different sequence variants that can form the same 3D structure and carry out similar functions.

Sequence differences can result from base substitutions or from insertions or deletions. When comparing two structures, insertions in the first structure relative to the second structure appear as deletions in the second relative to the first, so we refer to insertions and deletions collectively as "indels." Due to the flexibility of the RNA backbone, even large indels can be accommodated at certain positions to produce different versions of what is essentially the same motif. Comparison of different instances of recurrent motifs can help us understand the sequence variations compatible with the 3D structure, and thus facilitate the identification of motifs when all we have are RNA sequences. This is an important step in predicting RNA 3D structures and improving our ability to find non-coding RNA genes in genomes.

The take-home message from the 3D data is that to precisely define the 3D motifs of each hairpin, internal and multi-helix junction loop, the conserved interactions between motif nucleotides must be identified and classified.

1.1.4 Neutral Substitutions in Helices

The 3D structures of RNA double helices are very regular and largely independent of sequence, owing to the remarkable *isostericity* of the canonical *cis* Watson–Crick basepairs AU, UA, GC, and CG. "Isosteric" means "occupying the same space" and in the context of base-pairing, refers to the space between the sugar–phosphate backbones of the interacting strands of the helix. Because the canonical *cis* WC basepairs are isosteric, they can substitute for each other in RNA double helices without perturbing its structure. The key observation is that the RNA helix is defined by the type of interactions between the nucleotides, and not the specific sequence. It is usually not meaningful to speak of a "consensus" sequence for a helix because structure–neutral mutations can substitute one Watson–Crick base pair for another. The isostericity of the canonical *cis* WC basepairs is the physical basis for the comparative approach to RNA sequence analysis which led to accurate predictions of the secondary structures of large RNAs long before their 3D structures were determined (Pace et al. 1999). Of course, the exact thermodynamic stabilities of helices are sequence dependent due to variations in base-pairing and base-stacking free energies (Mathews and Turner 2006). Also, if the specific helix forms tertiary RNA interactions or binds a protein or other ligand, there may be additional base-specific constraints on the sequence. In fact many Watson–Crick basepairs in structured RNAs like the 16S and 23S rRNAs are very conserved, and this conservation correlates with the occurrence of specific tertiary RNA or RNA–protein interactions (Stombaugh et al. in preparation).

This idea of structure–neutral isosteric substitutions can be fruitfully applied to non-Watson–Crick basepairs- the basic building blocks of RNA 3D motifs- as explained in the next section.

1.2 Identifying, Classifying and Annotating Nucleotide Interactions that Stabilize RNA 3D Motifs

1.2.1 Reduced Representations of RNA 3D Structure

Atomic resolution 3D structures from X-ray crystallography provide detailed descriptions of RNA 3D structures and motifs in the form of sets of Cartesian coordinates for each atom. However, this description is too detailed for many applications, and in any case, the reported precision of crystallographic data, to thousandths of an Ångstrom, is misleading. To make the RNA structural data useful to bioinformatics applications, reduced representations of RNA structure are needed that capture the nature of the conserved interactions between the core nucleotides of each 3D motif. The interactions that interest us most are those that constrain the sequence and can therefore be used to identify motifs in genomic sequences. These interactions directly involve the bases. A lot of attention has been paid to classifying and annotating the non-WC basepairs, as they are the recurrent modular subunits of RNA 3D motifs, just as the Watson–Crick basepairs are for double helices. The crucial issue a classification should address is which basepairs substitute for each other in structure–neutral ways, without significantly perturbing the 3D structure of the motif.

1.2.2   Classification and Annotation of Base-Pairing Interactions in RNA Structures

RNA bases, purines and pyrimidines, present three edges for hydrogen-bonding interactions with other bases, the Watson–Crick, the Hoogsteen, and the Sugar Edges, illustrated for adenosine (A) in the left panel of Fig. 1.1 (Leontis and Westhof 2001). For pyrimidines, the Hoogsteen Edge is also called the "CH" edge. The RNA Sugar Edge includes the 2′-hydroxyl, a functional group that distinguishes RNA from DNA and plays an important role in RNA tertiary interactions and RNA chemistry. Bases can pair using any of the six combinations of the three edges, for example, the Watson–Crick Edge of one base with the Watson–Crick, Hoogsteen, or Sugar Edge of a second base. In addition, for each combination of edges, the bases can approach each other in two orientations, which are called *cis* and *trans*, by analogy to the geometric isomerism at carbon–carbon double bonds. As shown in the right panel of Fig. 1.1, in *cis* basepairs, the glycosidic bonds joining the bases to their respective sugar moieties are found on the same side of the axis shown in grey. This axis is defined by the hydrogen bonds joining the base edges. In the *trans*

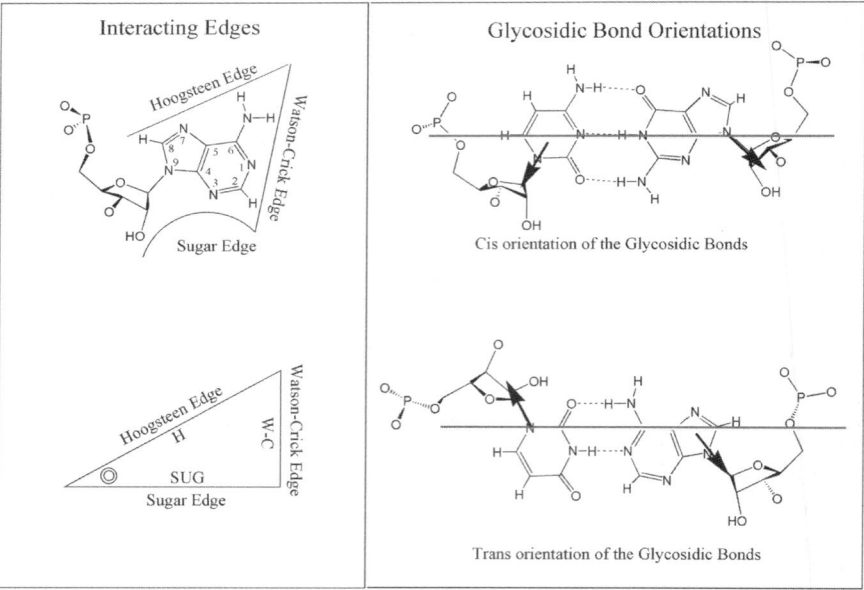

Fig. 1.1 Base edges and base-pair geometric isomerism. (*Upper left*) The structure of adenosine showing the three base edges (Watson–Crick, Hoogsteen and Sugar-edge) available for hydrogen–bonding interactions. (*Lower left*) Representation of RNA base as a triangle (see also Fig. 1.2). The position of the ribose is indicated by a circle in the corner defined by the Hoogsteen and Sugar edges. (*Right*) cis and trans base-pairing geometries, illustrated for two bases interacting with Watson–Crick edges (Leontis and Westhof 2001)

Table 1.1 The 12 geometric basepair families

No.	Glycosidic bond orientation	Interacting edges		Abbreviation	Symbol	Default local strand orientation
		NT1	NT2			
1	Cis	Watson–Crick	Watson–Crick	cWW		Anti-parallel
2	Trans	Watson–Crick	Watson–Crick	tWW		Parallel
3	Cis	Watson–Crick	Hoogsteen	cWH		Parallel
		Hoogsteen	Watson–Crick	cHW		
4	Trans	Watson–Crick	Hoogsteen	tWH		Anti-parallel
		Hoogsteen	Watson–Crick	tHW		
5	Cis	Watson–Crick	Sugar Edge	cWS		Anti-parallel
		Sugar Edge	Watson–Crick	cSW		
6	Trans	Watson–Crick	Sugar Edge	tWS		Parallel
		Sugar Edge	Watson–Crick	tSW		
7	Cis	Hoogsteen	Hoogsteen	cHH		Anti-parallel
8	Trans	Hoogsteen	Hoogsteen	tHH		Parallel

Table 1.1 (continued)

No.	Glycosidic bond orientation	Interacting edges		Abbreviation	Symbol	Default local strand orientation
		NT1	NT2			
9	Cis	Hoogsteen	Sugar Edge	cHS		Parallel
		Sugar Edge	Hoogsteen	cSH		
10	Trans	Hoogsteen	Sugar Edge	tHS		Anti-parallel
		Sugar Edge	Hoogsteen	tSH		
11	Cis	Sugar Edge (Priority)	Sugar Edge	cSs		Anti-parallel
		Sugar Edge	Sugar Edge (Priority)	csS		
12	Trans	Sugar Edge (Priority)	Sugar Edge	tSs		Parallel
		Sugar Edge	Sugar Edge (Priority)	tsS		

Each family is specified by the relative orientation of the glycosidic bonds (column 2) and the interacting edges of the bases (columns 3 and 4). Abbreviations and corresponding symbols for annotating basepairs in diagrams are given in columns 5 and 6. Column 7 defines the default local strand orientations for each base-pair family when both bases are in the default anti-configuration of the glycosidic bonds

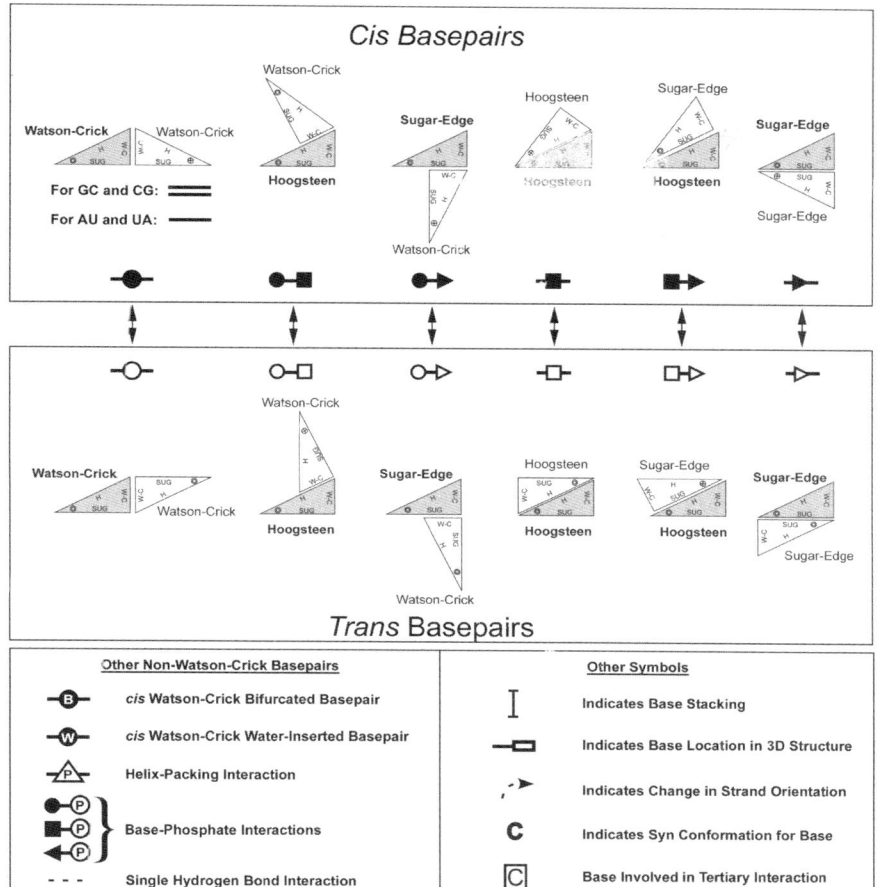

Fig. 1.2 Schematic representations of geometric families and symbols for annotating structures. *Upper panel:* The 12 geometric base pair families are shown using triangles to represent bases. Circles represent Watson–Crick edges, squares, Hoogsteen edges, and triangles, Sugar edges. Base pair symbols are composed by combining edge symbols, with solid symbols indicating *cis* basepairs and open symbol, *trans* basepairs. *Lower Left:* Symbols for other pairwise interactions. *Lower Right:* Additional symbols for base-stacking, reversal of chain direction in hairpin loops, *syn* bases, and bases forming tertiary interactions (Leontis et al. 2002)

orientation, the glycosidic bonds are on opposite sides of this axis. Thus, there are 12 basic geometric families of basepairs in RNA. Information regarding the base pair families, their abbreviations, and symbols for representing them in secondary structures are collected in Table 1.1. Each geometric family is shown schematically in the upper panel of Fig. 1.2, using right triangles to represent each base (Leontis and Westhof 2001; Leontis et al. 2002). The hypotenuse of each triangle represents the Hoogsteen Edge of the base. Circles or crosses are placed in the corner of the triangle defined by the Hoogsteen and Sugar Edges to indicate the direction of the sugar–phosphate backbone in the default case where all glycosidic bonds are in the *anti* configuration. A circle represents the sugar–phosphate backbone emerging 5′ to 3′ out of the plane toward the reader and the cross represents the opposite orientation (Leontis and Westhof 2001, 2002). The six basepairs in *cis* are shown in the upper half of Fig. 1.2 and the six basepairs in *trans* are shown immediately below the respective *cis* basepairs. Each of the 12 geometric base pair types is represented by a symbol to unambiguously annotate that pair in secondary structure diagrams, as described below (Leontis and Westhof 2001).

The *cis/trans* distinction for basepairs should not be confused with the designations *syn* and *anti* of rotational isomers of individual nucleotides that result from the rotation of the base about the glycosidic bond connecting the base to the sugar moiety.

Abbreviations. The geometric base pair families are abbreviated "cWW" for *cis* Watson–Crick/Watson–Crick, "tHS" for *trans* Hoogsteen/Sugar Edge, and so on, as summarized in Table 1.1. The cHH family is very rare and usually occurs with one nucleotide in the *syn* configuration of the glycosidic bond to minimize steric clash between the backbones of the interacting nucleotides. The cHS family usually occurs between adjacent nucleotides in the same strand to form platform motifs. The cWW, tWW, and tHH basepairs are generally symmetric – interchanging the bases produces equivalent basepairs – but the cSS and tSS pairs are not symmetric, so annotations are needed that reflect their asymmetry. In cSS pairs, the nucleotide that hydrogen bonds with its 2′-OH to both the 2′-OH and the base of the other nucleotide is assigned higher priority in the interaction and is indicated with an upper-case letter while the other nucleotide is indicated with a lower-case letter (i.e., cSs) Thus, the base pair shown in the lower right panel of Fig. 1.9, is an A/G cSs pair, as the A has higher priority than the G. For tSS basepairs, higher priority is assigned to the base that forms an H-bond with the 2′-OH of the other nucleotide, in addition to the base-to-base H-bonds (see Fig. 1.9).

Evidence Supporting the Triangle Abstraction: Base Triples and Quadruples. How realistic is the abstraction of RNA bases as triangles? It implies that a single RNA base can interact edge-to-edge in the same plane with up to three different bases, so as to produce base quadruples. Symbolic searching using the "Find RNA 3D" ("FR3D") RNA motif search program (Sarver et al. 2008) shows at least ten different base quadruples of this type, consistent with prediction (Nasalean et al. in preparation). Figure 1.3 shows an example of one of these quadruples from 16S rRNA (PDB: 1j5e), where the center base, G68 (blue), forms a cWW base pair with A101 (magenta), a tsS pair with A152 (orange), and a cHW pair with G64 (red).

Many different base triples and quadruples occur in RNA structures. As for the base quadruple in Fig. 1.3, almost all base triples and quadruples can be decomposed into combinations of the 12 geometric base pair families. In this way, it is straight-forward to classify these higher order groupings (Nasalean et al. in preparation). Most base triples comprise a central base interacting with two other bases using two distinct edges. However, a second type of base triple is also possible, in which one base, usually a purine, pairs with two other bases using the same edge-usually its Sugar Edge. This case is very frequent in tertiary interactions involving the minor groove. An example of such an interaction, which is also called a Type I A-minor motif in the literature, will be discussed below.

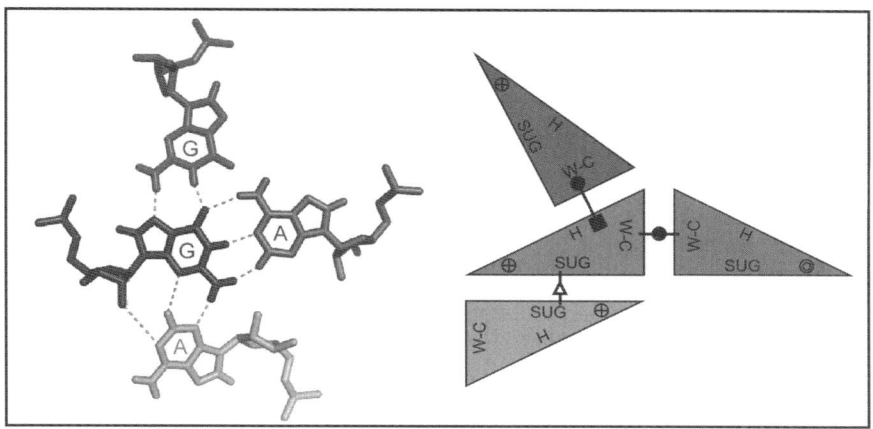

Fig. 1.3 Triangle abstraction for RNA bases. As implied by the triangle abstraction, RNA bases can interact with three different bases using their three edges, Watson–Crick (WC), Hoogsteen (H) and Sugar (Sug), forming "saturated" base quadruples. (*Left*) Example of a base quadruple of this type from *T. thermophilis* 16S rRNA (PDB file 1j5e) in which G68 (blue) forms a cWW pair with A101 (magenta), a cHW pair with G64 (red) and a tsS pair with A152 (yellow). The green dotted lines indicate Hydrogen bonds. (*Right*) Schematic representation showing each base as a triangle with edges labeled. The base-pairing type is given using the symbols from Fig. 1.2 (See figure insert for color reproduction)

Bifurcated and Water-Inserted Pairs. If one also allows for bifurcated and solvent-inserted pairs to extend the 12 base pair families, then the vast majority of basepairs can be classified within this framework (Leontis et al. 2002). In solvent-inserted basepairs, the base pair opens while maintaining one direct H-bond between the bases to allow a small molecule, usually water but sometimes an ion, to be inserted. The inserted molecule mediates additional interactions between the base edges.

Bifurcated basepairs involve H-bonds between an exocyclic functional group (amino or carbonyl oxygen) of one base and the base edge of the second base and can be accommodated in the framework of the 12 base pair families in the following way: The 12 families form two distinct groups of six each. Within each group, the six families are related by ~90° rotations in the base pair plane of one base relative to the other, without flipping either base. These rotations transform one base pair into another within each family by changing one interacting edge at a time. For example a cWW pair can be transformed in one step into a cWS, cSW, tWH or tHW pair, but not a cSS, tSH, tHS, or cHH pair. A 3 × 3 matrix represents each group of basepairs as shown in Table 1.2. The basepairs in neighboring horizontal or vertical cells in each matrix can be transformed by rotating one base with respect to the other ~90° without leaving the plane. Bifurcated basepairs result when this rotation is incomplete, so that an exocyclic functional group, G(O6), U(O6), A(N6), or C(N6) in the corner between the Watson–Crick and Hoogsteen edges, or G(N2), U(O2), or C(O2) in the corner between the Watson–Crick and Sugar Edges, interacts with one of the edges of the second base. The most common case involves the WC/H corner of one base and the WC edge of the second base. The bifurcated and

Table 1.2 Spatial relationship between the geometric base-pair families

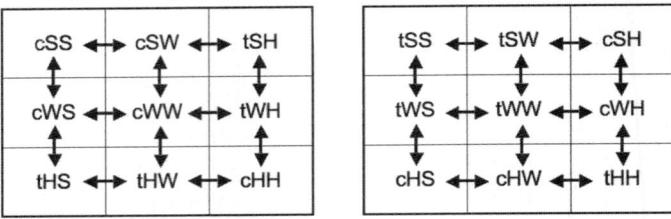

The geometric families form two distinct groups. Within each group, base pair types can be transformed by a ~90° rotation of one base in the base pair plane, changing one interacting edge at a time as shown by arrows connecting base pair families

water-inserted basepairs have been described previously in more detail (Leontis and Westhof 1998; Leontis et al. 2002; Auffinger and Hashem 2007).

1.2.3 Annotation of Secondary Structures

Annotations for 2D diagrams have been developed to communicate essential features of RNA 3D structures accurately and succinctly. In addition to the classical secondary structure, the annotations show (1) all non-Watson–Crick basepairs with unique symbols that specify the geometric family of the base pair; (2) all bases that are in the *syn* glycosidic configuration; (3) all points in the chain where the backbone reverses direction; (4) key base-stacking and base-phosphate interactions; and (5) sequential numbering of nucleotides in the 5′-to-3′ direction. Annotations of Group I introns, 16s rRNA and many aptamers and small ribozymes have been published (Adams et al. 2004; Lescoute and Westhof 2006a).

Annotation of BasePairs. The base pair symbols are derived in a simple way by associating a different symbol with each edge: the circle • with the Watson–Crick Edge, the square ■ with the Hoogsteen Edge and the triangle ▲ with < the Sugar Edge. Solid symbols indicate *cis* basepairs and open symbols *trans* basepairs. For bases pairing with different edges, the symbols indicate the edge used by each base to form the pair. When the same edge is used by both bases, the base pair type is indicated by a single symbol, filled or open, placed on a line joining the letters designating the bases. The base pair symbols with their respective pairing types are also shown in Fig. 1.2 and compiled in Table 1.1 and are used throughout this chapter to annotate diagrams representing 3D RNA motifs.

Symbols in common use that do not conflict with the new conventions can still be used,- notably "–" for AU or UA and " = " for GC or CG. "Wobble" GU or UG, being a type of cWW, is designated with a filled circle, •, not an open circle, to avoid confusion with *trans* basepairs. When only one hydrogen bond occurs between two bases or sugar atoms, a dashed line is used to denote the interaction. To denote cWW bifurcated or cWW water-inserted pairs, the letter "B" or "W" is added to the filled circle used to represent cWW pairs, as shown in the lower left corner of Fig. 1.2.

Helix Packing Interactions. Two nucleotides can interact by the interlocking of the Sugar Edges of two nucleotides without direct contact between the bases *per se.* This has been variously called "A-minor type 0" or "helix packing" motif and can be designated using the letter "P" placed in an open triangle (Nissen et al. 2001; Gagnon and Steinberg 2002; Mokdad et al. 2006).

Base-Stacking. Two RNA bases can stack face-to-face in four different ways, depending on the base faces that come in contact, the 5′-face or the 3′-face of each base. The 5′- and 3′-faces are defined by reference to the normal orientation of each base in the Watson–Crick helix, in which all bases are in the anti-glycosidic conformation; the 5′-face points toward the 5′-end of the strand and the 3′-face toward the 3′-end of the strand (Sarver et al. 2008). To show that two adjacent bases in the RNA chain are stacked, the letters representing them are drawn right above or below each other in the secondary structure. If one base is bulged out and not stacked on its neighbors in the chain, it is drawn to one side. In some motifs, "cross-strand stacking" occurs between bases in the same motif but on opposite strands. This can be indicated with an "I-beam" connecting the two stacked bases. When the stacked bases are far apart, one base can be represented by a rectangle placed above or below the base on which it stacks, and connected by a line to the letter representing it in the secondary structure.

Base–phosphate interactions are hydrogen bonds between the WC, Hoogsteen, or Sugar edges of a base and the phosphate oxygen atoms of a second nucleotide. Base–phosphate interactions are indicated by symbols comprising a circle containing a "P" to indicate the phosphate, connected by a line to a circle, square or triangle to indicate the interacting edge (Watson–Crick, Hoogsteen or Sugar) of the base. Different classes of the base–phosphate interactions can be proposed depending on the specific base and base edge interaction with the phosphate (Stombaugh et al. in preparation).

Additional Annotations. Some other descriptive symbols are used to denote changes in strand orientation (dashed line arrow or red solid line arrow) or to show that a base is in the syn conformation (bold nucleotide letter or red letter). A box is placed around nucleotides that participate in tertiary interactions and the box is connected with the appropriate interaction symbol to the interacting base(s).

1.2.4 Structure–Neutral Mutations in Recurrent RNA 3D Motifs

Structure–Neutral Mutations. Mutations in RNA sequence that disrupt the 3D structure of a functionally important motif are less likely to be passed on to subsequent generations as a result of the evolutionary process of natural selection. This is because the function of a molecule depends on its ability to fold into the functionally active 3D structure. Mutations that preserve 3D structure are called structure–neutral mutations. Two kinds of mutations need to be considered: substitutions and insertions or deletions (indels).

Insertions and Deletions. Indels can be structure neutral, depending on where they occur in a motif. A consequence of the high flexibility of the RNA backbone is that even a single nucleotide can be bulged out of an RNA motif without significantly perturbing its 3D structure. Sites that can accommodate one such insertion often allow two or more, as long as they do not interfere by steric clash with tertiary interactions the motif must form. Mutations that disrupt the structure of the motif and consequently impair its function will be selected against. By comparing instances of the same motif in 3D structures we can determine the nucleotide positions that tolerate insertions and thus improve our ability to predict the motif from sequences. This idea will be illustrated below for hairpin loop motifs.

Base Substitutions and Base Pair Isostericity. Base substitutions for basepairs are structure–neutral when they result in isosteric basepairs. The geometric basepair classification groups isosteric basepairs in the same geometric families. Basepairs from different geometric families are not isosteric. However, not all basepairs in the same geometric family are isosteric. Rather, each family comprises one or more subsets of isosteric basepairs (Leontis et al. 2002). This is illustrated in Fig. 1.4. Two basepairs are isosteric when they meet the following three criteria: (1) The C1′–C1′ distances are the same; (2) the paired bases are related by the same rotation in 3D space; and (3) H-bonds form between equivalent base positions. The cWW GC, CG, and AU basepairs (upper and lower left and upper center of Fig. 1.4) meet all three criteria and are isosteric to each other, as shown. The cWW AG pair (lower center) and GU pair (upper right) are in the same geometric family and so the paired bases are related by the same 3D rotation. However, the cWW AG pair has

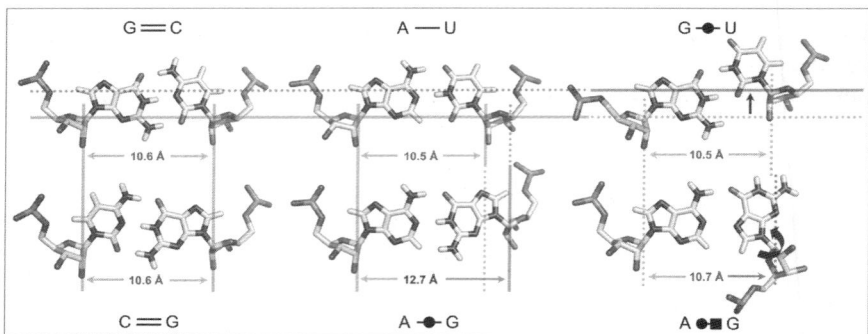

Fig. 1.4 Isosteric relationships between basepairs. Two basepairs are isosteric when they meet three criteria: (1) The C1′–C1′ distances are the same; (2) the paired bases are related by same rotations in 3D space; and (3) H-bonds are formed between equivalent base positions. The cWW GC, CG, and AU basepairs (upper and lower left and upper center) meet all three criteria and are isosteric to each other, as shown. The cWW AG pair (*lower center*) and GU pair (*upper right*) belong to the same geometric family and so the paired bases are related by the same 3D rotation. However, the cWW AG pair has a significantly longer C1′–C1′ distance (12.7 Å) and so is not isosteric to the other pairs, even though it meets the other two criteria. The C1′–C1′ distance in the cWW GU (wobble) pair is about the same, but the U is shifted toward the major groove, so H-bonding does not occur between the same positions as in the other cWW pairs. This change is more subtle and so GU is considered near isosteric to the canonical cWW pairs AU, UA, GC, and CG, consistent with its ability to substitute in Watson–Crick helices for these pairs. The last example, cWH AG (*lower right*), has about the same C1′–C1′ distance as the canonical cWW pairs, but belongs to a different geometric family. The bases are related by a very different 3D rotation so it is not isosteric or near isosteric to any of the cWW basepairs (See figure insert for colour reproduction)

a significantly longer C1'–C1' distance (12.7 Å) and so is not isosteric to the other pairs, even though it meets the other two criteria. While the C1'–C1' distance in the cWW GU (wobble) pair is about the same as for the GC, CG, AU, and UA (i.e., the canonical cWW pairs), the U in the GU pair is shifted toward the major groove so H-bonding does not occur between equivalent atomic positions compared to the canonical cWW pairs. This change is more subtle and so GU is considered *near isosteric* to the canonical cWW pairs AU, UA, GC, and CG, consistent with its ability to substitute for these pairs in Watson–Crick helices. The last example, cWH AG (Fig. 1.4, lower right), has about the same C1'–C1' distance as the canonical cWW pairs, but belongs to a different geometric family. The bases are related by a very different 3D rotation so it is not isosteric or near isosteric to any of the cWW basepairs shown in Fig. 1.4. For each of the 12 geometric families, isosteric and near isosteric subgroups have been identified (Leontis et al. 2002). These are applied to predict structure–neutral substitutions in 3D motifs, to supplement observed instances from the 3D database. The isosteric relations are summarized in Isostericity Matrices, as will be illustrated below (Sect. 1.3.2).

1.3 Defining Recurrent 3D Motifs and Identifying Them in Structures

Concise definitions of 3D motifs are needed to automatically search for them in 3D structures and to formulate algorithms to find them in RNA sequences. The definitions should be sufficiently precise to differentiate motifs with similar structures.

1.3.1 Classification of "Loop" Motifs

The "loop" motifs of secondary structure are classified according to their locations: (1) *Hairpin* (or terminal) *loops* are positioned at the ends of helices, (2) *internal loops*, are located within (or between) helices, and (3) *multi-helix junction loops* join three or more helices. Loop motifs have been further classified according to the number of nucleotides they contain: Hairpin loops have been classified as tri-loops (Lee et al. 2003; Lisi and Major 2007), tetra-loops (Woese et al. 1990), penta-loops (Stefl and Allain 2005) and so on, and internal loops as symmetric internal loops (2 × 2, 3 × 3, etc.) or asymmetric internal loops (1 × 2, 1 × 3, 2 × 3, etc.) depending on the number of nucleotides in the component strands. Likewise, junction loops have been classified according to the number of helices (3-way, 4-way or higher order junctions) and that of nominally unpaired bases linking the helices to each other (Altona 1996; Gan et al. 2004). These numerical classifications, however, can be misleading. On the one hand, nucleotides can be inserted or deleted at certain positions in motifs without significantly perturbing the rest of the structure. On the other hand, sequences of the same length can fold in very different ways. These effects are due to the high flexibility of the RNA backbone, and the sequence specific

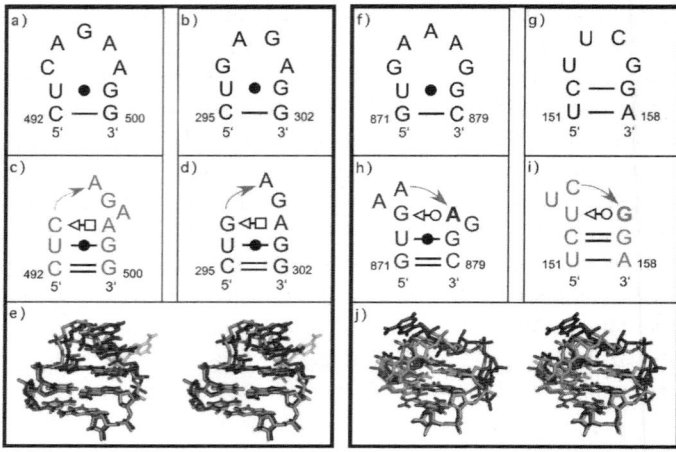

Fig. 1.5 Structurally similar hairpin loop motifs. **a–e** Comparison of sequence and structure annotations of two geometrically similar hairpin loop motifs, only one of which is a tetraloop and conforms to the consensus sequence "GNRA." **f–j** Comparison of sequence and structure annotations of geometrical hairpin loop motifs, where only one is a tetraloop and conforms to the consensus sequence "UNCG." **e** Stereo superpositions of motifs in **c** and **d**. **j** Stereo superpositions of motifs in **h** and **i** (See figure insert for colour reproduction)

folding of RNA. Consequently, the number of nucleotides in a "loop" is not a robust criterion for classification, as illustrated in the following examples.

Tetraloops and Pentaloops that form the same Motif. Figure 1.5 shows examples of hairpin loops that are classified differently at the level of sequence and secondary structure, yet form the same 3D structure. Figure 1.5a, b compare the pentaloop 5′-CAGAA-3′ with the tetraloop 5′-GAGA-3′. GAGA is an example of a "GNRA tetraloop" motif, so-named to indicate the consensus sequence identified by comparing secondary structures (Woese et al. 1990): G exclusively as the first base, A, C, G, or U ("N") as the second, A or G ("R") as the third, and A as the fourth base. The GAGA hairpin conforms to the "GNRA" consensus sequence and, as expected, forms the well-known 3D structure with the tSH closing base pair between the first and fourth bases of the loop (Ban et al. 2000). The strand reverses direction after the first nucleotide of the loop, as indicated by the curved arrow in Fig. 1.5c, and the second and third bases stack continuously on the fourth on the 3′-side of the loop. The CAGAA pentaloop sequence does not conform to the GNRA consensus, but the 3D structure shows that it forms the same 3D motif (Wimberly et al. 2000). The extra base of the penta-loop, A497, is bulged out, but this does not significantly perturb the 3D structure, as shown by superposition of the two motifs (Fig. 1.5e). Although the closing base pair is different, CA in the pentaloop and GA in the tetraloop, the *base pair type,* tSH, is the same in both structures. Moreover, tSH CA and tSH GA are isosteric (Leontis et al. 2002).

As a second example, the 5′-GAAAG-3′ pentaloop and the 5′-UUCG-3′ tetraloop appear unrelated at the level of sequence and secondary structure. UUCG conforms to the consensus "UNCG" sequence and not surprisingly, its 3D structure exhibits the characteristic features of these well-known motifs, including the *syn* configuration of the fourth base, the tSW closing base pair between the first and fourth base, the stacking of the third base on the first base and the chain reversal after the third base instead of the first base as in the GNRA tetraloops (Krasilnikov et al. 2003). The second base of UNCG loops is bulged out. The 3D structure of the GAAAG pentaloop (Ban et al. 2000), annotated in Fig. 1.5h and superposed on that of UUCG, shows that the GAAAG pentaloop and the UNCG tetraloop have very similar 3D structures. As in the previous example, the closing *base pair type* is the same (tSW), although the bases are different (GA vs. UG). The tSW GA and tSW UG pairs are *near isosteric*.

1.3.2 Defining and Naming 3D Motifs

The examples discussed above illustrate the confusion that results, especially for newcomers to RNA structural bioinformatics, with the use of names for motifs that are based on consensus sequences and number of nucleotides. These examples indicate the need for precise definitions and names for RNA motifs, to provide concise communication between humans and software agents, and to make automated reasoning about RNA possible. We demonstrate the process of constructing rigorous, structure-based definitions for 3D motifs, using "GNRA" loops as examples.

Defining the "GNRA" Hairpin Motif. There are instances of the same recurrent motif sharing a common set of core nucleotides and conserved interactions between them. The first step in constructing a structure-based definition is to identify all geometric instances in 3D structures to determine the core nucleotides and their interactions. We have written the "Find RNA 3D" (FR3D) suite of software tools to facilitate this process (Sarver et al. 2008). Using FR3D, we carried out a geometric search of the non-redundant RNA structure database using as query motif the centroid of a previous search. The query motif included the closing Watson–Crick base pair of the adjacent double helix, which, for the reasons discussed above, is treated as part of the motif (Sect. 1.1.3).

The search identified 108 instances with *geometric discrepancy* less than 0.75, as defined in Sarver et al. (2008). These instances correspond to 45 unique sequences and are listed in Fig. 1.6 with representative examples for each sequence. For each motif candidate, FR3D lists all base-pairing, base-stacking, and base-phosphate interactions between motif nucleotides and creates a structural alignment of all instances. The structural alignment identifies bases that superpose in 3D space as well as inserted bases not present in the query motif. Examination of the alignment shows that insertions occur between the 3rd and 4th, 4th and 5th and 5th and 6th nucleotide positions. The search reveals that none of the insertions occurs frequently, and so the core motif consists of the six nucleotide positions of the query motif.

"GNRA" Definition (upper left panel with annotations: 3 (1), 4 (1), 5 (1), 2, 1, 6, 5′ 3′, 10.1, 1.1)

Sugar Edge (NT 2) — tHS / Hoogsteen (NT 5)

tHS	A	C	G	U
A	---	2% (2)	89% (96)	1% (1)
C	1% (1)	1% (1)		3% (3)
G			4% (4)	
U	---			

Watson-Crick (NT 6) — cWW / Watson-Crick (NT 1)

cWW	A	C	G	U
A		---		6% (7)
C		1% (1)	56% (61)	
G		24% (26)		---
U	4% (4)		8% (9)	---

PDB File	1		2		3		4		5		6		1	2	3		4		5		6
2GIS	A	49	G	50	A	51	A	52	A	53	U	54	A	G	A	-	A	-	A	-	U
2AW4	A	1630	G	1631	A	1632	G	1633	A	1635	U	1636	A	G	A	-	G	A	A	-	U
2AVY	A	1012	G	1013	A	1014	G	1015	A	1016	U	1017	A	G	A	-	G	-	A	-	U
2AW4	A	2856	G	2857	C	2858	G	2859	A	2860	U	2861	A	G	C	-	G	-	A	-	U
1y0q	A	101	G	102	U	103	A	104	A	105	U	106	A	G	U	-	A	-	A	-	U
1s72	C	1275	U	1276	C	1277	A	1278	A	1280	C	1281	C	U	C	-	A	U	A	-	C
1j5e	C	522	A	523	G	524	C	525	C	526	G	527	C	A	G	-	C	-	C	-	G
1s72	C	217	C	218	G	219	G	221	A	222	G	223	C	C	G	C	G	-	A	-	G
1s72	C	804	G	805	A	806	A	807	A	808	G	809	C	G	A	-	A	-	A	-	G
1j5e	C	458	G	459	A	460	G	462	A	463	G	474	C	G	A	C	G	-	A	-	G
483d	C	2658	G	2659	A	2660	G	2661	A	2662	G	2663	C	G	A	-	G	-	A	-	G
2J01	C	955	G	956	A	957	A	959	A	960	G	962	C	G	A	U	A	-	A	C	G
1y0q	C	130	G	131	A	132	U	133	A	134	G	135	C	G	A	-	U	-	A	-	G
1mzp	C	25	G	26	C	27	A	28	A	29	G	30	C	G	C	-	A	-	A	-	G
1s72	C	89	G	90	C	91	G	92	A	93	G	94	C	G	C	-	G	-	A	-	G
1s72	C	1793	G	1794	G	1795	A	1796	A	1797	G	1799	C	G	G	-	A	-	A	C	G
1lng	C	208	G	209	G	210	A	211	A	212	G	213	C	G	G	-	A	-	A	-	G
2AW4	C	487	G	488	G	489	G	491	A	492	G	493	C	G	G	G	C	G	-	A	-	G
1s72	C	2248	G	2249	G	2250	G	2251	A	2252	G	2253	C	G	G	-	G	-	A	-	G
2J01	C	1806	G	1807	U	1808	A	1809	A	1810	G	1811	C	G	U	-	A	-	A	-	G
2J01	C	462	G	463	U	464	G	465	A	466	G	467	C	G	U	-	G	-	A	-	G
2AVY	C	862	U	863	A	864	A	865	C	866	G	867	C	U	A	-	A	-	C	-	G
1s72	G	1468	C	1469	A	1470	A	1471	C	1472	C	1474	G	C	A	-	A	-	C	U	C
1un6	G	82	G	83	A	84	A	85	A	86	C	94	G	G	A	-	A	-	A	-	C
1q93	G	13	G	14	A	15	G	16	A	17	C	18	G	G	A	-	G	-	A	-	C
1xjr	G	21	G	22	A	23	G	24	A	26	C	27	G	G	A	-	G	U	A	-	C
1u9s	G	204	G	205	C	206	A	207	A	208	C	209	G	G	C	-	A	-	A	-	C
2J01	G	86	G	87	C	88	G	89	A	90	C	91	G	G	C	-	G	-	A	-	C
2AVY	G	1515	G	1516	G	1517	A	1518	A	1519	C	1520	G	G	G	-	A	-	A	-	C
119a	G	146	G	147	G	148	A	149	G	150	C	151	G	G	G	-	A	-	G	-	C
1s72	G	2876	G	2877	U	2878	A	2879	A	2880	C	2881	G	G	U	-	A	-	A	-	C
1s72	G	1054	G	1055	U	1056	A	1057	A	1058	C	1060	G	G	U	-	A	-	A	G	C
2NZ4	G	108	G	109	U	110	G	111	A	112	C	113	G	G	U	-	G	-	A	-	C
1s72	U	1326	G	1327	A	1328	A	1329	A	1330	A	1331	U	G	A	-	A	-	A	-	A
1q96	U	13	G	14	A	15	G	16	A	17	A	18	U	G	A	-	G	-	A	-	A
2AVY	U	1165	G	1166	A	1167	A	1169	A	1170	A	1171	U	G	A	U	A	-	A	-	A
1s72	U	468	G	469	U	470	G	471	A	472	A	473	U	G	U	-	G	-	A	-	A
1s72	U	493	C	494	A	495	C	496	A	498	G	499	U	C	A	-	G	A	A	-	G
1kxk	U	33	G	34	A	35	A	36	A	37	G	38	U	G	A	-	A	-	A	-	G
2AW4	U	2356	G	2357	A	2358	C	2359	G	2360	G	2361	U	G	A	-	C	-	G	-	G
2AVY	U	296	G	297	A	298	G	299	A	300	G	301	U	G	A	-	G	-	A	-	G
2J01	U	1864	G	1865	C	1866	A	1876	A	1877	G	1878	U	G	C	-	A	-	A	-	G
2AW4	U	955	G	956	C	957	A	959	A	960	G	962	U	G	C	-	A	-	A	-	G
2AW4	U	1222	G	1223	U	1224	G	1225	A	1226	G	1227	U	G	C	U	A	-	A	C	G
1s72	U	733	U	734	C	735	A	736	A	737	G	738	U	U	C	-	A	-	A	-	G

Fig. 1.6 Structural definition of "GNRA" hairpin loop motif. *Upper left*: Annotations showing key features of structural definition of "GNRA" motif, including conserved base-pairing and base-stacking interactions and positions of insertions. *Right*: Unique instances of "GNRA" hairpin loop motif obtained by geometric search of non-redundant RNA structure database using FR3D. *Lower left*: Isostericity matrices for conserved basepairs in "GNRA" motif instances obtained by geometric search

The search also reveals two conserved base pair interactions – a cWW pair between the 1st and 6th bases and a tSH pair between the 2nd and 5th bases of the core motif. The strand always changes direction between the 2nd and 3rd nucleotides and stacking occurs between the two basepairs and also between the 3rd and 4th and the 4th and 5th bases. These structural features are summarized in the panel labeled "GNRA Definition" in Fig. 1.6. The positions where insertions are observed are also shown.

The search also returns valuable co-variation information for each base pair in the motif. This data is summarized as 4×4 contingency tables for each base pair superposed on the corresponding Isostericity Matrix for that base pair type (lower left panels of Fig. 1.6). The same background shading is used to indicate isosteric basepairs in each family. Similar shading indicate near isosteric relations while white boxes indicate basepairs that do not occur in that geometric family. This data shows that all canonical cWW pairs occur for the base pair between the 1st and 6th bases, but that CG and GC predominate. A significant fraction is the cWW UG pair, but no GU occurs. As shown by the background shading, the canonical cWW basepairs are isosteric to each other and near isosteric to UG. GU is not isosteric to UG.

The tHS base pair has two isosteric families. Almost all observed instances belong to the isosteric group 10.1 consisting of AN, CA, CC, and CU tHS basepairs. While most instances have the AG tHS base pair, a significant fraction do not. A small number of instances have GG tHS pairs, which belong to the second tHS isosteric group, indicated by different color. These loops form a similar, but not identical, hairpin loop, that forms specific tertiary interactions. While tHS AA is not observed in this set of structures, isostericity considerations indicate it may occur in new structures. The base pair information is included in the motif definition by indicating the geometric family and the preferred isosteric group within the family (see the upper left panel of Fig. 1.6).

Conserved Tertiary Interactions. The question arises whether motifs that match the structural definition for a motif but vary in length or sequence can still function in the same way. Again we use the example of the "GNRA" hairpin loops, which occur widely in RNA structures and function by mediating tertiary interactions. For example, the 3D structures of the 16S rRNAs of *E. coli* and *T. thermophilus* each contain 13 hairpin loops that meet the structural definition proposed above. Twelve of these mediate long-range tertiary interactions. Can structurally similar pentaloops also mediate these interactions? The answer is yes. Figure 1.7 shows an example of tertiary interactions involving homologous hairpin loops in the 23S rRNAs of *H. marismortui* and *T. thermophilus*, one of which is a pentaloop and the other a tetraloop. As shown by the annotations, they form identical tertiary interactions (Stombaugh 2004).

Fig. 1.7 Hairpin loops mediating tertiary interactions. Both the CAACU and GAAA hairpin loops meet the structural definition of a "GNRA" hairpin loop in Fig. 1.6. They occur at homologous sites in *H. marismortui* (*left*) and *T. thermophilus* 23S rRNA and mediate identical tertiary interactions (PDB files 1s72 and 2j01)

1.3.3   Defining Tertiary Interaction Motifs

Local vs. Composite Motifs: Similar procedures as outlined above for hairpin loop motifs are used to define internal and junction loop motifs. Again, it is important to find all instances of each motif to create accurate definitions. By definition, modular and recurrent internal loop motifs comprise two strand segments and are flanked by two helices. However, motifs first identified as internal loops are often found to also occur within multi-helix junction loops or more complex topologies involving pseudo-knots. The sarcin/ricin and kink-turn motifs have local and composite instances. When a motif that was first identified in an internal loop motif is also found in a junction or pseudo-knot composed of three or more different strand segments, it is called a composite motif. The original internal loop version is called a local motif. The search program FR3D was designed to find composite as well as local versions of recurrent motifs. In Fig. 1.8, local (left panel) and composite (right panel) versions of the sarcin/ricin motif are compared. The local version is the original sarcin/ricin motif of 23S rRNA and the composite is from a complex junction in Domain II of 23S rRNA. The annotated diagrams show that the two motifs comprise the same core nucleotides and have similar interactions between them.

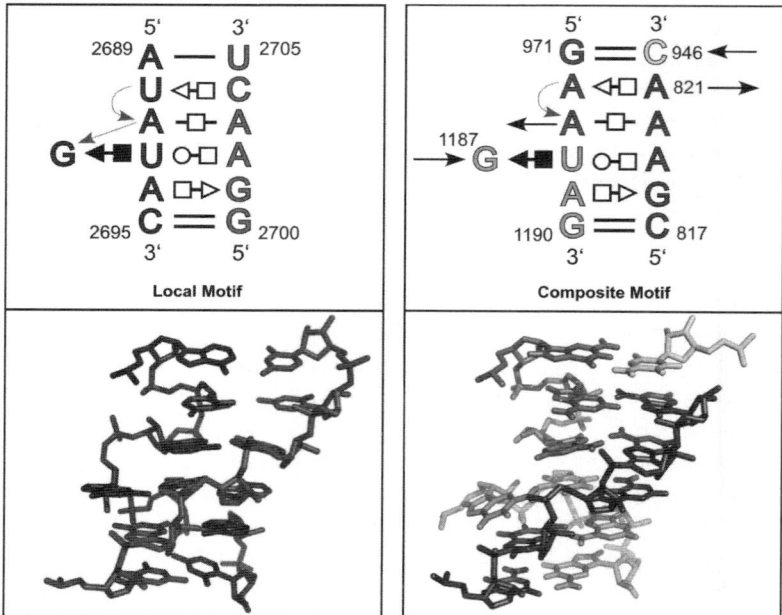

Fig. 1.8 Local vs. Composite motifs. *Left*: Local (internal loop) sarcin/ricin motif from *H. maris-mortui* 23S rRNA comprising two strand segments. *Right*: Composite sarcin/ricin motif from *E. coli* 23S rRNA comprising four different strand segments. The 3D structure of each motif is shown below each annotated diagram (PDB files 1s72 and 2aw4) (See figure insert for colour reproduction)

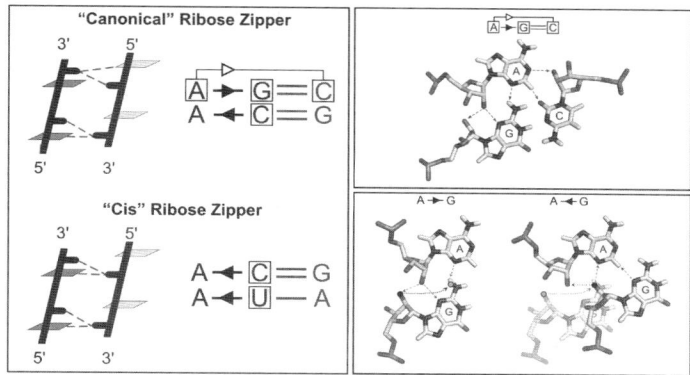

Fig. 1.9 Ribose zippers are tertiary interaction motifs composed of two Sugar-edge basepairs. *Left*: Schematic representation adapted from (Tamura and Holbrook 2002) and base-pair annotation (Leontis and Westhof 2001) of "canonical" and "*cis*" Ribose Zipper (RZ) tertiary motifs. *Upper Right*: Base triple composed of GC cWW, AG cSs and AC tSs basepairs. The A forms two pairs with its Sugar edge and is assigned higher priority in each interaction, as explained in the text. *Lower Right*: Comparison of A/G cSs (*left*) and A/G csS (*right*) basepairs. The dotted black arrow indicates the lateral shift that transforms one type into the other. In the A/G cSs, the A is the dominant base so the arrow points from the A to the G. The roles are reversed in the A/G csS pair. The dashed green arrows indicate hydrogen-bonds (See figure insert for colour reproduction)

Long-range tertiary interaction motifs form when different elements of the secondary structure dock to stabilize the native 3D structure of an RNA molecule. Many long-range tertiary motifs are recurrent. They are also defined by their core nucleotides and conserved interactions. Many are formed through the docking of hairpin or internal loop motifs in the minor grooves of helices or loop-receptor motifs. Some of these motifs have been given names, for example the "canonical" and "cis ribose zipper" (RZ) motifs shown in Fig. 1.9 (Cate et al. 1996a; Tamura and Holbrook 2002). In this figure, the schematic diagrams of Tamura & Holbrook and the corresponding base pair annotations are shown side-by-side for the canonical and cis ribose zipper. Each of these tertiary interaction motifs is a combination of sugar-edge basepairs formed when two adjacent Watson–Crick basepairs in a helix interact with two stacked "loop" nucleotides, usually adenosines, that (most often) belong to a hairpin or internal loop. In the canonical RZ, one of the loop nucleotides forms a cSs pair with one base of a cWW pair and a tSs pair with the other, as shown in Fig. 1.9. The second loop nucleotide forms a csS pair with one base of the second cWW base pair. The diagram in the upper right panel of Fig. 1.9 illustrates how a purine (A in this case) can pair with two different bases using its sugar edge to form a distinct kind of base-triple composed of cWW, cSs and tSs basepairs. In the cis RZ, both the loop nucleotides form cSs pairs with a cWW base pair. The lower right panel of Fig. 1.9 shows the difference between cSs and csS basepairs. The lateral shift that transforms cSs into csS is also shown. These examples show how tertiary interactions can be precisely defined in terms of the specific combinations of pairwise interactions of which they are composed. More complex tertiary interactions involve three and sometimes more pairwise interactions (Nasalean et al. in preparation).

3D Motifs and Sub-motifs. We have argued that it is useful to describe 3D motifs in terms of recurrent pairwise interactions because these interactions are conserved in geometrically similar 3D motifs and provide the means to precisely define recurrent motifs. Moreover, the pairwise interactions can be combined to describe more complex interactions, such as base triples and quadruples, and tertiary interaction motifs such as ribose zippers. Finally, software has been written to automatically classify pairwise interactions in 3D RNA structures and thus facilitate 3D searches for motifs.

In certain contexts it is useful to decompose motifs using more complex sub-motifs than basepairs or other pairwise interactions. For example, when predicting the thermodynamic stability of an RNA (or DNA), the free energies of proposed double helices are calculated using the nearest neighbor model, which requires decomposing each helix into overlapping pairs of neighboring basepairs. Each pair of stacked bases is assigned a free energy specific to the nucleotides composing the stacked pairs. A similar approach is used in the decomposition of 3D motifs into cycles of interacting nucleotides, as introduced by F. Major and co-workers (Lemieux and Major 2006; StOnge et al. 2007). These cycles are used to define graph grammars for predicting the 3D structures of RNA molecules (Lemieux and Major 2006).

1.4   Classification of Motifs According to Function

Structural vs. Functional Classifications. RNA motifs are classified structurally, to identify geometrically similar motifs, or functionally, to identify motifs that serve the same function. Recurrent RNA motifs often play the same or similar functional roles in different RNA molecules or in different places in the same RNA; so identifying them in sequence provides information about how the RNA folds and functions. For example, motifs that impart a sharp bend to helices toward the minor groove have been called kink-turn motifs (Klein et al. 2001; Strobel et al. 2004). The sequence signature of kink-turn motifs has been defined to facilitate their sequential finding (Lescoute et al. 2005).

The functional roles RNA motifs play can be roughly classified as architectural, structure-stabilizing, ligand-binding, or catalytic. Architectural motifs direct the organization and folding of the 3D structure. Kink-turns play architectural roles. Multi-helix junctions are key architectural motifs of RNA molecules. They create branch points in the secondary structure, making complex RNA structures possible. Junctions direct the folding by establishing specific co-axial stacking between pairs of helices at the junction thus organizing them in 3D space (Klosterman et al. 2004). For many junctions, non-Watson–Crick basepairs formed by junction "loop" nucleotides stabilize the native co-axial stacking (Lescoute and Westhof 2006b).

Structure stabilizing motifs include a variety of hairpin and internal loops that form 3D structures which stack two or more bases (usually A's) in appropriate geometries to form tertiary interactions. GNRA hairpin loops are the most common motifs of this kind. A number of internal loops mediate tertiary interactions very similar to those of GNRA loops (Gutell et al. 2000; Elgavish et al. 2001). While GNRA loops can interact with canonical Watson–Crick helices, more stable tertiary interactions can result, when they bind to loop-receptor motifs (Costa and Michel 1997). These are generally internal loops that use non-Watson–Crick basepairs to construct platforms on which GNRA loops can dock by base-stacking as well as base-pairing. The best-known motif of this type is the recurrent "11-nucleotide" GAAA loop receptor, first observed in the Group I intron (Cate et al. 1996b). Platforms usually project into the minor-groove side of helices.

Intercalation motifs "pinch" or "bulge out" a base that can then interact with a second motif by intercalation. The second motif creates a pocket for the intercalating base that consists of two bases, usually purines, that stack on either side of it and usually a third base that can base-pair with it, thus creating a stable tertiary interaction. T-loops, first observed in tRNA, are examples of recurrent motifs that have as one of their functions accepting an intercalating base (Nagaswamy and Fox 2002). T-loops occur in many different locations, mediate RNA–RNA or RNA–ligand interactions, and they often interact with other hairpin loops.

Different Motifs for the same Function. Different motifs can play the same role and can therefore substitute ("swap") for each other in the course of evolution. This is especially true of motifs that mediate long-range RNA–RNA interactions. Examples are shown in Fig. 1.10 of internal and hairpin loop motifs that occur

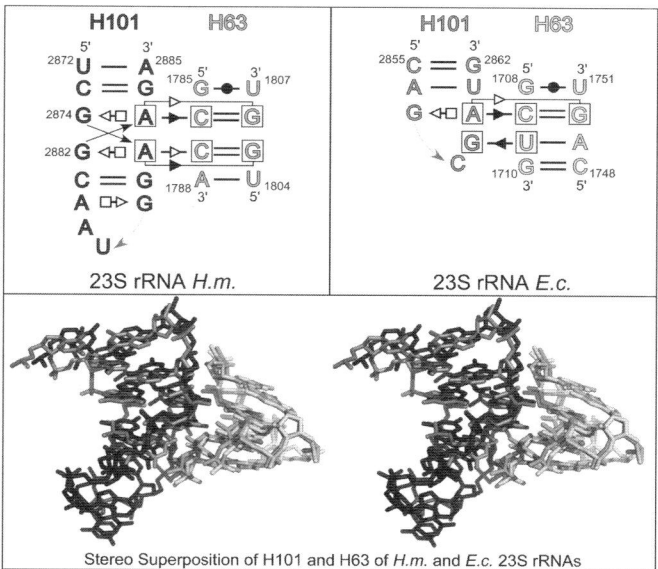

Fig. 1.10 Conserved tertiary interaction in 23S rRNA mediated different motifs. *Upper panels:* Annotated secondary structures of conserved interaction between Helices 101 (H101) and 63 in 23S rRNA of *H. marismortui* (*left*) and *E. coli* (*right*). In 23S of *H. marismortui,* the interaction is mediated by an internal loop in H101 (nucleotides 2,874; 2,875; 2,882; and 2,883), whereas in the *E. coli* structure it is mediated by a GNRA hairpin loop at the equivalent position of H101 (nucleotides 2,857–2,860). *Lower panel:* Stereo superposition of the 3D structures of Helices 101 and 63 from 23S rRNA of *H. marismortui* and *E. coli.* (PDB files 1s72 and 2aw4.) Color coding: *H. marismortui* Helix 101 (blue), Helix 63 (cyan), *E. coli* Helix 101 (orange), Helix 63 (yellow) (See figure insert for colour reproduction)

at equivalent locations in Helix 101 of evolutionarily distant *H. marismortui* and *E. coli* 23S rRNA. The motifs mediate corresponding, conserved tertiary interactions with Helix 63. Moreover, the geometry of the interaction is identical as shown by the 3D superposition of the interacting elements in the two structures.

1.5   Conclusions

Internal, junction, and hairpin loops that appear in secondary structures are, in many cases, instances of recurrent modular RNA motifs. Different sequences can form the same recurrent 3D motif as a result of structure–neutral mutations. RNA 3D motifs are defined by listing the conserved pairwise interactions between the core nucleotides (including base-pairing, stacking, and phosphate interactions). Definitions should include the geometric type of each conserved base pair, as well as the isosteric base pair groups represented in motif instances. All motif positions where insertions can occur without significantly perturbing the 3D structure should be identified and noted. For motifs that mediate RNA–RNA or RNA–protein interactions, the nucleotides that participate directly in these interactions are noted with the type of interaction formed, since these interactions may impose additional nucleotide-specific constraints that help identify them in sequence.

Motifs can be classified according to structural or functional similarity. During evolution, global structure changes more slowly than sequence or even local 3D structure; mutations can accumulate, including insertions, deletions, or substitutions that change the structure of a motif. However, if the motif is involved in crucial long-range interactions, the global function is preserved, resulting in a motif "swap" in which the tertiary or quaternary contact is mediated by geometrically distinct but functionally equivalent 3D motifs.

References

Adams PL, Stahley MR, Kosek AB, Wang J, Strobel SA (2004) Crystal structure of a self-splicing group I intron with both exons. Nature 430:45–50

Altona C (1996) Classification of nucleic acid junctions. J Mol Biol 263:568–581

Auffinger P, Hashem Y (2007) SwS: a solvation web service for nucleic acids. Bioinformatics 23:1035–1037

Ban N, Nissen P, Hansen J, Moore PB, Steitz TA (2000) The complete atomic structure of the large ribosomal subunit at 2.4 A resolution. Science 289:905–920

Cate JH, Gooding AR, Podell E, Zhou K, Golden BL, Kundrot CE, Cech TR, Doudna JA (1996a) Crystal structure of a group I ribozyme domain: principles of RNA packing. Science 273:1678–1685

Cate JH, Gooding AR, Podell E, Zhou K, Golden BL, Szewczak AA, Kundrot CE, Cech TR, Doudna JA (1996b) RNA tertiary structure mediation by adenosine platforms. Science 273:1696–1699

Costa M, Michel F (1997) Rules for RNA recognition of GNRA tetraloops deduced by in vitro selection: comparison with in vivo evolution. EMBO J 16:3289–3302

Elgavish T, Cannone JJ, Lee JC, Harvey SC, Gutell RR (2001) AA.AG@helix.ends: A:A and A:G base-pairs at the ends of 16 S and 23 S rRNA helices. J Mol Biol 310:735–753

Gagnon MG, Steinberg SV (2002) GU receptors of double helices mediate tRNA movement in the ribosome. RNA 8:873–877

Gan HH, Fera D, Zorn J, Shiffeldrim N, Tang M, Laserson U, Kim N, Schlick T (2004) RAG: RNA-as-graphs database – concepts, analysis, and features. Bioinformatics 20:1285–1291

Gutell RR, Cannone JJ, Shang Z, Du Y, Serra MJ (2000) A story: unpaired adenosine bases in ribosomal RNAs. J Mol Biol 304:335–354

Hendrix DK, Brenner SE, Holbrook SR (2005) RNA structural motifs: building blocks of a modular biomolecule. Q Rev Biophys 38:221–243

Holbrook SR (2005) RNA structure: the long and the short of it. Curr Opin Struct Biol 15:302–308

Klein DJ, Schmeing TM, Moore PB, Steitz TA (2001) The kink-turn: a new RNA secondary structure motif. EMBO J 20:4214–4221

Klosterman PS, Hendrix DK, Tamura M, Holbrook SR, Brenner SE (2004) Three-dimensional motifs from the SCOR, structural classification of RNA database: extruded strands, base triples, tetraloops and U-turns. Nucleic Acids Res 32:2342–2352

Krasilnikov AS, Yang X, Pan T, Mondragon A (2003) Crystal structure of the specificity domain of ribonuclease P. Nature 421:760–764

Lee JC, Cannone JJ, Gutell RR (2003) The lonepair triloop: a new motif in RNA structure. J Mol Biol 325:65–83

Lemieux S, Major F (2006) Automated extraction and classification of RNA tertiary structure cyclic motifs. Nucleic Acids Res 34:2340–2346

Leontis NB, Westhof E (1998) Conserved geometrical base-pairing patterns in RNA. Q Rev Biophys 31:399–455

Leontis NB, Westhof E (2001) Geometric nomenclature and classification of RNA basepairs. RNA 7:499–512

Leontis NB, Westhof E (2002) The annotation of RNA motifs. Comp Funct Genomics 3:518–524

Leontis NB, Westhof E (2003) Analysis of RNA motifs. Curr Opin Struct Biol 13:300–308

Leontis NB, Stombaugh J, Westhof E (2002) The non-Watson–Crick basepairs and their associated isostericity matrices. Nucleic Acids Res 30:3497–3531

Leontis NB, Lescoute A, Westhof E (2006) The building blocks and motifs of RNA architecture. Curr Opin Struct Biol 16:279–287

Leontis NB, Altman RB, Berman HM, Brenner SE, Brown JW, Engelke DR, Harvey SC, Holbrook SR, Jossinet F, Lewis SE, Major F, Mathews DH, Richardson JS, Williamson JR, Westhof E (2006) The RNA ontology consortium: an open invitation to the RNA community. RNA 12:533–541

Lescoute A, Leontis NB, Massire C, Westhof E (2005) Recurrent structural RNA motifs, isostericity matrices and sequence alignments. Nucleic Acids Res 33:2395–2409

Lescoute A, Westhof E (2006a) The interaction networks of structured RNAs. Nucleic Acids Res 34:6587–6604

Lescoute A, Westhof E (2006b) Topology of three-way junctions in folded RNAs. RNA 12:83–93

Lisi V, Major F (2007) A comparative analysis of the triloops in all high-resolution RNA structures reveals sequence structure relationships. RNA 13:1537–1545

Mathews DH, Turner DH (2006) Prediction of RNA secondary structure by free energy minimization. Curr Opin Struct Biol 16:270–278

Mokdad A, Krasovska MV, Sponer J, Leontis NB (2006) Structural and evolutionary classification of G/U wobble basepairs in the ribosome. Nucleic Acids Res 34:1326–1341

Nagaswamy U, Fox GE (2002) Frequent occurrence of the T-loop RNA folding motif in ribosomal RNAs. RNA 8:1112–1119

Nasalean L, Stombaugh J, Leontis NB (in preparation)

Nissen P, Ippolito JA, Ban N, Moore PB, Steitz TA (2001) RNA tertiary interactions in the large ribosomal subunit: the A-minor motif. Proc Natl Acad Sci U S A 98:4899–4903

Pace NR, Thomas BC, Woese CR (1999) Probing RNA structure, function, and history by comparative analysis. In: Gesteland RF, Cech TR, Atkins JF (eds.) The RNA World, Cold Spring Harbor Laboratory Press, Cold Spring Harbor, NY, pp. 113–141

Rangan P, Masquida B, Westhof E, Woodson SA (2003) Assembly of core helices and rapid tertiary folding of a small bacterial group I ribozyme. Proc Natl Acad Sci U S A 100:1574–1579

Richardson JS, Schneider B, Murray LW, Kapral GJ, Immormino RM, Headd JJ, Richardson DC, Ham D, Hershkovits E, Williams LD, Keating KS, Pyle AM, Micallef D, Westbrook J, Berman HM (2008) RNA backbone: consensus all-angle conformers and modular string nomenclature (an RNA ontology consortium contribution). RNA 14:465–481

Sarver M, Zirbel CL, Stombaugh J, Mokdad A, Leontis NB (2008) FR3D: finding local and composite recurrent structural motifs in RNA 3D structures. J Math Biol 56:215–252

St-Onge K, Thibault P, Hamel S, Major F (2007) Modeling RNA tertiary structure motifs by graph-grammars. Nucleic Acids Res 35:1726–1736

Stefl R, Allain FH (2005) A novel RNA pentaloop fold involved in targeting ADAR2. RNA 11:592–597

Stombaugh J (2004) Developing isostericity matrices: a tool for RNA structural alignment. MS Thesis

Stombaugh J, Zirbel CL, Westhof E, Leontis NB (submitted) Systematic evaluation of RNA base-pair isostericity matrices

Strobel SA, Adams PL, Stahley MR, Wang J (2004) RNA kink turns to the left and to the right. RNA 10:1852–1854

Sykes MT, Levitt M (2005) Describing RNA structure by libraries of clustered nucleotide doublets. J Mol Biol 351:26–38

Tamura M, Holbrook SR (2002) Sequence and structural conservation in RNA ribose zippers. J Mol Biol 320:455–474

Thirumalai D, Woodson SA (1996) Kinetics of folding of proteins and RNA. Acc Chem Res 29:433–439

Thirumalai D, Lee N, Woodson SA, Klimov D (2001) Early events in RNA folding. Annu Rev Phys Chem 52:751–762

Wimberly BT, Brodersen DE, Clemons WM Jr, Morgan-Warren RJ, Carter AP, Vonrhein C, Hartsch T, Ramakrishnan V (2000) Structure of the 30S ribosomal subunit. Nature 407:327–339

Woese CR, Winker S, Gutell RR (1990) Architecture of ribosomal RNA: constraints on the sequence of "tetra-loops". Proc Natl Acad Sci U S A 87:8467–8471

Zhuang X, Bartley LE, Babcock HP, Russell R, Ha T, Herschlag D, Chu S (2000) A single-molecule study of RNA catalysis and folding. Science 288:2048–2051

Chapter 2
Theory of RNA Folding: From Hairpins to Ribozymes

D. Thirumalai(⊠) and Changbong Hyeon

Abstract The rugged nature of the RNA folding landscape is determined by a number of conflicting interactions like repulsive electrostatic potential between the charges on the phosphate groups, constraints due to loop entropy, base stacking, and hydrogen bonding that operate on various length scales. As a result the kinetics of self-assembly of RNA is complex, but can be easily modulated by varying the concentrations, sizes, and shapes of the counterions. Here, we provide a theoretical description of RNA folding that is rooted in the energy landscape perspective and polyelectrolyte theory. A consequence of the rugged folding landscape is that, self-assembly of RNA into compact three-dimensional structures occurs by parallel routes, and is best described by the kinetic partitioning mechanism (KPM). According to KPM one fraction of molecules (Φ) folds rapidly while the remaining gets trapped in one of several competing basins of attraction. The partition factor Φ can be altered by point mutations as well as by changing the initial conditions such as ion concentration, size and valence of ions. We show that even hairpin formation, either by temperature or force quench, captures much of the features of folding of large RNA molecules. Despite the complexity of the folding process, we show that the KPM concepts from polyelectrolyte theory, and charge density of ions can be used to explain the stability, pathways and their diversity, and the plasticity of the transition state ensemble of RNA self-assembly.

2.1 Introduction

The landmark discovery that RNA molecules are ribozymes (RNA enzymes) (Guerriertakada et al. 1983; Kruger et al. 1982) has triggered an intense effort to decipher their folding mechanisms. In the intervening years an increasing repertoire of cellular functions has been associated with RNA (Doudna and Cech 2002). These

D. Thirumalai

Department of Chemistry and Biochemistry, University of Maryland, College Park, College Park, MD 20742, USA

e-mail: thirum@umd.edu

N.G. Walter et al. (eds.) *Non-Protein Coding RNAs*

doi: 10.1007/978-3-540-70840-7_2, © Springer-Verlag Berlin Heidelberg 2009

include their role in replication, translational regulation, viral propagation etc. Moreover, interactions of RNA with each other and with DNA and proteins are vital in many biological processes. Even, the central chemical activity of ribosomes, namely, the formation of the peptide bond in the biosynthesis of polypeptide chains by ribosomes near the peptidyl transfer center, involves only RNA, leading many to suggest that *ribosomes are ribozymes* (Nissen et al. 2000; Yusupov et al. 2001). The appreciation that RNA molecules play a major role in a number of cellular functions has made it important to establish the structure – function relationship. Thus, the need to understand, at the molecular level the ribozyme activity, inevitably leads to the question: How do RNA molecules fold?

In little over a decade great success has been achieved in an attempt to answer this question because of progress on a number of fronts. The number of experimentally determined high resolution RNA structures (Ban et al. 2000; Cate et al. 1996; Nissen et al. 2000; Yusupov et al. 2001) continues to increase which has enabled us to understand the interactions that stabilize the folded states. Single molecule (Ma et al. 2006; Onoa et al. 2003; Russell et al. 2002b; Woodside et al. 2006; Zhuang et al. 2000) and ensemble experiments (Zarrinkar and Williamson 1994; Koculi et al. 2006; Pan et al. 1999) using a variety of biophysical methods combined with theoretical techniques (Thirumalai and Woodson 1996; Thirumalai and Hyeon 2005) have led to a conceptual framework for predicting various processes by which RNA molecules fold.

There are two aspects to RNA folding. The first is the prediction of the folded structures from sequence (Hofacker 2003; Zuker and Stiegler 1981). The second problem concerns the mechanisms by which assembly of the three dimensional functionally competent structure forms, start from the unfolded conformations. In this chapter we describe the folding mechanisms from the energy landscape perspective with focus on the polyelectrolyte aspects of RNA.

At a first glance it might appear that the RNA folding problem should be simple at least in comparison to the better investigated problem of protein folding (Tinoco and Bustamante 1999). However, there are several reasons why RNA folding is a difficult problem.

1. The building blocks of RNA are the four nucleotides each with a base, ribose, and phosphate groups. The bases (two purines and two pyrimidines), that are chemically similar, interact with each other either through hydrogen bonding or base stacking. The secondary structural elements (helices, loops, bulges) are independently stable which gives the impression that the three dimensional assembly is built much the same way as complicated architecture using prefabricated building blocks. However, the difficulty arises not only because of the chemical similarity of the nucleotides but also due to the polyelectrolyte nature arising from the charged phosphate groups.

2. The bases, their ability to form hydrogen bonds through Watson–Crick (WC) pairing withstanding, are all hydrophobic. The uniformity of the hydrophilic backbone along with lack of diversity in the bases make RNA closer to a "homopolymer" than polypeptide chains (Thirumalai and Hyeon 2005). The "homopolymer" nature of nucleic acids results in RNA structures being able to adopt alternate structures i.e., the stability gap between the folded and the other

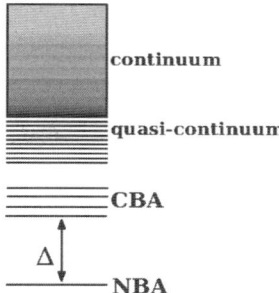

Fig. 2.1 View of the states of RNA as a free energy spectrum. The conformations in the NBA are separated from those in the competing basins of attraction (CBA) by the stability gap Δ. The structures in the CBA, while misfolded, can have many native-like features. Rapid folding without long pauses in the CBAs is likely if $\Delta/k_B T \gg 1$. Figure adapted from (Guo et al. 1992)

    misfolded structures is not large (Fig. 2.1). As a result, the energy landscape of RNA, even at the secondary structural level, is rugged containing many metastable conformations that serve as kinetic traps.

3. At some level, WC base pairing does simplify the prediction of RNA secondary structures. However, not all nucleotides are engaged in WC base pairing. Analysis of RNA secondary structures shows that the number of base-pairs (N_{BP}) varies with sequence length N as $N_{BP} = 0.27 \times N$. The linear growth of N_{BP} with N with slope 0.5 is expected if all the nucleotides are engaged in Watson–Crick base pairings. However, the slope is only 0.27 (Dima et al. 2005). This shows that 46% of the sequence, which is computed using $N_{BP}/N \approx (1-x)/2$, constitute non-pairing regions such as bulges, loops, dangling ends, and other motifs. The bulges and loops are important structural elements that glue the independent helices together to make the RNA structures compact.

4. Finally, the folding mechanisms can be greatly altered by changing the nature of counterions which makes it necessary to consider explicitly the polyelectrolyte nature of RNA. In particular, the important role of valence, shape and size of the counterions (Koculi et al. 2004, 2006, 2007) in modulating the secondary structures and possibly altering them during the course of tertiary structure formation, are difficult to predict (Chauhan and Woodson 2008; Thirumalai 1998; Wu and Tinoco 1998). The varying flexibilities of different regions of RNA, the homopolymer character of the building blocks, the key role of counterions in the folding process, and the presence of alternate structures render RNA folding a challenging problem.

2.2 Structural Characteristics of RNA

Determination of the size, shape, flexibility, and base-pairs statistics in RNA native structures, is important in understanding the nature of packing in folded structures and also in elucidating interaction between RNA and DNA or proteins. Analysis of the RNA native structures available in the Protein Data Bank (PDB) can be used to infer the general characteristics of the shapes and flexibility of folded RNA.

Native Structures are Compact: If RNA structures are compact then their volumes are expected to scale as $V \sim R_G^3 \sim a^3 N$, where R_G is the radius of gyration, a is an effective monomer length. More generally, Flory showed that $R_G \sim a N^\nu$ where the Flory exponent $\nu = 1/3$ for maximally compact structures, $\nu = 1/2$ for polymers in Θ- condition, and $\nu = 3/5$ for flexible polymers in good solvents. As RNA is a polyelec- trolyte valence, shape, and concentration (C) of counterions can alter solvent quality, and hence R_G. At low C, RNA is expanded and the transition to a compact structure occurs only when C exceeds the midpoint of the unfolded to folded transition.

Computation of the sizes of RNA structures using the PDB coordinates reveals that R_G, follows the Flory scaling law, namely, $R_G = a_N N^{1/3}$ Å (Hyeon et al. 2006). The pre-factor, $a_N = 5.5$ Å, corresponds approximately to the average distance between the phosphate groups (≈ 5.8 Å) along the ribose-phosphate backbone. For a given N, the approximate volume of RNA is larger than that of proteins whose R_G scales as $R_G = 3.1 N^{1/3}$ Å (Dima and Thirumalai 2004; Hyeon et al. 2006). In other words, RNA molecules are more loosely packed than proteins, which are probably linked to their folding being dependent on accommodation of counterions to form compact structures. The difference is due to the larger size of the nucleotides com- pared to amino acids and the nature of interactions that stabilize the folded states of RNA and proteins.

Folded RNAs are Prolate Ellipsoids: Even though folded RNAs are compact, as assessed by R_G, substantial deviations from sphericity have been found. When the shape of RNA molecules is characterized by the asphericity Δ and the shape param- eters S that are computed using the eigenvalues of the moment of inertia tensor (Aronovitz and Nelson 1986; Hyeon et al. 2006), we find that a large fraction of folded RNA structures are aspherical and the distribution of S values shows that RNA molecules are prolate. The prolate ellipsoid shape of RNA renders their dif- fusion intrinsically anisotropic. The observed difference between shapes of RNAs and globular proteins is primarily due to the nature of interactions that stabilize the folded structures of RNA and proteins. Packing in RNA is not only determined by the favorable interactions between nucleotides but also by counter-ion mediated long-range interactions. The volume excluded by counterions affects packing, and consequently the shape of RNA structures.

Persistence Length of RNA shows Similarity to Polyelectrolytes. From the poly- mer perspective, flexibility of RNA is best assessed by its persistence length, l_p, and its dependence on the changes in ionic strength. The overall compact RNA structure is formed by gluing together flexible (loops and bulges) and stiff helical regions. Despite the potential variations in the flexibility it is useful to obtain estimates of the global l_p. The total persistence length of RNA may be written as $l_p = l_p^0 + l_p^{el}$ where l_p^0 is the intrinsic persistence length and l_p^{el} is the electrostatic contribution. If RNA were a polyelectrolyte then $l_p^{el} = l_B/4\kappa^2 A^2$ where the Bjerrum length $l_B = e^2/4\pi\epsilon k_B T$ (e is the unit of charge, ϵ is the dielectric constant, k_B is the Boltzmann constant, and T is the temperature), for monovalent couterions $\kappa^2 = 8\pi l_B I$ (I is the ionic strength), and A is the average distance between the charges (Odijk 1977; Skolnick and Fixman

1977). The l_p values can be obtained from the distance distribution functions, which, for folded RNA molecules, can be directly computed using the PDB coordinates. The persistence length of the folded RNA can be extracted by fitting, for $r/R_G > 1$, the distance distribution function $P(r)$, which is computed using the coordinates of the folded RNA, to the wormlike chain model $P_{WLC}(r) \sim \exp\{-1/(1-(l_p r/R_G^2)^2)\}$ (Caliskan et al. 2005; Hyeon et al. 2006). The persistence length is scale-dependent and varies as $l_p = 1.5\ N^{0.33}$ Å (Hyeon et al. 2006). The dependence of l_p on N implies that the average length of helices with stacks should increases as N grows.

In principle, as the counterion concentration decreases the changes in l_p can be secured by obtaining $P(r)$ using Small Angle X-ray Scattering (SAXS) experiments. To date, SAXS data is available for only a few RNA molecules (*Azoarcus* ribozyme (Rangan et al. 2004), RNase P (Fang et al. 2002), and *Tetrahymena* ribozyme (Russell et al. 2002a)). Surprisingly, analysis of $P(r)$ for *Azoarcus* ribozyme and RNase *P* showed that the distance distribution function is well fit using $P_{WLC}(r)$ for the WLC model. As the concentration of Mg^{2+} and Na^+ decreases l_p increases (Caliskan et al. 2005) for *Azoarcus* ribozyme, $l_p \sim 21$ Å in the unfolded state, and $l_p \sim 10$ Å in the compact folded state. It is noteworthy that $l_p\ \kappa^{-2}$ which is predicted for polyelectrolytes (Odijk 1977; Skolnick and Fixman 1977) do not have globally compact folds like RNA molecules. Thus, not only does l_p change dramatically as RNA folds, but it also exhibits the characteristics of polyelectrolytes especially at low ionic strength. Thus, how the polyelectrolyte problem is solved in RNA remains a key problem.

2.3 Rugged Folding Landscape and the Kinetic Partitioning Mechanism

The observed multiple folding routes and the associated heterogeneity of folding pathways can be anticipated from the energy landscape perspective (Thirumalai and Woodson 1996). The states for RNA (or for proteins for that matter) can be represented as a free energy spectrum (Guo et al. 1992). If the free energy gap (Δ in Fig. 2.1) is large, then trapping in one of the many Competing Basins of Attraction (CBAs) is not very probable. The presence of many alternate structures implies that the stability gap (especially when scaled by N) for RNA is not very large. As a result, RNA folding landscape is rugged (Fig. 2.2a), and is characterized by the presence of multiple minima that are separated by free energy barriers of varying heights.

The rugged nature of the energy landscape arises due to the presence of several competing interactions. Favorable hydrophobic stacking, and tertiary interactions favor chain compaction while the negatively charged interactions are better accommodated by extended structures. As a result RNA molecules are "frustrated" because not all interactions involving a given nucleotide can be simultaneously satisfied. In addition, the polyelectrolyte nature of RNA also induces topological frustration. The formation of stable secondary structures is largely driven by interactions on "local" scales in which the persistence length is comparable to the Debye screening

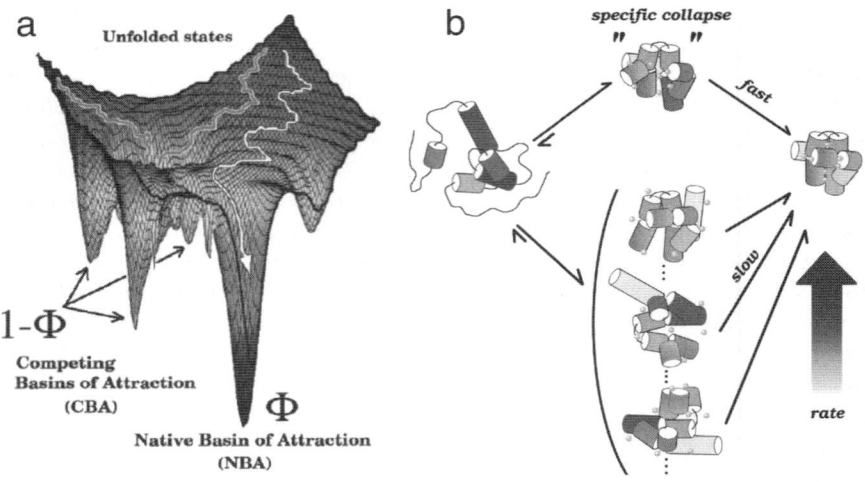

Fig. 2.2 (**a**) Schematic sketch of the rugged folding landscape of RNA. Conformational entropy and electrostatic repulsion between the phosphate groups favor the high free energy unfolded structures at low ionic strength. Under folding conditions a fraction of molecules (Φ) reach the NBA directly. A sketch of a trajectory for a fast track molecule that starts in a region of the energy landscape and which connects directly to the NBA is given in white. Trajectories (shown in green) that begin in other regions of the energy landscape can be kinetically trapped in the CBAs with probability ($1-\Phi$). The low dimensional representation of the complex energy landscape suggests that the initial conditions, which can be changed by counterions, stretching force, or denaturants, can alter the folding pathways. (**b**) Representation of RNA folding by KPM. Based on theory it is suggested that the fast track molecules specifically collapse into near native-like structures that rearrange to the native state without being trapped in the CBA. In contrast, the slow track molecules collapse to one of the manifold of misfolded structures. The collapse time scale, that depends on the nature of ions, for fast and slow track molecules, is similar. A spectrum of rates determine the transition from the CBAs to NBA (See figure insert for colour reproduction)

length. Compact folded structures result from the packing of locally formed secondary structures. Because there are multiple ways of assembling the stable secondary structures, several misfolded compact tertiary structures can form readily.

The incompatibility of the metastable misfolded structures that may share many of the correct secondary structures and the global stable fold, result in topological frustration. The folded structure may be thought of as the least frustrated and hence the most stable. From the perspective of topological frustration it follows that even the secondary structures can rearrange in the course of forming the global fold as was demonstrated in the context of P5abc formation (Wu and Tinoco 1998). In other words, organization of tertiary interactions might force the correct formation of even the secondary structures, as illustrated sometime ago using P5abc and more recently in the case of tertiary structure formation of a self-splicing group I intron in *Azoarcus* pre-tRNA (Chauhan and Woodson 2008).

The kinetic consequence of the rugged energy landscape is that folding is greatly impeded by long pausing in the CBAs. The structures in the CBAs could have many native-like features that make them long-lived under folding conditions. The diversity in the folding trajectories that leads to the kinetic partitioning mechanism (KPM) is best illustrated using the sketch of the energy landscape (Fig. 2.2b). Under folding conditions (excess Mg^{2+}) the heterogeneous population of unfolded molecules navigates the rugged energy landscape in search of the NBA (Fig. 2.2a). A fraction (Φ) of unfolded molecules reaches the NBA rapidly without being trapped in any of the CBAs (Fig. 2.2b). The precise value of Φ depends on the sequence as well as external conditions, and is an indicator of the size of the NBA that in turn is determined by the extent to which a given sequence under specific ionic condition is frustrated. The remaining fraction, $(1-\Phi)$, gets kinetically trapped in one of the many CBAs. The transitions from the CBAs to the NBA might require large conformational changes, and hence involve overcoming substantial free energy barriers. Consequently, the transition rate CBA → NBA might be extremely slow depending on the extent of structural rearrangement required to reach the folded state. Because there are many kinetic metastable states, several rate constants are needed to fully describe the CBA → NBA transition. Thus, with the multivalley structure of the free energy landscape, the initial ensemble of molecules kinetically partition into fast folders (Φ being their fraction) and slow folders. From the KPM it follows that the fraction of molecules that reach the NBA at time t is $f_{NBA} = 1 - \Phi \exp(-k_F t) - \Sigma a_i \exp(-k_i t)$ where k_F is the rate of reaching the NBA from the unfolded conformations for the fast folders, k_i is the rate of transition from the ith CBA to the NBA, and a_i is the corresponding amplitude.

Experimental Evidence. In key experiments, Zarrinkar and Williamson showed that the slow folding of *Tetrahymena* ribozyme is due to the presence of multiple long-lived metastable intermediates (Zarrinkar and Williamson 1994). This ribozyme, which has become the workhorse of group I intron folding, is roughly made up of two subdomains containing paired (P) regions P4–P6 and P3–P7 (Fig. 2.3). Using kinetics of oligonucleotide hybridization, two discrete intermediates along the presumed hierarchical folding pathway was identified. One of them is I_1 (folded P4–P6) and the other is I_2 in which both the major subdomains are nearly formed. Thus, in this picture, RNA folds through well-defined intermediates some of which are dependent on Mg^{2+}. The rate-limiting step is the association of the two major subdomains.

The possibility that $\Phi < 1$ implies that folding of RNA, regardless of the complexity of the fold, must occur by parallel pathways as predicted by KPM. The key prediction of KPM is that there must be a direct pathway from Unfolded Basin of Attraction (UBA) to the NBA. The evidence that *Tetrahymena* ribozyme folds by KPM was first provided by Pan et al. using a combination of theory and experiments (Pan et al. 2000; Thirumalai and Woodson 1996). Using native gel assay to measure the time-dependent increase in the population of the NBA under folding conditions and theoretical estimates for the rate of fast track molecules it was shown that $\Phi \approx 0.08$ for the precursor RNA. Thus, about 8% of the initially unfolded molecules reach the NBA without being kinetically trapped while the majority of the misfolded molecules fold through multiple intermediates. The results by Pan et al. also showed that addition of urea can modestly accelerate the rates of

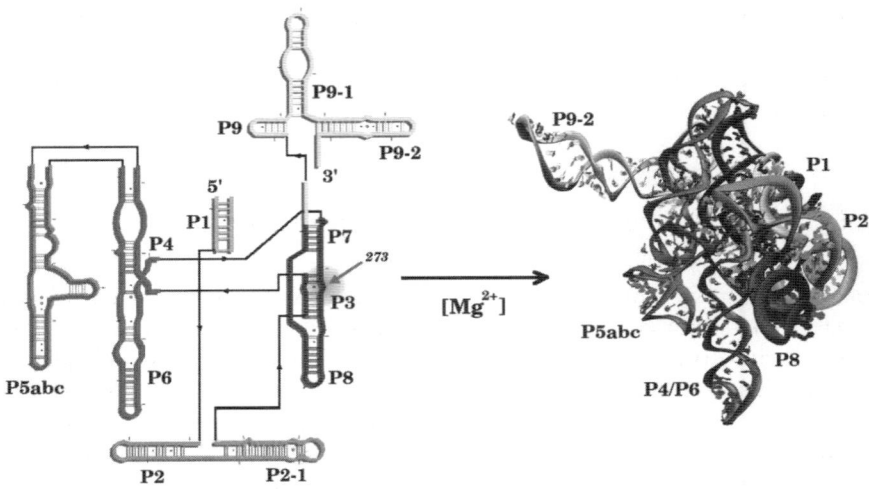

Fig. 2.3 Secondary structure of the most extensively studied group I intron from Tetrahymena. The secondary structure has a number of paired helices indicated by P1 through P9. Upon addition of excess Mg transition to compact tertiary structure, occurs (shown on the right) that is stabilized by the catalytic core formed by an interface involving the P5–P4–P6 and P3–P7–P8 helices. The structure of the independently folding P4–P6 domain is known in atomic detail (Cate et al. 1996). The structure on the right is a model proposed by Westhof and Michel (Lehnert et al. 1996) (See figure insert for colour reproduction)

escape from the misfolded conformations. Subsequent studies have used urea as an analytic probe of RNA stability (Sosnick and Pan 2003) in much the same way as it is done in protein folding studies. Another key prediction of the KPM is that point mutations can alter Φ. Remarkably, a single point mutation U273A in P3 increases Φ to about 80% (Pan et al. 2000). Thus, the mutation greatly reduces the kinetic possibility of being trapped in AltP3 that impedes folding of the wild type.

The most direct evidence for KPM was provided by using single molecule experiments that probes fluorescent energy transfer (FRET) efficiency (**E**) between two dyes attached to the 3′ and 5′ ends of the *Tetrahymena* ribozyme (Zhuang et al. 2000). The value of **E** is high (≈ 1) in the NBA whereas in the UBA **E** is low because the dyes are, on an average, far apart. Thus, under various folding or unfolding conditions, time-dependent changes in **E** in the FRET signal can be used as a reporter of the folding reaction. Addition of excess Mg^{2+} to initially unfolded molecules initiates the folding process. Under folding conditions **E** increases and the time needed to reach high **E** for the first time is the first passage time, τ_{1i} for the *i*th RNA molecule. From the distribution of first passage times, $P_{FP}(t)$, for an ensemble (in practice 100 molecules will suffice) of unfolded molecules, the probability that a molecule remains unfolded at time t is $P_u(t) = 1 - \int P_{FP}(s) \, ds$. Using the measured $P_{FP}(s)$ with single molecule FRET technique (Zhuang et al. 2000) the calculated $P_u(t)$ is best fit using a sum of two exponentials for the 400 nucleotide L-21 ribozyme (Thirumalai et al. 2001). The partition factor $\Phi \approx 0.06$. In other words, only 6% of the molecules fold rapidly by fast track without being kinetically trapped. It is worth noting that Φ for both L-21 is similar to the estimate for the pre-RNA, which suggests that the folding trajectories for the fast track molecules are similar.

2.4 Hairpin Formation Occurs by Multiple Routes

The relatively small stability gap between the native state and alternate misfolded or native-like conformations (Fig. 2.1) suggests that the folding landscape of even hairpins with a simple loop and a stem is rugged. The possibility of misfolding, at the secondary structural level, was already established in the context of tRNA folding, over 40 years ago (Lindhal et al. 1966). As a result, hairpin formation, when examined in detail, need not follow the classical two-state kinetics. Indeed, a series of recent experiments show that the kinetics of hairpin formation in RNA or ss-DNA is best described as a multi-step process (Jung and Van Orden 2006; Ma et al. 2006, 2007), thus challenging the conventional premise that small nucleic acid hairpins, fold in a two-state manner (Bloomfield et al. 2000; Tinoco et al. 2002; Turner et al. 1988).

The signatures of multi-state folding/unfolding are reflected in the kinetic data of ultra fast T-jump experiments that can discern the metastable intermediates. Multiple probes attached to the same molecule revealed that the folding is achieved through a series of dynamic steps that occur on vastly different time scales (Jung and Van Orden 2006; Ma et al. 2006, 2007). In contrast, single molecule force experiments (Liphardt et al. 2001; Woodside et al. 2006) showed that, when the ends of molecule are held at the transition mid-force (f_m), the hairpin stochastically hops between the two discrete values of end-to-end distance (R). The statistics of R exhibits a bimodal distribution without signatures of populated intermediates. However, when refolding is initiated by relaxing the applied force (f), metastable intermediates manifest themselves. By varying f, transitions from these misfolded structures to the folded structure can be facilitated – a process that is reminiscent of annealing by raising temperature.

To illustrate the consequences of the rugged folding landscape of nucleic acid hairpins, we simulated both thermodynamics and kinetics of RNA hairpin in detail by varying temperatures and mechanical forces using a coarse-grained Three Interaction Site (TIS) model (Hyeon and Thirumalai 2005, 2006). The TIS model simplifies the structural details of a nucleotide into the three coarse-grained interaction centers representing base, ribose, and phosphate group. Using the 22-nucleotide (nt) P5GA RNA hairpin (PDB ID: 1EOR) as a model system, we characterized the equilibrium ensemble of the RNA hairpin over the broad range of T and f conditions, and also simulated the relaxation dynamics of RNA hairpin under T and f-jump/quench conditions (Hyeon and Thirumalai 2008). The dynamics of RNA hairpins are monitored using two order parameters, i.e., the end-to-end distance (R) and the loop dihedral angles (φ) that can best describe the characteristics of the molecule. Here, $\varphi = 1 - \cos(\varphi_i - \varphi_i^0)$ where φ_i is the value of the ith dihedral angle in the GAAA tetraloop in the TIS representation of P5GA, and φ_i^0 is the corresponding value in the folded structure (Hyeon and Thirumalai 2008).

The equilibrium free energy surface expressed in terms of (R, φ) is characterized by two basins of attraction at the locus of critical points (T_m, f_m). Away from the critical condition, only one basin of attraction dominates. The free energy surface succinctly explains the origin of sharp bimodal transition between the folded and unfolded state when the RNA hairpin is subject to force. Thus, from thermodynamic consideration, hairpin formation can be described as a two-state system (see Hyeon and Thirumalai 2008 for details).

The refolding kinetics can be initiated by either a temperature (T) quench from high T to $T < T_m$ or by a force quench to $f < f_m$. Surprisingly, in both cases the kinetic folding pathways cannot be inferred from the free energy landscape. The RNA hairpin reaches the native state via multiple steps as observed in the recent kinetic experiments using high resolution T-jump experiments (Fig. 2.4). The expectation that kinetics can be gleaned from the free energy surface may be valid only if the RNA internal dynamics is rapid enough to establish quasi-equilibrium. For refolding induced by f or T-quench, such an assumption apparently breaks down. We find that the folding trajectories of different molecules are distinct which implies that there is diversity in the folding routes (Fig. 2.4).

The time-dependent changes in the order parameters R and ϕ show differences in folding pathways between T-quench and f-quench refolding. The ensemble of initially unfolded structures prepared by stretching the hairpin differs greatly from the thermally unfolded conformations. The initial ensemble of fully extended conformations, generated by forced-unfolding, is narrow and structurally homogeneous. The various conformations largely differ in the internal degrees of freedom while the overall end-to-end distance is large. Thus, the first step in the hairpin formation from the initially stretched conformations is the tetra-loop formation (Fig. 2.4), corresponding to the slow nucleation stage. Subsequent to the nucleation step the zipping of remaining base pairs leads to hairpin formation. Thus, hairpin is formed by this classic mechanism when folding is initiated by f-quench.

In contrast, upon T-quench, refolding commences from a broad thermal ensemble of unfolded conformations. As a result, nucleation can originate from regions other than near the tetra-loop. Consequently, the pathway diversity is larger when hairpin formation is initiated by T-quench rather than f-quench. The differences in the folding mechanism between these two methods are entirely due to the variations in the initial conformations. Just as folding trajectories in the self-assembly of ribozymes can be altered by pre-incubation with Na$^+$, here the routes to hairpin formation can be precisely controlled by applying mechanical force.

The simulations show that the complexity of energy landscape observed in ribozyme experiments is already reflected in the formation of simple RNA hairpin (Chen and Dill 2000; Thirumalai and Hyeon 2005; Treiber and Williamson 2001; Woodson 2005). Exploring the details of the heterogeneous kinetics requires multiple probes that control the conformations in the ensemble of unfolded states.

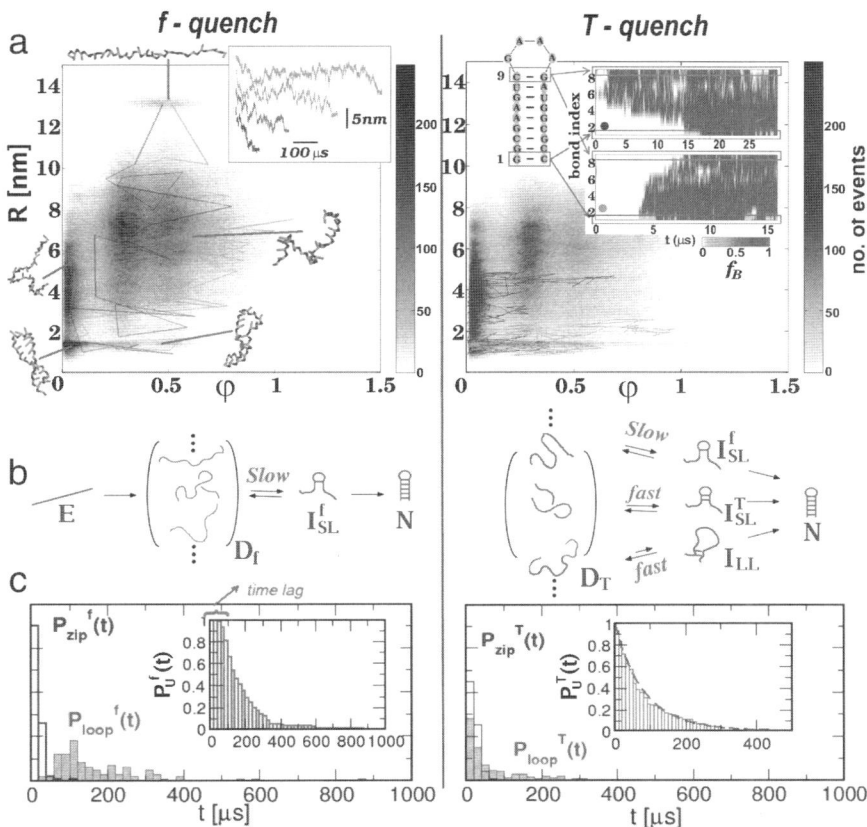

Fig. 2.4 Kinetic analysis of the refolding trajectories upon f-quench and T-quench. (**a**) Conformational space navigated by the refolding trajectories projected onto the (R, φ) plane. The trajectories of individual molecules are overlapped onto the (R, φ) plane. The corresponding trajectories monitored using a single parameter are shown in the insets. (**b**) Summary of the pathways to the NBA inferred from the dynamics depicted in (**a**). (**c**) Statistical analysis of refolding kinetics. The refolding time for each molecule is decomposed into looping and zipping time as $\tau_{FP} = \tau_{loop} + \tau_{zip}$. The fraction of unfolded molecules ($P_u(t) = 1 - \int_0^t d\tau P_{FP}(\tau)$) where $P_{FP}(\tau)$ is the refolding or first passage time distribution) is plotted in the inset. The probability of the hairpin remaining unfolded upon f-quench $P_u^f(t)$ shows a lag phase (left hand side of C) suggesting the presence of an intermediate, while $P_u^T(t)$ is well fit using $P_u^T(t) = 0.4\exp(-t/62\mu s) + 0.6\exp(-t/100\mu s)$ (See figure insert for colour reproduction)

2.5 Ion–RNA Interactions Affects Stability, Pathway Diversity and Transition States

To fold, RNA must overcome the large electrostatic repulsion between the negatively charged phosphate groups. At high temperatures ion–RNA interactions are weak, and the gain in translational entropy makes the ions disperse homogeneously in solution without condensing onto RNA. As a result RNA is relatively extended with $R_G \sim aN^\nu$ ($\nu \approx 1$). A naive estimate of the electrostatic repulsion is $E_R \approx (Ne)^2/\varepsilon R_G \approx Nk_BT(l_B/a)$ (Thirumalai et al. 2001). Since $(l_B/b) > 1$ it follows that $E_R/k_BT >> 1$ even when N is small. Therefore, under folding conditions substantial softening of the electrostatic interactions must be achieved through the screening of the electrostatic repulsion or counterion condensation.

Although a complete theoretical treatment of the interaction of counterions and RNA (or other polyelectrolytes for that matter) is lacking, the qualitative aspects of RNA–ion interactions can be understood using the Manning picture (Manning 1978). Charge neutralization is thought to result from the condensation of counterions onto the charged polyanion resulting in overall minimization of the free energy of RNA. Because folded RNA is aperiodic with irregular grooves the electrostatic potential is non-uniform. As a result, the condensed ions can be grouped into distinct classes. Examination of crystal structures of RNA, biophysical and theoretical analysis shows that ions in the vicinity of the strong electrostatic RNA molecule can be considered as (a) diffuse ions that are localized within the volume of RNA or (b) discrete ions that interact specifically with certain sites in the folded structure (Draper 2004).

The theories based on the Manning picture as well as solution to the non-linear Poisson–Boltzmann (NLPB) equation (Draper 2004) show that bulk of the charge neutralization is due to the non-specific association of the diffuse ions on RNA (Heilman-Miller et al. 2001). Counterion-condensation occurs at low temperatures because the loss in the translational entropy of the ions (viewed as unstructured species) is compensated by a gain in the association energy between ions and RNA. As a result of the condensation of the ions there is a substantial reduction in the overall average charge per phosphate group. For highly charged rod ion, condensation occurs if $l_B/A > 1/Z$ where Z is the counterion valence, and A, the distance between charges which is about 3 Å for poly A and 1.3 Å for A-form double helix. The estimate based on charged rods also provides a useful measure of the charge renormalization for RNA.

A few key consequences of the Manning theory follow by treating the condensed and free (in solution) counterions as two equilibrium phases. The chemical potential of the free ions is $\mu_F = -k_BT \log \phi$ where ϕ is the volume fraction of the counterions, while the chemical potential of the diffuse condensed ions is $\mu_C = Ne_RZk_BT \times (l_B/R_G)$ where N is the number of nucleotides, e_R ($<e$) is the fraction of net charge on the phosphate upon condensation of the ions, and Z is the valence of counterion. By equating $\mu_F = \mu_C$ we obtain $Ne_R = -(R_G/l_BZ)\log \phi$. Using an appropriate R_G value for *Tetrahymena* ribozyme it turns out that nearly 90% of the charge is neutralized with $\phi = 0.01$ for monovalent ions. A similar value is obtained for $[CO(NH3)_6]^{3+}$ with $\phi \approx 10^{-6}$ (Heilman-Miller et al. 2001).

The relationship between the extent of charge neutralization and the size of the compact structures lead to several qualitative predictions. (a) Multivalent cations are more efficient than monovalent ions in reducing Ne. As a result the concentration of ions needed for inducing compact RNA structures decreases as Z increases. (b) Compared to monovalent ions RNA structures formed by multivalent ions are more compact with R_G decreasing as $\approx 1/Z^2$. (c) The more compact misfolded structures have lower free energies than those formed in the presence of monovalent ions. As a result the folding rates from the compact misfolded structures to the native state should decrease as Z increases. Below we briefly discuss experiments that have provided support to these predictions.

Stability and Valence. From the simple theoretical picture we infer that as the valence increases the efficiency of inducing compact conformation must also increase. Indeed, experiments show that the midpoint of transition (C_m) decreases as Z increases. In particular, for *Tetrahymena* ribozyme $C_m = 0.46 M$ in Na^+ whereas in $[CO(NH3)_6]^{3+}$, $C_m = 12 \mu M$ (Heilman-Miller et al. 2001). Cooperativity of the folding transition, and hence the free energy of stability of the native state typically increases with Z although other factors such as size and shape of ions also play an important role (see below). The crucial role played by the ion valence can be understood by considering the electrostatic attraction between the residual charge on the counterions and the renormalized total charge (Ne_R) on RNA after ion condensation. In the monovalent case there is only a weak dipole–dipole interaction between ion pairs that form when $Z = 1$ ion interacts with the negatively charged phosphate groups. In contrast, multivalent ions induce attractive bridging interactions between regions of RNA that are well separated. Such long range interactions ($\propto 1/r$) in the presence of multivalent ions stabilize compact structures more effectively than monovalent ions because they can bridge two or more phosphate groups.

Charge Density of Ions and RNA Stability. The extent of compaction depends on size as well, and we expect variations in stability at a fixed Z but differing size. Excluded volume interactions between condensed counterions result in spatial correlations that position any two metal ions at distances greater than the sum of their ionic radii. As a result ion size not only determines the distance of closest approach to the negatively charged phosphate group but also affects the spatial location of other diffuse ions. The effect of ion–ion correlation is difficult to include in theoretical treatments within the Manning picture and the NLPB approach (see however Ha and Thirumalai 2003; Tan and Chen 2005; Chen 2008). However, qualitative effects of correlations between condensed ions, due to excluded volume, on RNA stability can be obtained using simple arguments.

As both Z and the volume (V) of the ions contribute to the distribution of ions around the RNA, they also both determine the nature of the counterion-induced compact states of RNA. Thus, to a first approximation, the natural variable that should control RNA stability is the charge density $\zeta = Ze/V$ where Ze is the charge and V is the volume of the cation. In accordance with this expectation, the changes in stability of in *Tetrahymena* ribozyme in various Group II metal ions (Mg^{2+}, Ca^{2+}, Ba^{2+}, and Sr^{2+}) showed a remarkable linear variation with ζ (Fig. 2.5). The extent of stability is largest for ions with the largest ζ (smallest V). Brownian dynamics simulations showed that this effect could be captured solely by non-specific interaction of ions with polyelectrolytes (Fig. 2.5) in the absence of any site-specific ion–RNA interactions. These findings and similar variations of stability in different sized diamines (Koculi et al. 2004) show that (a) the bulk of the stability arises from non-specific association of ions with RNA, and (b) stability can be greatly altered by valence, shape, and size of the counterions.

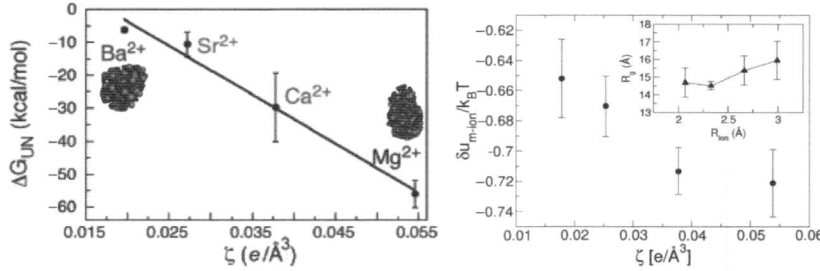

Fig. 2.5 (*Left*) Stability of *Tetrahymena* L-21Sca ribozyme (ΔG_{UF}) vs. cation charge density (ζ). (*Right*) The Brownian Dynamics simulations of polyelectrolyte collapse for the average stabilization energy between monomer of polyelectrolyte and ion as a function of ζ. The remarkable linear dependence on the left is captured by ion-induced collapse of flexible polyelectrolytes. The radius of gyration of the collapsed polyelectrolyte ($N = 120$) is plotted as a function of group II metalion size in the inset (See figure insert for colour reproduction)

Diversity of Folding Routes Depends on Initial Conditions and ζ. Single molecule FRET experiments indicate that the time-dependent changes ($\mathbf{E}(t)$) in FRET efficiencies vary from molecule to molecule (Zhuang et al. 2000). While the averages over an ensemble of such single molecule measurements are consistent with bulk experiments, the substantial variations might be indicative of inherent pathway diversity. Indeed, a consequence of the KPM is that the partition factor, Φ, can be altered not only by sequence variations but can also be changed by altering initial conditions. Analysis of single molecule FRET data of L-21 Sca I construct of the group I intron shows that preincubation in excess Na^+ before initiating folding alters Φ. Monovalent ions induce compaction in the initial unfolded ensemble of structures. As a result, folding commences from a region of the rugged energy landscape that restricts the starting unfolding ensemble which differs from the more expanded ensemble of structures in the absence of Na^+. Thus, upon initiation of folding the ribozyme assembles via different routes. The partition factor is a global measure of the pathway diversity because Φ is proportional to the number of molecules that fold rapidly through a restricted channel in the folding landscape (Fig. 2.2) without being kinetically trapped. Thus, the nature of pathways traversed depends critically on the starting RNA structures that can be manipulated by pre-incubation with monovalent cations.

Varying ζ can also change the diversity of folding routes. Just as pre-incubation with Na^+ leads to a more compact ensemble of initial structures the extent of collapsed structures can be altered by varying ζ. As a result, ions of differing ζ can modulate the diversity of folding routes. From a suitable generalization of the Manning picture it follows that as ζ increases, the extent of compaction of RNA increases. From the folding landscape (Fig. 2.6) it follows that the low ζ-ensemble of structures $\{\mathbf{I}_{NS}\}$ is less compact than those formed in ions with high ζ. As a result the higher the entropy associated with low ζ ions, the number of conformations in the $\{\mathbf{I}_{NS}\}$ ensemble is greater than in ions with high ζ. Thus, we expect that folding pathway diversity should increase when the ion charge density is low – prediction that can be tested using single molecule measurements.

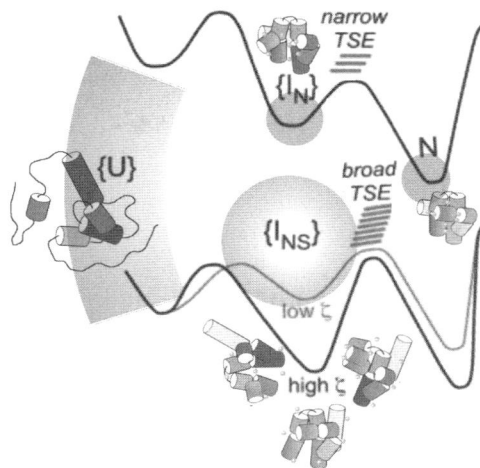

Fig. 2.6 Coupling of diversity of the folding pathways and heterogeneity of the transition state structures of *Tetrahymena* ribozyme to the charge density of counterions (ζ). The majority of the ribozyme folds through intermediates in which the core P3 helix is replaced by a non-native helix alt-P3. The pathway diversity increases as ζ decreases (lower part of the figure). For the fraction $((1-\Phi))$ slow track molecules, the transition state ensemble (TSE) along the $\{U\}\rightarrow\{I_{NS}\}\rightarrow N$ pathway becomes broader and less structured as the ζ decreases. Thus, the system is more dynamic in polyamines (low ζ) than Mg^{2+} (high ζ). The fast track molecules that fold via $\{U\}\rightarrow\{I_N\}\rightarrow N$ (Φ) (upper part of the figure) first form specifically collapsed compact structure that becomes increasingly native-like as the folding reaction proceeds. For $\{U\}\rightarrow\{I_N\}\rightarrow N$ we suggest that the TSE is narrow with little structural heterogeneity. The pathway diversity is expected to increase as ζ decreases (See figure insert for colour reproduction)

Role of ζ on the Plasticity of the Transition State Ensemble (TSE). In contrast to protein folding much less is known about the nature of TSE in self-assembly of large RNA molecules. Only recently has there been concerted efforts to decipher the nature of TSE in RNA folding (Bokinsky et al. 2003; Fang et al. 2002; Koculi et al. 2006). Although it is tempting to propose a very general picture of the TSE or the rate-limiting step in RNA self-assembly, it should be kept in mind that, just as the nature of intermediates in RNA folding can be easily altered by changing the properties of ions so too can the location and plasticity of the TSE (Koculi et al. 2006). Using single molecule FRET experiments of docking and undocking in hairpin ribozyme it has been suggested that the TSE is compact in Mg^{2+}, and perhaps share much of the structural characteristics of collapsed native-like intermediates. By using a combination of biophysical methods it has been suggested that the TSE for the C domain of RNase P involves reorganization of metal-ion binding sites late in the folding process.

A much more general analysis of the TSE and its variations requires studies that alter the valence, size, and shape of the ions. In a recent study, the variations in folding kinetics and TSE movements were probed using concentrations of polyamines ($^+H_3N(CH_2)_nNH_3^+$ with $n = 2–5$) as a natural perturbation of the RNA folding landscape. Several key observations were made: (a) The TSE is much broader in polyamines than the Mg^{2+} which has a larger ζ. (b) Ions with larger ζ give rise to higher free energy barriers to folding. (c) By using the Tanford β parameter it was surmised that the average location of the TSE is closer to the native state (β closer to unity) when folded in ions with small ζ. These observations allowed us to describe the general changes in the TSE as ζ is changed (Fig. 2.6). At low ζ the $\{I_{NS}\}$ ensemble is less compact with higher entropy than when ζ is high. Thus, the free energy barriers are largely determined by entropy changes at low ζ. By contrast, when ζ is high (n = 2 in polyamine for example) the $\{I_{NS}\}$ ensemble is more compact, and the free energy barrier separating the intermediates and the NBA is largely enthalpic. Thus, by modulating the charge density of ions one can modulate the interplay between entropy and enthalpy and control the very nature of transition state structures in RNA folding.

2.6 Folding Rates, free Energy Barriers and Kramers Prefactor for RNA

Besides the nature of ions, sequence length (N) also plays an important role in determining the folding time, τ_F. Given that non-coding RNAs are "evolved" heteropolymers, it is not surprising that the N should play a crucial role in controlling the folding rate (k_F). For minimally frustrated sequences, $\log(k_F/k_0) \sim \alpha \log N$ with $\alpha \approx 4$ (Thirumalai 1995) at $T < T_F$ where the prefactor k_0 can be obtained using Kramers' theory. Because biopolymers are topologically frustrated, there is residual roughness even in two-state folders. As a result, the folding kinetics characterized with a single barrier crossing event follows the relation

$$\log (\tau_F/\tau_0) = \Delta G^{\ddagger}_{UF}/k_B T$$

where the prefactor τ_0, that is often estimated using transition state theory (TST), has to be determined using Kramers theory. If we assume a Gaussian distribution for the free energy barriers with dispersion $< (\Delta G^{\ddagger}_{UF})^2 > \sim N$ then $\Delta G^{\ddagger}_{UF}/k_B T \sim N^{\beta}$ with $\beta = 1/2$ (Thirumalai 1995; Thirumalai and Hyeon 2005). Other arguments predict that $\beta = 2/3$ (Finkelstein and Badretdinov 1997; Wolynes 1997). The sublinear scaling of the effective barrier height with N naturally explains both rapid folding (kinetics) and marginal stability (thermodynamics) of single domain proteins and RNA.

In contrast to proteins (Li et al. 2004), the number of experiments for RNA molecules that report τ_F as a function of N is small; hence, the variation of k_F with N has not been examined. Experiments on hairpin formation in oligonucleotides and helix-coil transition theories already showed that k_F must be sensitive to N. We have analyzed the N dependence on RNA folding kinetics using the available data from the literature (Thirumalai and Hyeon 2005). Here, we extend these calculations using a slightly larger dataset. Surprisingly, the rates that vary over 7 orders of

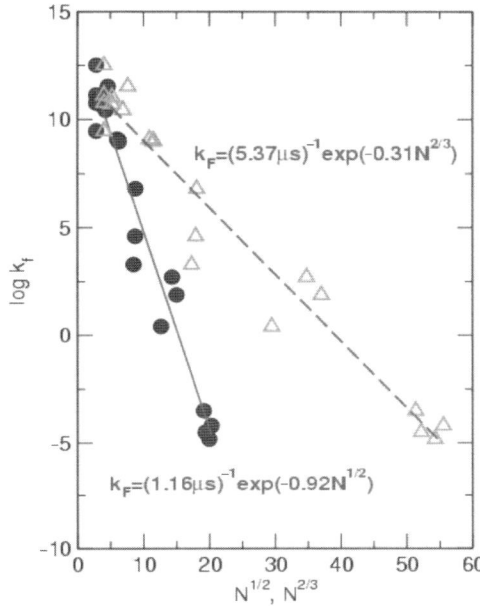

Fig. 2.7 Dependence of RNA folding rates as a function of N, the number of nucleotides. Fits of log k_f as a function of N^β with $\beta = 1/2$ or $\beta = 2/3$ are also shown

magnitude depend on N as predicted by theory. The correlation coefficients for both values of β are in excess of 0.9. In contrast to proteins, the predicted N dependence of k_F is more closely obeyed (Finkelstein and Badretdinov 1997; Galzitskaya et al. 2003, 2004). Using the results in Fig. 2.7, the difficult-to-measure prefactor τ_0, which should be estimated by using Kramers' theory, can be calculated. From the scaling plots in Fig. 2.7 we find that $\tau_0 \approx 1.2\,\mu s$ for $\beta = 1/2$ and $\tau_0 \approx 5.4\,\mu s$ for $\beta = 2/3$. Both these estimates for the RNA folding prefactor are nearly *six orders of magnitude larger than the TST value* $(=h/k_B T \sim 0.2\,ps)$. The large value of τ_0 implies that the effective free energy barriers from the measurements of rates alone using TST prefactor, overestimates the activation free energies by $\sim 15\,k_B T$. The TST prefactor is applicable only if breakage of a single bond is involved at the transition state. While this may be appropriate for gas phase reactions it cannot describe folding that is determined by collective events. The prefactor τ_0 represents the time scale in which folding can occur without barriers, i.e., by diffusion limited process. An estimate for the most elementary event in folding (for example base pairing in RNA) leads to the Kramers' estimate of $\sim 1\,\mu s$ for τ_0. Our estimate is in accord with the typical base pairing rate (Porschke and Eigen 1971; Porschke et al. 1973).

2.7 Conclusions

A number of factors, such as the lack of diversity of the building blocks, sequence variations, polyelectrolyte character of the phosphate backbone, and the subtle roles played by the ions, contribute to the complexity of RNA folding. The interplay of these factors are evident in the emergence of astounding variety of structures with each fold having both regions of flexibility and rigidity – features that lend themselves to RNA molecules being able to execute wide-ranging cellular functions. However, from a biophysical perspective the following features make it hard to provide a molecular understanding of RNA folding. (a) It would seem that the constraint of Watson–Crick base pairing and the inherent stability of RNA secondary structures would make RNA folding relatively simpler than the protein-folding problem. However, nearly half of the base pairs are involved in non-WC structures, which makes it difficult to predict even the RNA secondary structures especially when the number of nucleotides exceeds about 50. (b) The inherent complexity of RNA folding kinetics can be better appreciated by comparisons to the better-studied protein folding problem. To a large extent, folded proteins are stabilized by favorable interactions between hydrophobic residues that are buried in the interior. The interactions between all the residues are short-range, and are in the order of the size of the residues themselves (~6 Å). In contrast, the ranges of interactions between the nucleotides or the structural motifs that drive RNA folding vary greatly. The ion-mediated interactions occur on the persistence length scale that varies from about (1–2) nm depending on the ion concentration. Other interactions using hydrogen bonds between the bases and stacking interaction that stabilize various elements of the RNA structure are shorter range in distance. The interplay of the interactions on distinct length scales that can be altered by changing valence and size of ions gives rise to multiple scenarios for folding. Despite these difficulties it is remarkable that, at some global level, the principles based on KPM, polyelectrolyte theory, and ion–RNA interactions allow us to qualitatively rationalize many puzzling aspects of RNA folding. Developments in single molecule experiments and novel theoretical tools will be needed to quantitatively understand the richness of RNA folding.

Acknowledgments One of us (DT) is grateful to Sarah A. Woodson for pleasurable collaboration on all aspects of RNA folding for over 12 years. We are pleased to acknowledge useful discussions with her and Eda Koculi on ion-RNA interactions. This work was supported in part by a grant from the National Science Foundation (CHE 05-14056).

References

Aronovitz JA, Nelson DR (1986) Universal features of polymer shapes. J Phys 47(9):1445–1456

Ban N, Nissen P, Hansen J, Moore PB, Steitz TA (2000) The complete atomic structure of the large ribosomal subunit at 2.4 angstrom resolution. Science 289(5481):905–920

Bloomfield VA, Crothers DM, Tinoco I Jr (2000) Nucleic acids, structures, properties and functions. University Science Books, Sausalito, CA

Bokinsky G, Rueda D, Misra VK, Rhodes MM, Gordus A, Babcock HP et al. (2003) Single-molecule transition-state analysis of RNA folding. Proc Natl Acad Sci U S A 100(16):9302–9307

Caliskan G, Hyeon C, Perez-Salas U, Briber RM, Woodson SA, Thirumalai D (2005) Persistence length changes dramatically as RNA folds. Phys Rev Lett 95(26):268–303

Cate JH, Gooding AR, Podell E, Zhou KH, Golden BL, Kundrot CE et al. (1996) Crystal structure of a group I ribozyme domain: principles of RNA packing. Science 273(5282):1678–1685

et al. 2005; Woodson 2005a). RNA secondary structure forms in a wide range of concentrations of K^+, Na^+ and Mg^{2+}. In contrast, formation of tertiary interactions often requires the presence of divalent metal ions. In kinetic studies, Mg^{2+} ions are often used to trigger tertiary folding. In the absence of Mg^{2+}, group I and II intron ribozymes both form secondary structure. Addition of Mg^{2+} quickly induces structure compaction of these RNAs, followed by a slow step that rearranges packing of helices and folds the ribozymes into the native state (Pyle et al. 2007; Woodson 2005b). In a recent study, several different conformations of a hairpin ribozyme were distinguished by flushing the enzyme with a series of buffers containing different concentrations of Mg^{2+} (Liu et al. 2007). Mg^{2+} is also often used to affect RNA conformational dynamics in NMR and single-molecule fluorescence studies (Al-Hashimi et al. 2003; Bokinsky and Zhuang 2005).

3.2.2 Methods to Monitor RNA Folding

UV/Optical Melting. In UV or optical melting, absorption of nucleic acids at around 260 nm (OD260) is monitored with gradually increased temperature. As base pairs are disrupted, OD260 increases about 25% (Bloomfield et al. 1999). This UV hyperchromicity results from loss of base stacking upon denaturation of double strands (Tinoco 1960). UV melting has been extensively used to measure thermodynamics of small duplex and hairpin structures of nucleic acids. Using a nearest-neighbor approximation (Tinoco et al. 1971), data from these experiments were used to compile free energy and enthalpy tables for base pairs in double helices, for loops and for other secondary structures (Mathews et al. 1999). Such thermodynamic information is the foundation for successful RNA secondary structure prediction programs based on energy minimization methods (Zuker 2003).

Although UV melting is a straightforward method, high quality data is difficult to obtain. Mergny and Lacroix (2003) reviewed good practices for performing this experiment and for proper analysis of data. The melting curve, absorbance (A) as a function of temperature (T), can be transformed to a differential melting curve, which plots dA/dT vs. T (Fig. 3.2). The differential melting curve is particularly powerful in distinguishing unfolding signals of different domains (Theimer et al. 2005), or in distinguishing unfolding of secondary and tertiary structures (Bukhman and Draper 1997; Lorenz et al. 2006; Shiman and Draper 2000).

Although most UV melting experiments track only OD260, the broad UV absorbance spectrum from 220 to 320 nm provides key information on nucleic acid structure. A thermal difference spectrum is the difference between two spectra at temperatures above and below T_m and its spectral shape can be used to interpret RNA folds (Mergny et al. 2005). Besides absorbance, other spectral signals, such as fluorescence or circular dichroism, can also be used to monitor RNA folding (Bloomfield et al. 1999).

Calorimetry. Heat associated with structural transitions of nucleic acids has long been measured by calorimetry, yielding changes in enthalpy during denaturation

a   b

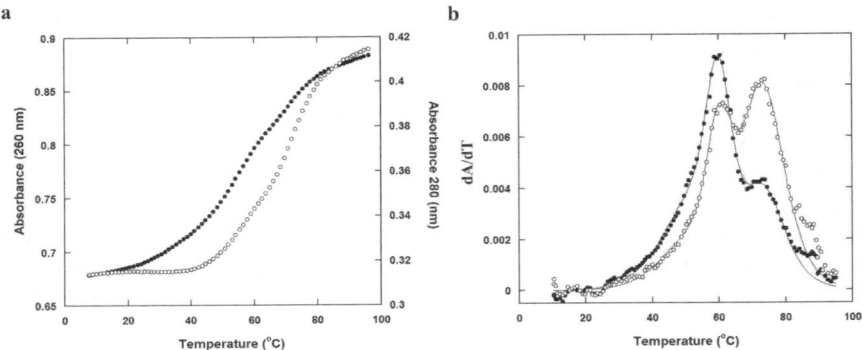

Fig. 3.2 Thermal melting of a human telomere pseudoknot. (**a**) UV absorbance at 260 nm (●) and 280 nm (○) are plotted as a function of temperature. (**b**) Differential melting curve, dA/dT vs. T, reveals unfolding transitions of two helices of the pseudoknot. Courtesy of Carla A Theimer, University at Albany, State University of New York

(Sturtevant and Geiduschek 1958). In differential scanning calorimetry (DSC), difference in heat flow to the sample and a reference is measured, as temperature is changed at a constant rate. Heat capacity change, ΔC_p, is measured as a function of temperature (Privalov and Dragan 2007). Structural transitions of RNA are characterized by heat absorption peaks. The area under such a peak yields ΔH^0 directly. ΔS^0 and ΔG^0 can be obtained indirectly, but they are less reliable than ΔH^0. Isothermal titration calorimetry (ITC) measures release or absorption of heat when two solutions are quickly mixed; it is widely used to study ligand binding to macromolecules (Buurma and Haq 2007). It can also be used to study RNA folding that is coupled to ligand binding, such as riboswitches (Gilbert et al. 2007). ΔH^0 can be calculated from ITC data using assumed thermodynamic models, whereas in DSC experiments, ΔH^0 is obtained independently.

DSC is the most direct and thermodynamically rigorous measurement of ΔH^0 of RNA folding. ΔH^0 can also be extracted from melting experiments using van't Hoff plots (Puglisi and Tinoco 1989). This treatment depends on a two-state equilibrium hypothesis, and assumes that ΔH^0 of nucleic acid structures is temperature independent. Factors causing the difference between $\Delta H^0_{\text{van't Hoff}}$ and $\Delta H^0_{\text{calorimetry}}$ have been discussed (Chaires 1997; Mergny and Lacroix 2003; Mikulecky and Feig 2006). With recent improvement in instrumentation sensitivity, DSC and ITC use significantly less amount of sample than before. A surge of application of calorimetry, particularly ITC, appears in recent literature (Feig 2007).

RNA Footprinting of Chemical Modification. Most biochemical and biophysical methods measure the overall change in RNA folding. In contrast, RNA footprinting and NMR techniques have potential to monitor each nucleotide simultaneously during folding. RNA footprinting is based on the principle that different regions of a folded RNA have different solvent accessibility. Various chemical reagents and nucleases are more likely to react with solvent-exposed single strands than base paired helices or protein bound regions. The footprinting assays can be used to map structured and unstructured regions in an RNA with single-nucleotide resolution. Such information can be used as constraints to improve prediction of RNA structure (Mathews et al. 2004).

Hydroxyl radicals (Brenowitz et al. 2002; Tullius and Greenbaum 2005) cleave the backbone of RNA at positions of all four nucleotides. Hydroxyl radicals can be produced by the Fenton reaction of $Fe(EDTA)^{2-}$ and H_2O_2 (Price and Tullius 1992) or radiation of water by a synchrotron beam (Sclavi et al. 1997). When partially cleaved RNA samples are run on a denaturing polyacrylamide gel, the frequency of cleavage at each nucleotide indicates levels of folding. Time-resolved synchrotron footprinting can be used to study folding kinetics of large RNAs (Sclavi et al. 1998). Several software packages have been developed to semi-automatically analyze footprinting gel patterns (Das et al. 2005; Takamoto et al. 2004). This development not only allows quantitative extrapolation of folding kinetics from data, but also makes it possible to rigorously compare folding rates of different domains in large RNAs. Hydroxyl footprinting has been used to elucidate folding pathways of a group I intron ribozyme, revealing local folding kinetics and parallel folding pathways (Fig. 3.3) (Laederach et al. 2006, 2007). Furthermore, a new method, multiplexed hydroxyl radical cleavage analysis has been developed to map long range interactions in ribozymes (Das et al. 2008).

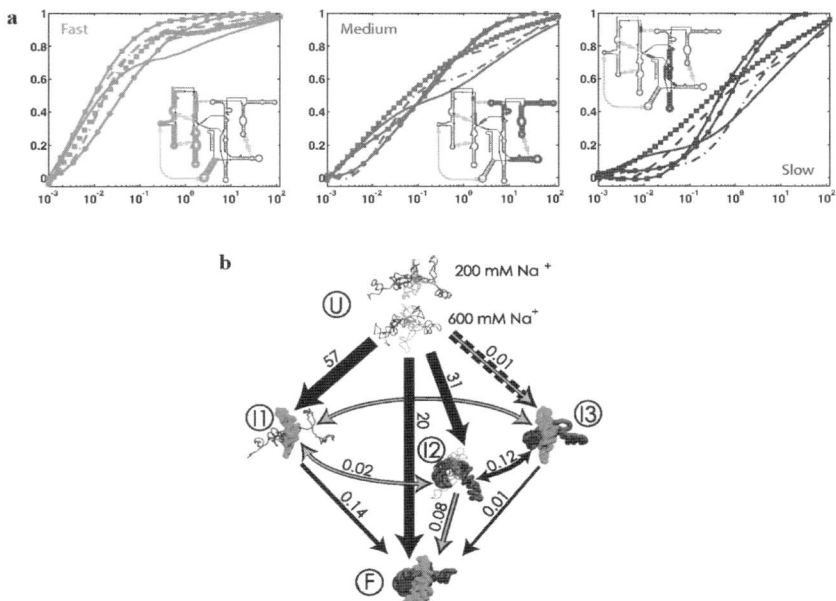

Fig. 3.3 Hydroxyl footprinting shows different local folding of L-21 group I intron ribozyme from *Tetrahymena thermophila*. (**a**) Fast, medium and slow folding of domains. Each curve represents one ionic condition. (**b**) A multi-pathway kinetic folding model involves three intermediates (I1, I2 and I3) from unfolded (U) to folded (F) state. Adapted from (Laederach et al. 2007); (See figure insert for colour reproduction)

Many chemicals, such as dimethylsulfate (DMS), selectively modify certain types of bases (Tijerina et al. 2007). A primer extension reaction on a modified RNA will reveal the positions of reactive (single-stranded) bases because the added bulky chemical groups can stop reverse transcriptase. Recently, a new method called Selective 2′-Hydroxyl Acylation Analyzed by Primer Extension (SHAPE) has been developed to target 2′-hydroxyl groups of ribose by *N*-methylisatoic anhydride (Merino et al. 2005). SHAPE is specific to RNA and has the advantage of resolving all four ribonucleotides simultaneously. Using SHAPE chemistry, Weeks and colleagues revisited folding of tRNA and observed non-hierarchical folding, i e., tertiary interaction forms before secondary structure (Wilkinson et al. 2005). Although the non-hierarchical folding of tRNA has been long suspected, as the most stable secondary structure of tRNA should be a long hairpin (Gralla and DeLisi 1974), the SHAPE experiment provides direct evidence of this unusual folding pathway with single-nucleotide resolution. SHAPE results were further used to interpret multiple UV melting transitions of tRNA, another perennial problem. SHAPE chemistry has also been employed to study structure and dynamics of dimerization domains of several retroviral RNAs (Badorrek et al. 2006; Badorrek and Weeks 2005, 2006). Interestingly, SHAPE was used to examine local RNA dynamics in crystals and in the crystallization process (Vicens et al. 2007). This effort should be very helpful to solve the puzzling question of why some point mutations make RNA so crystalizable while others do not.

Another distinct advantage of RNA footprinting is that it can be used to study RNA folding in vivo. A recent study compared structures of tmRNA in vitro and in a few *E. coli* cell lines (Ivanova et al. 2007). Another study tested binding of aminoglycoside antibiotics to the HIV DIS kissing complex in *E. coli* (Ennifar et al. 2006).

NMR. NMR methods have long been used to determine structures of small RNAs (Latham et al. 2005). Several new NMR techniques have been developed in recent years to probe RNA conformational dynamics (Fürtig et al. 2007; Getz et al. 2007; Shajani and Varani 2007). The residual dipolar coupling (RDCs) technique is particularly promising in monitoring domain motions of RNA. A series of studies have been conducted on HIV TAR RNA and TAR-like RNAs, all of which have two helices connected by a bulge (Al-Hashimi et al. 2003; Hansen and Al-Hashimi 2007; Zhang et al. 2007, 2006). Similar Mg^{2+}-dependent domain motion has also been observed by single-molecule fluorescence techniques (Bokinsky and Zhuang 2005). A comparison between the two techniques to study RNA dynamics in the future will be very important. In principle, NMR covers a wide range of time scale from picoseconds to seconds whereas single-molecule fluorescence covers a range from milliseconds to seconds.

Biochemical Function. RNA starts to fold while being transcribed by RNA polymerase. It is difficult to apply most techniques described above to study RNA folding during transcription. Instead, catalytic activity of ribozymes has been used as an indicator for completion of native folding (Pan and Sosnick 2006). In addition, ribozyme activity has also been utilized to study RNA folding in vivo (Mahen et al. 2005).

3.3 Single-Molecule Measurements of Folding Thermodynamics and Kinetics

We are used to thinking about making measurements on samples that have large numbers of molecules; it may be helpful to see how many molecules we actually deal with in an experiment. Consider a microliter of water containing 0.1 μM solute; it has about 10^{19} water molecules and 10^{11} solute molecules. A spectroscopic measurement of the solution, such as fluorescence, will depend on the properties of all the 10^{11} solute molecules. If the solute is RNA in a solvent where half the molecules are unfolded, the fluorescence will correspond to the mixture of folded and unfolded molecules. However, in a single-molecule experiment each molecule will show either the fluorescence of the folded or the unfolded species. If the kinetics of unfolding are in an experimentally accessible range, we can see "hopping"; the fluorescence changes with time from one species to another. Clearly, if there are multiple species we will see multiple fluorescent spectra. Measuring 1,000 molecules is sufficient to have a very high probability of detecting species that are only present as 1% of the molecules. They would be very difficult to observe in the ensemble mixture. Fluorescence was used as an example, but absorption, scattering, or any measurable property can be substituted. For an RNA molecule, the end-to-end distance of the molecule can be measured using laser tweezers or atomic force microscopy; the molecular extension is indicative of RNA folding.

Single-molecule methods are most advantageous for characterization of kinetic mechanisms because kinetics is stochastic and several reaction pathways can coexist. This means that reactions do not occur synchronously; they occur randomly. We are all familiar with radioactive decay in that it is impossible to predict when a nucleus will react. We can measure a half-life for an ensemble of nuclei, but an individual nucleus can react before a 0.1 half-life or still not react after 10 half-lives. The distribution of lifetimes is exponential with 46% of actual lifetimes being between 0.5 half-life and 2 half-lives. For a reaction with intermediates, the stochastic nature of kinetics means that all species are present throughout the reaction. For a reaction with two intermediates, although we start with pure reactant, soon there will be four species: reactant, intermediate 1, intermediate 2, and product. It will be difficult to count the number of intermediates, to measure their concentrations as a function of time to establish a mechanism, and to obtain the rate constants of the substeps. In a single-molecule experiment there is only one species in the reaction at any time. Its lifetime and the new species it forms can be measured. Repeating the reaction many times characterizes the mechanism of the reaction and the distribution of lifetimes for each species. The transformation of one species into another following first-order kinetics gives an exponential distribution of lifetimes, with the mean lifetime of the species equal to the reciprocal of the rate constant for its reaction. However, if there are hidden intermediates, second-order reactions, or other mechanisms, the shape of the distribution can reveal them (McKinney et al. 2006). Single-molecule experiments provide information not available otherwise.

3.3.1 Force

Single-molecule experiments allow force to be applied to a molecule; thus mechanical unfolding, rather than the more familiar thermal and chemical denaturation, can be studied (Tinoco et al. 2006). However there are important differences in – and advantages of – the application of force. Force is a local perturbation, whereas temperature and solutes are global. This means we can pull or push on one molecule, or one part of a molecule, without disturbing any of the others. The four main methods that have been used to apply force to single DNA and RNA molecules are atomic force microscopy (AFM), optical tweezers, magnetic tweezers and flow stretching (Williams and Rouzina 2002). In AFM a sharp tip on a small cantilever picks up a molecule attached to a surface. Moving the cantilever relative to the surface applies a force on the molecule. With optical tweezer experiments, micron-size beads are attached to the molecule, and the beads are manipulated by two laser traps, or one trap and a micropipette. In magnetic tweezers and flow stretching, one end of the molecule is attached to a surface and the other end is linked to a (magnetic) bead. Application of a magnetic field or flow stretches the molecule. In all these types of experiments, applied force and extension of the molecule are measured as a function of time and are often presented as force-extension curve. Structure of nucleic acids is indicated by the extension of the molecule, and its stability is interpreted from the mechanical work of unfolding.

We will now focus on the concept of applying force to studying RNA structure and function. There have been many reviews in the last 3 years describing single-molecule studies of RNA (Bokinsky and Zhuang 2005; Cornish and Ha 2007; Li et al. 2008; Tinoco et al. 2006; Tinoco and Onoa 2005; Zhuang 2005); therefore, we will emphasize understanding and applying the methods.

Experimental Methods and Interpretation. Typical experimental designs for using force to study unfolding and folding of polynucleotides and polypeptides are shown in Fig. 3.4. The atomic force microscope has been mainly applied to proteins (Forman and Clarke 2007), but some work has been done on RNA (Bonin et al. 2002; Green et al. 2004). Optical tweezers have been mainly applied to RNA (Chen et al. 2007; Green et al. 2008; Greenleaf et al. 2008; Li et al. 2006a, b, 2007; Liphardt et al. 2001; Onoa et al. 2003; Vieregg et al. 2007; Wen et al. 2007), but some work has also been done on proteins (Cecconi et al. 2005). Currently, optical tweezers are the most convenient way to unfold structures adopted by single-stranded RNA. The forces typically applied are in the range of 1–100 piconewtons (pN), and changes in extension of the molecules between one and a few thousand nanometers (nm) can be measured with nm precision. The RNA of interest is synthesized by transcription of a plasmid coding for the RNA and for an extra 0.5–5 kb nucleotides on each end. DNA strands (obtained from the plasmid) complementary to the extra RNA nucleotides are added to make handles. One DNA strand has attached biotins, the other has digoxigenins. Thus the RNA can be held between two micron-sized beads with either streptavidin or anti-digoxigenin on their surfaces; the beads are controlled by a laser trap and a piezo-driven micropipette (Liphardt et al. 2001).

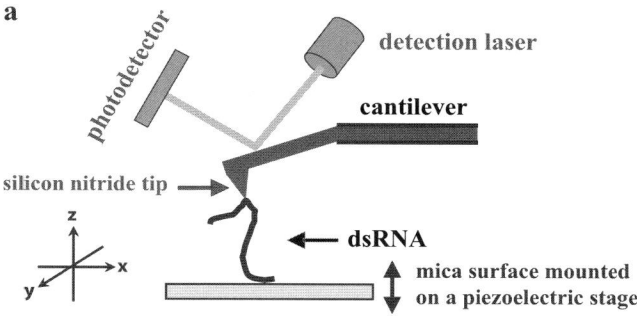

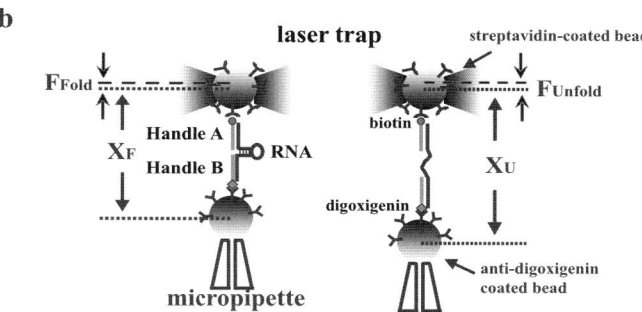

Fig. 3.4 Application of force to RNA. (**a**) An atomic force microscope cantilever is used to measure a force-extension curve for a double-stranded RNA attached to a surface. (**b**) Laser tweezers are used to measure a force-extension curve for an RNA held between two beads. The drawings are not to scale. Adapted from (Tinoco et al. 2006)

The simplest experiment is a pulling curve in which the force and the molecular extension (distance between beads) are measured, as one bead is smoothly moved away from the other and back. For an RNA hairpin of 48 bp (Fig. 3.5a) the pulling curve (Fig. 3.5b) is a result of (1) the straightening of the RNA·DNA hybrid handles, plus (2) the cooperative unfolding of the RNA hairpin, plus (3) the straightening of the single-stranded RNA and the RNA·DNA hybrid handles. The handles of 1–10 kbp are coiled like spaghetti at zero force, but straighten out as the force on their ends is increased. The extent of coiling is characterized by the ratio of the end-to-end distance (x) (the straight-line distance between the ends) and the contour length (L) (the distance between the ends measured by moving along the molecule). The ratio (x/L) varies from approximately 0 at zero force to one when the pulling force has fully straightened the handles. The dependence of the (x/L) ratio on force depends on one parameter: the persistence length, P, that measures the stiffness of the chain; the corresponding equation is the worm-like-chain (WLC) model (Bustamante et al. 1994).

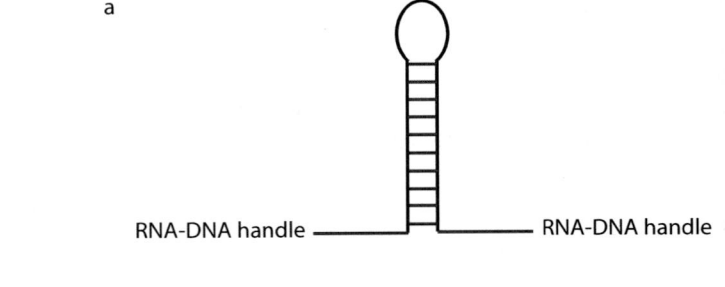

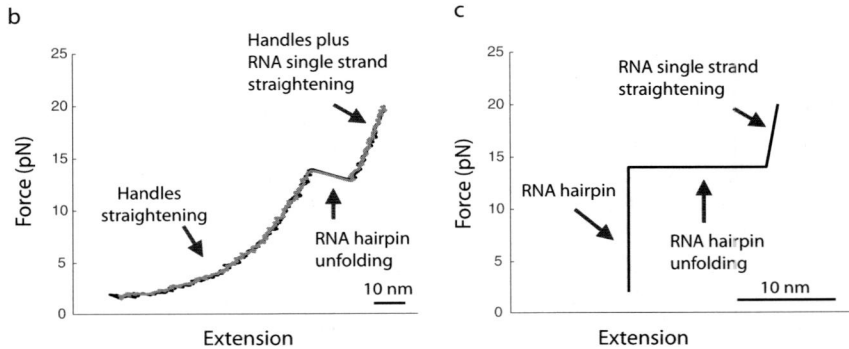

Fig. 3.5 Pulling RNA structures with optical tweezers. (**a**) A schematic RNA hairpin attached to RNA·DNA handles. The drawings are not to scale. (**b**) A typical force unfolding curve showing the stretching of the handles, the abrupt unfolding of the RNA, and later the continued stretching of the single-stranded RNA plus the handles. Refolding trajectory is shown in grey. (**c**) An idealized force unfolding curve for the RNA hairpin without handles that assumes the RNA unfolds in an all-or-none reaction

$$F = \frac{kT}{P} \left[\frac{1}{4(1-x/L)^2} + x/L - 1/4 \right]$$

in which k = Boltzmann constant; T = temperature. For nucleic acids the persistence length ranges from 50 nm for double-stranded DNA, to 1 nm for single-stranded RNA. The curved regions of the pulling curve are fitted well by the WLC with P = 10 nm for the RNA·DNA handles and P = 1 nm for the single-stranded RNA.

As the force increases, eventually the hairpin will unfold. In Fig. 3.5b the hairpin unfolds cooperatively to a single strand at a force of 15 pN; the abrupt increase in extension is termed a rip. The rip force and rip size reveal how stable the structure was and how many nucleotides were unfolded.

3.3.2   Thermodynamics

Reversible Work and Gibbs Free Energy. While noting that the force-extension curve (Fig. 3.5b) is the same when the force is raised as well as lowered, we learn that the transition is reversible. More complex structures, as we shall see, usually unfold in several steps and are not reversible. The reason reversibility is important is because reversible work (at constant temperature and pressure) is equal to the Gibbs free energy change, ΔG.

$$\Delta G = W_{rev} = \int_{x_1}^{x_2} F dx$$

Thus measuring the area under a reversible pulling curve – the integral of force times distance between extensions x_1 and x_2 – gives the reversible work and thereby the free energy necessary to stretch and unfold the construct from x_1 to x_2. In general this will be a combination of straightening the handles and unfolding the RNA.

To obtain the free energy for unfolding the RNA we subtract the contribution from straightening the handles; this is equivalent to measuring the area under the rectangle in Fig. 3.5c. In Fig. 3.5b the negative slope of the rip is caused by the laser trap (Liphardt et al. 2001). In Fig. 3.5c we assume that the hairpin does not partially open before 15 pN; a cooperative transition occurs. The vertical line corresponds to the constant distance between the ends of the stem of the folded hairpin. At 15 pN an all-or-none transition occurs (250 mM Na$^+$, 25°C) to give a single strand of RNA; the change in extension is the difference in end-to-end distance between the single strand and the ends of the stem (~2 nm). The single strand then straightens as the force increases. The area under the rectangle (15 pN × 20.1 nm) is the measured free energy; it is the free energy difference between fully formed hairpin and single strand at 15 pN. The value is 301.5 pN·nm = 181.6 kJ mol^{-1} = 43.4 kcal mol^{-1}.

Comparison with Zero-Force Measurements. To compare free energy of a transition measured by two different methods, it is essential that the initial and final states of the transition be the same for the different measurements. Here the initial state is the hairpin at our chosen temperature and solvent; the final state is the single strand at the same temperature and solvent. Both hairpin and single strand are under a tension of 15 pN. To obtain the free energy change at zero force, we assume that the free energy of the folded hairpin is independent of force and that the reaction is all-or-none. The free energy of the single strand does depend on force because the higher the force, the more stretched out the single strand is; the RNA has less entropy and higher free energy. To calculate the change in free energy of the single strand when it is straightened by a force of 15 pN, we integrate the WLC equation from $x_1 = 0$ to x_2 at $F = 15$ pN for a single strand contour length of 52 nucleotides (~0.59 nm per nucleotide). The result is that the stretched

RNA is 54.7 kJ mol^{-1} or 13.1 kcal mol^{-1} higher in free energy than it is at zero force. We conclude that at zero force the change in free energy for the unfolding transition is 126.9 kJ mol^{-1} or 30.3 kcal mol^{-1}.

To measure the free energy change of the transition in the usual bulk experiments, we would do a thermal melting experiment in the same solvent. However, folding free energies of some RNAs are difficult to obtain from this approach. For instance, a TAR hairpin from HIV-1 and the same hairpin, TARdb, with the three-base bulge deleted, have very high melting temperatures and their melting profiles are not two-state (Li, PTX, unpublished data). Instead, we can compare force unfolding with nearest-neighbor calculations of RNA free energies. For TARdb the measured values differ by 4% from the calculated results in 1 M NaCl from Mfold and agree within experimental error (Vieregg et al. 2007). For TAR they differ by 11%, outside the estimated error of 5%. Free energy values of other hairpins measured by force unfolding, agree with Mfold values with less than 10% difference (Collin et al. 2005; Dumont et al. 2006; Liphardt et al. 2001).

Irreversible Work and Gibbs Free Energy. It is rare that pulling curves are reversible as the one seen in Fig. 3.5b. The integral of force times distance (the area under the pulling curve) is still the work, but the work is not equal to the Gibbs free energy change. In principle by pulling the RNA slowly enough, the unfolding could be done reversibly, but slow pulling eventually becomes impractical, or even impossible because of drift in the instrument. When a process is not reversible, the transition is controlled by kinetics. That means it occurs stochastically; there is a distribution of transition forces, and therefore of work values when the process is repeated. Various amounts of work is dissipated – released as heat to the surroundings; also, rarely, the work may be less than the reversible work, as heat from the surroundings is converted to useful work. The second law states that heat cannot be converted to work, averaged over many molecules, at constant temperature. It does allow fluctuations to occur that convert heat into work at constant temperature. For macroscopic systems the fluctuations are negligible compared to the mean value, but for single-molecule systems the fluctuations, as seen in the distribution of work values, are measurable and provide useful new information.

The width of the work distribution approaches zero (within experimental error) for a reversible process; this width increases as the process becomes less reversible. During unfolding the reversible work will be near the minimum of the distribution (zero dissipated work). During refolding the reversible work will be near the maximum of the distribution (again zero dissipated work). To obtain the free energy, the irreversible work is measured many times on unfolding and refolding to obtain the corresponding distributions. The crossing of the two distributions gives the reversible work, as proven in the Crooks fluctuation theorem (Crooks 1999) which relates the distributions for folding and unfolding.

$$\frac{P_{un}(w)}{P_{re}(-w)} = \exp\left(\frac{w - \Delta G}{kT}\right)$$

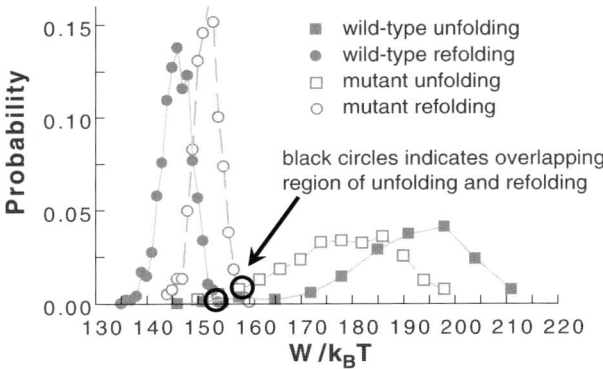

Fig. 3.6 Application of Crooks theorem to measure the Gibbs free energy change for unfolding a wild-type and mutant RNA. Over a thousand irreversible unfolding and refolding trajectories are measured; the crossover between the two distributions is the reversible Gibbs free energy. Note that the effect of changing one base pair out of 34 bps can be measured. Adapted from (Collin et al. 2005)

$P_{un}(w)$ is the probability of measuring unfolding work, w; $P_{re}(-w)$ is the probability of measuring refolding work, $-w$. Work done on the system, unfolding, is positive; work done by the system, refolding, is negative. When work, w, is equal to free energy, ΔG, the probabilities are equal – the distributions cross. A plot of the logarithm of the ratio of probabilities vs. w/kT gives a line of slope 1 with y-intercept = $-\Delta G/kT$ and x-intercept = $\Delta G/kT$. An example of its application is shown in Fig. 3.6 where work distributions are plotted for unfolding and refolding a three-helix junction from *E. coli* ribosomal RNA that binds the S15 ribosomal protein (Collin et al. 2005). The unfolding/refolding curves were repeated about 1,000 times to obtain the work distributions for the wild-type sequence and for a mutant with one base pair changed out of 34 bps. Clearly the process is very irreversible; the average work dissipated is 50–100 kJ mol^{-1}. However, the intersection of the curves is apparent, and a precise value can be obtained by plotting the logarithm of the ratio of probabilities vs. w/kT to obtain $\Delta G = 381.8 \pm 1$ kJ mol^{-1} and $\Delta G\ 391.3 \pm 0.5$ kJ mol^{-1} for unfolding the wild-type and mutant, respectively. It would be extremely difficult to measure this small difference by other means. The reversible work can also be estimated using Jarzynski's equality (Jarzynski 1997; Liphardt et al. 2002).

3.3.3 Kinetics

When a reaction reaches equilibrium, concentrations no longer change, but individual molecules keep switching between the reactive species present in the sample, a folded molecule unfolds and an unfolded molecule refolds, as depicted in the mechanism below.

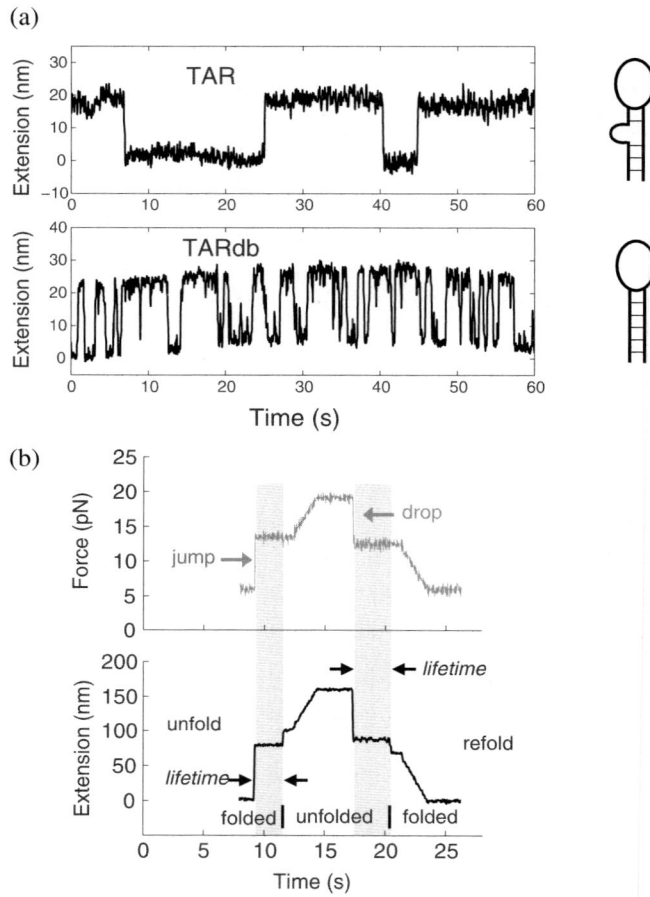

Fig. 3.7 Kinetics at constant forces. (**a**) Hopping of TAR and TARdb between folded and unfolded species at a force near the critical force where the species have equal stabilities. (**b**) Force jump and force drop protocols used to measure kinetics at forces where only one of the species is stable

$$\text{Folded} \rightleftarrows \text{Unfolded}$$

Single-molecule kinetics can thus be measured by observing this hopping between states at constant force (Li et al. 2006a, b; Liphardt et al. 2001; Manosas et al. 2007; Wen et al. 2007). At the critical force where the rate constants for forward and reverse reactions are the same, a molecule has equal probability to be folded and unfolded. Figure 3.7a shows unfolding/refolding hopping for the TAR hairpin (TAR) and the TAR hairpin with deletion of the three-base bulge (TARdb) near the critical force for each (Li et al. 2007). The mean lifetime, $\langle \tau \rangle$ of each state equals the reciprocal of the rate constant.

$$\langle \tau_{\text{folded}} \rangle = 1/k_1 \quad \langle \tau_{\text{unfolded}} \rangle = 1/k_{-1}$$

Because kinetics is stochastic there is a distribution of lifetimes, as seen in Fig. 3.7a. Different folding/unfolding kinetics of the two RNAs is also apparent. For simple two-state (first-order) kinetics the distribution of lifetimes is exponential. If intermediates exist between the two states, each should be visible if the lifetime is 2–3 times longer than the time resolution of the instrument. If intermediates can not be detected, their presence is still revealed in the distribution, which is no longer exponential. For N steps (N–1 intermediates) with equal rate constants the probability distribution density, $dP(\tau)$ is a Poisson equation.

$$\frac{dP(\tau)}{d\tau} = \left(\frac{k^N \tau^{N-1}}{(N-1)!} \right) (e^{-k\tau})$$

If the rate constants of each step are not equal the probability distribution depends on sums and differences of the rate constants.

Figure 3.8 shows the distributions of lifetimes for a reactant going to product with 0, 1, and 2 hidden intermediates with identical rate constants. The presence of one or more intermediates is clearly indicated if the distribution is not exponential. Curve fitting and statistical analysis of the distribution is required to quantitate the number of intermediates and their rate constants (McKinney et al. 2006).

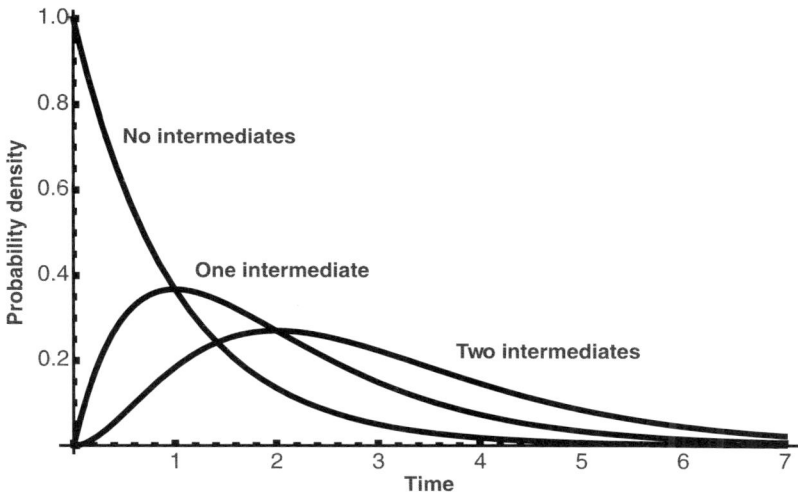

Fig. 3.8 Probability densities for the distribution of lifetimes for a reaction with no (the distribution is exponential), one, and two intermediates. Each step in the reaction has the same rate constant, $k = 1$

Hopping does not occur at forces away from the critical force – where two species have equal populations. The lifetime of one of the species increases, and eventually the other species is not detectable. To measure the kinetics under these conditions, the force is quickly jumped or dropped to a force that destabilizes the reactant, and its lifetime is measured (Li et al. 2006a, b). Figure 3.7b shows the protocol. The kinetics are measured at constant force (as in hopping), but the forces are different for unfolding and refolding.

Increasing force favors the longer species; its equilibrium concentration and its lifetime increases. For a two-state (no intermediates) reaction, the equilibrium constant, K, and rate constants, k, depend exponentially on force.

$$K_{(F)} = K_{(F=0)} e^{F \Delta X / k_B T}$$

$$k_{(F)} = k_{(F=0)} e^{F \Delta X^{\ddagger} / k_B T}$$

with ΔX, the difference in end-to-end extension between the two species, and $\Delta X^{\ddagger}$, the distance to the transition state in the reaction. $\Delta X^{\ddagger}$ is positive in unfolding and negative in refolding. Both ΔX and $\Delta X^{\ddagger}$ may depend on force. Importantly, $\Delta X^{\ddagger}$ indicates the position of the transition state along the reaction coordinate and can be used to interpret molecular structure at the transition state (Li et al. 2006b; Liphardt et al. 2001; Woodside et al. 2006a, b).

If there is no change in extension ΔX, then force has no effect on the equilibrium constant. The effect of force on the kinetics is the magnitude of the distance to the transition state, $\Delta X^{\ddagger}$, reflecting whether the transition state is similar to reactant or product. A compliant reactant, such as an RNA hairpin, has its transition state 5–10 nm away from the initial conformation (Liphardt et al. 2001). A brittle reactant involving the tertiary interactions in kissing hairpins or a pseudoknot, has a distance to the transition state of order 1 nm (Chen et al. 2007; Li et al. 2006a).

Comparison with Zero Force Kinetics. We discussed how to compare free energies (and therefore equilibrium constants) at zero force with those measured experimentally at non-zero forces. The fact that energies depend only on initial and final states allows this. However, kinetics depends on detailed mechanisms for reactions, so if different mechanisms occur, the rate constants will also be different. In unfolding a hairpin the force is applied to the ends of the molecule and the base pairs break sequentially from the end of the stem. In thermal or denaturant unfolding, base pairs can break from both ends of the stem, as well as internally. Similarly, when refolding under force, base pairs form, to close the loop before the end of the stem forms. Changes in mechanism mean that no simple extrapolation or correction to the rate constants measured under force can give rate constants measured thermally.

The important question is, which rates are more relevant to understanding the biological functions of the RNA? Clearly, RNAs in cells do not unfold and refold at high temperatures, or in 8 M urea. Instead, an RNA will fold during its transcription from DNA. Maybe force-drop experiments that identify kinetics as a function of force can be extrapolated to zero force to estimate rates of folding of RNA during its synthesis. Similarly, unfolding of RNA must occur during its translation by ribosomes, or transcription by RNA-dependent RNA polymerase. The ribosome or polymerase pulls on one end opening base pairs sequentially. This may be analogous to the force applied to the ends of an RNA by laser tweezers. Notably, a series of single-molecule mechanical studies have been carried out to elucidate mechanisms by which helicases (Cheng et al. 2007; Dumont et al. 2006; Johnson et al. 2007), and ribosomes unwind RNA (Wen et al. 2008).

3.4 Conclusion

Better understanding of how RNA folds and unfolds will provide better opportunities to understand, predict and control RNA function. Our knowledge of RNA folding is still limited, but it is improving with advances in existing methods and the advent of new techniques. This review serves as a short guide to common methods used for studying the thermodynamics and kinetics of RNA folding. A conspicuous omission is fluorescence techniques at both ensemble and single-molecule levels. However, this topic is discussed in great detail in another chapter of this book.

References

Al-Hashimi HM, Pitt SW, Majumdar A, Xu W, Patel DJ (2003) Mg2+-induced variations in the conformation and dynamics of HIV-1 TAR RNA probed using NMR residual dipolar couplings. J Mol Biol 329:867–873

Auton M, Bolen DW (2007) Application of the transfer model to understand how naturally occurring osmolytes affect protein stability. Methods Enzymol 428:397–418

Badorrek CS, Weeks KM (2005) RNA flexibility in the dimerization domain of a gamma retrovirus. Nat Chem Biol 1:104–111

Badorrek CS, Weeks KM (2006) Architecture of a gamma retroviral genomic RNA dimer. Biochemistry 45:12664–12672

Badorrek CS, Gherghe CM, Weeks KM (2006) Structure of an RNA switch that enforces stringent retroviral genomic RNA dimerization. Proc Natl Acad Sci U S A 103:13640–13645

Baird SD, Turcotte M, Korneluk RG, Holcik M (2006) Searching for IRES. RNA 12:1755–1785

Bartley LE, Zhuang X, Das R, Chu S, Herschlag D (2003) Exploration of the transition state for tertiary structure formation between an RNA helix and a large structured RNA. J Mol Biol 328:1011–1026

Bloomfield VA, Crothers DM, Tinoco I Jr (1999) Electronic and vibrational spectroscopy. Nucleic acids: structures, properties, and functions. University Science Book, Sausalito, CA

Blount KF, Breaker RR (2006) Riboswitches as antibacterial drug targets. Nat Biotechnol 24:1558–1564

Bokinsky G, Zhuang X (2005) Single-molecule RNA folding. Acc Chem Res 38:566–573

Bonin M, Zhu R, Klaue Y, Oberstrass J, Oesterschulze E, Nellen W (2002) Analysis of RNA flexibility by scanning force spectroscopy. Nucleic Acids Res 30:e81

Brenowitz M, Chance MR, Dhavan G, Takamoto K (2002) Probing the structural dynamics of nucleic acids by quantitative time-resolved and equilibrium hydroxyl radical "footprinting". Curr Opin Struct Biol 12:648–653

Brierley I, Pennell S, Gilbert RJ (2007) Viral RNA pseudoknots: versatile motifs in gene expression and replication. Nat Rev Microbiol 5:598–610

Buchmueller KL, Webb AE, Richardson DA, Weeks KM (2000) A collapsed non-native RNA folding state. Nat Struct Biol 7:362–366

Bukhman YV, Draper DE (1997) Affinities and selectivities of divalent cation binding sites within an RNA tertiary structure. J Mol Biol 273:1020–1031

Bustamante C, Marko JF, Siggia ED, Smith S (1994) Entropic elasticity of lambda-phage DNA. Science 265:1599–1600

Buurma NJ, Haq I (2007) Advances in the analysis of isothermal titration calorimetry data for ligand-DNA interactions. Methods 42:162–172

Cecconi C, Shank EA, Bustamante C, Marqusee S (2005) Direct observation of the three-state folding of a single protein molecule. Science 209:2057–2060

Chaires JB (1997) Possible origin of differences between van't Hoff and calorimetric enthalpy estimates. Biophys Chem 64:15–23

Chapman EJ, Carrington JC (2007) Specialization and evolution of endogenous small RNA pathways. Nat Rev Genet 8:884–896

Chen G, Wen JD, Tinoco I Jr (2007) Single-molecule mechanical unfolding and folding of a pseudoknot in human telomerase RNA. RNA 13:2175–2188

Cheng W, Dumont S, Tinoco I Jr, Bustamante C (2007) NS3 helicase actively separates RNA strands and senses sequence barriers ahead of the opening fork. Proc Natl Acad Sci U S A 104:13954–13959

Collin D, Ritort F, Jarzynski C, Smith SB, Tinoco I Jr, Bustamante C (2005) Verification of the Crooks fluctuation theorem and recovery of RNA folding free energies. Nature 437:231–234

Condon A, Davy B, Rastegari B, Tarrant F, Zhao S (2004) Classifying RNA pseudoknotted structures. Theor Comput Sci 320:35–50

Cornish PV, Ha T (2007) A survey of single-molecule techniques in chemical biology. ACS Chem Biol 2:53–61

Crooks GE (1999) Entropy production fluctuation theorem and the nonequilibrium work relation for free-energy differences. Phys Rev E 60:2721–2726

Das R, Laederach A, Pearlman SM, Herschlag D, Altman RB (2005) SAFA: semi-automated footprinting analysis software for high-throughput quantification of nucleic acid footprinting experiments. RNA 11:344–354

Das R, Kudaravalli M, Jonikas M, Laederach A, Fong R, Schwans JP, Baker D, Piccirilli JA, Altman RB, Herschlag D (2008) Structural inference of native and partially folded RNA by high-throughput contact mapping. Proc Natl Acad Sci U S A 105:4144–4149

De Rose VJ (2003) Metal ion binding to catalytic RNA molecules. Curr Opin Struct Biol 13:317–324

Dirks RM, Pierce NA (2004) An algorithm for computing nucleic acid base-pairing probabilities including pseudoknots. J Comput Chem 25:1295–1304

Dock AC, Lorber B, Moras D, Pixa G, Thierry JC, Giégé R (1984) Crystallization of transfer ribonucleic acids. Biochimie 66:179–201

Doty P, Boedtker H, Fresco JR, Haselkorn R, Litt M (1959) Secondary structure in ribonucleic acids. Proc Natl Acad Sci U S A 45:482–499

Draper DE, Grilley D, Soto AM (2005) Ions and RNA folding. Annu Rev Biophys Biomol Struct 34:221–243

Dumont S, Cheng W, Serebrov V, Beran RK, Tinoco I Jr, Pyle AM, Bustamante C (2006) RNA translocation and unwinding mechanism of HCV NS3 helicase and its coordination by ATP. Nature 439:105–108

Ennifar E, Paillart JC, Bodlenner A, Walter P, Weibel JM, Aubertin AM, Pale P, Dumas P, Marquet R (2006) Targeting the dimerization initiation site of HIV-1 RNA with aminoglycosides: from crystal to cell. Nucleic Acids Res 34:2328–2339

Feig AL (2007) Applications of isothermal titration calorimetry in RNA biochemistry and biophysics. Biopolymers 87:293–301

Forman JR, Clarke J (2007) Mechanical unfolding of proteins: insights into biology, structure and folding. Curr Opin Struct Biol 17:58–66

Fürtig B, Buck J, Manoharan V, Bermel W, Jäschke A, Wenter P, Pitsch S, Schwalbe H (2007) Time-resolved NMR studies of RNA folding. Biopolymers 86:360–383

Getz M, Sun X, Casiano-Negroni A, Zhang Q, Al-Hashimi HM (2007) NMR studies of RNA dynamics and structural plasticity using NMR residual dipolar couplings. Biopolymers 86:384–402

Giedroc DP, Theimer CA, Nixon PL (2000) Structure, stability and function of RNA pseudoknots involved in stimulating ribosomal frameshifting. J Mol Biol 298:167–185

Gilbert SD, Montange RK, Stoddard CD, Batey RT (2006) Structural studies of the purine and SAM binding riboswitches. Cold Spring Harb Symp Quant Biol 71:259–268

Gilbert SD, Love CE, Edwards AL, Batey RT (2007) Mutational analysis of the purine riboswitch aptamer domain. Biochemistry 46:13297–13309

Gollnick P, Babitzke P (2002) Transcription attenuation. Biochim Biophys Acta 1577:240–250

Gralla J, DeLisi C (1974) mRNA is expected to form stable secondary structures. Nature 248:330–332

Green NH, Williams PM, Wahab O, Davies MC, Roberts CJ, Tendler SJ, Allen S (2004) Single-molecule investigations of RNA dissociation. Biophys J 86:3811–3821

Green L, Kim CH, Bustamante C, Tinoco I, Jr (2008) Characterization of the mechanical unfolding of RNA pseudoknots. J Mol Biol 375:511–528

Greenleaf WJ, Frieda KL, Foster DA, Woodside MT, Block SM (2008) Direct observation of hierarchical folding in single riboswitch aptamers. Science 319:630–633

Grimson A, Farh KK, Johnston WK, Garrett-Engele P, Lim LP, Bartel DP (2007) MicroRNA targeting specificity in mammals: determinants beyond seed pairing. Mol Cell 27:91–105

Hannon GJ, Rivas FV, Murchison EP, Steitz JA (2006) The expanding universe of noncoding RNAs. Cold Spring Harb Symp Quant Biol 71:551–564

Hansen AL, Al-Hashimi HM (2007) Dynamics of large elongated RNA by NMR carbon relaxation. J Am Chem Soc 129:16072–16082

Henkin TM, Grundy FJ (2006) Sensing metabolic signals with nascent RNA transcripts: the T box and S box riboswitches as paradigms. Cold Spring Harb Symp Quant Biol 71:231–237

Ivanova N, Lindell M, Pavlov M, Holmberg Schiavone L, Wagner EG, Ehrenberg M (2007) Structure probing of tmRNA in distinct stages of trans-translation. RNA 13:713–722

Jan E (2006) Divergent IRES elements in invertebrates. Virus Res 119:16–28

Jang SK (2006) Internal initiation: IRES elements of picornaviruses and hepatitis c virus. Virus Res 119:2–15

Jarzynski C (1997) Nonequilibrium equality for free energy differences. Phys Rev Lett 78:2690–2693

Johnson DS, Bai L, Smith BY, Patel SS, Wang MD (2007) Single-molecule studies reveal dynamics of DNA unwinding by the ring-shaped T7 helicase. Cell 129:1299–1309

Jossinet F, Ludwig TE, Westhof E (2007) RNA structure: bioinformatic analysis. Curr Opin Microbiol 10:279–285

Koculi E, Hyeon C, Thirumalai D, Woodson SA (2007) Charge density of divalent metal cations determines RNA stability. J Am Chem Soc 129:2676–2682

Komar AA, Hatzoglou M (2005) Internal ribosome entry sites in cellular mRNAs: mystery of their existence. J Biol Chem 280:23425–23428

Laederach A, Shcherbakova I, Liang MP, Brenowitz MA, Altman RB (2006) Local kinetic measures of macromolecular structure reveal partitioning among multiple parallel pathways from the earliest steps in the folding of a large RNA molecule. J Mol Biol 358:1179–1190

Laederach A, Shcherbakova I, Jonikas MA, Altman RB, Brenowitz M (2007) Distinct contribution of electrostatics, initial conformational ensemble, and macromolecular stability in RNA folding. Proc Natl Acad Sci U S A 104:7045–7050

Lambert D, Draper DE (2007) Effects of osmolytes on RNA secondary and tertiary structure stabilities and RNA-Mg2+ interactions. J Mol Biol 370:993–1005

Latham MP, Brown DJ, McCallum SA, Pardi A (2005) NMR methods for studying the structure and dynamics of RNA. Chembiochem 6:1492–1505

Lemay JF, Penedo JC, Tremblay R, Lilley DM, Lafontaine DA (2006) Folding of the adenine riboswitch. Chem Biol 13:857–868

Li PTX, Bustamante C, Tinoco I Jr (2006a) Unusual mechanical stability of a minimal RNA kissing complex. Proc Natl Acad Sci U S A 103:15847–15852

Li PTX, Collin D, Smith SB, Bustamante C, Tinoco I Jr (2006b) Probing the mechanical folding kinetics of TAR RNA by hopping, force-jump and force-ramp methods. Biophys J 90:250–260

Li PTX, Bustamante C, Tinoco I Jr (2007) Real-time control of the energy landscape by force directs the folding of RNA molecules. Proc Natl Acad Sci U S A 104:7039–7044

Li PTX, Vieregg J, Tinoco I Jr (2008) How RNA unfolds and refolds. Annu Rev Biochem 77:27.1–27.24

Liphardt J, Onoa B, Smith SB, Tinoco I Jr, Bustamante C (2001) Reversible unfolding of single RNA molecules by mechanical force. Science 292:733–737

Liphardt J, Dumont S, Smith SB, Tinoco I Jr, Bustamante C (2002) Equilibrium information from nonequilibrium measurements in an experimental test of Jarzynski's equality. Science 296:1832–1835.

Liu S, Bokinsky G, Walter NG, Zhuang X (2007) Dissecting the multistep reaction pathway of an RNA enzyme by single-molecule kinetic "fingerprinting". Proc Natl Acad Sci U S A 104:12634–12639

Long D, Lee R, Williams P, Chan CY, Ambros V, Ding Y (2007) Potent effect of target structure on microRNA function. Nat Struct Mol Biol 14:287–294

Lorenz C, Piganeau N, Schroeder R (2006) Stabilities of HIV-1 DIS type RNA loop-loop interactions in vitro and in vivo. Nucleic Acids Res 34:334–342

Lu M, Draper DE (1995) On the role of rRNA tertiary structure in recognition of ribosomal protein L11 and thiostrepton. Nucleic Acids Res 23:3426–3433

Mahen EM, Harger JWC, Calderon EM, Fedor MJ (2005) Kinetics and thermodynamics make different contributions to RNA folding in vitro and in yeast. Mol Cell 19:27–37

Manosas M, Wen JD, Li PTX, Smith SB, Bustamante C, Tinoco I, Jr, Ritort F (2007) Force unfolding Kinetics of RNA using Optical Tweezers. II. Modeling Experiments. Biophys J 92:3010–3021

Mathews DH, Turner DH (2006) Prediction of RNA secondary structure by free energy minimization. Curr Opin Struct Biol 16:270–278

Mathews DH, Sabina J, Zuker M, Turner DH (1999) Expanded sequence dependence of thermodynamic parameters provides improved prediction of RNA secondary structure. J Mol Biol 288:911–940

Mathews DH, Disney MD, Childs JL, Schroeder SJ, Zuker M, Turner DH (2004) Incorporating chemical modification constraints into a dynamic programming algorithm for prediction of RNA secondary structure. Proc Natl Acad Sci U S A 101:7287–7292

McKinney SA, Joo C, Ha T (2006) Analysis of single-molecule FRET trajectories using hidden Markov modeling. Biophys J 91:1941–1951

Mergny JL, Lacroix L (2003) Analysis of thermal melting curves. Oligonucleotides 13:515–537

Mergny JL, Li J, Lacroix L, Amrane S, Chaires JB (2005) Thermal difference spectra: a specific signature for nucleic acid structures. Nucleic Acids Res 33:e138

Merino EJ, Wilkinson KA, Coughlan JL, Weeks KM (2005) RNA structure analysis at single nucleotide resolution by selective 2'-hydroxyl acylation and primer extension (SHAPE). J Am Chem Soc 127:4223–4231

Mikulecky PJ, Feig AL (2004) Heat capacity changes in RNA folding: application of perturbation theory to hammerhead ribozyme cold denaturation. Nucleic Acids Res 32:3967–3976

Mikulecky PJ, Feig AL (2006) Heat capacity changes associated with nucleic acid folding. Biopolymers 82:38–58

Noller HF (2005) RNA structure: reading the ribosome. Science 309:1508–1514

Onoa B, Dumont S, Liphardt J, Smith SB, Tinoco I, Jr., Bustamante C (2003) Identifying kinetic barriers to mechanical unfolding of the T. thermophila ribozyme. Science 299:1892–1895

Pan T, Sosnick T (2006) RNA folding during transcription. Annu Rev Biophys Biomol Struct 35:161–175

Parisien M, Major F (2008) The MC-Fold and MC-Sym pipeline infers RNA structure from sequence data. Nature 452:51–55

Price MA, Tullius TD (1992) Using hydroxyl radical to probe DNA structure. Methods Enzymol 212:194–219

Privalov PL, Dragan AI (2007) Microcalorimetry of biological macromolecules. Biophys Chem 126:16–24

Puglisi JD, Tinoco I Jr (1989) Absorbance melting curves of RNA. Methods Enzymol 180:304–325

Pyle AM, Fedorova O, Waldsich C (2007) Folding of group II introns: a model system for large, multidomain RNAs? Trends Biochem Sci 32:138–145

Ralston CY, He Q, Brenowitz M, Chance MR (2000) Stability and cooperativity of individual tertiary contacts in RNA revealed through chemical denaturation. Nat Struct Biol 7:371–374

Reeder J, Giegerich R (2004) Design, implementation and evaluation of a practical pseudoknot folding algorithm based on thermodynamics. BMC Bioinformatics 5:104

Ren J, Rastegari B, Condon A, Hoos HH (2005) HotKnots: heuristic prediction of RNA secondary structures including pseudoknots. RNA 11:1494–1504

Rivas E, Eddy SR (1999) A dynamic programming algorithm for RNA structure prediction including pseudoknots. J Mol Biol 285:2053–2068

Rösgen J (2007) Molecular basis of osmolyte effects on protein and metabolites. Methods Enzymol 428:459–486

Ruan J, Stormo GD, Zhang W (2004) An iterated loop matching approach to the prediction of RNA secondary structures with pseudoknots. Bioinformatics 20:58–66

Sashital DG, Butcher SE (2006) Flipping off the riboswitch: RNA structures that control gene expression. ACS Chem. Biol. 1:341–345

Schuwirth BS, Borovinskaya MA, Hau CW, Zhang W, Vila-Sanjurjo A, Holton JM, Cate JH (2005) Structures of the bacterial ribosome at 3.5 A resolution. Science 310:827–834

Sclavi B, Woodson S, Sullivan M, Chance MR, Brenowitz M (1997) Time-resolved synchrotron X-ray "footprinting", a new approach to the study of nucleic acid structure and function: application to protein-DNA interactions and RNA folding. J Mol Biol 266:144–159

Sclavi B, Sullivan MC, Chance MR, Brenowitz M, Woodson SA (1998) RNA folding at millisecond intervals by synchrotron hydroxyl radical footprinting. Science 279:1940–1943

Shajani Z, Varani G (2007) NMR studies of dynamics in RNA and DNA by 13C relaxation. Biopolymers 86:348–359

Shapiro BA, Yingling YG, Kasprzak W, Bindewald E (2007) Bridging the gap in RNA structure prediction. Curr Opin Struct Biol 17:157–165

Shiman R, Draper DE (2000) Stabilization of RNA tertiary structure by monovalent cations. J Mol Biol 302:79–91

Stark H, Lührmann R (2007) Cryo-electron microscopy of spliceosomal components. Annu Rev Biophys Biomol Struct 35:435–457

Street TO, Bolen DW, Rose GD (2006) A molecular mechanism for osmolyte-induced protein stability. Proc Natl Acad Sci U S A 103:13977–14002

Sturtevant JM, Geiduschek EP (1958) The heat denaturation of DNA. J Am Chem Soc 80:2911

Su LJ, Brenowitz M, Pyle AM (2003) An alternative route for the folding of large RNAs: apparent two-state folding by a group II intron ribozyme. J Mol Biol 334:639–652

Takamoto K, Chance MR, Brenowitz M (2004) Semi-automated, single-band peak-fitting analysis of hydroxyl radical nucleic acid footprint autoradiograms for the quantitative analysis of transitions. Nucleic Acids Res 32:E119

Theimer CA, Feigon J (2006) Structure and function of telomerase RNA. Curr Opin Struct Biol 16:307–318

Theimer CA, Blois CA, Feigon J (2005) Structure of the human telomerase RNA pseudoknot reveals conserved tertiary interactions essential for function. Mol Cell 17:671–682

Tijerina P, Mohr S, Russell R (2007) DMS footprinting of structured RNAs and RNA-protein complexes. Nat Protoc 2:2608–2623

Tinoco I Jr (1960) Hypochromism in polynucleotides. J Am Chem Soc 82:4785–4790

Tinoco I Jr, Bustamante C (1999) How RNA folds. J Mol Biol 293:271–281

Tinoco I Jr, Onoa B (2005) Folding, unfolding, and dynamics of RNA. One molecule at a time. In: Gesteland R, Cech T, Atkins J (eds.) The RNA World, 3rd edn. Cold Spring Harbor Laboratory, Cold Spring Harbor, pp. 723–745

Tinoco I Jr, Uhlenbeck OC, Levine MD (1971) Estimation of secondary structure in ribonucleic acids. Nature 230:362–367

Tinoco I Jr, Li PTX, Bustamante C (2006) Determination of thermodynamics and kinetics of RNA reactions by force. Q Rev Biophys 39:325–360

Tullius TD, Greenbaum JA (2005) Mapping nucleic acid structure by hydroxyl radical cleavage. Curr Opin Chem Biol 9:127–134

Vicens Q, Gooding AR, Laederach A, Cech TR (2007) Local RNA structural changes induced by crystallization are revealed by SHAPE. RNA 13:536–548

Vieregg J, Cheng W, Bustamante C, Tinoco I Jr (2007) Measurement of the effect of monovalent cations on RNA hairpin stability. J Am Chem Soc 129:14966–14973

Wen JD, Manosas M, Li PTX, Smith SB, Bustamante C, Ritort F, Tinoco I Jr. (2007) Force unfolding kinetics of RNA using optical tweezers. I. Effects of experimental variables on measured results. Biophys J 92:2996–3009

Wen JD, Lancaster L, Hodges C, Zeri AC, Yoshimura SH, Noller HF, Bustamante C, Tinoco I Jr (2008) Following translation by single ribosomes one codon at a time. Nature 452:598–603

Wickiser JK, Cheah MT, Breaker RR, Crothers DM (2005a) The kinetics of ligand binding by an adenine-sensing riboswitch. Biochemistry 44:13404–13414

Wickiser JK, Winkler WC, Breaker RR, Crothers DM (2005b) The speed of RNA transcription and metabolite binding kinetics operate an FMN riboswitch. Mol Cell 18:49–60

Wilkinson KA, Merino EJ, Weeks KM (2005) RNA SHAPE chemistry reveals nonhierarchical interactions dominate equilibrium structural transitions in tRNA(Asp) transcripts. J Am Chem Soc 127:4659–4667

Williams MC, Rouzina I (2002) Force spectroscopy of single DNA and RNA molecules. Curr Opin Struct Biol 12:330–336

Woodside MT, Anthony PC, Behnke-Parks WM, Larizadeh K, Herschlag D, Block SM (2006a) Direct measurement of the full, sequence-dependent folding landscape of a nucleic acid. Science 314:1001–1004

Woodside MT, Behnke-Parks WM, Larizadeh K, Travers K, Herschlag D, Block SM (2006b) Nanomechanical measurements of the sequence-dependent folding landscapes of single nucleic acid hairpins. Proc Natl Acad Sci U S A 103:6190–6195

Woodson SA (2005a) Metal ions and RNA folding: a highly charged topic with a dynamic future. Curr Opin Chem Biol 9:104–109

Woodson SA (2005b) Structure and assembly of group I introns. Curr Opin Struct Biol 15:324–330

Yanofsky C (2007) RNA-based regulation of genes of tryptophan synthesis and degradation, in bacteria. RNA 13:1141–1154

Zhang Q, Sun X, Watt ED, Al-Hashimi HM (2006) Resolving the motional modes that code for RNA adaptation. Science 311:653–656

Zhang Q, Stelzer AC, Fisher CK, Al-Hashimi HM (2007) Visualizing spatially correlated dynamics that directs RNA conformational transitions. Nature 450:1263–1267

Zhuang X (2005) Single-molecule RNA science. Annu Rev Biophys Biomol Struct 34:399–414

Zuker M (2003) Mfold web server for nucleic acid folding and hybridization prediction. Nucleic Acids Res 31:3406–3415

Chapter 4
Ribozyme Catalysis of Phosphodiester Bond Isomerization: The Hammerhead RNA and Its Relatives

William G. Scott

Abstract The hammerhead ribozyme is a comparatively small, self-cleaving RNA that has served as a prototype for understanding ribozyme catalysis. It has been intensively investigated using a variety of biochemical and biophysical techniques, yet for a simple ribozyme, it continues to yield surprises. A new structure of a full-length hammerhead ribozyme now reconciles over a decade of experimental discord and has helped to formulate a unified understanding of how it and other phosphodiester isomerase ribozymes function as catalysts. This whole family of ribozymes appears to exploit the chemistry of acid-base catalysis in a manner reminiscent of protein enzymes, such as RNase A, that catalyze similar reactions. Specifically, the roles of general base and general acid are often filled by the nucleotide functional groups themselves, in contrast to the originally anticipated ancillary structural role that RNA was thought to play, wherein the RNA was believed to be a passive scaffold upon which catalytically indispensable divalent metal ions might bind.

4.1   Introduction

The hammerhead ribozyme (Fig. 4.1) is representative of a class of small self-cleaving and ligating ribozymes that catalyze phosphodiester bond isomerization chemistry. Nucleic acids almost always comprise phosphodiester backbone linkages between the 3'-oxygen of one nucleotide ribose and the 5'-oxygen of an adjacent nucleotide ribose. While the phosphodiester backbone of DNA is extremely stable, the backbone of RNA is somewhat less so, due to the presence of the 2'-hydroxyl on the ribose. If a ribose 2'-hydroxyl should become deprotonated, it becomes a potent nucleophile that may attack the adjacent phosphodiester linkage, inducing a phosphodiester bond isomerization reaction and resulting in backbone cleavage. This reaction is accelerated in a basic solution that favors deprotonation of the

W.G. Scott
Department of Chemistry, University of California, 1156 High Street, Santa Cruz, CA 95064, USA
e-mail: wgscott@chemistry.ucsc.edu

N.G. Walter et al. (eds.) *Non-Protein Coding RNAs*   73
doi: 10.1007/978-3-540-70840-7_4, © Springer-Verlag Berlin Heidelberg 2009

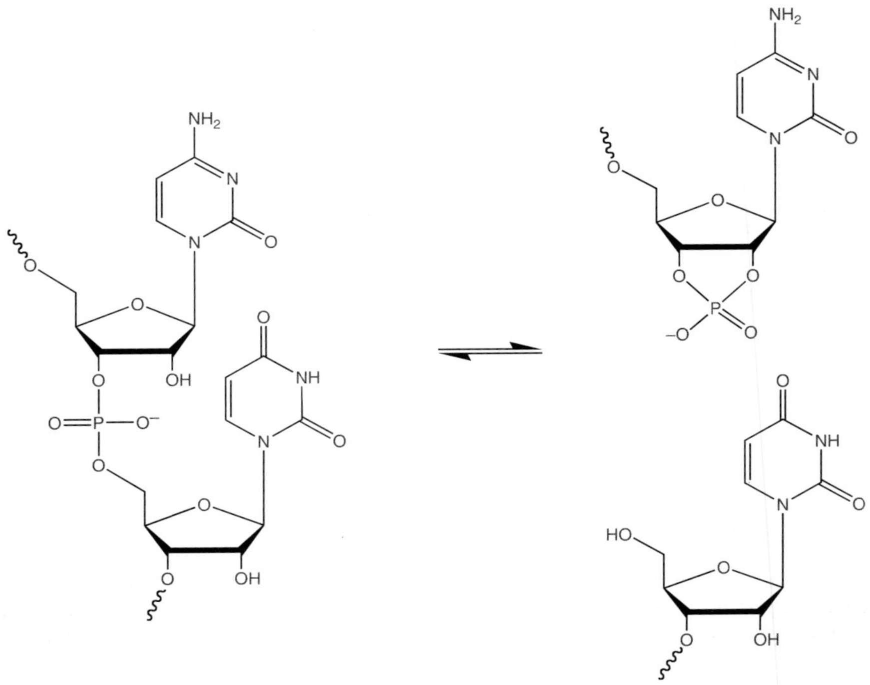

Fig. 4.1 RNA degradation via phosphodiester bond isomerization. The reactant and product are isomers, as the number and identity of all of the atoms in the RNA molecule remains constant. (Water is not added unless the 2′,3′-cyclic phosphate in the product subsequently hydrolyzes.) The phosphate remains in the diester state in both species, thus making the back (ligation) reaction possible without the input of ATP or other exogenous energy sources. The reaction rate is enhanced in a basic solution, and is suppressed when the phosphate conformation is restricted to the anti-periplanar double-gauche configuration compatible with A-form helices and similar secondary structures

2′-hydroxyl and thereby generation of the nucleophile. It is thus often referred to as base-catalyzed RNA degradation (and sometimes, wrongly, RNA hydrolysis). Since neither water nor hydroxide ion is actually added to the RNA, and the phosphate remains in the diester state, the reaction is simply an isomerization of the phosphodiester. It is arguably the simplest reaction that RNA can undergo.

4.1.1   Transition-State Structural Constraints

The phosphodiester isomerization reaction (Fig. 4.2) that degrades RNA is dependent upon not only the pH of the solution, but also the conformation of the RNA. The reaction proceeds through a trigonal bipyramidal oxyphosphorane transition-state, and inversion of configuration takes place as the reaction proceeds (Slim and Gait 1991; van Tol et al. 1990). This observation, coupled with the principle of microscopic reversibility, dictates that the 2′-O and 5′-O atoms must occupy the axial

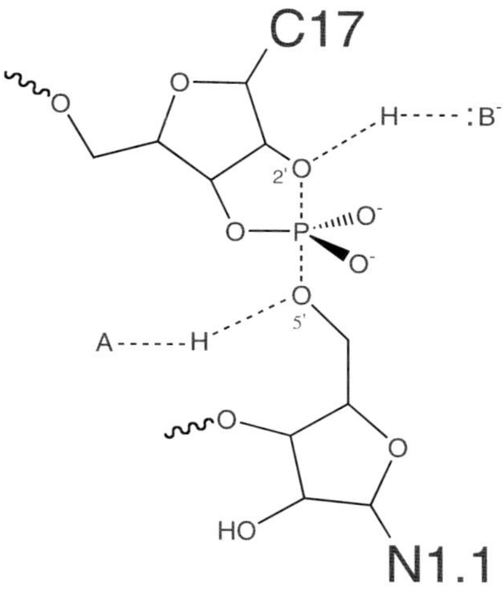

Fig. 4.2 The transition-state geometry for the reaction depicted in Fig. 4.1. Because both the spontaneous and enzyme-catalyzed versions of the phosphodiester isomerization reaction (Fig. 4.1) proceed via inversion of configuration of the non-bridging phosphate oxygen atoms, by far the simplest explanation is that the reaction is concerted, and proceeds through a single transition-state in which the attacking nucleophile, the 2′-O atom, is aligned with the phosphorus atom and the 5′-O leaving group. The principle of microscopic reversibility dictates that the transition-states for forward and reverse concerted reactions are indistinguishable, which entails that the 2′-O and 5′-O atoms must occupy the axial positions of a trigonal bypiramidal transition-state. The transition-state is thought to be associative, and partial axial bonds are indicated as dotted lines. Abstraction of a proton from the 2′-O atom initiates the cleavage (forward) reaction, and acquisition of a proton balances the accumulating negative charge on the 5′-O atom as the bond between it and the phosphorus atom breaks. A general base (indicated as: B⁻) is most likely responsible for abstracting the proton from the 2′-O, and a general acid (indicated as A–H) can donate a proton to the 5′-O atom as a negative charge begins to accumulate. Partial proton dissociation and association are also indicated with dotted lines. In the context of the hammerhead ribozyme, the reaction is greatly enhanced at residue C17, where the phosphate between it and the 3′-adjacent residue, N1.1, isomerizes

positions in the bipyramidal transition-state configuration, and that the axial positions are occupied by the 3′-O and the two non-bridging phosphate oxygen atoms. This in turn requires that the 2′-O atom, the phosphorus atom, and the 5′-O atom be approximately co-linear for the reaction to take place.

4.1.2 Phosphate Configuration and Reactivity

Random RNA sequences, or RNAs heated above their melting temperatures, tend to degrade rather more quickly than RNA sequestered within A-form helices. The helical geometry restrains the phosphodiester linkage to an anti-periplanar double-gauche-(−) configuration, thus minimizing repulsion between the electron lone pairs on the bridging phosphate oxygens (Govil 1976). This configuration (Fig. 4.3)

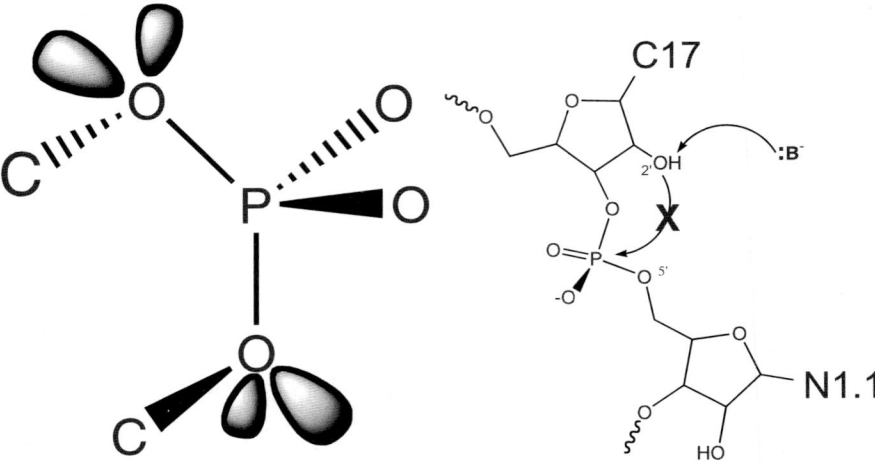

Fig. 4.3 The non-bonding orbitals that contain the bridging oxygen electron lone pairs minimize overlap (and therefore electrostatic repulsion) when the phosphodiester adopts the anti-periplanar double-gauche conformation shown in panel (**a**). This conformation is always found in A-form and B-form nucleic acid helices, and is incompatible with the geometry required for the phosphodiester bond isomerization reaction whose transition-state is depicted in Fig. 2, as shown in panel (**b**). A-form RNA helices thus lock the phosphodiester linkage into a conformation that is incompatible with the formation of the required in-line transition-state, and therefore suppress spontaneous RNA cleavage via phosphodiester isomerization

positions the 2'-O atom more than 90° away from the collinear orientation that would be most compatible with the potential formation of the trigonal bipyramidal transition-state required for the isomerization reaction. Therefore, RNA sequences that are sequestered in helical or helix-like secondary structures are protected from spontaneous (uncatalyzed) degradation, and the lability of RNA is thus quite context-dependent on the RNA structure. It is therefore reasonable to expect that non-coding RNAs will have well-defined three-dimensional structures with high helical content based upon natural selective pressures favoring long-lived RNA sequences.

4.2 Catalysis of RNA Phosphodiester Isomerization Reactions

Because the anti-periplanar double-gauche phosphodiester conformation is that of an energetic minimum, most of the phosphates in RNAs, which tend to be largely helical, quite possibly due to selective pressure against instability, are found in this conformation. This renders helical RNA resistant to the spontaneous cleavage isomerization reaction. Ribozyme enhancement of this reaction substantially above the background rate therefore requires several catalytic strategies that work in consonance to achieve efficient rate-enhancement. These include optimal orientation of the substrate, acid-base catalysis, transition-state stabilization, and possibly additional catalytic components.

Highly active ribozymes, such as the full-length hammerhead ribozyme, employ all of these strategies (and probably others that are more poorly understood).

4.2.1 Substrate Orientation

The spontaneous phosphodiester isomerization reaction, whose transition-state is shown in Fig. 4.2, requires proper alignment of the attacking nucleophile (the 2′-O atom in the cleavage reaction), the adjacent phosphorus atom, and the leaving group (the 5′-O atom in the cleavage reaction) in order to proceed. Hence it is expected that a ribozyme that catalyzes a site-specific cleavage will somehow favor distortion of its target substrate substantially from an A-form helix-like structure. The backbone of the substrate RNA will appear to be kinked, changing the configuration of the phosphate to one that favors formation of the bipyramidal transition-state, and the attacking nucleophile will be aligned with the leaving group, as shown in Fig. 4.4. Thus the first requirement (and opportunity) for catalytic enhancement is substrate alignment.

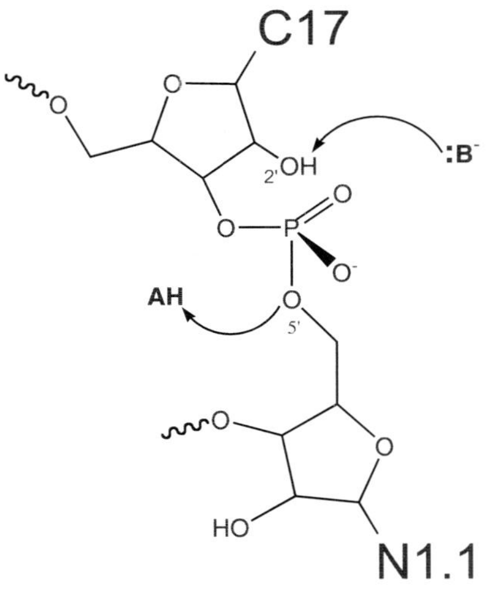

Fig. 4.4 In contrast, when the phosphodiester conformation becomes deformed and deviates from the gauche conformation, it may become more susceptible to in-line attack. Ribozymes, such as the hammerhead, that accelerate the cleavage reaction, are observed to distort the substrate substantially from the A-form conformation into one in which the attacking nucleophile approaches near-perfect alignment with the phosphorus and leaving-group. When the substrate is bound in a conformation that permits an in-line attack to occur, abstraction of the 2′-proton by a general base (:B⁻ in the figure) is likely to result in phosphodiester isomerization, especially if a general acid (AH in the figure) is simultaneously present to protonate the 5′-O leaving group. Although general acid-base catalysis is illustrated, other mechanisms involving water (specific acid-base catalysis) and Lewis acid-base catalysis, are also possible and are discussed in the text

4.2.2   Base Catalysis

The reaction is initiated by abstraction of a proton from the 2′-hydroxyl (Fig. 4.4), generating the nucleophile, a charged oxygen atom. The pK_a is on the order of 12 or 13, and spontaneous de-protonation is thus a rare occurrence. Base catalysis is thus the second opportunity for rate-enhancement, and several possible mechanisms exist. Specific base catalysis, i.e., de-protonation catalyzed by abstraction of the 2′-proton by a hydroxide ion, is thought to be involved in spontaneous RNA degradation. General Brønsted base catalysis can involve other entities, including potential metal hydroxides (such as magnesium hydroxide) or enzyme functional groups (histidines in the context of RNase A, or nucleotide bases in the context of ribozymes). Lewis bases are also potential participants, such as an inner-sphere interaction of a Mg^{2+} ion with the 2′-O atom, favoring its de-protonation.

4.2.3   Acid Catalysis

As the bond between the 2′-O atom, the attacking nucleophile, and the adjacent phosphorus atom forms, the bond between the phosphorus atom and the 5′-O leaving group breaks, resulting in an accumulating negative charge on the 5′-O atom, as is illustrated in the transition-state structure depicted in Fig. 4.2. The negative charge becomes neutralized in an aqueous solution when the 5′-O atom acquires a proton Analogous with base catalysis, acid catalysis can enhance the reaction rate by stabilizing the leaving group, providing the third opportunity for catalysis (Fig. 4.4). Specific acid catalysis, i.e., donation of a proton from a water molecule, will always occur in an aqueous solution, due to the very high pK_a of a primary alkoxide. However, general acid catalysis, in the form of a Brønsted acid like a fully hydrated $Mg (H_2O)_6^{2+}$ complex that can donate a proton, or an enzyme functional group (histidine, or a somewhat acidic nucleotide base), or in the form of a Lewis acid (again, an inner-sphere interaction with a Mg^{2+} ion) is also a possibility.

4.2.4   Transition-State Stabilization

The pentacoordinated trigonal bipyramidal oxyphosphorane transition-state (Fig. 4.2) will possess not one but two negative charges localized on the non-bridging phosphate oxygen atoms. The resulting electrostatic repulsion raises the potential energy of the transition-state structure substantially, and thus any mechanism that might help to dissipate the excess negative charge that accumulates in the transition-state, would lower its potential energy and thus accelerate the reaction. Transition-state stabilization via electrostatic screening or a similar effect thus provides a forth potential opportunity for catalytic enhancement. A positively-charged lysine in RNase A, for example, has been invoked as a participant in transition-state charge stabilization

(Raines 1998). In ribozymes, divalent (and other) cations have been implicated (Dahm et al. 1993; Dahm and Uhlenbeck 1991), as well as in some cases, the nucleic acid functional groups (Rupert et al. 2002).

4.2.5   Additional Effects

Other potential contributions to catalysis, such as orbital steering (Scott 2001), entropy effects (Hertel et al. 1994 1997), and so forth, have also been suggested, and quite likely make some contribution to enhancing catalysis. The above four effects are however reasonably well-understood, uncontroversial (at least by the standards of enzymology), and are demonstrably important in ribozyme catalysis of phosphodiester isomerization reactions in the case of several different small self-cleaving and self-ligating ribozymes (Breaker et al. 2003; Emilsson et al. 2003).

4.3   The Small Phosphodiester Isomerase Ribozymes

The first two ribozymes that were discovered, RNase P (Guerrier-Takada et al. 1983) and the Group I intron (Zaug and Cech 1986), catalyze fairly complex reactions (precursor tRNA processing and exon splicing, respectively). The third ribozyme to be discovered was the hammerhead ribozyme (Prody et al. 1986), now known to be a member of a class of several small self-cleaving and self-ligating ribozymes, each of unique sequence and structure, that mediate rolling-circle replication of satellite virus RNAs and similar molecules. Although differing in structure, the hammerhead, HDV, hairpin, VS and several other such ribozymes catalyze the simple phosphodiester isomerization reaction described above. The *glm*S ribozyme, a riboswitch that regulates gene expression in bacteria, is also a member of this class in that it catalyzes the same chemical reaction. The reaction is identical to the first step of the RNase A catalyzed reaction as well. Because of its simplicity, it offers the best hope for understanding and elucidating the most fundamental features of ribozyme catalysis.

4.3.1   Biological Context

Satellite RNAs may accompany viral infections (Symons 1997). Examples include the satellite RNA of tobacco ringspot virus, from which the hammerhead and hairpin ribozymes were first discovered, and the hepatitis delta virus (HDV) RNA, a satellite of hepatitis B. These are typically small (fewer than 400 nt) single stranded RNA molecules that are covalently closed circles. As templates for the host cell's replicative machinery, they are copied as long linear concatamers that must subsequently cleave into monomeric fragments, and these fragments must then recircularize to form new (complementary) templates for subsequent rounds of replication. The cleavage must be site-specific and must be reversible in that the ligation is required for template circulation. Various non-coding structural RNA motifs have now been identified that specifically catalyze site-specific cleavage and ligation. The cleavage reaction generates a 2′, 3′-cyclic phosphate, as is typical of non-catalyzed base-mediated RNA degradation. The ligation reaction reverses the cleavage reaction, with formation of a 3′ to 5′ phosphodiester linkage, using a 2′, 3′-cyclic phosphate as a substrate.

4.3.1.1 Rolling Circle Replication

Virusoid and satellite RNAs are small circular, single-stranded RNAs that are virus-like entities (Symons 1997) found in association with several types of plant RNA viruses (such as tobacco ringspot virus) and, in the case of the hepatitis delta virus (HDV), with hepatitis B. These small circular RNAs rely upon the cellular machinery of the host as well as products of viral infection to replicate via a rolling-circle mechanism (Fig. 4.5). The covalently-closed single strand of RNA is a template for

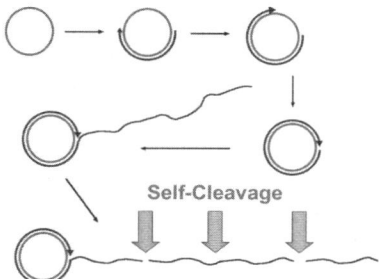

Fig. 4.5 Rolling-circle replication. Satellite RNAs are virus-like RNA genomes associated with several types of viruses. Two different examples that contain ribozyme sequence include the satellite RNA of tobacco ringspot virus (sTRSV), which is associated with an RNA virus called tobacco ringspot virus, an RNA virus that infects tobacco plants, and hepatitis delta virus (HDV), which is associated with hepatitis B (a DNA virus that infects humans). Although the sequence and structure of each self-cleaving ribozyme motif are unique, all catalyze the same chemical reaction and are functionally quite similar. Genomes of satellite RNAs are typically small (about 400 nucleotides) and are covalently closed circles that are replicated by the host cell's RNA polymerase. The polymerase copies the circular template processively, generating a long linear complementary concatomeric copy of the circular genome. This must be processed into linear monomers (catalyzed by a ribozyme self-cleavage reaction), each linear monomer must then become circular (catalyzed by a ribozyme self-ligation reaction), and the circular complementary copy of the original genomic strand must then serve as a template for additional replication via the rolling-circle mechanism, to generate copies of the original (sense) genomic strand of RNA, also requiring ribozyme-mediated cleavage and ligation reactions. In the case of the sTRSV, the hammerhead motif is found in the sense strand, and the hairpin motif is found in the anti-sense or intermediate strand. In the case of HDV, two separate but very similar HDV ribozyme sequences carry out the analogous processing reactions

an RNA polymerase that creates a complementary copy of the circular molecule. However, this molecule will be linear, and as the polymerase travels along the RNA for several revolutions, a long linear concatameric complementary copy of the circular template is produced. To complete the replication cycle, the linear concatamer must be separated into linear monomers, and these monomeric complementary copies of the original circular RNA must then close up to form circular molecules. These can then undergo the same sort of rolling-circle replication, with concomitant production of linear concatameric copies of the original circular template. These must be divided into linear monomeric fragments which again will circulate and ligate to form covalently closed circular copies of the original satellite RNA. The linear concatamers are cleaved into monomeric fragments autolytically, i.e., without the intervention of any enzymes or other intermolecular species with the possible exception of divalent cations. (A protein has been identified that may aid in this process by binding to the RNA (Luzi et al. 1997), but its presence is not essential for the self-cleavage reaction to take place in vitro).

4.3.1.2 Cleavage and Ligation

A relatively small, autonomously folding motif of RNA found at the cleavage-site junction is responsible for catalyzing a highly sequence-specific self-cleavage event in each case. In the case of the satellite RNA of tobacco ringspot virus, for example, an approximately 60 nucleotide sequence that has been dubbed the "hairpin" self-cleaving RNA is found at the junction of two monomeric sequences in the linear concatameric complementary copies of the original circular satellite RNA (Buzayan et al. 1986; Hampel and Tritz 1989; Hampel et al. 1990). A different sequence of approximately 50 nucleotides, called the "hammerhead" self-cleaving RNA, is found at the analogous positions in the concatameric copy of the original sequence, produced in the second phase of the rolling circle replication (Prody et al. 1986). These self-cleaving motifs reappear in a variety of other satellite RNA species. Similarly, HDV is a single-stranded satellite RNA virus associated with hepatitis B, and the HDV self-cleaving RNA, again consisting of an autonomously folded region of about 80 nucleotides, is involved in the rolling-circle replication of the hepatitis delta virus (Kuo et al. 1988; Wu et al. 1989). The VS self-cleaving RNA is a motif of about 160 nucleotides involved in the rolling-circle replication of a retro-plasmid in Neurospora (Beattie et al. 1995). In each case, the self-cleaving RNA catalyzes a highly sequence-specific phosphodiester bond cleavage reaction, that yields monomeric fragments having 5′-hydroxyl and 2′,3′-cyclic phosphate termini. Each monomeric fragment can then re–circulate when the two ends of the monomer approach one another and the complete folding motif is regenerated. The ends are ligated when the self-cleaving RNA catalyzes the reverse chemical reaction, that is, ligation of the phosphodiester backbone. Hence the RNA is catalytic in the sense that the cleavage is highly specific, greatly accelerated over the background rate of the reaction, and is reversible. However, these are not true enzymatic catalysts in the technical sense because they are not regenerated in such a way that true multiple turn over in the presence of an excess of substrate occurs. The natural biological reaction is one of, or a succession of, a single-substrate turnover cleavage, and a single-turnover ligation event.

The hammerhead, hairpin, VS and HDV self-cleaving RNAs can be made into true RNA enzymes, however, by a trivial alteration of their phosphodiester bond connectivity in such a way that a single-strand of RNA corresponding to the autonomous folding motif is divided into two strands, one of which (the substrate strand) gets cleaved by the other. When this is done, these four small self-cleaving RNAs become true ribozymes that catalyze multiple turnover cleavage reactions with the kinetic properties typically observed in true protein enzymes.

4.3.2   Other Contexts

The hammerhead ribozyme has also been discovered in the context of RNA transcripts within non-coding repetitive DNA in eukaryotic organisms including the newt (Forster et al. 1988) and schistosome (Ferbeyre et al. 1998). The function of these RNA transcripts is unknown, but they are thought to be replicated via a rolling-circle mechanism similar to that of satellite virus RNAs. The hammerhead motifs perform the same function.

The *glm*S ribozyme is unique in the world of naturally-occurring ribozymes in two respects. It is a riboswitch, and the regulatory effecter. Glucosamine-6-phosphate (GlcN6P), participates in the acid/base catalysis of RNA self-cleavage. The ribozyme is derived from a self-cleaving RNA sequence found in the 5′-UTR of the *glm*S message; it cleaves itself, inactivating the message, when the co-factor GlcN6P binds. GlcN6P production is thus regulated in many Gram-positive bacteria *via* this ribozyme-mediated negative feedback mechanism. The *glm*S ribozyme is thus both a riboswitch and a self-cleaving RNA (Winkler et al. 2004).

4.4   The Hammerhead Ribozyme

The hammerhead ribozyme in many respects is the model "RNase A of ribozymes" in that it is a comparatively simple and well-studied prototype ribozyme that in principle should be capable of revealing the secrets of its catalytic potential – if we are able to pose the right questions and carry out useful and informative experiments. Much attention has been focused upon this particular ribozyme with the hope that with a good understanding of its catalytic properties, our grasp of the phenomenon of RNA catalysis in general will become more comprehensive, leading to generalizations that are applicable to the larger ribozymes, to RNA splicing and peptidyl transfer, and perhaps even beyond – to a unified understanding of RNA and protein enzymology.

The hammerhead ribozyme is arguably the most intensively studied ribozyme, if one normalizes the number of experiments with respect to molecular weight. Its small size, thoroughly-investigated cleavage chemistry (McKay 1996; Nelson and Uhlenbeck 2006), various crystal structures (Dunham et al. 2003; Murray et al. 1998b, 2000, 2002; Pley et al. 1994; Scott et al. 1996; Winkler et al. 2004), and its biological relevance make the hammerhead ribozyme particularly well-suited for biochemical and biophysical investigations into the fundamental nature of RNA catalysis. Despite the extensive structural and biochemical characterization of the hammerhead ribozyme, the relationship between hammerhead ribozyme structure, biochemistry and catalytic mechanism was a source of considerable discord until 2006 (Blount and Uhlenbeck 2005; Nelson and Uhlenbeck 2006), when the structure of a full-length hammerhead ribozyme (Martick and Scott 2006) helped to resolve most of the seemingly irreconcilable experimental results.

4.4.1   Hammerhead Ribozyme Biochemistry

The minimal hammerhead ribozyme consists of a core region of 15 conserved (mostly invariant) nucleotides flanked by three helical stems. In 2003 it finally became clear that optimal activity required the presence of a tertiary interaction between stems I and II. Although there is little apparent sequence variation, the contact appears to be present in most if not all hammerhead sequences. Although the minimal hammerhead has a turnover rate of approximately $1 \, \text{min}^{-1}$, full-length sequences that include the tertiary contact are up to 1,000-fold more active (de la Pēna et al. 2003; Khvorova et al. 2003).

4.4.1.1   Rate Enhancement

The rate of non-site-specific, spontaneous decay of RNA is highly dependent upon the secondary structural context, but is on average about $10^{-6} \, \text{min}^{-1}$. (Soukup and Breaker 1999). Hence the rate enhancement enjoyed by an optimized minimal hammerhead is in the order of 10^6, and for the full-length natural hammerhead, can be as much as 10^9. To achieve this magnitude of rate-enhancement, not to mention site-specificity, the hammerhead ribozyme must adopt several effective catalytic strategies simultaneously. Each of these are separated (perhaps somewhat artificially) and analyzed below.

4.4.1.2   Metal Ions and Catalysis

Originally it was believed that all ribozymes, including the hammerhead ribozyme, were obligate metalloenzymes (Pyle 1993; Steitz and Steitz 1993). Mg^{2+} ion is assumed to be the biologically relevant divalent cation, although the hammerhead is active in the presence of a variety of divalent cations (Dahm and Uhlenbeck 1991). Proposed roles for Mg^{2+} ion in catalysis included both acid and base catalysis components (Dahm et al. 1993; Steitz and Steitz 1993) (with Brønsted and Lewis variants of this proposal articulated) as well as direct coordination of the pro-R non-bridging phosphate oxygen of the scissile phosphate for transition-state stabilization. Mg^{2+} ion has also been implicated in structural roles that facilitate formation of the active ribozyme (Bassi et al. 1995, 1996, 1997, 1999; Hammann et al. 2001a, b; Hammann and Lilley 2002; Lilley 1998, 1999; Penedo et al. 2004; Zhou et al. 2002).

In 1998 it was demonstrated that the hammerhead, along with the hairpin and VS ribozymes (but not the HDV ribozyme) could also function in the absence of divalent metal ions as long as a high enough concentration of positive charge was present (molar quantities of Li^+, Na^+, or even the non-metallic NH_4^+ ion allow cleavage to take place), permitting suggestion that ribozymes were not strictly metalloenzymes (Murray et al. 1998a).

Because of the volume of research devoted to understanding the mechanistic roles of divalent metal ions in hammerhead ribozyme catalysis, and because a fundamental tenet of ribozyme enzymology has been that all ribozymes are metallo-enzymes, it was unexpected to find that at least three of the four small, naturally-occurring ribozymes can function reasonably efficiently in the absence of divalent metal ions. This was discovered in the course of performing experimental controls for time-resolved crystallographic freeze-trapping experiments in crystals of the minimal hammerhead ribozyme (Murray et al. 1998a, 2002).

This result is dramatically illustrated in Fig. 4.6 (Scott 1999), which shows that EDTA can abolish cleavage activity by sequestering divalent cations, as one would expect, but in the cases of the hammerhead, hairpin and *Neurospora* VS ribozymes (i.e., three of the four naturally occurring small self-cleaving RNAs), the activity returns when the concentration of EDTA, and therefore Na^+, is increased further. High concentrations of Li^+, Na^+, NH_4^+ and other monovalent cations apparently enable the RNA to fold in much the same way that divalent metal ions allow it to. (The crystal structures of the minimal hammerhead ribozyme in the presence of $1.8\,M$ Li_2SO_4 and in the presence of $10\,mM$ $MgCl_2$ at low ionic strength are identical within experimental error.) It therefore appears that RNA folding and non specific electrostatic transition-state stabilization accounts for much, if not all, of the catalytic enhancement over background rates found with these ribozymes (Murray et al. 1998a). For example, hammerhead 16.1, which is considered to be an optimized hammerhead ribozyme sequence for single-turnover reactions, cleaves only threefold faster in the presence of $10\,mM$ $MgCl_2$ and $2\,M$ Li_2SO_4 than it does in the presence of $2\,M$ Li_2SO_4 alone (Murray et al. 1998a). The rates of hairpin and VS ribozymes in $2\,M$ Li_2SO_4 actually exceed those measured under "standard" low ionic strength conditions, and the rate of cleavage for the non-optimized hammerhead sequence used for crystallization is enhanced fivefold in $2\,M$ Li_2SO_4 alone, vs. standard reaction conditions. The non-optimized sequence used for crystallization tends to form alternative, inactive structures in solution, such as a dimer of the enzyme strand, that dominates at lower ionic strength.

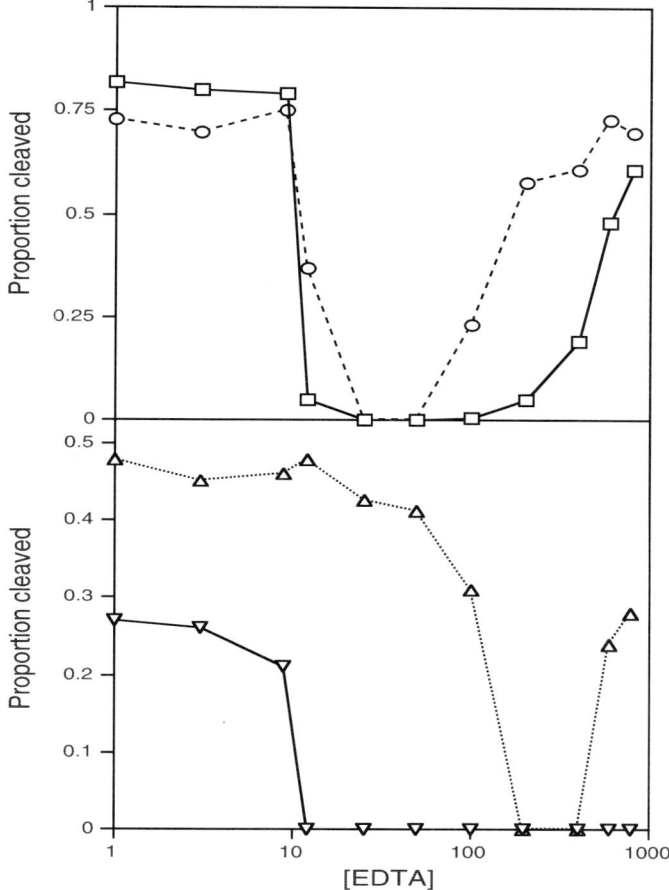

Fig. 4.6 Na$_3$EDTA titrations demonstrate that magnesium-dependent ribozyme-catalyzed RNA cleavage reactions of the HH$_{16.1}$(□), hairpin (○) and VS (Δ) ribozymes but not the HDV ribozyme (∇) are quenched by EDTA and stimulated by monovalent cations. In three of the four cases, the extent of cleavage, suppressed to zero by a stoichiometric excess of EDTA, is almost completely restored by the presence of about 3M Na$^+$ in the absence of divalent cations. This demonstrates that the hairpin, hammerhead and VS ribozymes do not require divalent metal ions for catalysis, but the HDV ribozyme appears to (and thus serves as an internal positive control) (Figure courtesy of John Burke)

This result also implied that any chemical role of Mg^{2+} ion in the ribozyme reaction was likely to be one of comparatively nonspecific electrostatic stabilization, rather than more direct participation in the chemical step of catalysis. It also suggested that, if acid/base catalysis takes place in ribozymes, the RNA itself, rather than serving as a passive scaffold for binding metal ions that served the roles of general acid and base catalysts, was an active participant in the chemistry of catalysis. Subsequently, with the structural elucidation of the hairpin (Ferre-D'amare and Rupert 2002; Rupert and Ferre-D'Amare 2001; Rupert et al. 2002) and full-length hammerhead (Martick and Scott 2006) structures, it was in fact revealed that RNA bases and other functional groups were positioned to provide the moieties likely responsible for acid-base catalysis.

4.4.1.3   Acid–Base Chemistry

Originally, hydrated Mg^{2+} and other hydrated divalent metal ions were thought to play a direct chemical role of general base and general acid in ribozyme catalysis, with the RNA itself serving as an ancillary and passive scaffold upon which metal ions would bind and would be positioned in the active site.

With the discovery that the hairpin, hammerhead and VS ribozymes were not strictly metalloenzymes (Murray et al. 1998a), it became apparent that in at least these three cases, the RNA, rather than serving merely as a metal ion-binding scaffold, must itself be an active participant in the chemistry of catalysis. The crystal structure of the hairpin ribozyme (Rupert and Ferre-D'Amare 2001) (as well as the HDV ribozyme (Ferre-D'Amare et al. 1998; Ke et al. 2004), which is in fact a metalloenzyme) soon validated this prediction, but it was not apparent from that of the minimal hammerhead (Pley et al. 1994; Scott et al. 1995, 1996), what functional groups might be involved in acid-base catalysis. So the focus of biochemical mechanistic investigations in the hammerhead turned to this problem.

The invariant core residues G12 and G8 in the hammerhead ribozyme were finally identified in 2005 as likely candidates for participation in acid-base chemistry, through careful purine modification studies conducted by John Burke and coworkers (Han and Burke 2005; Heckman et al. 2005). Substitution of G12 (pK$_a$ 9.5) with inosine (pK 8.7), 2, 6-diaminopurine (pK 5.1), or 2-aminopurine (pK 3.8) shifts the reaction rate profile in a manner consistent with G12's suggested role in general base (or acid) catalysis without significantly perturbing ribozyme folding (Han and Burke 2005). Similar substitutions at G8 also implicated this invariant residue in acid-base catalysis, but in this case (as well as with the invariant G5), the modifications partially inhibited ribozyme folding as well (Han and Burke 2005). These experiments could not determine specifically whether an individual nucleotide, such as G12, was the general acid or the general base, but clearly implicated G12 and G8 in acid-base catalysis.

4.4.1.4   Kinetics

The minimal hammerhead ribozyme, under "standard" reaction conditions (10 mM Tris, pH 7.5, 10 mM $MgCl_2$) has a turnover rate in the order of 1 min^{-1} and a K_m of about 10 μm, and a log-linear dependence of rate on pH with a slope of 0.7. Above pH 8.5–9.0 (depending upon reaction conditions), the rate becomes pH-independent, suggesting an apparent kinetic pK$_a$ of about 8.5–9.0 (Dahm and Uhlenbeck

1991; Hertel et al. 1994; Stage-Zimmermann and Uhlenbeck 1998b). This observation is consistent with both Mg^{2+}-mediated and guanine-mediated acid–base chemistry. The full-length hammerhead ribozyme shows similar pH dependence, but the cleavage rate is enhanced up to about 1,000-fold (i.e., about $15\,s^{-1}$) (Canny et al. 2004). There exists no compelling evidence that the reaction is sequential rather than concerted, although this remains an issue for debate. It is perplexing that the pH-dependence of the rate-limiting step is similar in both the minimal and full-length ribozymes, despite the remarkable difference in reaction rate.

4.4.1.5 Internal Equilibria

Catalysts enhance both the forward and reverse reaction rates by the same magnitude, and thus cannot alter the equilibrium constant of the reaction they catalyze (i.e., the ratio of products to reactants is not changed by an enzyme). However, in the case of the hammerhead ribozyme, the division of the complex into enzyme and substrate is artificial, and it is more meaningful to examine the internal equilibrium within the enzyme-substrate complex (Hertel et al. 1994; Hertel and Uhlenbeck 1995). In this case, it has been found that in the minimal hammerhead, the internal equilibrium can be perturbed specifically in the direction of ligation by altering the relative helical orientations of stems I and II by means of chemical cross linking and other tethers (Blount and Uhlenbeck 2002; Stage-Zimmermann and Uhlenbeck 1998a). Recently, differences in internal equilibria have been reported in the context of the natural full-length hammerhead ribozyme sequence as well, in which one class of hammerheads (represented by the sTRSV) favors virtually complete cleavage, and another (represented by the smα hammerhead) possesses an internal equilibrium in which about 1/3 of the RNA is in the ligated form (Canny et al. 2007). (The two classes of hammerhead are described in more detail in Sect. 5.3.)

4.4.2 Hammerhead Ribozyme Structure

The crystal structure of a minimal hammerhead ribozyme was the first near-atomic resolution structure of a ribozyme to be determined. The first example, solved by McKay and coworkers in 1994 (Pley et al. 1994), was that of a minimal hammerhead RNA enzyme strand bound to a DNA substrate-analogue inhibitor, and in 1995 a different all-RNA hammerhead construct having a 2′-OMe inhibitory substitution of the nucleophilic 2′-OH of C17 appeared (Scott et al. 1995). Subsequently, structures of minimal hammerheads without modified nucleophiles appeared in various states (Scott et al. 1996) of pre-catalytic conformational changes, and finally a structure of the cleavage product appeared (Murray et al. 2000) in 2000, providing the opportunity to construct the first "molecular movie" of ribozyme catalysis.

4.4.2.1   Minimal vs. full-Length Hammerhead Ribozymes

It was immediately apparent from the first hammerhead crystal structure (Pley et al. 1994) that a conformational change would need to take place, to position the attacking nucleophile in line for activation of the cleavage reaction. The desired conformation corresponds to the configuration illustrated schematically in Fig. 4.4, and the actual orientation observed in the initial crystal structures corresponds to that depicted schematically in Fig. 4.3. The requirement for this conformational change motivated the subsequent crystallographic freeze-trapping experiments (Murray et al. 1998b).

Meanwhile, a growing list of discrepancies between the minimal hammerhead ribozyme structure and mechanistic biochemical experiments designed to probe transition-state interactions began to accumulate (Blount and Uhlenbeck 2005). The observed hydrogen-bonding patterns within the minimal hammerhead crystal structures could not explain many of the in variances, including the immutability of G8, G12, G5, C3 and a number of other core residues (McKay 1996). Even more concerning was evidence that the phosphate of A9 and the scissile phosphate, separated by 18 Å in the minimal hammerhead crystal structures, might bind a single metal ion in the transition-state of the reaction (Wang et al. 1999). Such an interaction would require the two phosphates to approach within about 4.4 Å, but this requirement is incompatible with the minimal hammerhead crystal structure, unless significant unwinding or base unpairing were to take place in one or more of the helices (Murray and Scott 2000).

When the hammerhead RNA was first discovered, it was observed to be embedded within a ~370 nucleotide single-stranded genomic satellite RNA, most of which could be deleted while preserving the RNA's catalytic properties (Prody et al. 1986). Eventually, it was found that about 13 core nucleotides and a minimal number of flanking helical nucleotides were all that was required for a respectable catalytic turnover rate of $1\,min^{-1}$ to $10\,min^{-1}$, and this "minimal" hammerhead construct became the focus of attention (Ruffner et al. 1990; Uhlenbeck 1987). It thus came as a great surprise to most in the field when, in 2003, it was finally pointed out that for optimal activity, the hammerhead ribozyme in reality requires the presence of sequence in stems I and II, that interact to form tertiary contacts that were removed in the process of eliminating seemingly superfluous sequences from the hammerhead ribozyme, in the standard reductionist approach often employed in molecular biology (Lilley 2003). Once the full ramifications of this revelation became apparent, i.e., that the entire field had been studying the residual catalytic activity of an over-zealously truncated version of the full-length ribozyme, attention shifted away from the minimal constructs. It also quickly became apparent that a crystal structure of the full-length hammerhead ribozyme, in which these distal tertiary contacts were present, might be of considerable interest. The crystal structure of a full-length hammerhead appeared in 2006, and it was indeed found to reconcile most of the significant experimental discrepancies (Martick and Scott 2006; Nelson and Uhlenbeck 2006).

Secondary structures of the minimal and full-length hammerhead ribozymes are presented in Fig. 4.7, oriented to reflect the corresponding tertiary structures. Comparison of the folds of the minimal and full-length hammerhead ribozyme structures is likewise illustrated in Fig. 4.8.

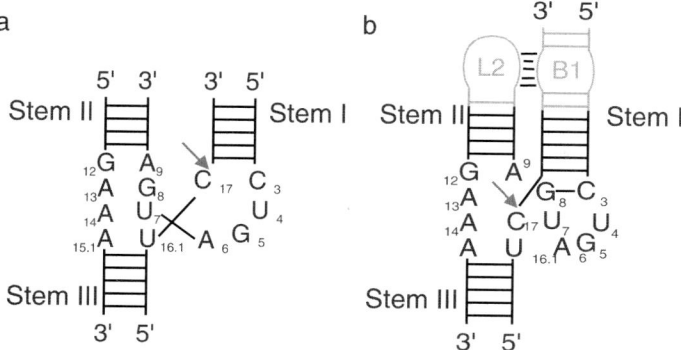

Fig. 4.7 The minimal and full-length hammerhead ribozyme secondary structures. Figure A shows the minimal hammerhead ribozyme sequence, the focus of study between about 1987 and 2003, when it became apparent that an additional tertiary contact having very limited sequence conservation and no readily discernible covariance is also present in natural hammerhead ribozyme sequence. The presence of the contact can enhance catalysis up to 1,000-fold. The crystal structure of the full-length hammerhead, when compared to that of the minimal hammerhead, as shown in Fig. 4.8, reveals that the presence of the tertiary contact stabilizes an active site conformation, not observed in the minimal hammerhead structures, in which several of the invariant residues shown explicitly in the above figure are arranged to orient the substrate for in-line attack and to position it for acid-base catalysis, as shown in Fig. 4.9 (See figure insert for colour reproduction)

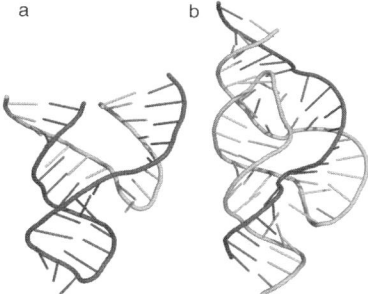

Fig. 4.8 Minimal and full-length hammerhead ribozyme tertiary structures. Backbone representations of the minimal (**a**) and full-length (**b**) hammerhead ribozymes are shown together for comparison. The substrate strands are shown in dark grey, and the enzyme strands in light grey. Although the overall fold is similar within the regions of sequence that both ribozymes possess, there is a pronounced kink in the substrate at the active site corresponding to a localized conformational change that rearranges the phosphodiester backbone conformation at the cleavage site from one that resembles that shown in Fig. 4.3 to that shown in Fig. 4.4. The tertiary contact between stems I and II is apparent in Fig. B, and profoundly bends and distorts stem I in such a way that it becomes co- linear with stems II and III

4.4.2.2 Substrate Orientation

The backbone folds of the minimal and full-length hammerhead ribozymes, illustrated in Fig. 4.8, are rather similar within the subset of nucleotides that both constructs share in common (an observation that helps to explain how the minimal hammerhead ribozyme could be catalytically active in the crystal). Within the common region that both have, the most striking difference in the backbone structure occurs at the cleavage site, where the substrate strand makes a sharp U-turn like bend in the full-length hammerhead. This sharp bend or kink is absent in the minimal hammerhead structure.

Examination of the structural details in this region quickly reveals the reason for the observed backbone kink. The cleavage-site base, C17, has rotated almost 180° relative to the minimal hammerhead and in such a way that the 2′-O becomes almost completely aligned (within 17°) with the adjacent phosphate, as illustrated in Fig. 4.9. In the crystal structure, the cleavage reaction is inhibited by the presence of a 2′-OMe modification of the nucleophile; the extra methyl group is omitted for clarity in the illustration.

4.4.2.3 The Active Site and Acid–Base Catalysis

The structure of the active site, shown in Fig. 4.9, also immediately explains why G12 is invariant. The 2′-O nucleophile is within hydrogen bonding distance of the O6 and N1 of G12, whose Watson–Crick face makes no other contact within

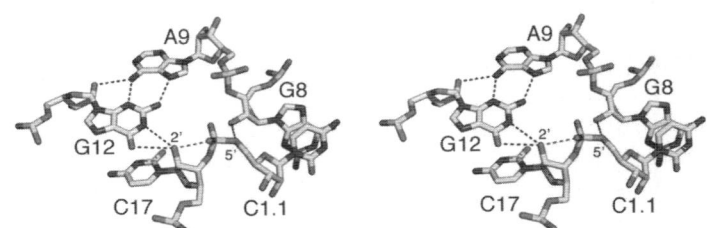

Fig. 4.9 Cross-eyed stereo view of the hammerhead ribozyme active site (from 2GOZ). The 2′-O of the cleavage-site nucleotide, C17, is oriented for in-line attack. G12 is positioned for general base catalysis, where O6 and N1 are within hydrogen-bonding distance of the 2′-O nucleophile. The leaving-group 5′-O of C1.1 accepts a hydrogen bond from the 2′-OH of G8 (whose nucleotide base forms a Watson–Crick base-pair with C3, not shown). The A9 and scissile phosphate non-bridging oxygen atoms approach within 4.3 Å. The grey-scale depicts atom identity, with nitrogen as darkest, then oxygen, then phosphorus, and finally carbon as the lightest grey. Dark dotted lines indicate hydrogen bonds, and the light dotted line between the 2′-O and the scissile phosphate indicates the direction of potential in-line attack. The pre-cleavage state is captured using a 2′-OMe on C17 (omitted from the figure for clarity, but present in the coordinates deposited as 2GOZ)

the RNA. The position of G12 with respect to the cleavage site strongly suggests G12 is the general base in the cleavage reaction. This suggestion is in fact in complete agreement with mechanistic biochemical experiments performed by Burke and coworkers in 2005 (Han and Burke 2005; Heckman et al. 2005), described in Sect. 5.1.

One of the most remarkable structural differences between the minimal and full-length hammerhead involves repositioning of the invariant residue G8. In the minimal structure, G8 forms a sheared reverse-Hoogsteen base-pair with A13 within the augmented stem II helix (Fig. 4.12c). In the full-length structure, G8 abandons its position in the stem II helix, and forms a Watson–Crick base-pair with the invariant C3. Burke in 2005 observed the modification of G8 perturbed ribozyme folding as well as acid-base catalysis (Han and Burke 2005). The observed effect on folding is consistent with the observed base-pairing interaction. The observed base-pair has since been experimentally corroborated as critical to catalysis (Nelson and Uhlenbeck 2008b; Przybilski and Hammann 2007). Single point-mutations of G8 or C3 kill catalytic activity, but compensatory double-mutants restore Watson–Crick base-pairing rescue activity. The ribose, rather than the base, of G8 appears to be involved in acid catalysis. The 2'-OH of G8 donates a hydrogen bond to the O5' leaving group of residue 1.1.

4.4.2.4 Transition-State Stabilization

The rearrangement of pairing between stem II and stem I induces a conformational change that has the effect of positioning the A9 and scissile phosphates within 4.3 Å of one another, consistent with the idea that both phosphates might coordinate a single metal ion in the transition-state. The need to shield the close approach of two negative charges on two non-bridging phosphate oxygens from electrostatic repulsion therefore exists even in the pre-catalytic conformation trapped in the crystal structure. However, no divalent metal ions have yet been observed to bridge the two phosphates. In the original structure, obtained in the presence of molar quantities of NH_4^+ and 1 mM Mg^{2+}, this is not terribly surprising, as the high concentration of monovalent cations is likely to provide a sufficient charge screening to permit the ribozyme to fold into the observed conformation, and may inhibit binding of the 1 mM Mg^{2+}. However, crystals soaked with 50 mM Mn^{2+} reveal unambiguously several metal binding sites, thanks to the X-ray absorption properties of Mn^{2+} (Martick et al. 2008). Although no Mn^{2+} is observed to bind the scissile phosphate, a single Mn^{2+} binds with full occupancy the A9 phosphate and makes an inner-sphere contact with N7 of G10.1, exactly as observed in the crystal structures of the minimal hammerhead. Hence, the role of this cation (or others) in transition-state stabilization remains obscure. It is possible that as an additional negative charge accumulates in the oxyphosphorane transition-state, the observed or other divalent metal ion, might be recruited to bridge the two phosphates as predicted.

4.4.3   Structure-Function Correlates

The minimal hammerhead crystal structure was unable to explain the observed invariance of many of the core nucleotides, including the immutability of C3, G5, G8 and G12, based on the observed hydrogen bonding interactions. In addition, the nucleophile was not positioned for an in-line attack, and the A9 and scissile phosphates were much further separated than would be allowed if they conspire to bind a single metal ion in the transition-state of the reaction. The full-length hammerhead structure successfully answers each of these concerns, and a fairly exhaustive study has been conducted that concludes that most, if not all, of the significant concerns have finally been reconciled (Nelson and Uhlenbeck 2006, 2008a). Elsewhere, the conformational equilibrium between minimal and full-length hammerheads has been investigated via adiabatic morphing (Scott 2007), and a rationalization for the observed catalytic activity in crystals of unmodified minimal hammerhead ribozymes has been articulated. Hence the means by which the hammerhead catalyzes acid-base chemistry, has, for the most part now been elucidated, apart from electrostatic transition-state stabilization (Martick et al. 2008). The other outstanding problem is how the hammerhead, in the natural biological context of satellite RNA rolling-circle replication, switches between nuclease and ligase activities.

4.4.3.1   Two Classes of Hammerhead Ribozyme

One of the many reasons why the distal sequence of the full-length hammerhead ribozyme involved in formation of the tertiary interaction between stems I and II, evaded detection until 2003 is that there are in fact two separate classes of contacts. One is found in the *Schistosomal* hammerhead (smα), and the other in the first hammerhead discovered- that of the satellite RNA of tobacco ringspot virus (sTRSV) (Khvorova et al. 2003). The sequence and secondary structural representations of these two ribozymes are shown in Fig. 4.10 in a format that reflects the tertiary structures shown in Fig. 4.11.

A recent crystal structure of a slowly cleaving sTRSV hammerhead, in which G12 is replaced by A, has been obtained in both ligated and cleaved forms (Chi et al. 2008). The catalytic core of the uncleaved form is virtually identical to that of the smα hammerhead, despite the G12A mutation and the absence of a 2′-OMe on C17. The cleaved form reveals a 2′,3′-cyclic phosphate and hints at a transition-state stabilization interaction with the exocyclic amine of A9. The tertiary contact between Stems I and II that stabilizes the active site conformation, in contrast, is strikingly different from that observed in the smα hammerhead. Figure 4.10 illustrates the sequence differences and the individual base tertiary interactions using Westhof's notation. The one tertiary base-pairing interaction that the two classes of hammerhead share is a Hoogsteen pair between a U in stem I and an A in stem II, as shown in Fig. 4.11.

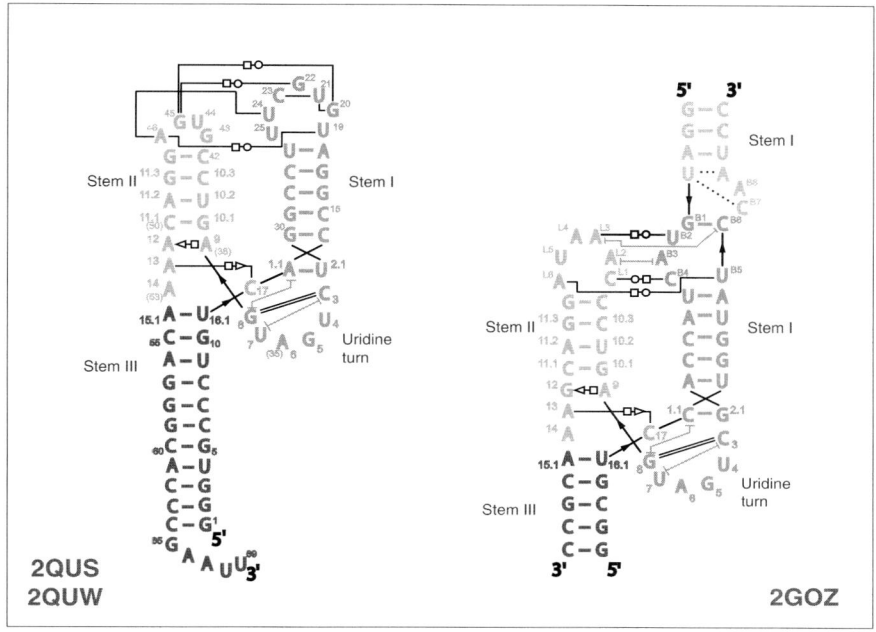

Fig. 4.10 The two classes of full-length hammerhead ribozymes. The sTRSV hammerhead is on the left, and the smα1 hammerhead is on the right. Although the primary sequence elements of the tertiary contacts have only one interaction (an AU Hoogsteen pair between a U in stem-loop I and an A in stem-loop II), the contacts have very similar effects upon the conformation of the hammerhead ribozyme active site

4.4.3.2 Internal Equilibrium, Switches, and the Hammerhead Ribozyme

Rolling circle replication of satellite RNAs requires processing of the genomic and anti-genomic linear concatamers. The multimers must be cleaved into monomeric fragments, and those fragments must then recirculate and the ends must ligate. The hammerhead must therefore be capable of both catalytic cleavage and catalytic ligation reactions at different stages of the replicative cycle. For this to happen efficiently, a switching mechanism, by which the internal equilibrium might become shifted more toward cleavage or more toward ligation, is required. Is there a structural basis for ribozyme switching?

The minimal hammerhead sequence tends to favor cleavage over ligation rather strongly; the full-length sequence less so. The full-length smα hammerhead exists in an internal equilibrium in which 1/3 of the RNA is in the ligated state (Canny et al. 2007). Hence a particularly simple switching mechanism could involve the conformational change required to bring the minimal hammerhead conformation in line with what is observed for the full-length hammerhead. Two significant conformational changes, one in the tertiary contact region, and the other in the active site,

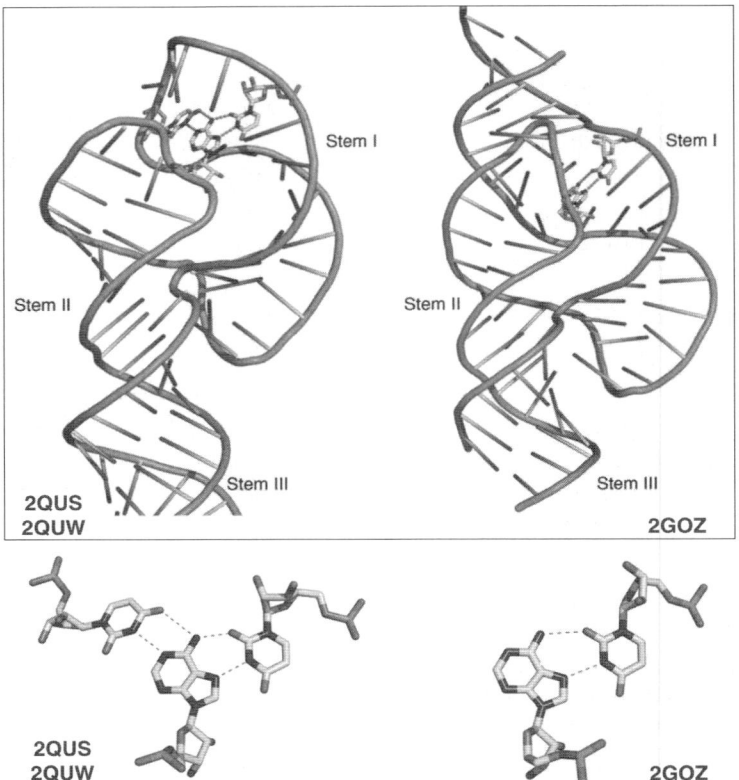

Fig. 4.11 The conserved AU Hoogsteen pair that forms between stem-loop I and stem-loop II in both hammerhead ribozyme classes. The top frame shows the details of the interaction, and the bottom frame shows these interactions in the context of the backbone structures. In the case of the sTRSV hammerhead ribozyme, the conserved AU Hoogsteen pair is part of a base triple; a Watson–Crick AU pair forms within the loop-loop interaction as well. It is noteworthy that the internal equilibrium of the smα1 hammerhead is such that about 1/3 of the molecules remain ligated, whereas the sTRSV hammerhead internal equilibrium strongly favors complete cleavage

are observed in the full-length hammerhead structure relative to the minimal structure, and could form the basis for a molecular switch.

4.4.3.3 GNRA Tetraloop

The GNRA tetraloop is among the most stable non-helical secondary structural elements commonly found in RNA. The first hammerhead structure solved included a GAAA tetraloop capping stem II (Pley et al. 1994). The GAAA tetraloop is a specific example of the general GNRA tetraloop, in which the first residue is

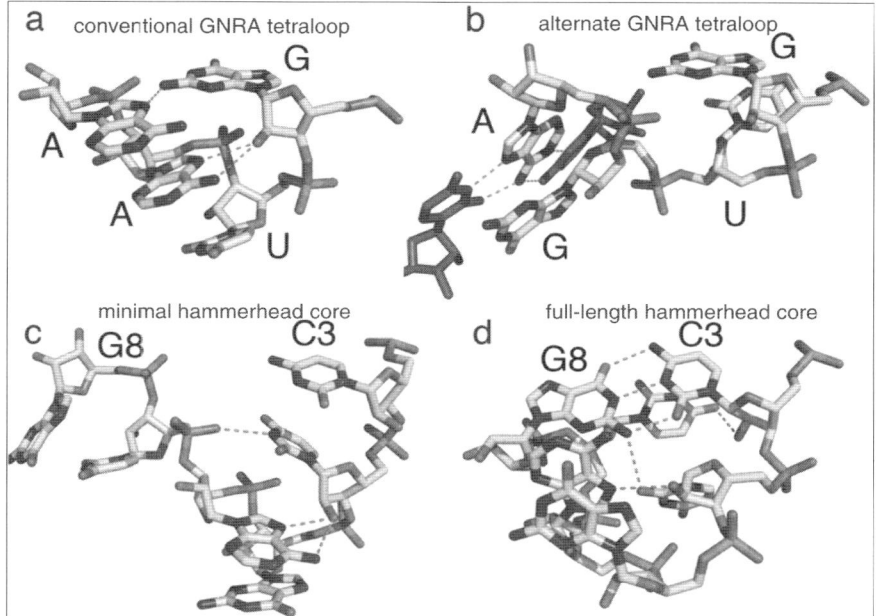

Fig. 4.12 There are two components observed in the hammerhead conformational switch between that favored in the minimal hammerhead and that favored in the full-length hammerhead. The first occurs within the context of the tertiary contact. In its absence, in the sTRSV ribozyme, stem-loop II adopts the conventional GNRA tetraloop structure shown in a. In the presence of the tertiary contact, the GNRA tetraloop of stem II substantially rearranges so that the final A can participate in the conserved AU Hoogsteen pairing interaction. Stem II residues are shaded according to atom identity as in Fig. 4.9 and the U residues from stem I are shown in solid dark grey. The second conformational switch, as noted in the text, involves base-pairing of G8 with C3, absent in the minimal hammerhead structure. Whether these interactions form in a sequential or concerted manner is presently unknown

always a G, the second can be anything, the third is restricted to purine, and the fourth is always A. The thermodynamically stable structure typically formed is one in which the second, third and forth residues form a 3′ stacking interaction, and the exocyclic amine of G forms a hydrogen bond with the N7 of the final A (as shown in Fig. 4.12a).

Although the minimal hammerhead stem II GNRA tetraloop structure adheres to this expectation, the conformation of the GUGA tetraloop in the sTRSV hammerhead is rather different. The first G adopts a similar position, the final three nucleotides are completely unstacked, and both the third G and the final, invariant, A, are involved in tertiary contact base-pairing interactions. Notably, the final A forms the conserved Hoogsteen interaction observed in both classes of hammerheads (as shown in Fig. 4.12b).

4.4.3.4 The G8-C3 Pair

As described in Sect. 4.2, a Watson–Crick base-pair forms between G8 and C3 in the full-length hammerhead structure that is absent in the minimal hammerhead (Figs. 4.12c and 4.12d (Dunham et al. 2003)). Establishment of this base-pair has been shown to be critical for catalysis, and alterations that perturb the base-pairing capacity of G8 affect both folding and catalysis (Nelson and Uhlenbeck 2008b; Przybilski and Hammann 2007). Hence formation of the G8-C3 Watson–Crick pair is likely to constitute a second molecular switch that can turn catalysis on and off.

4.4.3.5 Stem I Helical Pitch

In addition, the internal equilibrium between cleavage and ligation of bound substrate may also be fine-tuned by the orientation of Stem I in relation to Stem II. Previously, it has been observed that cross linking Stems I and II in a minimal hammerhead construct can differentially alter the internal equilibrium of the hammerhead as a function of relative helix orientation (Rueda et al. 2003; Sigurdsson et al. 1995; Stage-Zimmermann and Uhlenbeck 2001). Superposition of the sequence shared between the smα hammerhead and the sTRSV hammerhead (Fig. 4.13) in fact reveals the sTRSV hammerhead Stem I to be significantly more tightly wound than the smα hammerhead, with the sTRSV substrate strand being more tightly

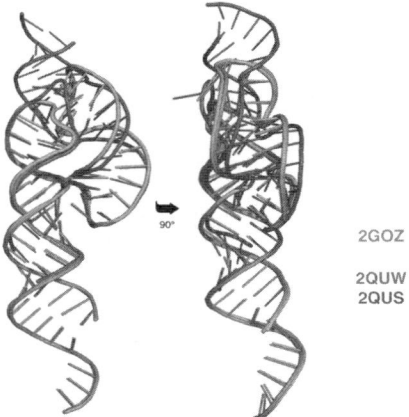

Fig. 4.13 The net effect of the different tertiary contacts in the two classes of hammerhead ribozyme. Residues shared in common between the smα1 hammerhead ribozyme (2GOZ) and the sTRSV ribozyme (2QUW, 2QUS) have been superimposed, and with the exception of stem I, starting 3′ to the cleavage site, these superimpose within experimental error. The most significant deviation is a pronounced unwinding of stem I in 2GOZ relative to the others. The internal equilibrium of the hammerhead ribozyme has previously been demonstrated to be perturbed by the presence of a tether or chemical crosslink between stems I and II. Hence it is likely that the greater unwinding of stem I in 2GOZ correlates with a shift in internal equilibrium toward ligation

kinked. Hence it is likely that the observed differences between the sTRSV and smα hammerhead Stem I orientations correlate with the observed differences in internal equilibria. It is noteworthy that the cleaved and uncleaved forms of the sTRSV are far more similar to each other (Fig. 4.13) than either is to the smα hammerhead that, as noted, tends to favor ligation more than the sTRSV.

4.5 Other Examples of Phosphodiester Isomerase Ribozymes

4.5.1 Hairpin Ribozyme

The crystal structure of a hairpin ribozyme transition-state analogue (1M5O) reveals several active-site interactions with a vanadate mimic of the penta- coordinated oxy-phosphorane transition-state (Rupert et al. 2002). Although the geometry is not a perfectly symmetrical trigonal bipyramid, the observed interactions are suggestive. G8 (no relation to the hammerhead residue) in this case appears positioned to be a general base in both the ligated and the transition-state analogue structure (Rupert and Ferre-D'Amare 2001; Rupert et al. 2002). Unique to the transition-state analogue structure are additional interactions, including a hydrogen bond between the 5′-O (the leaving group in the cleavage reaction, and attacking nucleophile in the ligation reaction) and the N1 of A57, thus suggesting A57 is the general acid in the cleavage reaction. In the hairpin ribozyme, exocyclic amines of both A57 and A9 each hydrogen bond to the pro-R oxygen of the vanadate, suggesting that one or both of these residues participates in transition-state stabilization. No metal ions are found in the active site of the hairpin ribozyme.

4.5.2 HDV Ribozyme

Crystal structures of the HDV ribozyme before (Ke et al. 2004) and after (Ferre-D'Amare et al. 1998) cleavage suggest a role for C75 in acid/base catalysis, although there remains some controversy as to whether C75 is the general base, rather than the general acid, in the cleavage reaction. A divalent metal ion is also required for the HDV ribozyme to function, and it is also thought to play a complementary role in acid-base catalysis (i.e., if C75 is the base, the divalent metal ion is the acid, and vice-versa).

4.5.3 GlmS Ribozyme

The fold of the glmS ribozyme is that of a double pseudoknot. The GlcN6P co-factor binding site is positioned immediately adjacent to the scissile phosphate. The C2-NH_2 amine in GlcN6P, and the analogous C2-OH in a Gly6P inhibitor, are positioned within hydrogen-bonding distance of the 5'-oxygen leaving group, together suggesting that GlcN6P is the general acid catalytic component in the self-cleavage reaction (Cochrane et al. 2007; Klein and Ferre-D'Amare 2006). G40 in turn is positioned such that its N1 is within hydrogen-bonding distance of the nucleophilic 2'-OH at the ribozyme cleavage site, suggesting G40 may be the general base component (similar to what is seen in the hammerhead ribozyme structure).

Structures of the uncleaved RNA in the absence of the cofactor reveal that the substrate is positioned for in-line attack in a pre-formed active site, and binding of the cofactor then initiates the cleavage reaction by providing the acidic component to the catalyst. From the structural perspective, it does not appear that any metal ions are involved directly in the chemistry of catalysis.

4.5.4 The Group I Intron

It is worth noting that the group I intron also catalyzes a series of phosphodiester isomerization reactions that result in intron excision with concomitant ligation of two adjacent exons (Stahley and Strobel 2006). The details of the reaction mechanism are however far more complex and are not directly related to ribozymes that produce a 2',3'-cyclic phosphate product.

4.6 Concluding Remarks

The hammerhead ribozyme and its relatives in the family of small self-cleaving RNAs are an important class of non-coding RNAs with highly specific functions that arise from unique structures. These were originally discovered in the context of the rolling-circle replicative cycle of virus-like satellite RNAs and viroids, but have also been found in some eukaryotes, hinting that they, or others like them who are yet to be discovered, may play a wider role in regulatory pathways. We have, in fact, recently discovered several hammerhead ribozymes embedded within the 3'UTRs of various mammalian mRNAs that appear to regulate gene expression (Martick et al. 2008b). Each member of the family of small ribozymes that catalyze phosphodiester isomerizations appears to use a unique catalytic strategy; yet all catalyze the same simple chemical reaction. Evolution has thus given rise to a variety of RNA structures endowed with catalytic potential that have survived natural selection. One can only wonder at the potential richness of the RNA catalytic repertoire in a postulated pre-biotic RNA world.

References

Bassi GS, Mollegaard NE, Murchie AI, von Kitzing E, Lilley DM (1995) Ionic interactions and the global conformations of the hammerhead ribozyme. Nat Struct Biol 2:45–55

Bassi GS, Murchie AI, Lilley DM (1996) The ion-induced folding of the hammerhead ribozyme: core sequence changes that perturb folding into the active conformation. RNA (New York) 2:756–768

Bassi GS, Murchie AI, Walter F, Clegg RM, Lilley DM (1997) Ion-induced folding of the hammerhead ribozyme: a fluorescence resonance energy transfer study. EMBO J 16:7481–7489

Bassi GS, Mollegaard NE, Murchie AI, Lilley DM (1999) RNA folding and misfolding of the hammerhead ribozyme. Biochemistry 38:3345–3354

Beattie TL, Olive JE, Collins RA (1995) A secondary-structure model for the self-cleaving region of Neurospora VS RNA. Proc Natl Acad Sci USA 92:4686–4690

Blount KF, Uhlenbeck OC (2002) Internal equilibrium of the hammerhead ribozyme is altered by the length of certain covalent cross-links. Biochemistry 41:6834–6841

Blount KF, Uhlenbeck OC (2005) The structure-function dilemma of the hammerhead ribozyme. Annu Rev Biophys Biomol Struct 34:415–440

Breaker RR, Emilsson GM, Lazarev D, Nakamura S, Puskarz IJ, Roth A, Sudarsan N (2003) A common speed limit for RNA-cleaving ribozymes and deoxyribozymes. RNA (New York) 9:949–957

Buzayan JM, Hampel A, Bruening G (1986) Nucleotide sequence and newly formed phosphodiester bond of spontaneously ligated satellite tobacco ringspot virus RNA. Nucleic Acids Res 14:9729–9743

Canny MD, Jucker FM, Kellogg E, Khvorova A, Jayasena SD, Pardi A (2004) Fast cleavage kinetics of a natural hammerhead ribozyme. J Am Chem Soc 126:10848–10849

Canny MD, Jucker FM, Pardi A (2007) Efficient ligation of the Schistosoma hammerhead ribozyme. Biochemistry 46:3826–3834

Chi YI, Martick M, Kim R, Scott WG, Kim SH (2008) Capturing hammerhead ribozyme structures in action by modulating general base catalysis. (in press, PLoS Biology, 2008)

Cochrane JC, Lipchock SV, Strobel SA (2007) Structural investigation of the GlmS ribozyme bound to its catalytic cofactor. Chem Biol 14:97–105

Dahm SC, Derrick WB, Uhlenbeck OC (1993) Evidence for the role of solvated metal hydroxide in the hammerhead cleavage mechanism. Biochemistry 32:13040–13045

Dahm SC, Uhlenbeck OC (1991) Role of divalent metal ions in the hammerhead RNA cleavage reaction. Biochemistry 30:9464–9469

De la Pēna M, Gago S, Flores R (2003) Peripheral regions of natural hammerhead ribozymes greatly increase their self-cleavage activity. EMBO J 22:5561–5570

Dunham CM, Murray JB, Scott WG (2003) A helical twist-induced conformational switch activates cleavage in the hammerhead ribozyme. J Mol Biol 332:327–336

Emilsson GM, Nakamura S, Roth A, Breaker RR (2003) Ribozyme speed limits. RNA (New York) 9:907–918

Ferbeyre G, Smith JM, Cedergren R (1998) Schistosome satellite DNA encodes active hammerhead ribozymes. Mol Cell Biol 18:3880–3888

Ferre-D'amare AR, Rupert PB (2002) The hairpin ribozyme: from crystal structure to function. Biochem Soc Trans 30:1105–1109

Ferre-D'Amare AR, Zhou K, Doudna JA (1998) Crystal structure of a hepatitis delta virus ribozyme. Nature 395:567–574

Forster AC, Davies C, Sheldon CC, Jeffries AC, Symons RH (1988) Self-cleaving viroid and newt RNAs may only be active as dimers. Nature 334:265–267

Govil G (1976) Conformational structure of polynucleotides around the O-P bonds: refined parameters for CPF calculations. Biopolymers 15:2303–2307

Guerrier-Takada C, Gardiner K, Marsh T, Pace N, Altman S (1983) The RNA moiety of ribonuclease P is the catalytic subunit of the enzyme. Cell 35:849–857

Hammann C, Cooper A, Lilley DM (2001a) Thermodynamics of ion-induced RNA folding in the hammerhead ribozyme: an isothermal titration calorimetric study. Biochemistry 40:1423–1429

Hammann C, Lilley DM (2002) Folding and activity of the hammerhead ribozyme. Chembiochem 3:690–700

Hammann C, Norman DG, Lilley DM (2001b) Dissection of the ion-induced folding of the hammerhead ribozyme using 19F NMR. Proc Natl Acad Sci U S A 98:5503–5508

Hampel A, Tritz R (1989) RNA catalytic properties of the minimum (-)sTRSV sequence. Biochemistry 28:4929–4933

Hampel A, Tritz R, Hicks M, Cruz P (1990) 'Hairpin' catalytic RNA model: evidence for helices and sequence requirement for substrate RNA. Nucleic Acids Res 18:299–304

Han J, Burke JM (2005) Model for general acid-base catalysis by the hammerhead ribozyme: pH-activity relationships of G8 and G12 variants at the putative active site. Biochemistry 44:7864–7870

Heckman JE, Lambert D, Burke JM (2005) Photocrosslinking detects a compact, active structure of the hammerhead ribozyme. Biochemistry 44:4148–4156

Hertel KJ, Herschlag D, Uhlenbeck OC (1994) A kinetic and thermodynamic framework for the hammerhead ribozyme reaction. Biochemistry 33:3374–3385

Hertel KJ, Peracchi A, Uhlenbeck OC, Herschlag D (1997) Use of intrinsic binding energy for catalysis by an RNA enzyme. Proc Natl Acad Sci U S A 94:8497–8502

Hertel KJ, Uhlenbeck OC (1995) The internal equilibrium of the hammerhead ribozyme reaction. Biochemistry 34:1744–1749

Ke A, Zhou K, Ding F, Cate JH, Doudna JA (2004) A conformational switch controls hepatitis delta virus ribozyme catalysis. Nature 429:201–205

Khvorova A, Lescoute A, Westhof E, Jayasena SD (2003) Sequence elements outside the hammerhead ribozyme catalytic core enable intracellular activity. Nat Struct Biol 10:708–712

Klein DJ, Ferre-D'Amare AR (2006) Structural basis of glmS ribozyme activation by glucosamine-6-phosphate. Science 313:1752–1756

Kuo MY, Sharmeen L, Dinter-Gottlieb G, Taylor J (1988) Characterization of self-cleaving RNA sequences on the genome and antigenome of human hepatitis delta virus. J Virol 62:4439–4444

Lilley DM (1998) Folding of branched RNA species. Biopolymers 48:101–112

Lilley DM (1999) RNA folding and catalysis. Genetica 106:95–102

Lilley DM (2003) Ribozymes–a snip too far? Nat Struct Biol 10:672–673

Luzi E, Eckstein F, Barsacchi G (1997) The newt ribozyme is part of a riboprotein complex. Proc Natl Acad Sci U S A 94:9711–9716

Martick M, Scott WG (2006) Tertiary contacts distant from the active site prime a ribozyme for catalysis. Cell 126:309–320

Martick M, Lee TS, York DM, Scott WG (2008a) Solvent structure and hammerhead ribozyme catalysis chemistry and biology 15:332–342

Martick M, Horan LH, Noller HF, Scott WG (2008b) A discontinuous hammerhead ribozyme embedded in a mammalian mRNA. Nature (in press)

McKay DB (1996) Structure and function of the hammerhead ribozyme: an unfinished story. RNA (New York) 2:395–403

Murray JB, Dunham CM, Scott WG (2002) A pH-dependent conformational change, rather than the chemical step, appears to be rate-limiting in the hammerhead ribozyme cleavage reaction. J Mol Biol 315:121–130

Murray JB, Scott WG (2000) Does a single metal ion bridge the A-9 and scissile phosphate groups in the catalytically active hammerhead ribozyme structure? J Mol Biol 296:33–41

Murray JB, Seyhan AA, Walter NG, Burke JM, Scott WG (1998a) The hammerhead, hairpin and VS ribozymes are catalytically proficient in monovalent cations alone. Chem Biol 5:587–595

Murray JB, Szoke H, Szoke A, Scott WG (2000) Capture and visualization of a catalytic RNA enzyme-product complex using crystal lattice trapping and X-ray holographic reconstruction. Mol Cell 5:279–287

Murray JB, Terwey DP, Maloney L, Karpeisky A, Usman N, Beigelman L, Scott WG (1998b) The structural basis of hammerhead ribozyme self-cleavage. Cell 92:665–673

Nelson JA, Uhlenbeck OC (2006) When to believe what you see. Mol Cell 23:447–450

Nelson JA, Uhlenbeck OC (2008a) Hammerhead redux: does the new structure fit the old biochemical data? RNA (New York) 14:605–615

Nelson JA, Uhlenbeck OC (2008b) Minimal and extended hammerheads utilize a similar dynamic reaction mechanism for catalysis. RNA (New York) 14:43–54

Penedo JC, Wilson TJ, Jayasena SD, Khvorova A, Lilley DM (2004) Folding of the natural hammerhead ribozyme is enhanced by interaction of auxiliary elements. RNA (New York) 10:880–888

Pley HW, Flaherty KM, McKay DB (1994) Three-dimensional structure of a hammerhead ribozyme. Nature 372:68–74

Prody GA, Bakos JT, Buzayan JM, Schneider IR, Breuning G (1986) Autolytic processing of dimeric plant virus satellite RNA. Science 231:1577–1580

Przybilski R, Hammann C (2007) The tolerance to exchanges of the Watson Crick base pair in the hammerhead ribozyme core is determined by surrounding elements. RNA (New York) 13:1625–1630

Pyle AM (1993) Ribozymes: a distinct class of metalloenzymes. Science 261:709–714

Raines RT (1998) Ribonuclease A. Chem Rev 98:1045–1066

Rueda D, Wick K, McDowell SE, Walter NG (2003) Diffusely bound Mg^{2+} ions slightly reorient stems I and II of the hammerhead ribozyme to increase the probability of formation of the catalytic core. Biochemistry 42:9924–9936

Ruffner DE, Stormo GD, Uhlenbeck OC (1990) Sequence requirements of the hammerhead RNA self-cleavage reaction. Biochemistry 29:10695–10702

Rupert PB, Ferre-D'Amare AR (2001) Crystal structure of a hairpin ribozyme-inhibitor complex with implications for catalysis. Nature 410:780–786

Rupert PB, Massey AP, Sigurdsson ST, Ferre-D'Amare AR (2002) Transition state stabilization by a catalytic RNA. Science 298:1421–1424

Scott WG (1999) RNA structure, metal ions, and catalysis. Curr Opin Chem Biol 3:705–709

Scott WG (2001) Ribozyme catalysis via orbital steering. J Mol Biol 311:989–999

Scott WG (2007) Morphing the minimal and full-length hammerhead ribozymes: implications for the cleavage mechanism. Biol Chem 388:727–735

Scott WG, Finch JT, Klug A (1995) The crystal structure of an all-RNA hammerhead ribozyme: a proposed mechanism for RNA catalytic cleavage. Cell 81:991–1002

Scott WG, Murray JB, Arnold JR, Stoddard BL, Klug A (1996) Capturing the structure of a catalytic RNA intermediate: the hammerhead ribozyme. Science 274:2065–2069

Sigurdsson ST, Tuschl T, Eckstein F (1995) Probing RNA tertiary structure: interhelical crosslinking of the hammerhead ribozyme. RNA (New York) 1:575–583

Slim G, Gait MJ (1991) Configurationally defined phosphorothioate-containing oligoribonucleotides in the study of the mechanism of cleavage of hammerhead ribozymes. Nucleic Acids Res 19:1183–1188

Soukup GA, Breaker RR (1999) Relationship between internucleotide linkage geometry and the stability of RNA. RNA (New York) 5:1308–1325

Stage-Zimmermann TK, Uhlenbeck OC (1998a) Circular substrates of the hammerhead ribozyme shift the internal equilibrium further toward cleavage. Biochemistry 37:9386–9393

Stage-Zimmermann TK, Uhlenbeck OC (1998b) Hammerhead ribozyme kinetics. RNA (New York) 4:875–889

Stage-Zimmermann TK, Uhlenbeck OC (2001) A covalent crosslink converts the hammerhead ribozyme from a ribonuclease to an RNA ligase. Nat Struct Biol 8:863–867

Stahley MR, Strobel SA (2006) RNA splicing: group I intron crystal structures reveal the basis of splice site selection and metal ion catalysis. Curr Opin Struct Biol 16:319–326

Steitz TA, Steitz JA (1993) A general two-metal-ion mechanism for catalytic RNA. Proc Natl Acad Sci U S A 90:6498–6502

Symons RH (1997) Plant pathogenic RNAs and RNA catalysis. Nucleic Acids Res 25:2683–2689

Uhlenbeck OC (1987) A small catalytic oligoribonucleotide. Nature 328:596–600

van Tol H, Buzayan JM, Feldstein PA, Eckstein F, Bruening G (1990) Two autolytic processing reactions of a satellite RNA proceed with inversion of configuration. Nucleic Acids Res 18:1971–1975

Wang S, Karbstein K, Peracchi A, Beigelman L, Herschlag D (1999) Identification of the hammerhead ribozyme metal ion binding site responsible for rescue of the deleterious effect of a cleavage site phosphorothioate. Biochemistry 38:14363–14378

Winkler WC, Nahvi A, Roth A, Collins JA, Breaker RR (2004) Control of gene expression by a natural metabolite-responsive ribozyme. Nature 428:281–286

Wu HN, Lin YJ, Lin FP, Makino S, Chang MF, Lai MM (1989) Human hepatitis delta virus RNA subfragments contain an autocleavage activity. Proc Natl Acad Sci U S A 86:1831–1835

Zaug AJ, Cech TR (1986) The intervening sequence RNA of Tetrahymena is an enzyme. Science 231:470–475

Zhou J M, Zhou D M, Takagi Y, Kasai Y, Inoue A, Baba T, Taira K (2002) Existence of efficient divalent metal ion-catalyzed and inefficient divalent metal ion-independent channels in reactions catalyzed by a hammerhead ribozyme. Nucleic Acids Res 30:237ϵ–2382

Chapter 5
The Small Ribozymes: Common and Diverse Features Observed Through the FRET Lens

Nils G. Walter(✉) **and Shiamalee Perumal**

Abstract The hammerhead, hairpin, HDV, VS and *glmS* ribozymes are the five known, naturally occurring catalytic RNAs classified as the "small ribozymes." They share common reaction chemistry in cleaving their own backbone by phosphodiester transfer, but are diverse in their secondary and tertiary structures, indicating that Nature has found at least five independent solutions to a common chemical task. Fluorescence resonance energy transfer (FRET) has been extensively used to detect conformational changes in these ribozymes and dissect their reaction pathways. Common and diverse features are beginning to emerge that, by extension, highlight general biophysical properties of non-protein coding RNAs.

5.1   Introduction

Since the discovery in the early 1980s that certain biological catalysts involved in the processing of genetic information are composed of RNA (Kruger et al. 1982; Guerrier-Takada et al. 1983), a number of such natural ribozymes have been discovered, and research in the field has focused on elucidating their enzymatic mechanisms and secondary and tertiary structures. In recent years, the spotlight has been on emerging high-resolution crystal structures that illustrate the precise manner in which ribozymes orient and align reactive groups. The main challenge now lies in linking these static snapshots to the dynamical features of RNA structure to answer the outstanding question of how chemical catalysis arises. This chapter summarizes how the current application of fluorescence resonance energy transfer (FRET) has helped dissect the reaction mechanisms of the small ribozymes. Common and distinct features are beginning to emerge under the magnifying lens of FRET.

N.G. Walter
Department of Chemistry, University of Michigan, Ann Arbor, 93. N University, MI 48109–1055, USA
e-mail: nwalter@umich.edu

N.G. Walter et al. (eds.) *Non-Protein Coding RNAs*
doi: 10.1007/978-3-540-70840-7_5, © Springer–Verlag Berlin Heidelberg 2009

5.2   The Class of Small Ribozymes

5.2.1   *Common Mechanism and Catalytic Strategies*

Biological evolution has produced and preserved five known, structurally distinct ribozymes that promote non-hydrolytic phosphodiester backbone cleavage in RNA, the hammerhead, hairpin, hepatitis delta virus (HDV), Varkud satellite (VS), and *glmS* ribozymes. Given their relatively small size (<200 nt) and common reaction mechanism, these self-cleaving RNAs are grouped as the class of "small ribozymes" (Doudna and Lorsch 2005; Fedor and Williamson 2005; Bevilacqua and Yajima 2006; Scott 2007). Unlike their larger counterparts, such as group I and II intron ribozymes or RNase P, the small ribozymes do not require an external nucleophile but site-specifically activate one of their own 2′-OH moieties by de-protonation to attack the adjacent 3′, 5′-phosphodiester and substitute the 5′-oxygen in an S_N2-type phosphodiester transfer. As a result, the 5′- and 3′-reaction products carry 2′, 3′-cyclic phosphate and 5′-OH termini, respectively (Fig. 5.1).

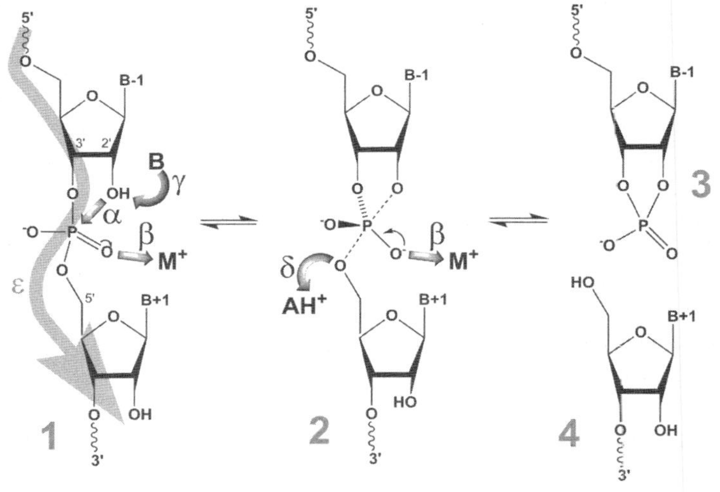

Fig. 5.1 Mechanism of RNA cleavage by the small ribozymes. The backbone (**1**) passes through a penta coordinate transition state (**2**) that degrades into the 5′- and 3′-cleavage products carrying 2′, 3′-cyclic phosphate (**3**) and 5′-hydroxyl termini (**4**), respectively. Five catalytic strategies that can promote the reaction mechanism are highlighted by broad shaded arrows and Greek letters: *α*, in-line nucleophilic attack of the 2′-oxygen on the scissile phosphate at an idealized 2′O-P-5′O angle of 180°; *β*, neutralization of the negative charge on a non-bridging phosphate oxygen; *γ*, deprotonation of the 2′-hydroxyl group by a base B; *δ*, protonation of the developing negative charge on the 5′-oxygen leaving group by an acid AH⁺; and *ε*, ground state structure destabilization (substrate straining)

Although many protein-based ribonucleases catalyze the same reaction, the small ribozymes afford superior site-specificity by using base pairing and other interactions, to align the cleavage site for in-line nucleophilic attack. Arguably, evolution chose these RNA-based catalysts not for their rate acceleration and capacity to process, which is higher for the protein enzymes, but for their compact usage of genetic information to accomplish unique sequence specificity. This sequence specificity is all the more surprising since the mechanism catalyzed is the same as that of the non-enzymatic degradation of RNA (Zhou and Taira 1998), which implies that a ribozyme has to avoid non-specific backbone cleavage while promoting the site-specific path.

The small ribozymes, though very different in chemical makeup, are turning out to use the same repertoire of catalytic strategies as their protein counterparts (Emilsson et al. 2003; Doudna and Lorsch 2005; Fedor and Williamson 2005; Bevilacqua and Yajima 2006; Scott 2007; Walter 2007), including (Fig. 5.1): positioning of the reacting functional groups in an optimal in-line attack configuration (α); electrostatic catalysis to stabilize the enhanced negative charge on the phosphate oxygens in the transition state (β); general base catalysis by removing the proton from the attacking 2′-OH nucleophile (γ); general acid catalysis by donating a proton to the 5′-oxygen leaving group (δ); and destabilization of the ground state structure (ε).

In protein-based enzymes, the chemical versatility of the amino-acid side chains contributes polar, charged or uncharged polar side chains as obvious participants in general acid–base and electrostatic catalysis. Initially, it was therefore less clear how ribozymes with their more limited chemical makeup may affect general acid–base catalysis at neutral pH, especially since ionization of the nucleobases and riboses in isolation only occurs at considerably acidic or basic pH. Consequently, the first proposals suspected external cofactors such as liganded, partially hydrated Mg^{2+}-ions as effectors of ribozyme chemistry (Pyle 1993); but the field has now embraced the idea of chemistry catalyzed by pK_a-shifted nuclecbases (Doudna and Lorsch 2005; Fedor and Williamson 2005; Bevilacqua and Yajima 2006) and even considers mechanisms involving proton relays through structural water molecules (Walter 2007). Intriguingly, the field has come full circle through discovery of the *glmS* ribozyme, since this ribozyme indeed requires an external, small-molecule co-factor as an essential reaction participant (McCarthy et al. 2005; Hampel and Tinsley 2006; Klein and Ferre-D'Amare 2006; Tinsley et al. 2007).

The three questions to be specifically addressed in this review are: (i) How can FRET dissect the folding and reaction pathways responsible for positioning the participants of small ribozyme catalysis? (ii) How do small ribozymes direct catalytic power towards one specific bond without sacrificing their overall backbone integrity? (iii) Which common and diverse features emerge from a comparison of the five known, naturally occurring small ribozymes? We will address these questions by first summarizing what insights FRET has helped reveal for each small ribozyme.

5.3 The Hammerhead Ribozyme

The hammerhead ribozyme was first discovered in viroids and satellite RNAs of plant viruses, where it is thought to be essential for RNA self-cleavage and -ligation during double-rolling circle replication of the parasitic pathogen (Prody et al. 1986; Forster and Symons 1987; Flores et al. 2004). It was subsequently also identified in the genomes of the newt (*Notophthalamus viridescens*), schistosome trematodes, cave crickets (*Dolichopoda* species), and even mammals (Martick et al. 2008b) and emerges as the smallest and most common self-cleaving RNA motif from *in vitro* evolution experiments, suggesting that nature may have created it multiple times by convergent evolution (Salehi-Ashtiani and Szostak 2001).

The hammerhead motif consists of a catalytic core of 11 conserved nucleotides flanked by three helical stems that are arranged in a "Y-shape" by a sharp uridine (U-)turn in the backbone that juxtaposes Stems I and II (Ruffner et al. 1990; Wedekind and McKay 1998) (Fig. 5.2a). The ribozyme is catalytically active in the presence of either millimolar divalent or molar monovalent cations with similar sequence requirements, suggesting that divalents are not obligatory participants in phosphodiester transfer (Murray et al. 1998; Curtis and Bartel 2001; O'Rear et al. 2001). Most of the functional groups in the catalytic core are essential for catalytic activity (McKay 1996), making it initially difficult to distinguish moieties important for folding from those directly involved in catalysis. In addition, a controversy ensued between early crystal structures and biochemical evidence that suggested significant conformational changes from these structures were necessary to reach the catalytically active state (Blount and Uhlenbeck 2005).

Only in 2003 did it become clear that all naturally occurring hammerhead ribozymes have non-conserved (and therefore initially overlooked) loop–loop interactions between Stems I and II that further stabilize the Y-shape (Fig. 5.2b) and significantly enhance catalysis, especially at near-physiologic concentrations of Mg^{2+} (0.5–1 mM free Mg^{2+}) (De la Pena et al. 2003; Khvorova et al. 2003; Penedo et al. 2004).The controversy was lifted by new crystal structures including such loop–loop interactions by showing how these distal tertiary contacts rearrange the catalytic core, in a way much more consistent with the biochemical data, where the scissile phosphate is poised for general acid–base catalysis by functional groups of guanosines G8 and G12 (Martick and Scott 2006; Nelson and Uhlenbeck 2006) and potentially a structural water molecule (Martick et al. 2008a) (please note that these structures are the focus of the preceding chapter by Scott). The current, unifying model suggests that both the "minimal" and "extended" hammerhead ribozymes dynamically adopt both inactive and active conformations similar to the two types of crystal structures, albeit with different bias, leading to the observed difference in catalytic activity (Martick and Scott 2006; Nelson and Uhlenbeck 2008b; Nelson and Uhlenbeck 2008a).

The hammerhead ribozyme thus emerges as a small, yet very dynamic RNA motif, which has largely eluded high-resolution solution-phase structure determination for example by, NMR spectroscopy (Simorre et al. 1997; Simorre et al. 1998; Bondensgaard et al. 2002; Furtig et al. 2008). Fluorescence spectroscopy is

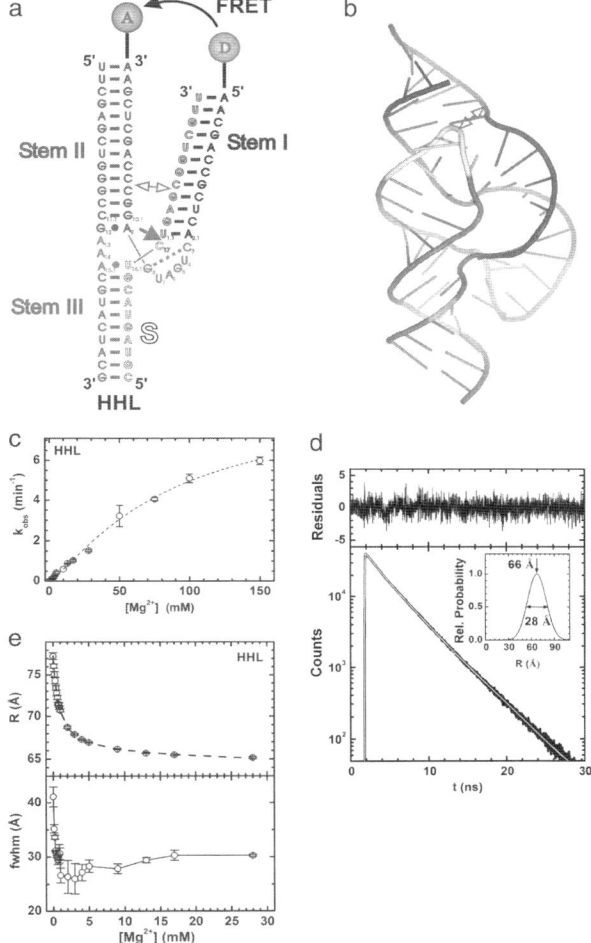

Fig. 5.2 Time-resolved FRET studies of the hammerhead ribozyme. (**a**) Secondary structure of the hammerhead ribozyme HHL utilized in these studies. Helices are color coded, the catalytic core is highlighted in green, and the cleavage site in the substrate (outlined, S) is marked by a closed red arrow. Stems I and II are labeled with donor (D) and acceptor (A) fluorophores, as indicated, which undergo FRET. In extended forms of the ribozyme, Stems I and II interact in the regions marked by an open double-arrow. (**b**) Cartoon representation of the crystal structure of the extended *Schistosoma* hammerhead ribozyme (PDB ID: 2GOZ) (Martick and Scott 2006), color-coded as in panel A except for the silver capping loop that is present only in the extended form. The interaction of an internal and a capping loop in Stems I and II, respectively, is marked by an open double-arrow. (**c**) Mg^{2+} dependence of the observed cleavage rate constant of HHL under standard conditions (50 mM Tris-HCl, pH 8.0, 25°C). (**d**) Representative HHL donor fluorescence decay (bottom, black line) with corresponding fit (white line) in the presence of acceptor under standard conditions with 9 mM Mg^{2+}. Top, fit residuals; inset, resulting single-distance distribution with the indicated mean and fwhm values. (**e**) Mean distance (*top*) and fwhm (*bottom*) of the Stem I–II distance distribution of HHL as a function of Mg^{2+} concentration under standard conditions, as derived by time-resolved FRET analysis similar to panel D. In part modified with permission from (Rueda et al. 2003)

a biophysical technique capable of probing conformational ensembles in solution. FRET, in particular, is a photon-less process that occurs at nanometer distances between a donor (D) fluorophore in the excited electronic state and an acceptor (A) in the ground electronic state. The rate of energy transfer k_T is dependent on the distance r between D and A, as well as the D excited state lifetime τ_D (Lakowicz 2006)

$$k_t(r) = \frac{1}{\tau_D} \left(\frac{R_0}{r} \right)^6$$

The distance R_0 at which the FRET efficiency is 50% is called the Förster distance, after Theodor Förster who first described the theory behind the phenomenon (Förster 1946; Förster 1948). The Förster distance typically ranges from 20 to 70 Å and depends primarily on the spectral overlap of the D emission and A excitation, as well as the relative orientation of the transition dipole moments of the fluorophores. The FRET efficiency is most strongly dependent on the D–A distance around the Förster distance, making it a well suited molecular ruler in biology (Stryer 1978):

$$E_T = \frac{R_0^6}{R_0^6 + r^6}$$

Given the "Y-shaped" structure of the hammerhead ribozyme, its two proximal Stems I and II are suitable placement sites for a D–A pair. Most FRET-based structure probing so far was reported for "minimal" hammerhead ribozyme constructs (Tuschl et al. 1994; Bassi et al. 1997, 1999; Rueda et al. 2003), but their similar global structures and apparent conformational exchange on the local structural level make these studies also relevant to the less studied "extended" versions (Penedo et al. 2004).

A commonly employed "minimal" hammerhead ribozyme termed HHL is shown in Fig. 5.2a. Lilley and co-workers used steady-state FRET (i.e., continuous fluorophore excitation combined with inspection of the emission spectra (Walter 2001, 2002)) to establish that, upon Mg^{2+} titration, two sequential metal ion-induced structural transitions lead to adoption of the "Y-shaped" structure. The first transition leads to coaxial stacking of Stems II and III with an apparent half-titration point for Mg^{2+} of ~0.1 mM, while the second transition presumably organizes the U-turn motif in the catalytic core such that Stems I and II become juxtaposed with an apparent Mg^{2+} half-titration point of ~1 mM (Bassi et al. 1997). In an extended hammerhead ribozyme, these two transitions collapse into a single one at the lower Mg^{2+} concentration (Penedo et al. 2004), suggesting that the loop–loop interactions of the extended hammerhead ribozyme reduce the Mg^{2+}-requirement by an order of magnitude for adopting the Y-shaped structure with folded U-turn of the core.

We utilized the minimal HHL ribozyme to reveal, by combining cleavage assays (Fig. 5.2c) and time-resolved FRET (Figs. 5.2d, e), a third folding transition at even higher Mg^{2+} concentrations (half-titration point of 90 mM), which finally activates the ribozyme (Rueda et al. 2003). For time-resolved FRET we site-specifically labeled Stems I and II with fluorescein donor (D) and tetramethylrhodamine acceptor (A), respectively, as indicated in Fig. 5.2a. To avoid flurophore quenching by adjacent guanines, two A:U base pairs were placed next to the fluorophores at the ends of the labeled stems.

Time-resolved FRET measures the donor fluorescence decay lifetimes (typically on the order of a few nanoseconds) in both a donor-only and a donor-acceptor doubly-labeled sample to derive the fluorophore distance distribution from the relative decay acceleration in the presence of the acceptor (Walter 2001). Relative to the timescale of the donor fluorescence decay, most molecular motions are slow so that time-resolved FRET takes a snapshot of the donor-acceptor distance distribution arising (at least in part) from structural flexibility among the ensemble of molecules in the sample.

Given the placement of the fluorophores in the HHL construct, the mean Stem I–stem II distance and its full width at half-maximum (fwhm) were measured in standard buffer (50 mM Tris-HCl, pH 8.0) at 25°C as a function of Mg^{2+} concentration (Figs. 5.2d, e). A large global structural transition with a Mg^{2+} dissociation constant in the physiological range of ~1 mM was observed which brings Stems I and II closer together and sharpens their distance distribution (i.e., lowers the fwhm) (Fig. 5.2e), essentially consistent with the earlier studies (Bassi et al. 1997; Hammann and Lilley 2002). Na^+ ions impair this Mg^{2+}-induced transition. However, a previously undetected, relatively subtle global rearrangement coincides with catalytic activation at ~100-fold lower Mg^{2+} affinity. This transition broadens the Stem I–II distance distribution and is not impaired by Na^+ (Fig. 5.2e). Notably, a catalytically more active hammerhead ribozyme (termed HHα) exhibits the same general behavior with higher Mg^{2+} affinity and a larger fwhm than HHL, further supporting the notion that a shortening of the mean Stem I–II distance combined with larger flexibility at high Mg^{2+}-concentrations is important for catalysis (Rueda et al. 2003).

5.4   The Hairpin Ribozyme

The hairpin ribozyme is derived from the negative-polarity strand of the satellite RNA associated with the tobacco ringspot virus, where it complements a hammerhead ribozyme motif in the positive-polarity strand to affect double-rolling circle replication of the satellite (Haseloff and Gerlach 1989; Feldstein et al. 1990; Walter and Burke 1998). A minimal hairpin ribozyme of ~50 nucleotides can be segmented into two separate strands (termed RzA and RzB) that bind a substrate to form domain A, comprising helices H1 and H2 and the symmetric internal loop A (Fig. 5.3a). Domain A is connected by a flexible hinge to domain B, which comprises

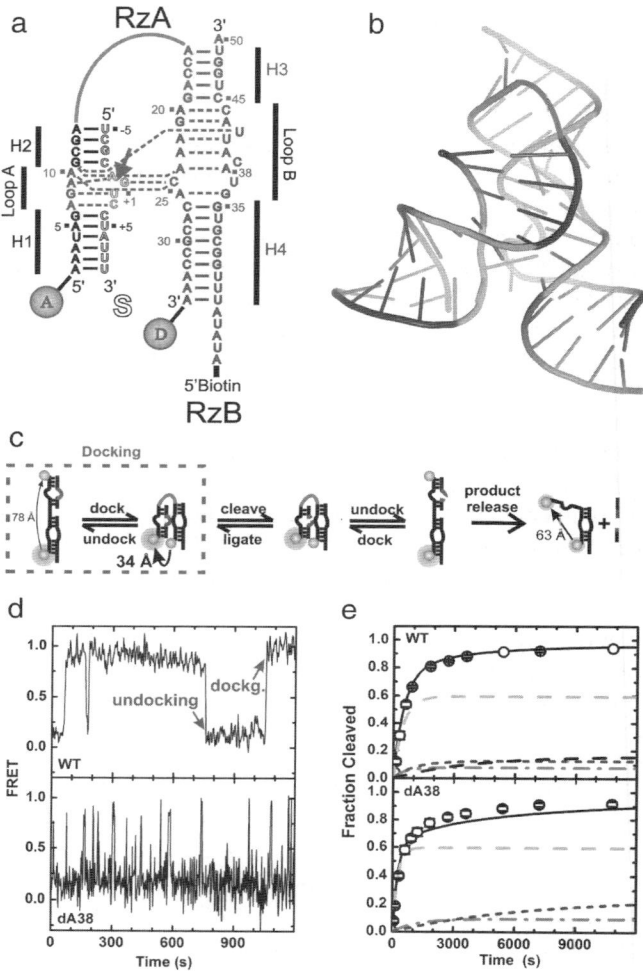

Fig. 5.3 Steady-state FRET studies of the hairpin ribozyme. (**a**) Secondary structure of the hairpin ribozyme from the tobacco ringspot satellite RNA utilized in these studies. Helices are color coded, the catalytic core is highlighted in green, and the cleavage site in the substrate (outlined, S) is marked by a closed red arrow. The RzA strand is labeled with donor (D) and acceptor (A) fluorophores, as indicated, which undergo FRET. (**b**) Cartoon representation of the crystal structure of a junctionless, all-RNA hairpin ribozyme (PDB ID: 2OUE) (Salter et al. 2006), color-coded as in panel A; a gray line was added to indicate the connectivity of the two domains. (**c**) Reaction pathway of the hairpin ribozyme as revealed by ensemble and single molecule FRET. Fluorophore distances measured by time-resolved FRET are indicated. (**d**) Representative single molecule FRET time trace of wild-type (WT) and dA38 mutant ribozymes after binding of non-cleavable (2′-OMe-A$_{-1}$) substrate analog under standard conditions (50 mM Tris-HCl, pH 7.5, 12 mM Mg^{2+}, 25°C). One undocking and one docking event are indicated. (**e**) Experimental (symbols) and simulated (lines) cleavage time courses of the WT and dA38 mutant ribozymes under standard conditions. The predicted contributions from four and three experimentally detected molecule sub-populations, respectively, are indicated by gray and dashed lines. In part modified with permission from (Rueda et al. 2004) (See figure insert for color reproduction)

H3 and H4 and the asymmetric internal loop B. Site-specific cleavage and ligation occur in the substrate strand of domain A (Fig. 5.3a) and require the presence of either millimolar concentrations of divalent or molar concentrations of monovalent cations (Hampel and Cowan 1997; Nesbitt et al. 1997; Young et al. 1997; Murray et al. 1998). A number of crystal structures have highlighted the intricate network of hydrogen bonding and stacking interactions that docks loops A and B to form the catalytic core, and single out a specific phosphodiester bond for cleavage (Fig. 5.3b) (Rupert and Ferre-D'Amare 2001; Rupert et al. 2002; Alam et al. 2005; Salter et al. 2006; Torelli et al. 2007), presumably involving (electrostatic) transition state stabilization (Rupert et al. 2002; Kuzmin et al. 2004; Kuzmin et al. 2005) and/or general acid–base catalysis by RNA functional groups of A38 and perhaps G8 (Pinard et al. 2001; Bevilacqua 2003; Wilson et al. 2006) and/or structural water molecules (Rhodes et al. 2006; Torelli et al. 2007).

To dissect steps on the reaction pathway of the hairpin ribozyme, the ensemble relaxation kinetics from the undocked to the docked conformational state upon substrate addition were monitored by steady-state FRET between domain-terminal fluorophores (Walter et al. 1998c). The results revealed that substrate binding (secondary structure formation) and domain docking (tertiary structure formation) are two distinct steps on the reaction pathway. Upon docking, the ribozyme reversibly cleaves the substrate, followed by undocking and product release (Fig. 5.3c). Docking of domains A and B, as monitored by FRET, is thus required and at least partially rate-limiting for both cleavage and ligation. Ensemble steady-state FRET assays also revealed that most modifications to the RNA, or reaction conditions known to inhibit catalysis do in fact prevent the necessary domain docking (Walter et al. 1998c).

To further characterize the folding intermediates, the undocked and docked conformers of the hairpin ribozyme were examined under varying Mg^{2+} concentrations by time-resolved FRET (Walter et al. 1999). This analysis yielded the interdomain distance distributions, their mean distances, and their fractional contributions, which define differences in docking free energy. The catalytically active tertiary structure was found to be stabilized by both specific docking interactions between domains A and B and the topology of the intervening helical junction. In particular, at near-physiological Mg^{2+} concentrations, the naturally occurring four-way junction thermodynamically favors docking, whereas a nicked or connected two-way junction and a three-way junction favor the undocked conformer. These findings highlighted the importance of the four-way junction as part of the hairpin ribozyme motif found in the satellite RNA to bring domains A and B into proximity, independent of specific docking interactions between them (Walter et al. 1999). This view is qualitatively supported by ensemble steady-state FRET assays (Murchie et al. 1998; Walter et al. 1998a, b).

Over the past few years, single molecule FRET has become increasingly popular for illuminating pathways of RNA folding and catalysis (Zhuang 2005; Ditzler et al. 2007). Using prism-based total internal reflection fluorescence microscopy (TIRFM) (Walter et al. 2008), low background noise levels can be obtained during the observation of single, surface immobilized hairpin ribozyme molecules that reproduce the catalytic rate constants of free solution conditions (Zhuang et al. 2002). Since single molecules exist in either the undocked or the docked conformation, single molecule FRET can detect (given sufficient time resolution) individual transitions between the two states (Fig. 5.3d). Statistics of the dwell or residency times in a specific state then yield the kinetics associated with the inter-conversion of states (Zhuang et al. 2002; Ditzler et al. 2007), unlike in ensemble behavior where the individual molecule behavior is averaged out and only non-equilibrium relaxation kinetics can be measured.

Three single molecule FRET approaches have been used to dissect the reaction pathway of two- and four-way hairpin ribozymes (Fig. 5.3c). They provide access to the chemical turnover rates (Ditzler et al. 2007): (i) Probing of inactivated ribozyme-substrate and -product complexes in combination with ensemble activity assays and mechanistic modeling (Rueda et al. 2004). This approach depends on the accuracy of assumptions necessary for the kinetic modeling, such as that of identical folding behavior of the inactivated and active ribozyme complexes. (ii) Direct observation of catalytic turnover due to associated changes in FRET or dwell time constant (Nahas et al. 2004). This approach depends on the ability to clearly distinguish such changes from unrelated fluctuations. (iii) Indirect observation of the probability for catalytic events using a succession of buffer exchanges to produce distinct time sequences of the single molecule FRET signal that serve as kinetic "fingerprints" of specific catalytic intermediates (Liu et al. 2007). This approach requires a very careful "sorting" of the resulting time traces by behavioral classes.

These three approaches showed that the rate of substrate cleavage is rate-limited by a combination of conformational transitions and reversible chemical equilibrium (Zhuang et al. 2002; Rueda et al. 2004). Adoption of the four-way junction and shortening of the substrate accelerate docking and product dissociation, respectively, shifting the rate limitation largely toward reaction chemistry (Nahas et al. 2004). Strikingly, all studies consistently found evidence for non-interchanging sub-populations of the hairpin ribozyme that are readily distinguished by their dwell time in the docked state (Fig. 5.3d) (Zhuang et al. 2002; Bokinsky et al. 2003; Nahas et al. 2004; Okumus et al. 2004; Rueda et al. 2004; Liu et al. 2007). These complex structural dynamics quantitatively explains the heterogeneous cleavage kinetics common to many catalytic RNAs (Zhuang et al. 2002; Rueda et al. 2004), as evident from the fact that only the sum of the predicted contributions of each sub-population can fully reproduce the biphasic kinetics of product formation. Figure 5.3e illustrates this point on exemplary wild-type (WT) and dA38 mutant cleavage time courses.

These observations suggest that single molecule approaches are critical in delineating the complexities of RNA folding and function (Zhuang 2005; Ditzler et al. 2007). Furthermore, in conjunction with site-specific mutations, metal ion titrations, and computational modeling they reveal that the domains of the hairpin ribozyme are in near-contact in the docking transition state even though the native tertiary contacts are at most partially formed (Bokinsky et al. 2003). More in-depth single molecule FRET analysis indicate that most site-specific RNA modifications affect the rate constants of docking, undocking, and chemistry even when they are distant from any direct docking and catalytic interactions (Rueda et al. 2004). This effect is likely due to a long-range network of hydrogen bonding and stacking interactions that involves several structural water molecules in the catalytic core (Rhodes et al. 2006; Walter 2007).

5.5   The HDV Ribozyme

The hepatitis delta virus (HDV) is a human pathogen and satellite of the hepatitis B virus; co-infection with both leads to enhanced liver damage. HDV contains a viroid-like genomic RNA that is thought to undergo replication via a double rolling-circle mechanism wherein both the genomic and anti-genomic RNAs self-cleave and re-ligate into circular monomers (Taylor 2006). The cleavage activities map to ~85-nucleotide RNA motifs, the genomic and anti-genomic ribozymes, that have ~75% sequence homology and share a common secondary and tertiary structure (Been 2006). The HDV ribozyme is comprised of five helices P1, P1.1, P2, P3, and P4 that are tight-knit into a double-nested pseudoknot (Fig. 5.4a) (Perrotta and Been 1991; Ferre-D'Amare et al. 1998), making this fold thermodynamically highly stable, perhaps as an adaptation to the conditions in mammalian cells (Perrotta and Been 1991). A structurally and biochemically closely related ribozyme motif of unknown biological function resides in an intron of the human *CPEB3* gene, suggesting that HDV may have parasitically arisen from the human transcriptome (Salehi-Ashtiani et al. 2006).

The cleavage site of the HDV ribozyme is located at the junction of a single-stranded 5′-sequence and the G1:U37 wobble pair that closes the P1 helix (Fig. 5.4a). The HDV ribozyme was the first (and arguably is still the clearest) example of an RNA enzyme where structural and kinetic data suggest a specific role for an RNA side chain in catalysis; the exact nature of this role however remains controversial. In particular, two kinetically equivalent roles – those of general base and general acid catalyst – have been alternatively proposed for nucleobase C75 in the genomic ribozyme (or the equivalent C76 in the antigenomic form) (Perrotta et al. 1999; Nakano et al. 2000; Ke et al. 2004; Das 2005).

The crystal structures of the reaction precursor and product show that C75 is jutting from its location in the joiner between helices P4 and P2 (J4/2) toward the cleavage site (Ferre-D'Amare et al. 1998; Ferre-D'Amare and Doudna 2000; Ke et al. 2004) (Fig. 5.4b). Controversially, the conformation of the catalytic pocket in the precursor crystal led to a model of general base involvement of C75 in the reaction mechanism (Ke et al. 2004), while the hydrogen bond between C75(N3) and the 5′-OH leaving group in the product crystal (Ferre-D'Amare et al. 1998; Ferre-D'Amare and Doudna 2000) is suggestive of general acid catalysis. ^{13}C-NMR spectroscopy yielded no clear evidence for the significant pK_a shift expected for C75(N3) to play a direct role in catalysis (Luptak et al. 2001), whereas a recent Raman crystallography approach did (Gong et al. 2007). A hydrated Mg^{2+} ion,

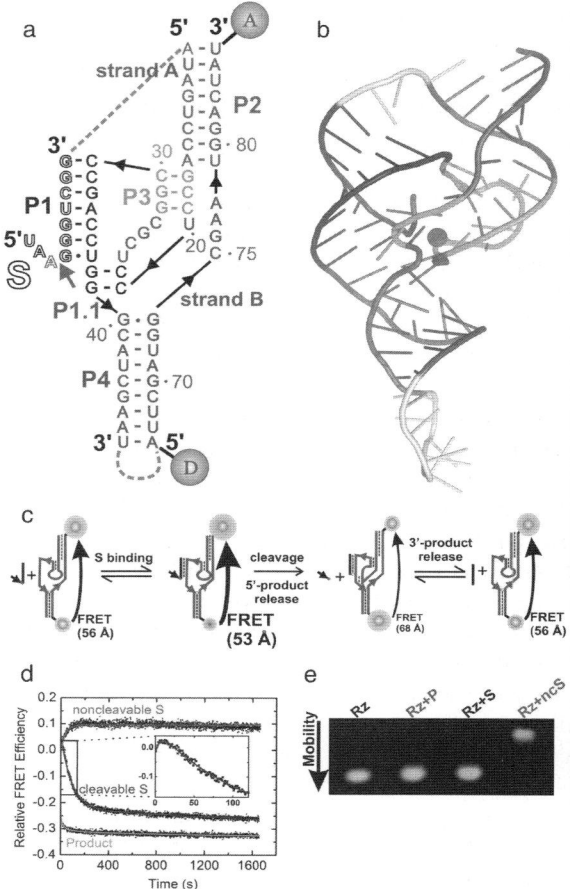

Fig. 5.4 FRET studies of the HDV ribozyme. (**a**) Secondary structure of the HDV ribozyme D1 utilized in these studies. Helices are color coded, the catalytic core is highlighted in green, and the cleavage site in the substrate (outlined, S) is marked by a closed red arrow. Strand B of the ribozyme is labeled with donor (D) and acceptor (A) fluorophores, as indicated, which undergo FRET. Dashed gray lines indicate connections that are present in the naturally occurring, cis-acting, genomic and antigenomic HDV ribozymes. (**b**) Cartoon representation of the crystal structure of the genomic HDV ribozyme (PDB ID: 1SJ3) (Ke et al. 2004), color-coded as in panel A with connections removed for the trans-acting D1 construct shown in silver. Red sphere, presumably catalytically involved Mg^{2+} ion. (**c**) Reaction pathway of the HDV ribozyme as revealed by ensemble steady-state FRET. Fluorophore distances measured by time-resolved FRET are indicated. (**d**) Relative FRET efficiency (calculated as ratio of the acceptor:donor fluorescence signals) over time of the doubly labeled D1 construct upon addition of a fivefold excess of non-cleavable substrate analogue (ncS3), cleavable substrate (S3; which binds, leading to an increase [inset], followed by a decrease upon cleavage), or 3′ product (3′P), as indicated, under standard conditions (40 mM Tris-HCl, pH 7.5, 11 mM Mg^{2+}, 25 mM DTT, 25°C). (**e**) Non-denaturing gel electrophoresis of the doubly labeled ribozyme alone (Rz) and in complex with 3′-product (Rz + P), cleavable substrate (Rz + S; this complex undergoes catalysis and is indistinguishable from the 3′-product complex), and non-cleavable substrate (Rz + ncS), as indicated. Detection is by FRET in the gel. In part modified with permission from (Pereira et al. 2002) (See figure insert for color reproduction)

located near the cleavage site (Ke et al. 2004) is, while not absolutely obligatory for activity (Nakano et al. 2003), thought to assist C75 by providing the complementary function in general acid–base catalysis (Nakano et al. 2000; Nakano and Bevilacqua 2007). However, it has been difficult to discern by either mechanistic or structural studies which specific role is played by these two components (Bevilacqua and Yajima 2006).

Studies of the HDV ribozyme mechanism are complicated by the fact that a conformational change accompanies catalysis (Fig. 5.4c). Probing a synthetic three-strand form (Fig. 5.4a) by a combination of steady-state FRET (Fig. 5.4d), time-resolved FRET, and electrophoretic mobility shift FRET assays (Fig. 5.4e) revealed that the distance between the P2 and P4 termini is significantly shorter (by ~15 Å) in the reaction precursor than in the product (Pereira et al. 2002). Complementary assays based on 2-aminopurine fluorescence (Harris et al. 2002) and terbium(III)-mediated footprinting (Jeong et al. 2003) highlighted local conformational changes that accompany the global conformational change detected by FRET. Subsequently, the crystal structure of the genomic HDV ribozyme precursor (Ke et al. 2004) showed a similar, if less pronounced P2–P4 contraction relative to the product structure (Ferre-D'Amare et al. 1998), validating the FRET observations on the three-strand form. Time-resolved FRET studies of the latter RNA in dependence of Mg^{2+} found that the precursor shortens while the product expands with increasing divalent metal ion concentration, thereby amplifying the structural differences observed in the crystal structures (Tinsley et al. 2004). Such amplification of the conformational change may contribute to systematically lower cleavage rate constants observed for multi-strand ("trans-acting") constructs relative to the contiguous ("cis-acting") genomic and anti- genomic HDV ribozymes (Tinsley and Walter 2007). Finally, the Mg^{2+} affinity of the C75 wild-type as monitored by time-resolved FRET is slightly (~2-fold) lower than that of a C75U mutant, consistent with the notion that C75 binds in proximity to and competes with a divalent metal ion (Tinsley et al. 2004), as also suggested by X-ray crystallography (Ke et al. 2004) and mechanistic studies (Nakano et al. 2000).

5.6   The VS Ribozyme

The Varkud Satellite (VS) RNA is located within the mitochondria of the bread mold *Neurospora*, where it replicates a retrotransposon by reverse transcription into a (circular) DNA plasmid, transcription into multimeric VS RNA copies, and self-cleavage and re-ligation back into circular, monomeric VS RNA (Collins and Saville 1990; Saville and Collins 1990). The VS ribozyme is the largest and most complex of the small ribozymes, and accordingly, least structurally understood. The secondary structure of the VS ribozyme consists of six helical segments: Stem-loop I forms the substrate domain with the cleavage site and stem-loops II through VI comprising the catalytic domain. The catalytic domain is organized into two three-way junctions (II-III-VI) and (III-IV-V) that share helix III (Fig. 5.5a). It recognizes the substrate predominantly through tertiary interactions, particularly the

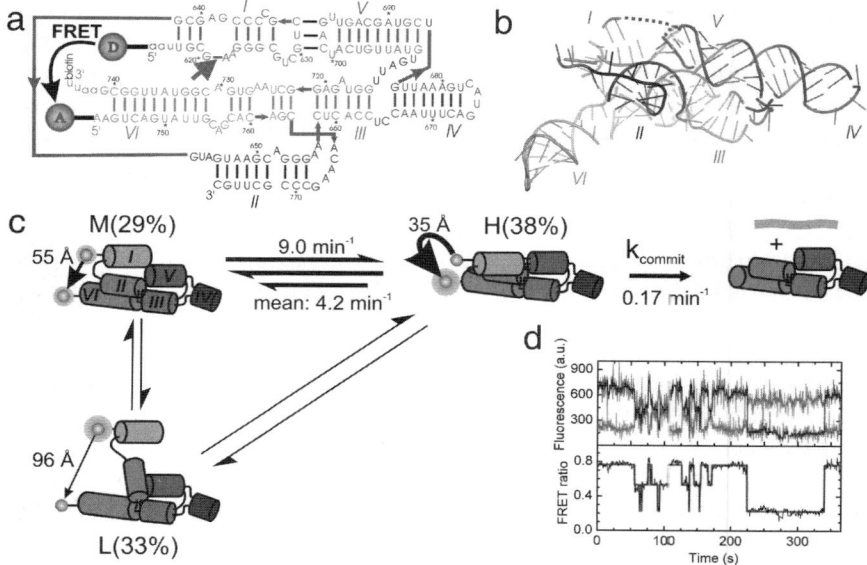

Fig. 5.5 Single molecule FRET studies of the VS ribozyme. (**a**) Secondary structure of the VS ribozyme G11 utilized in these studies. Helices are color coded, the cleavage site is marked with a closed red arrow, and the FRET donor (D) and acceptor (A) and biotin labeling sites are indicated. (**b**) Tertiary structure model of the VS ribozyme, using the same color code as in panel A. Some strand connectivities are incomplete in the model; the ones in the I–V kissing loop are indicated by dashed lines. (**c**) Structural and kinetic model of the reaction pathway of the WT VS ribozyme as derived from single molecule FRET. The helix colors match those in panel A. (**d**) Single molecule FRET time trace of a VS ribozyme (Hiley and Collins 2001). The raw Cy3 donor and Cy5 acceptor fluorescence signals are green and red, respectively (upper panel). Superimposed in gray are the data after applying a non-linear filter (Haran 2004). The FRET ratio (lower panel, black trace) is calculated from the filtered data as acceptor/(donor + acceptor) and reveals three distinct states by Hidden Markov modeling (red line). In part modified with permission from (Pereira et al. 2008) (See figure insert for color reproduction)

Mg²⁺-dependent kissing-loop interaction between stems I and V (Rastogi et al. 1996; Andersen and Collins 2000). This exposes the substrate to the catalytic core around the 730 loop of helix VI (Hiley et al. 2002) (Fig. 5.5a), wherein functional groups of G638 and A756 appear to be involved in catalysis (Wilson et al. 2007; Jaikaran et al. 2008). Cleavage activity is observed in the absence of divalent metal ions, consistent with the sufficiency of RNA residues for catalysis (Murray et al. 1998). Models of the global fold of the VS ribozyme have been derived from ensemble steady-state FRET of the isolated three-way junctions (Lafontaine et al. 2001, 2002), mutagenesis, native gel electrophoresis, hydroxyl radical foot printing, and UV-induced cross linking (Beattie et al. 1995; Hiley and Collins 2001; Hiley et al. 2002).

To assess the folding dynamics of the VS ribozyme, single molecule FRET was performed on the well-studied G11 construct (Pereira et al. 2008). The 5′ end of

G11 was labeled with a Cy3 FRET donor by a ligation approach, while the non-essential closing loop of helix VI was opened to attach a 5′-Cy5 FRET acceptor and a 3′-biotin for surface immobilization. This wild-type construct is designed to monitor global distance changes between the substrate stem-loop I and the catalytic core in helix VI (Fig. 5.5a). It exhibits dynamic three-state folding, where especially a mid (M) and high FRET (H) state inter-convert with rapid and heterogeneous kinetics (Fig. 5.5c), showing occasional excursions into a long-lived low FRET (L) state (Fig. 5.5d). Disruption of the kissing-loop interaction upon mutation of a single base pair completely eliminates both the H state and catalytic activity, while a second-site mutation to invert the base pair restores both, suggesting that the H state is required for catalytic activity (Pereira et al. 2008). Kinetic modeling showed, however, that formation of the H state is not rate-limiting, suggesting that a slow and local conformational change, hidden from FRET observation, must be traversed before the ribozyme undergoes self-cleavage (Pereira et al. 2008).

Mutation of the II–III–VI junction leads to reduced catalytic activity as a consequence of less frequent H state access, suggesting that the II–III–VI junction acts as an important structural scaffold onto which the I–V kissing-loop interaction is built. These observations provide evidence for hierarchical folding of the VS ribozyme as an example of a more complex ribozyme with multiple structural motifs (Pereira et al. 2008). Notably, a change in topology that connects stem-loop I with the 3′- rather than the 5′-end of the catalytic core leads to considerably faster cleavage activity, ratelimited by proton transfer (Smith and Collins 2007), and more rapid and stable docking into the high FRET state (Pereira et al. 2008).

5.7 The *glmS* Ribozyme

A recently discovered class of gene regulatory RNAs, termed riboswitches, are commonly found in the 5′-untranslated regions of mRNAs in Gram-positive bacteria such as *Bacillus*, where they typically change conformation upon binding a specific metabolite, thus refolding an adjacent expression platform that may be sometimes up or down – and which regulates expression of the downstream gene involved in biosynthesis or transport of the metabolite (Winkler and Breaker 2005; Coppins et al. 2007; Edwards et al. 2007; Al-Hashimi and Walter 2008) (please note that riboswitches are the focus of the accompanying chapters by Batey and Schwalbe and coworkers). Given that riboswitches are mostly bacterial, highly selective receptors for small, drug-like metabolites, they may represent a new class of RNA targets for the development of antibiotics (Blount and Breaker 2006).

The riboswitch known as the *glmS* ribozyme is unique in that binding of its ligand glucosamine-6-phosphate (GlcN6P) induces self-cleavage (Winkler et al. 2004) and subsequent intracellular degradation of the embedding *glmS* mRNA, which encodes GlcN6P synthase (Collins et al. 2007). While ligand is absolutely required for catalytic activity and specificity for GlcN6P is high, several structurally related amine-containing compounds were found to partially activate the riboswitch, suggesting that the aminogroup participates in general acid–base catalysis rather than functions as an allosteric activator (McCarthy et al. 2005). Crystal structures of the *glmS* ribozyme (Figs. 5.6a, b) revealed three parallel helical stacks with a doubly pseudo knotted core that binds GlcN6P in a manner consistent with general acid–base catalysis by GlcN6P, in conjunction with G40 of the ribozyme and possibly structural water molecules (Klein and Ferre-D'Amare 2006; Cochrane et al. 2007; Walter 2007). Notably, the *glmS* ribozyme can fold and function in the absence of divalent metal ions (Roth et al. 2006).

To address with maximal sensitivity the question of whether any conformational rearrangements accompany ligand binding by the *glmS* ribozyme, the closing loop of P1 was removed to obtain a trans-acting ribozyme with external substrate for

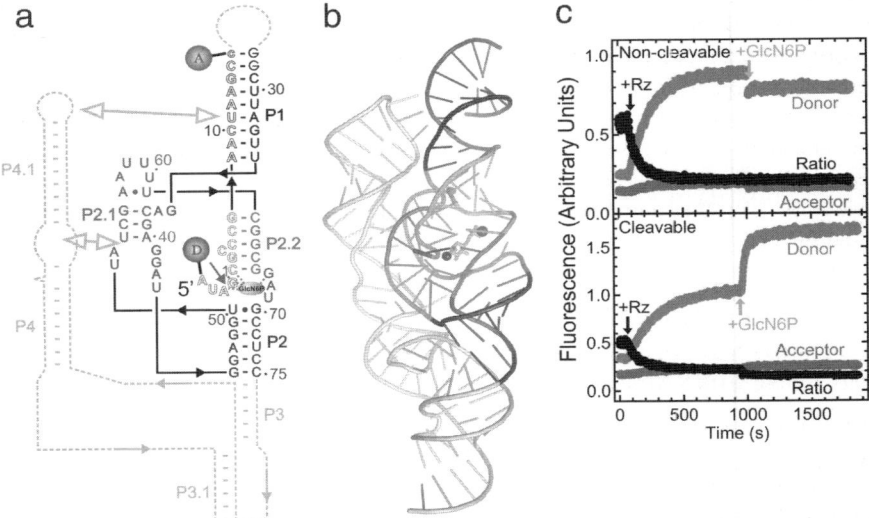

Fig. 5.6 Steady-state FRET studies of the *glmS* ribozyme. (**a**) Secondary structure of the trans-acting *glmS* ribozyme utilized in these studies. Helices are color coded, the catalytic core is highlighted in green, the cleavage site in the substrate (outlined, S) is marked by a closed red arrow, and the cofactor GlcN6P is shown as an orange oval interacting with several core residues (dashed orange lines). The substrate strand is labeled with donor (D) and acceptor (A) fluorophores, as indicated, which undergo FRET. Dashed gray lines indicate connections and a downstream pseudoknot that are present in the naturally occurring, cis-acting ribozyme. (**b**) Cartoon representation of the crystal structure of the *Thermoanaerobacter tengcongensis glmS* ribozyme (PDB ID: 2H0Z) (Klein and Ferre-D'Amare 2006), color-coded as in panel A with removed parts shown in silver. Red sphere, ligand-chelating Mg^{2+} ion; blue sphere, amino group of the ligand presumably involved in catalysis. (**c**) Changes over time in donor, acceptor, and acceptor:donor ratio signals upon addition of a tenfold excess of *glmS* ribozyme (Rz) and subsequently saturating (10 mM) GlcN6P ligand to either the non-cleavable or cleavable substrate under standard conditions (50 mM HEPES-KOH, pH 7.5, 200 mM K$^+$, 10 mM Mg^{2+}, 25 mM DTT, 25°C), as indicated. In part modified with permission from (Tinsley et al. 2007) (See figure insert for color reproduction)

FRET labeling, while the downstream P3–P4.1 was truncated to obtain a minimal, slightly destabilized ribozyme core (Tinsley et al. 2007) (Figs. 5.6a, b). Fluorescein donor and tetramethylrhodamine acceptor fluorophores were attached to the substrate 5′- and 3′-termini, respectively, to detect changes in distance along the central P1:P2.2 helical axis that encompasses the GlcN6P binding and cleavage sites.

Addition of an excess of ribozyme to the FRET labeled non-cleavable substrate analog resulted in a significant increase in donor fluorescence and corresponding decrease in acceptor:donor signal ratio (Fig. 5.6c top). These changes are expected as the binding of the ribozyme will cause an extension of the initially random-coil substrate in the complex. Upon further addition of 10 mM GlcN6P, however, no significant changes in fluorescence signal and FRET efficiency were observed (Fig. 5.6c top). By contrast, a considerable increase in donor signal and resulting decrease in FRET efficiency were evident upon addition of the GlcN6P ligand to the ribozyme-cleavable substrate complex (Fig. 5.6c bottom), as expected from ligand-induced substrate cleavage and dissociation of the 5′-product with attached donor fluorophore. These results are corroborated by time-resolved FRET analysis, where the donor-acceptor distance in the absence and presence of GlcN6P was both found to be 52 Å (Tinsley et al. 2007). Taken together, these results support the notion that the *glmS* ribozyme is fully folded in solution prior to binding its activating ligand (cofactor), which is consistent with precursor and product crystal structures (Klein and Ferre-D'Amare 2006; Cochrane et al. 2007) and observations from hydroxyl radical foot printing in solution (Hampel and Tinsley 2006).

5.8 Common and Diverse Features: Attaining Site-Specificity in Cleavage

The RNA backbone can spontaneously undergo acid-base catalyzed self-cleavage via the same reaction chemistry utilized by the small ribozymes (Fig. 5.1). Under physiologic conditions, spontaneous RNA degradation occurs at a rate constant estimated to average $\sim 10^{-7}\,min^{-1}$ per bond (Emilsson et al. 2003), with a strong dependence on both sequence and structure context (Kaukinen et al. 2002). In double-stranded (Watson–Crick base paired) RNA, the angle between the 2′O-P-5′O atoms is 65–70°, which is significantly different from the optimal angle for in-line attack of close to 180° (Min et al. 2007). This renders the backbone of double-stranded RNA chemically quite inert. Strikingly, all crystal structures of small ribozymes that are thought to be catalytically relevant find the cleavage site in a partially not Watson–Crick paired (or single-stranded) RNA segment featuring a backbone kink (Rupert and Ferre-D'Amare 2001; Ke et al. 2004; Klein and Ferre-D'Amare 2006; Martick and Scott 2006). In fact, the absence of such a backbone kink was used as an indication that the early crystal structures of the hammerhead ribozyme could only be activated by a conformational change after which the cleavage site would adopt non-double-helical characteristics (Wedekind and McKay 1998; Blount and Uhlenbeck 2005).

The small ribozymes have evolved in a convergent manner to prepare a specific bond for a common reaction chemistry, and while they utilize different strategies and structural contexts, in all cases the two nucleotides immediately upstream and downstream of the scissile phosphate are splayed apart in distinct hydrogen bonding and stacking interactions, enforcing unusual torsion angles that kink the backbone (Sefcikova et al. 2007; Walter 2007). Estimates suggest that adoption of an ideal in-line attack configuration accelerates spontaneous backbone cleavage only by <100-fold (Emilsson et al. 2003; Min et al. 2007), supporting the notion that backbone kinking is likely to serve more than one role: to facilitate in-line attack, to specify a bond near the kink for cleavage by exposure to the general acid and base that affect catalysis, and to destabilize the ground state structure (Fig. 5.1).

Specific nucleo bases, structural water molecules, and – in the case of the HDV ribozyme – a hydrated divalent metal ion are generally proximal to the exposed scissile phosphate of the small ribozymes, placed there by the intricacies of the global and local RNA fold. The nucleo bases so far implicated in general acid–base (or electrostatic) catalysis are mostly the purines G and A, perhaps due to the larger number of their functional groups and more extended aromatic system with larger surface area for stacking. The only exception to this rule is the HDV ribozyme that utilizes a cytosine (C75) for catalysis and is, perhaps not coincidentally, also the only small ribozyme strongly dependent on divalent metal ions for activity (Murray et al. 1998). This and other roles of metal ions are the focus of the following segment.

5.9   Common and Diverse Features: The Roles of Metal Ions

All small ribozymes show at least residual catalytic function in the complete absence of divalent metal ions, as long as sufficient countercharge in the form of monovalents is provided to neutralize the negatively charged sugar–phosphate backbone (Hampel and Cowan 1997; Nesbitt et al. 1997; Young et al. 1997; Murray et al. 1998; Nakano et al. 2000, 2003; Curtis and Bartel 2001; O'Rear et al. 2001; Roth et al. 2006). This observation contrasts with ribozymes catalyzing more complex phosphodiester transfers, which consistently appear to require a pair of Mg^{2+} ions for catalysis (Walter 2007) (see also accompanying chapters by Woodson, Pyle, and Harris and coworkers). The simpler chemistry executed by the small ribozymes typically requires neither an external nucleophile (as do most larger ribozymes) nor an external catalytic cofactor (except perhaps for ubiquitous water molecules), making small ribozymes surprisingly self-sufficient and versatile, perhaps as an adaptation to their function in parasite replication. Exceptions are the *glmS* ribozyme, which requires GlcN6P as a cofactor and is not a parasite-derived ribozyme but is embedded in bacterial mRNA to control gene expression (Winkler et al. 2004), and the HDV ribozyme, which not only is part of the HDV viroid but is also embedded in the human genome (Salehi-Ashtiani et al. 2006).

The role of metal ions in folding of the small ribozymes is much more unequivocal, especially under physiologic (low-ionic strength) conditions, where Mg^{2+} is the preferred, highest-affinity folding cofactor. Mg^{2+} is thought to bind RNA in three different modes, by specific inner-sphere coordination (chelation), by outer-sphere binding where a single water layer influences the ion position through steric packing and hydrogen bonding, and by diffuse association with the long-range RNA electrostatic field through multiple water layers (Draper et al. 2005). FRET studies have shown that tertiary structure folding is generally accelerated (and thermodynamically stabilized) and unfolding decelerated by increasing Mg^{2+} concentrations, although differential saturation behavior of the two Mg^{2+} dependency curves is observed for the hairpin ribozyme (Bokinsky et al. 2003). Resulting modeling studies interpret this behavior as an indication that the folding transition state is compacted by Mg^{2+}-based bridging of the two ribozyme domains, even before tertiary hydrogen bond and stacking interactions are extensively formed (Bokinsky et al. 2003). How general this behavior is for the small ribozymes remains to be seen.

5.10   Common and Diverse Features: RNA Structural Dynamics

The catalytic transition state of any enzyme is, by definition, a high-energy state along the reaction pathway that only exists on the timescale of a bond vibration. By contrast, structures observed in a crystalline and solution state by, for example, X-ray crystallography and NMR, respectively, are ensemble-averaged, low-energy ground states, even if perturbed by employing transition state analogs. As a consequence, the available high-resolution crystal structures of ribozymes need to be animated with their dynamics to understand the inherently dynamic processes underlying catalysis.

Ensemble and single molecule FRET has extensively contributed to our understanding of how and why the small ribozymes fold or show intrinsic dynamics. With one exception, global structural changes that rearrange helical alignments for catalytic activation (as in the hammerhead ribozyme), "dock" substrate and catalytic core domains (as in the hairpin and VS ribozymes), or accompany and possibly control catalysis (as in the HDV ribozyme) have readily been detected by FRET in all small ribozymes. It is tempting to speculate that such large-scale conformational rearrangements from inactive to active states serve a biological role by controlling the activity of the catalytic motif during satellite replication in dependence of intracellular cues. Notably, the *glmS* ribozyme is devoid of any such significant conformational change, but is catalytically triggered by a different kind of intracellular cue in the form of a threshold concentration of its catalytic cofactor GlcN6P.

Protein enzymes are thought to exploit a global network of coupled molecular motions to facilitate the chemical reaction they catalyze. This network comprises fast thermal motions that are at equilibrium while the reaction progresses along the reaction pathway; and these motions lead to slower, larger-scale conformational changes that directly facilitate reaction chemistry (Hammes-Schiffer and Benkovic 2006). Whether ribozymes employ a similar strategy to accelerate catalysis is not known, but single molecule FRET probing in combination with molecular dynamics simulations has revealed that at least the hairpin ribozyme exhibits analogous networks of coupled molecular motions throughout its catalytic core (Rueda et al. 2004; Rhodes et al. 2006). Reaction chemistry in an enzymatic reaction is largely of a local nature, suggesting that the dynamics contributing directly to catalysis in RNA would mostly entail vibrations, torsional librations, sugar re-puckering, and longitudinal and lateral motions of bases, all of which take place in the tens-of-femtoseconds to low-nanosecond time regime (Al-Hashimi and Walter 2008). It remains to be determined whether global structural changes at the nanosecond to minute timescale, as observable by FRET, couple with these local dynamics and contribute to the catalytic power of the small ribozymes.

In conclusion, many common features are emerging from a comparison of the five naturally occurring, structurally very distinct small ribozymes. Diverseness between them probably highlights evolutionary adaptation to specific functional requirements. Future studies using FRET and other biophysical techniques promise to elucidate more about the fundamental principles underlying RNA catalysis, enabling an even broader comparison of enzymes, including those based on protein, and a potential utilization in human technology.

References

Al-Hashimi HM, Walter NG (2008) RNA dynamics: it's about time. Curr Opin Struct Biol 18:321–329

Alam S, Grum-Tokars V, Krucinska J, Kundracik ML, Wedekind JE (2005) Conformational heterogeneity at position U37 of an all-RNA hairpin ribozyme with implications for metal binding and the catalytic structure of the S-turn. Biochemistry 44:14396–14408

Andersen AA, Collins RA (2000) Rearrangement of a stable RNA secondary structure during VS ribozyme catalysis. Mol Cell 5:469–478

Bassi GS, Murchie AI, Walter F, Clegg RM, Lilley DM (1997) Ion-induced folding of the hammerhead ribozyme: a fluorescence resonance energy transfer study. EMBO J 16:7481–7489

Bassi GS, Mollegaard NE, Murchie AI, Lilley DM (1999) RNA folding and misfolding of the hammerhead ribozyme. Biochemistry 38:3345–3354

Beattie TL, Olive JE, Collins RA (1995) A secondary-structure model for the self-cleaving region of Neurospora VS RNA. Proc Natl Acad Sci U S A 92:4686–4690

Been MD (2006) HDV ribozymes. Curr Top Microbiol Immunol 307:47–65

Bevilacqua PC (2003) Mechanistic considerations for general acid-base catalysis by RNA: revisiting the mechanism of the hairpin ribozyme. Biochemistry 42:2259–2265

Bevilacqua PC, Yajima R (2006) Nucleobase catalysis in ribozyme mechanism. Curr Opin Chem Biol 10:455–464

Blount KF, Breaker RR (2006) Riboswitches as antibacterial drug targets. Nat Biotechnol 24:1558–1564

Blount KF, Uhlenbeck OC (2005) The structure-function dilemma of the hammerhead ribozyme. Annu Rev Biophys Biomol Struct 34:415–440

Bokinsky G, Rueda D, Misra VK, Rhodes MM, Gordus A, Babcock HP, Walter NG, Zhuang X (2003) Single-molecule transition-state analysis of RNA folding. Proc Natl Acad Sci U S A 100:9302–9307

Bondensgaard K, Mollova ET, Pardi A (2002) The global conformation of the hammerhead ribozyme determined using residual dipolar couplings. Biochemistry 41:11532–11542

Cochrane JC, Lipchock SV, Strobel SA (2007) Structural investigation of the GlmS ribozyme bound to its catalytic cofactor. Chem Biol 14:97–105

Collins RA, Saville BJ (1990) Independent transfer of mitochondrial chromosomes and plasmids during unstable vegetative fusion in Neurospora. Nature 345:177–179

Collins JA, Irnov I, Baker S, Winkler WC (2007) Mechanism of mRNA destabilization by the glmS ribozyme. Genes Dev 21:3356–3368

Coppins RL, Hall KB, Groisman EA (2007) The intricate world of riboswitches. Curr Opin Microbiol 10:176–181

Curtis EA, Bartel DP (2001) The hammerhead cleavage reaction in monovalent cations. RNA 7:546–552

Das S, Piccirilli J (2005) General acid catalysis by the hepatitis delta virus ribozyme. Nat Chem Biol 1:45–52

De la Pena M, Gago S, Flores R (2003) Peripheral regions of natural hammerhead ribozymes greatly increase their self-cleavage activity. EMBO J 22:5561–5570

Ditzler MA, Aleman EA, Rueda D, Walter NG (2007) Focus on function: single molecule RNA enzymology. Biopolymers 87:302–316

Doudna JA, Lorsch JR (2005) Ribozyme catalysis: not different, just worse. Nat Struct Mol Biol 12:395–402

Draper DE, Grilley D, Soto AM (2005) Ions and RNA folding. Annu Rev Biophys Biomol Struct 34:221–243

Edwards TE, Klein DJ, Ferre-D'Amare AR (2007) Riboswitches: small-molecule recognition by gene regulatory RNAs. Curr Opin Struct Biol 17:273–279

Emilsson GM, Nakamura S, Roth A, Breaker RR (2003) Ribozyme speed limits. RNA 9: 907–918

Fedor MJ, Williamson JR (2005) The catalytic diversity of RNAs. Nat Rev Mol Cell Biol 6: 399–412

Feldstein PA, Buzayan JM, van Tol H, deBear J, Gough GR, Gilham PT, Bruening G (1990) Specific association between an endoribonucleolytic sequence from a satellite RNA and a substrate analogue containing a 2'-5' phosphodiester. Proc Natl Acad Sci U S A 87: 2623–2627

Ferre-D'Amare AR, Doudna JA (2000) Crystallization and structure determination of a hepatitis delta virus ribozyme: use of the RNA-binding protein U1A as a crystallization module. J Mol Biol 295:541–556

Ferre-D'Amare AR, Zhou K, Doudna JA (1998) Crystal structure of a hepatitis delta virus ribozyme. Nature 395:567–574

Flores R, Delgado S, Gas ME, Carbonell A, Molina D, Gago S, De la Pena M (2004) Viroids: the minimal non-coding RNAs with autonomous replication. FEBS Lett 567:42–48

Förster T (1946) Energiewanderung Und Fluoreszenz. Naturwissenschaften 33:166–175

Förster T (1948) Intermolecular energy migration and fluorescence. Ann Phys (Leipzig) 2:55–75

Forster AC, Symons RH (1987) Self-cleavage of virusoid RNA is performed by the proposed 55-nucleotide active site. Cell 50:9–16

Furtig B, Richter C, Schell P, Wenter P, Pitsch S, Schwalbe H (2008) NMR-spectroscopic characterisation of phosphodiester bond cleavage catalysed by the minimal hammerhead ribozyme. RNA Biol 5:41–48

Gong B, Chen JH, Chase E, Chadalavada DM, Yajima R, Golden BL, Bevilacqua PC, Carey PR (2007) Direct measurement of a pK(a) near neutrality for the catalytic cytosine in the genomic HDV ribozyme using Raman crystallography. J Am Chem Soc 129:13335–13342

Guerrier-Takada C, Gardiner K, Marsh T, Pace N, Altman S (1983) The RNA moiety of ribonuclease P is the catalytic subunit of the enzyme. Cell 35:849–857

Hammann C, Lilley DM (2002) Folding and activity of the hammerhead ribozyme. Chembiochem 3:690–700

Hammes-Schiffer S, Benkovic SJ (2006) Relating protein motion to catalysis. Annu Rev Biochem 75:519–541

Hampel A, Cowan JA (1997) A unique mechanism for RNA catalysis: the role of metal cofactors in hairpin ribozyme cleavage. Chem Biol 4:513–517

Hampel KJ, Tinsley MM (2006) Evidence for preorganization of the glmS ribozyme ligand binding pocket. Biochemistry 45:7861–7871

Haran G (2004) Noise reduction in single-molecule fluorescence trajectories of folding proteins. Chem Phys 307:137–145

Harris DA, Rueda D, Walter NG (2002) Local conformational changes in the catalytic core of the trans-acting hepatitis delta virus ribozyme accompany catalysis. Biochemistry 41: 12051–12061

Haseloff J, Gerlach WL (1989) Sequences required for self-catalysed cleavage of the satellite RNA of tobacco ringspot virus. Gene 82:43–52

Hiley SL, Collins RA (2001) Rapid formation of a solvent-inaccessible core in the Neurospora Varkud satellite ribozyme. EMBO J 20:5461–5469

Hiley SL, Sood VD, Fan J, Collins RA (2002) 4-thio-U cross-linking identifies the active site of the VS ribozyme. EMBO J 21:4691–4698

Jaikaran D, Smith MD, Mehdizadeh R, Olive J, Collins RA (2008) An important role of G638 in the cis-cleavage reaction of the Neurospora VS ribozyme revealed by a novel nucleotide analog incorporation method. RNA 14:938–949

Jeong S, Sefcikova J, Tinsley RA, Rueda D, Walter NG (2003) Trans-acting hepatitis delta virus ribozyme: catalytic core and global structure are dependent on the 5′ substrate sequence. Biochemistry 42:7727–7740

Kaukinen U, Lyytikainen S, Mikkola S, Lonnberg H (2002) The reactivity of phosphodiester bonds within linear single-stranded oligoribonucleotides is strongly dependent on the base sequence. Nucleic Acids Res 30:468–474

Ke A, Zhou K, Ding F, Cate JH, Doudna JA (2004) A conformational switch controls hepatitis delta virus ribozyme catalysis. Nature 429:201–205

Khvorova A, Lescoute A, Westhof E, Jayasena SD (2003) Sequence elements outside the hammerhead ribozyme catalytic core enable intracellular activity. Nat Struct Biol 10:708–712

Klein DJ, Ferre-D'Amare AR (2006) Structural basis of glmS ribozyme activation by glucosamine-6-phosphate. Science 313:1752–1756

Kruger K, Grabowski PJ, Zaug AJ, Sands J, Gottschling DE, Cech TR (1982) Self-splicing RNA: autoexcision and autocyclization of the ribosomal RNA intervening sequence of Tetrahymena. Cell 31:147–157

Kuzmin YI, Da Costa CP, Fedor MJ (2004) Role of an active site guanine in hairpin ribozyme catalysis probed by exogenous nucleobase rescue. J Mol Biol 340:233–251

Kuzmin YI, Da Costa CP, Cottrell JW, Fedor MJ (2005) Role of an active site adenine in hairpin ribozyme catalysis. J Mol Biol 349:989–1010

Lafontaine DA, Norman DG, Lilley DM (2001) Structure, folding and activity of the VS ribozyme: importance of the 2- 3–6 helical junction. EMBO J 20:1415–1424

Lafontaine DA, Norman DG, Lilley DM (2002) The global structure of the VS ribozyme. EMBO J 21:2461–2471

Lakowicz JR (2006) Principles of fluorescence spectroscopy, 3rd edn. Springer, New York

Liu S, Bokinsky G, Walter NG, Zhuang X (2007) Dissecting the multistep reaction pathway of an RNA enzyme by single-molecule kinetic "fingerprinting". Proc Natl Acad Sci U S A 104:12634–12639

Luptak A, Ferre-D'Amare AR, Zhou K, Zilm KW, Doudna JA (2001) Direct pK(a) measurement of the active-site cytosine in a genomic hepatitis delta virus ribozyme. J Am Chem Soc 123:8447–8452

Martick M, Scott WG (2006) Tertiary contacts distant from the active site prime a ribozyme for catalysis. Cell 126:309–320

Martick M, Lee TS, York DM, Scott WG (2008a) Solvent structure and hammerhead ribozyme catalysis. Chem Biol 15:332–342

Martick M, Horan LH, Noller HF, Scott WG (2008b) A discontinuous hammerhead ribozyme embedded in a mammalian messenger RNA. Nature 454:899–902

McCarthy TJ, Plog MA, Floy SA, Jansen JA, Soukup JK, Soukup GA (2005) Ligand requirements for glmS ribozyme self-cleavage. Chem Biol 12:1221–1226

McKay DB (1996) Structure and function of the hammerhead ribozyme: an unfinished story. RNA 2:395–403

Min D, Xue S, Li H, Yang W (2007) 'In-line attack' conformational effect plays a modest role in an enzyme-catalyzed RNA cleavage: a free energy simulation study. Nucleic Acids Res 35:4001–4006

Murchie AI, Thomson JB, Walter F, Lilley DM (1998) Folding of the hairpin ribozyme in its natural conformation achieves close physical proximity of the loops. Mol Cell 1:873–881

Murray JB, Seyhan AA, Walter NG, Burke JM, Scott WG (1998) The hammerhead, hairpin and VS ribozymes are catalytically proficient in monovalent cations alone. Chem Biol 5:587–595

Nahas MK, Wilson TJ, Hohng S, Jarvie K, Lilley DM, Ha T (2004) Observation of internal cleavage and ligation reactions of a ribozyme. Nat Struct Mol Biol 11:1107–1113

Nakano S, Bevilacqua PC (2007) Mechanistic characterization of the HDV genomic ribozyme: a mutant of the C41 motif provides insight into the positioning and thermodynamic linkage of metal ions and protons. Biochemistry 46:3001–3012

Nakano S, Chadalavada DM, Bevilacqua PC (2000) General acid-base catalysis in the mechanism of a hepatitis delta virus ribozyme. Science 287:1493–1497

Nakano S, Cerrone AL, Bevilacqua PC (2003) Mechanistic characterization of the HDV genomic ribozyme: classifying the catalytic and structural metal ion sites within a multichannel reaction mechanism. Biochemistry 42:2982–2994

Nelson JA, Uhlenbeck OC (2006) When to believe what you see. Mol Cell 23:447–450

Nelson JA, Uhlenbeck OC (2008a) Hammerhead redux: does the new structure fit the old biochemical data? RNA 14:605–615

Nelson JA, Uhlenbeck OC (2008b) Minimal and extended hammerheads utilize a similar dynamic reaction mechanism for catalysis. RNA 14:43–54

Nesbitt S, Hegg LA, Fedor MJ (1997) An unusual pH-independent and metal-ion-independent mechanism for hairpin ribozyme catalysis. Chem Biol 4:619–630

O'Rear JL, Wang S, Feig AL, Beigelman L, Uhlenbeck OC, Herschlag D (2001) Comparison of the hammerhead cleavage reactions stimulated by monovalent and divalent cations. RNA 7(4):537–545

Okumus B, Wilson TJ, Lilley DM, Ha T (2004) Vesicle encapsulation studies reveal that single molecule ribozyme heterogeneities are intrinsic. Biophys J 87:2798–2806

Penedo JC, Wilson TJ, Jayasena SD, Khvorova A, Lilley DM (2004) Folding of the natural hammerhead ribozyme is enhanced by interaction of auxiliary elements. RNA 10:880–888

Pereira MJ, Harris DA, Rueda D, Walter NG (2002) Reaction pathway of the trans-acting hepatitis delta virus ribozyme: a conformational change accompanies catalysis. Biochemistry 41: 730–740

Pereira MJB, Nikolova EN, Hiley SL, Collins RA, Walter NG (2008) Single VS ribozyme molecules reveal dynamic and hierarchical folding toward catalysis. J Mol Biol 382:496–509

Perrotta AT, Been MD (1991) A pseudoknot-like structure required for efficient self-cleavage of hepatitis delta virus RNA. Nature 350:434–436

Perrotta AT, Shih I, Been MD (1999) Imidazole rescue of a cytosine mutation in a self-cleaving ribozyme. Science 286:123–126

Pinard R, Hampel KJ, Heckman JE, Lambert D, Chan PA, Major F, Burke JM (2001) Functional involvement of G8 in the hairpin ribozyme cleavage mechanism. EMBO J 20:6434–6442

Prody GA, Bakos JT, Buzayan JM, Schneider IR, Bruening G (1986) Autolytic Processing of Dimeric Plant Virus Satellite RNA. Science 231:1577–1580

Pyle AM (1993) Ribozymes: a distinct class of metalloenzymes. Science 261:709–714

Rastogi T, Beattie TL, Olive JE, Collins RA (1996) A long-range pseudoknot is required for activity of the Neurospora VS ribozyme. EMBO J 15:2820–2825

Rhodes MM, Reblova K, Sponer J, Walter NG (2006) Trapped water molecules are essential to structural dynamics and function of a ribozyme. Proc Natl Acad Sci U S A 103:13380–13385

Roth A, Nahvi A, Lee M, Jona I, Breaker RR (2006) Characteristics of the glmS ribozyme suggest only structural roles for divalent metal ions. RNA 12:607–619

Rueda D, Wick K, McDowell SE, Walter NG (2003) Diffusely bound Mg2+ ions slightly reorient stems I and II of the hammerhead ribozyme to increase the probability of formation of the catalytic core. Biochemistry 42:9924–9936

Rueda D, Bokinsky G, Rhodes MM, Rust MJ, Zhuang X, Walter NG (2004) Single-molecule enzymology of RNA: essential functional groups impact catalysis from a distance. Proc Natl Acad Sci U S A 101:10066–10071

Ruffner DE, Stormo GD, Uhlenbeck OC (1990) Sequence requirements of the hammerhead RNA self-cleavage reaction. Biochemistry 29:10695–10702

Rupert PB, Ferre-D'Amare AR (2001) Crystal structure of a hairpin ribozyme-inhibitor complex with implications for catalysis. Nature 410:780–786

Rupert PB, Massey AP, Sigurdsson ST, Ferre-D'Amare AR (2002) Transition state stabilization by a catalytic RNA. Science 298:1421–1424

Salehi-Ashtiani K, Szostak JW (2001) In vitro evolution suggests multiple origins for the hammerhead ribozyme. Nature 414:82–84

Salehi-Ashtiani K, Luptak A, Litovchick A, Szostak JW (2006) A genomewide search for ribozymes reveals an HDV-like sequence in the human CPEB3 gene. Science 313:1788–1792

Salter J, Krucinska J, Alam S, Grum-Tokars V, Wedekind JE (2006) Water in the active site of an all-RNA hairpin ribozyme and effects of Gua8 base variants on the geometry of phosphoryl transfer. Biochemistry 45:686–700

Saville BJ, Collins RA (1990) A site-specific self-cleavage reaction performed by a novel RNA in Neurospora mitochondria. Cell 61:685–696

Scott WG (2007) Ribozymes. Curr Opin Struct Biol 17:280–286

Sefcikova J, Krasovska MV, Sponer J, Walter NG (2007) The genomic HDV ribozyme utilizes a previously unnoticed U-turn motif to accomplish fast site-specific catalysis. Nucleic Acids Res 35:1933–1946

Simorre JP, Legault P, Hangar AB, Michiels P, Pardi A (1997) A conformational change in the catalytic core of the hammerhead ribozyme upon cleavage of an RNA substrate. Biochemistry 36:518–525

Simorre JP, Legault P, Baidya N, Uhlenbeck OC, Maloney L, Wincott F, Usman N, Beigelman L, Pardi A (1998) Structural variation induced by different nucleotides at the cleavage site of the hammerhead ribozyme. Biochemistry 37:4034–4044

Smith MD, Collins RA (2007) Evidence for proton transfer in the rate-limiting step of a fast-cleaving Varkud satellite ribozyme. Proc Natl Acad Sci U S A 104:5818–5823

Stryer L (1978) Fluorescence energy transfer as a spectroscopic ruler. Annu Rev Biochem 47:819–846

Taylor JM (2006) Structure and replication of hepatitis delta virus RNA. Curr Top Microbiol Immunol 307:1–23

Tinsley RA, Walter NG (2007) Long-range impact of peripheral joining elements on structure and function of the hepatitis delta virus ribozyme. Biol Chem 388:705–715

Tinsley RA, Harris DA, Walter NG (2004) Magnesium dependence of the amplified conformational switch in the trans-acting hepatitis delta virus ribozyme. Biochemistry 43:8935–8945

Tinsley RA, Furchak JR, Walter NG (2007) Trans-acting glmS catalytic riboswitch: locked and loaded. Rna 13:468–477

Torelli AT, Krucinska J, Wedekind JE (2007) A comparison of vanadate to a 2'-5' linkage at the active site of a small ribozyme suggests a role for water in transition-state stabilization. RNA 13:1052–1070

Tuschl T, Gohlke C, Jovin TM, Westhof E, Eckstein F (1994) A three-dimensional model for the hammerhead ribozyme based on fluorescence measurements. Science 266:785–789

Walter NG (2001) Structural dynamics of catalytic RNA highlighted by fluorescence resonance energy transfer. Methods 25:19–30

Walter NG (2002) Probing RNA structural dynamics and function by fluorescence resonance energy transfer (FRET). Curr Protoc Nucleic Acid Chem 11.10:11.0.1–0.23

Walter NG (2007) Ribozyme catalysis revisited: is water involved? Mol Cell 28:923–929

Walter NG, Burke JM (1998) The hairpin ribozyme: structure, assembly and catalysis. Curr Opin Chem Biol 2:24–30

Walter F, Murchie AI, Thomson JB, Lilley DM (1998a) Structure and activity of the hairpin ribozyme in its natural junction conformation: effect of metal ions. Biochemistry 37:14195–14203

Walter F, Murchie AIH, Lilley DMJ (1998b) Folding of the four-way RNA junction of the hairpin ribozyme. Biochemistry 37:17629–17636

Walter NG, Hampel KJ, Brown KM, Burke JM (1998c) Tertiary structure formation in the hairpin ribozyme monitored by fluorescence resonance energy transfer. EMBO J 17:2378–2391

Walter NG, Burke JM, Millar DP (1999) Stability of hairpin ribozyme tertiary structure is governed by the interdomain junction. Nat Struct Biol 6:544–549

Walter NG, Huang C, Manzo AJ, Sobhy MA (2008) Do-it-yourself guide: how to use the modern single molecule toolkit. Nat Methods 5:475–489

Wedekind JE, McKay DB (1998) Crystallographic structures of the hammerhead ribozyme: relationship to ribozyme folding and catalysis. Annu Rev Biophys Biomol Struct 27:475–502

Wilson TJ, Ouellet J, Zhao ZY, Harusawa S, Araki L, Kurihara T, Lilley DM (2006) Nucleobase catalysis in the hairpin ribozyme. RNA 12:980–987

Wilson TJ, McLeod AC, Lilley DM (2007) A guanine nucleobase important for catalysis by the VS ribozyme. Embo J 26:2489–2500

Winkler WC, Breaker RR (2005) Regulation of bacterial gene expression by riboswitches. Annu Rev Microbiol 59:487–517

Winkler WC, Nahvi A, Roth A, Collins JA, Breaker RR (2004) Control of gene expression by a natural metabolite-responsive ribozyme. Nature 428:281–286

Young KJ, Gill F, Grasby JA (1997) Metal ions play a passive role in the hairpin ribozyme catalysed reaction. Nucleic Acids Res 25:3760–3766

Zhou DM, Taira K (1998) The hydrolysis of RNA: from theoretical calculations to the hammerhead ribozyme-mediated cleavage of RNA. Chem Rev 98:991–1026

Zhuang X (2005) Single-molecule RNA science. Annu Rev Biophys Biomol Struct 34:399–414

Zhuang X, Kim H, Pereira MJ, Babcock HP, Walter NG, Chu S (2002) Correlating structural dynamics and function in single ribozyme molecules. Science 296:1473–1476

Chapter 6
Structure and Mechanism of the *glmS* Ribozyme

Juliane K. Soukup and Garrett A. Soukup(⊠)

Abstract The self-cleaving *glmS* ribozyme is a mechanistically unique functional RNA in the category of riboswitches and RNA catalysts. Its catalytic activity provides the basis of genetic regulation and depends upon glucosamine-6-phosphate (GlcN6P) as a coenzyme. Substantial biochemical and biophysical data relating to the structure and function of the *glmS* ribozyme has been amassed in a relatively short period of time since its discovery. A precise and comprehensive mechanistic understanding of coenzyme function in *glmS* ribozyme self-cleavage has however not been elaborated. Here, evidence regarding the structure and function of the *glmS* ribozyme is carefully weighed to provide a comprehensive mechanistic model of coenzyme action in acid-base catalysis at the enzyme's active site.

6.1 Introduction

The discovery of bacterial riboswitches expanded the knowledge of genetic control where RNA functions as a molecular switch regulating messenger RNA transcription, translation, or processing through direct interaction with specific cellular metabolite. (Nudler and Mironov 2004; Winkler and Breaker 2005). A stunning array of riboswitches has been characterized; these respond to eleven different cellular metabolites ultimately regulating ~4% of protein coding genes in gram positive bacteria (Irnov et al. 2006). Riboswitches are often found in genes constituting biosynthetic pathways which produce their cognate metabolites; riboswitches thus afford an elegant feedback regulation mechanism that allows bacterial cells to appropriately respond to metabolic supply and demand. Riboswitches therefore represent pharmaceutical targets significant for the development of novel antibiotics.

G.A. Soukup
Department of Biomedical Sciences, Creighton University School of Medicine, 2500 California Plaza, Omaha, NE 68178, USA
e-mail: garrettsoukup@creighton.edu

N.G. Walter et al. (eds.) *Non-Protein Coding RNAs*
doi: 10.1007/978-3-540-70840-7_6, © Springer–Verlag Berlin Heidelberg 2009

To exert control over gene expression, riboswitches couple the task of metabolite recognition with that of modulating a requisite aspect of gene expression. Consequently, riboswitches typically reside in 5′ untranslated regions of bacterial messenger RNAs. They are generally composed of two interdependent domains that include a natural ligand-binding or "aptamer" domain and an "expression platform" of which the precise conformation impacts gene expression (Winkler and Breaker 2005). Two common mechanisms of riboswitch function involve ligand-dependent conformational changes that modulate (a) terminator/antiterminator stem formation impacting transcription, or (b) Shine–Dalgarno sequence accessibility impacting translation (Nudler and Mironov 2004; Winkler and Breaker 2005).

A third mechanism is uniquely represented by the catalytic *glmS* riboswitch (Winkler et al. 2004) as a ribozyme that undergoes self-cleavage in response to binding glucosamine-6-phosphate (GlcN6P), the metabolic product of the GlmS enzyme required for cell wall biosynthesis. *glmS* ribozyme self cleavage via internal RNA transesterification substantially reduces messenger RNA translation (Winkler et al. 2004) by destabilizing the transcript (Collins et al. 2007). Therefore, artificial agonists of *glmS* ribozyme activity might function as antibiotic agents for number of pathogens that harbor the riboswitch (Barrick et al. 2004). This, and the fact that the *glmS* ribozyme is a natural curiosity among catalytic RNAs, has propelled intensive study of the RNA's structure and function.

6.2 *glmS* Ribozyme Architecture

Insight regarding the general function of the *glmS* ribozyme is provided by biochemical and biophysical analyses that convey the overall three-dimensional organization of the RNA catalyst and its bound ligand. Such studies have enabled a vital distinction between potential mechanistic roles of GlcN6P as an allosteric effector or a coenzyme.

The *glmS* ribozyme consists of eight paired regions (P1–P4) that organize a phylogenetically conserved catalytic core (Barrick et al. 2004; Winkler et al. 2004) between P1 and P2 as exemplified by the *B. cereus* ribozyme (Fig. 6.1a). The P3–P4 domain is dispensable for core GlcN6P-dependent self-cleavage (Winkler et al. 2004), but enhances activity particularly under low magnesium ion conditions (Wilkinson and Been 2005; Soukup 2006). The overall tertiary structure of the RNA is derived largely by the formation of pseudoknots. The pseudoknot forming P3.1 resides in the P3–P4 domain and mildly affects core ribozyme activity (Wilkinson and Been 2005; Roth et al. 2006). The double pseudoknot forming P2.1 and P2.2 comprises highly conserved segments in the catalytic core and partially resembles a pseudoknot interaction proposed to organize the core segments (Soukup 2006) and supported by UV-cross linking analysis (Hampel and Tinsley 2006). The true complexity of the pseudoknot interaction was only revealed in detail by X-ray crystallography (Klein and Ferré-D'Amaré 2006). Further, the GlcN6P-binding site of the *glmS* ribozyme has been demonstrated to reside within the proposed pseudoknot core by nucleotide analog interference mapping (NAIM) and suppression (NAIS) experiments utilizing GlcN6P and glucosamine (GlcN) as ligands (Jansen et al. 2006).

The detailed three-dimensional structure of the *glmS* ribozyme was first determined by X-ray crystallographic analysis of the *T. tengcongensis* ribozyme in unliganded pre- and post-cleavage states and bound to glucose-6-phosphate (Glc6P) (Klein and Ferré-D'Amaré 2006), where Glc6P serves as a competitive inhibitor of GlcN6P (McCarthy et al. 2005). The crystal structure of the *B. anthracis* ribozyme bound to GlcN6P was later solved by replacing the 2′-hydroxyl adjacent to scissile phosphodiester with a 2′-methoxy group to prevent self-cleavage (Cochrane et al. 2007). Importantly, each structure conveys identical architecture exemplified by the *B. anthracis* ribozyme (Fig. 6.1b). However, the structure of the *B. anthracis glmS* ribozyme is exclusively used throughout this chapter because (a) it portrays GlcN6P binding, and (b) the numbering of nucleotide identities throughout the catalytic core is identical to that of the *B. cereus* ribozyme for which considerable biochemical data exist.

The crystal structures of *glmS* ribozymes demonstrate that ligand binds within the catalytic core adjacent to the cleavage site, where the P4 domain hugs the core through minor groove interactions with P2.1 (Fig. 6.1b). The catalytic core contains four metal ion-binding sites, two of which are separated from the ligand-binding and cleavage sites. Two additional and fully hydrated metal ions facilitate ligand phosphate recognition (Fig. 6.1c) similar to thiamine pyrophosphate (TPP) riboswitch recognition of ligand (Serganov et al. 2006; Thore et al. 2006) and possibly representing a general mode for RNA recognition of anionic ligands. Moreover, GlcN6P binds the catalytic core of the RNA with its amine group adjacent to the scissile phosphoester (Fig. 6.1c) and its phosphate group and associated metal ions distal to the cleavage site.

The main theme throughout such biophysical and additional biochemical analyses is that the *glmS* ribozyme does not exhibit conformational changes upon binding ligand. The use of "in-line" probing (Winkler et al. 2004), chemical and enzymatic probing (Hampel and Tinsley 2006; Tinsley et al. 2007), and FRET analysis (Tinsley et al. 2007) agree that the solution structure of the *glmS* ribozyme is unaffected by ligand binding. The crystal structures of the pre- and post-cleavage *T. tengcongensis* ribozymes, and those of the Glc6P- and GlcN6P-bound *T. tengcongensis* and *B. anthracis* ribozymes are also superimposable and essentially indistinguishable (Klein and Ferré-D'Amaré 2006; Cochrane et al. 2007; Klein et al. 2007b). Taken together, such data strongly support a mechanism by which the *glmS* ribozyme is fully pre-folded prior to binding GlcN6P at the active site, ruling out any allosteric activation mechanism. However, consideration of how GlcN6P functions as a coenzyme to support *glmS* ribozyme self-cleavage requires a basic appreciation of catalytic strategies effecting RNA transesterification.

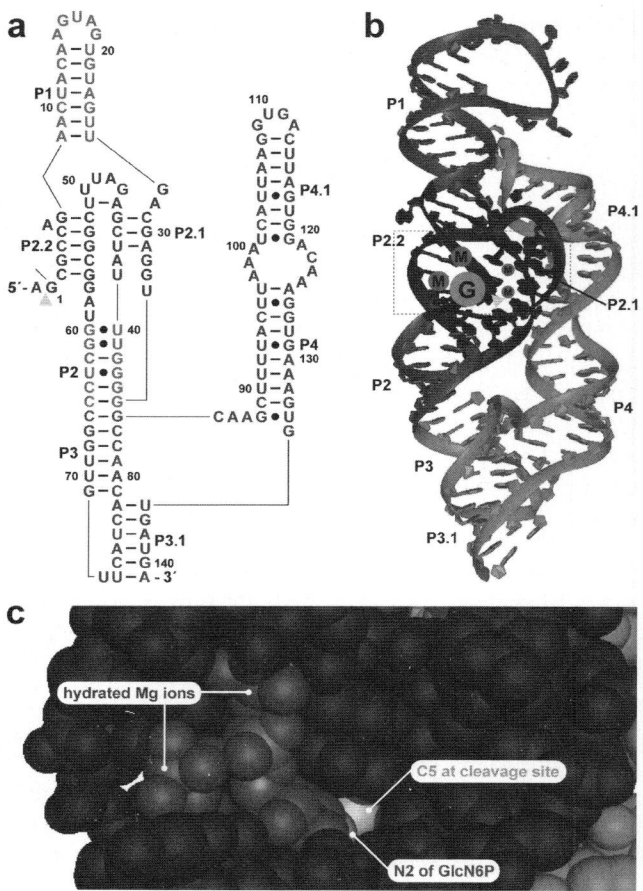

Fig. 6.1 Overview of *glmS* ribozyme structure. (**a**) Secondary structure. Depicted is the secondary structure of the *B. cereus* ribozyme based upon phylogenetic comparison (Winkler et al. 2004) and crystal structures of the *T. tengcongensis* (Klein and Ferré-D'Amaré 2006) and *B. anthracis* ribozymes (Cochrane et al. 2007). The highly conserved catalytic core (blue) comprises nucleotides adjacent to and including paired regions P2.1 and P2.2. The minimal portion that supports GlcN6P-dependent self-cleavage additionally includes P1 and P2 (red). The remaining portion including P3–P4 (green) enhances ribozyme activity, but is not required. Nucleotides are numbered relative to the site of cleavage indicated by the yellow arrowhead. (**b**) Tertiary structure. The ribbon model depicts the structure of the *B. anthracis* ribozyme determined by X-ray crystallography (Cochrane et al. 2007) in complex with U1A protein (not shown) bound to a site inserted into the loop of P1. Importantly, nucleotide identities and numbering within the catalytic core (blue) are identical to that of *B. cereus* ribozyme. The scissile phosphodiester bond is denoted by the yellow arrowhead, and the approximate site of GlcN6P binding (orange encircled G) and four sites of divalent metal ion binding (each magenta encircled M) are indicated. (**c**) GlcN6P binding site. The space-filling model shows the approximate region indicated in Fig. 6.1b (dashed box) with two hydrated metal ions and GlcN6P bound adjacent to the scissile phosphate. Ligands are colored by atom (Mg, magenta; O, red; P, orange; C, gray; and N, cyan), whereas C5′ forming the scissile phosphate ester is colored yellow (See figure insert for color reproduction)

6.3 Catalytic Strategies Promoting Internal RNA Transesterification

There are limited means by which internal RNA transesterification can be accelerated by self-cleaving ribozymes. The uncatalyzed rate constant for transesterification and cleavage at any given phosphodiester linkage is projected to be approximately $10^{-8}\,min^{-1}$ (Li and Breaker 1999). However, rate constants for transesterification catalyzed by many self-cleaving ribozymes far exceed this (>1 min^{-1}), relying upon the combined contributions of up to four individual strategies to accelerate the reaction. The individual contribution of each catalytic strategy to the overall rate enhancement observed for self-cleaving RNA catalysts has been previously elaborated (Emilsson et al. 2003) and is briefly reviewed here.

Internal RNA transesterification resulting in 2′,3′-cyclic phosphodiester and 5′ hydroxyl products must proceed through a nucleophilic substitution (S_N2) reaction. (Fig. 6.2). Attack of the 2′ oxygen on the adjacent phosphorus center (Fig. 6.2a) forms a trigonal bipyrimidal pentacoordinate phosphate intermediate (Fig. 6.2b) with the 5′ oxygen leaving group geometrically opposite the 2′ oxygen. Therefore, an "in-line" conformation of 2′ oxygen, phosphorus, and 5′ oxygen within the scissile phosphodiester linkage represents a geometrical prerequisite, the constraint of which provides as much as a 100-fold increase in the rate constant for transesterification, exclusive of additional strategies for catalysis (Soukup and Breaker 1999). More substantial rate enhancements are achieved by the three remaining strategies of (1) general base catalysis deprotonating and substantially increasing the nucleophilicity of the 2′ oxygen, (2) general acid catalysis protonating the 5′ hydroxyl leaving group, and (3) stabilization of the transition state by alleviating the

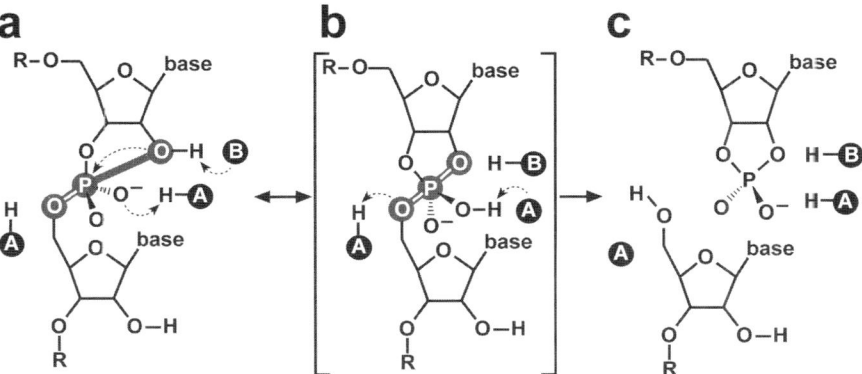

Fig. 6.2 Internal RNA transesterification. (**a**) Scissile phosphodiester bond. (**b**) Transition state. (**c**) Reaction products. The reaction is promoted by an in-line conformation of the highlighted (gray) 2′ oxygen, phosphorus, and 5′ oxygen; deprotonation of the 2′ hydroxyl by base catalysis (B); protonation of the 5′ oxygen by acid catalysis (A); and transition state stabilization, for example, by protonation of a non-bridging phosphate oxygen by another acid (A) to alleviate negative charge

formation of negative charge on non-bridging phosphate oxygens (Fig. 6.2b). Importantly, nucleobase functional groups or metal ions may support any of these three mechanisms (Bevilacqua and Yajima 2006).

How the *glmS* ribozyme mechanistically accomplishes GlcN6P-dependent self-cleavage is fascinating from the perspective that its coenzyme dependence is unique among catalytic RNAs. It must however employ some combination of these four strategies to achieve ligand-dependent catalysis. In this chapter, we will further consider available biochemical and biophysical data to rationalize a comprehensive mechanistic model for *glmS* ribozyme function that adheres to (a) the framework of conventional catalytic strategy, and (b) the chemical properties of nucleobases and the coenzyme, GlcN6P.

6.4   Coenzyme Recognition and Contribution to Catalysis

The first evidence that ligand participates directly in catalysis of the *glmS* ribozyme came from biochemical analysis of various GlcN6P analogs (McCarthy et al. 2005). Importantly, the ribozyme was shown to be activated by the buffering agent tris(hydroxymethyl)aminomethane (TRIS) which mimics the amine-containing portion of GlcN6P, and was otherwise devoid of activity with buffering agents that lack primary amines. Moreover, other GlcN6P analogs including GlcN, serinol, and ethanolamine (Fig. 6.3a) were shown to support catalysis whereas non-amine-containing analogs (e.g., Glc6P, glucose, trimethylene glycol, and ethanol) do not. Thus, the *glmS* ribozyme strictly requires an amine-containing ligand that minimally possesses an ethanolamine moiety for activity. Further, pH-reactivity profiles for *glmS* ribozyme self-cleavage revealed that catalysis precisely reflects the acid dissociation constant (pK_a) of GlcN6P, GlcN, or serinol, where the proximity effect of phosphate on the solution pK_a of GlcN6P appears to be masked by association with the ribozyme. These studies concluded that deprotonated GlcN6P functions as a coenzyme to activate catalysis, where proton transfer is mediated by the amine functionality to promote general acid–base catalysis (McCarthy et al. 2005).

Additional analysis of *glmS* ribozyme activation by various GlcN6P analogs echoes the need for an amine functionality and demonstrate the effect of other functional group modifications (Lim et al. 2006). Interestingly, virtually every functional group of GlcN6P is found to impact binding/catalysis of the *glmS* ribozyme (Fig. 6.3b). The basis for these results is readily apparent from analysis of the crystal structures of *glmS* ribozymes (Klein and Ferré-D'Amaré 2006; Cochrane et al. 2007; Klein et al. 2007b), where direct contacts are observed between each of the functional groups of GlcN6P and various nucleobase and backbone functional groups of the catalytic core or hydrated metal ions (Fig. 6.3c). These studies suggest that the ribozyme's catalytic core is highly tuned to the binding and utilization of GlcN6P as a coenzyme; a point that is further illustrated by an apparent inability to expand the functionality of the *glmS* ribozyme to include ligand analog usage through mutagenesis and in vitro selection (Link et al. 2006).

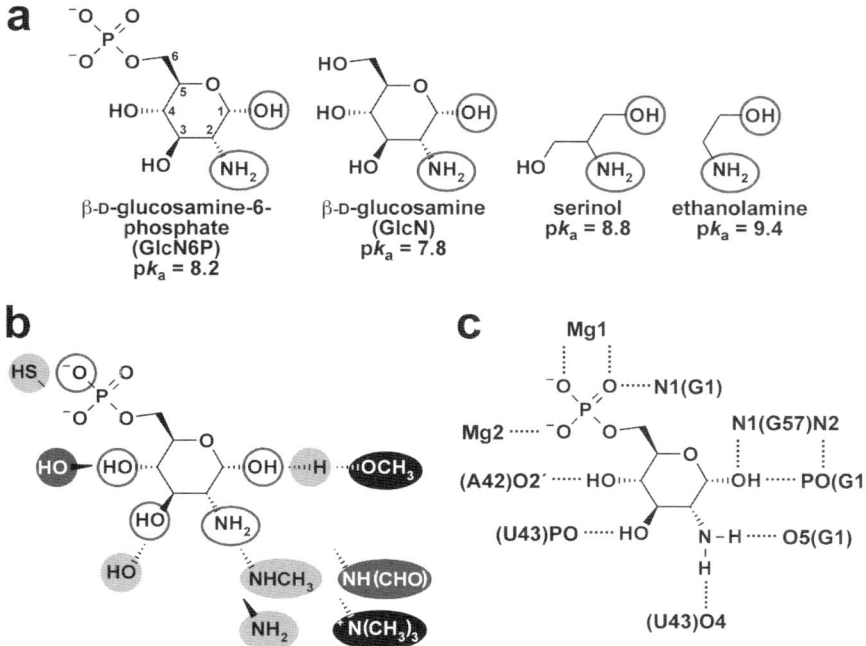

Fig. 6.3 GlcN6P functional group requirements and recognition. (**a**) Structure of GlcN6P and related analogs that support *glmS* ribozyme catalysis (McCarthy et al. 2005). Each ligand is depicted with requisite hydroxyl and amine groups indicated (encircled). The pk_a corresponding to ionization of each amine group is provided. (**b**) Additional functional group requirements for GlcN6P-dependent *glmS* ribozyme self-cleavage. Functional group modifications of GlcN6P (encircled) that diminish activity to fractions within 1/100, greater than 1/100, or completely are respectively highlighted light gray, dark gray, or black (Lim et al. 2006). (**c**) GlcN6P recognition by the *glmS* ribozyme. Depicted are contacts to each of two divalent metal ions (Mg1 and Mg2) and to nucleotide functional groups within the core of the *glmS* ribozyme determined by X-ray crystallography (Cochrane et al. 2007)

The most revealing aspect of crystal structure analysis of *glmS* ribozymes is that the amine functionality of GlcN6P is perfectly positioned to serve as a general acid catalyst (Klein and Ferré-D'Amaré 2006; Cochrane et al. 2007). The coenzyme's amine may serve to protonate the 5′ oxygen leaving group of G1, but its C1-hydroxyl might also function to stabilize the transition state through a network of interactions with a non-bridging phosphate oxygen at the scissile linkage -and with nucleobase functional groups on G57 (Fig. 6.3c). These findings lead to the conclusion that the coenzyme functions primarily as a general acid catalyst in *glmS* ribozyme self-cleavage (Klein and Ferré-D'Amaré 2006; Cochrane et al. 2007). However, if the identity and nature of any general base catalyst is to be appreciated in the *glmS* ribozyme self-cleavage mechanism, further consideration of the evidence is warranted.

6.5   Other Interactions and Contributions to Catalysis

The crystal structures of *glmS* ribozymes in conjunction with biochemical effects of nucleotide functional group modifications provide substantial additional insight regarding the structure and function of the RNA catalyst. Although the *glmS* ribozyme requires divalent metal ions for efficient catalysis, activity has been demonstrated at high concentrations of monovalent metal ion, and phosphorothioate substitution of the scissile phosphodiester linkage fails to exhibit a thiophilic metal ion (i.e., manganese) rescue effect (Roth et al. 2006). These studies suggest that metal ions serve only structural roles in *glmS* ribozyme self-cleavage. This point is corroborated by crystal structure analyses that provide no evidence of metal ion binding at the cleavage site (Klein and Ferré-D'Amaré 2006; Cochrane et al. 2007). Therefore, general acid and general base catalysis appear to be properties of the coenzyme and RNA alone.

Further scrutiny of the results of nucleotide analog interference mapping (NAIM) and suppression (NAIS) experiments (Jansen et al. 2006) reveal a number of other interesting features of *glmS* ribozyme structure and function. Phosphorothioate effects exhibiting manganese rescue identify the two metal ion binding sites that are not associated with coenzyme binding on the back side of the ribozyme along P2.1 (Fig. 6.4). Other backbone and nucleobase modification interferences and interference suppressions that are associated with GlcN6P- and GlcN-dependent function, respectively, reveal few direct contacts with the coenzyme's phosphate moiety and associated metal ions but which nonetheless lie at the top of P2.1 most visible from the front side of the ribozyme (Fig. 6.4). This analysis suggests that the relatively more important metal ion binding sites are those along P2.1 on the back side of the ribozyme. A number of interferences are also observed for nucleotide modifications that lie along or within P2.1, many of which demonstrate indirect suppressions when the coenzyme lacks a phosphate group (Fig. 6.4). This analysis suggests that although the coenzyme's phosphate group and associated metal ions are not required for catalysis, their presence most affects the fine structure of P2.1. Furthermore, 2'-deoxy interferences that lie along P2.1 (Figs. 6.4 and 6.5a) reveal the importance of a ribose zipper interaction with the P4 domain (Fig. 6.5b). Interestingly, these interferences are indirectly suppressed when manganese is substituted for magnesium, suggesting that its tighter binding along P2.1 alleviates any additional requirement for P4 domain interactions to aid in P2.1 organization. Altogether, this analysis underscores the importance of P2.1 organization affected by coenzyme phosphate, metal ion binding, and P4 domain interactions with regard to the formation of the ribozyme's active site, which lies at the base of P2.1 (Fig. 6.5a).

Although metal ions do not appear to play a direct role in catalysis of the *glmS* ribozyme, it should be noted that metal ions need not directly contact the scissile phosphodiester linkage to influence catalysis. A recent review by Sigel and Pyle (2007) details alternative roles for metal ions in catalysis. Namely, the positive charge provided by metal ions can alter pk_a values of RNA functional groups at a substantial distance, and direct coordination of transition state metal ions with the

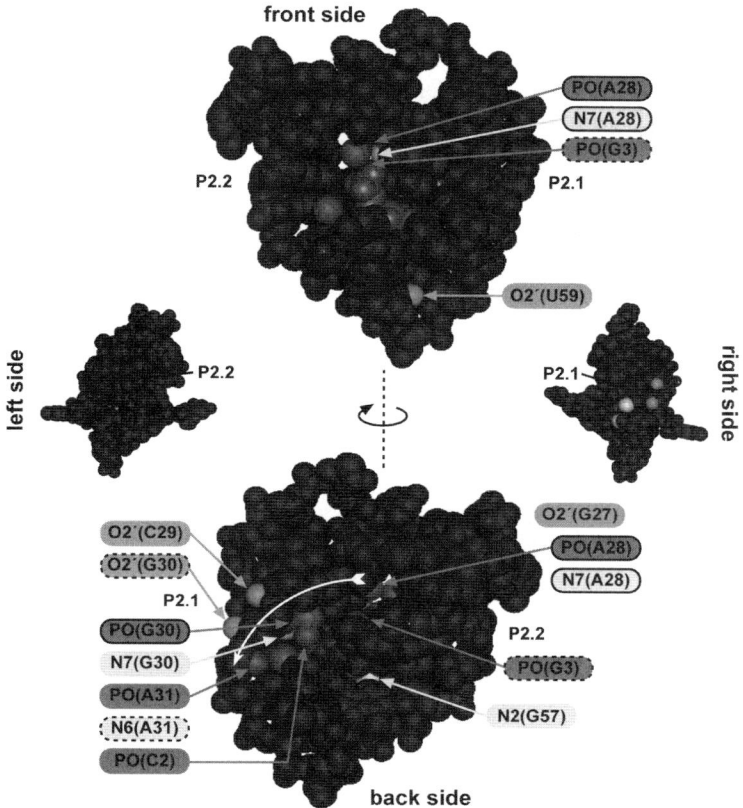

Fig. 6.4 Functional group requirements for metal ion binding and GlcN6P recognition determined by NAIM and NAIS. Depicted is the catalytic core only of the *glmS* ribozyme (blue) shown from each side relative to the "front" or GlcN6P (gray)-bound side (Cochrane et al. 2007). Functional groups within the catalytic core for which chemical mutagenesis results in interference (Jansen et al. 2006) are colored indicating requisite R_p phosphate oxygens (red), 2′ hydroxyls (green), or nucleobase functional groups (yellow). Each group is labeled in detail to provide nucleotide identity and number. Sites that cannot be observed in the space-filling model lack arrows, and sites at which interference suppression are observed with GlcN (Jansen et al. 2006) are outlined with solid or dashed lines to illustrate full or partial suppression, respectively. Magnesium ions (magenta) are shown as unhydrated for simplicity, and the white arrow denotes the backbone from A28 to A31 along which interference suppressions with GlcN are largely observed (See figure insert for color reproduction)

N7 of G can lower its N1 pk_a by 2.3 pH units. One cannot therefore exclude the possibility that metal ion binding near the base of P2.1 might influence adjacent active site nucleobase functional groups. In support of this notion, the *glmS* ribozyme exhibits 7-deazaguanine interference at the active site G33 in the presence of manganese ions (Soukup and Soukup, unpublished observations) which is not apparent in the presence of magnesium ions (Jansen et al. 2006). Consequently, alternative roles for metal ions near the *glmS* ribozyme active site require further scrutiny.

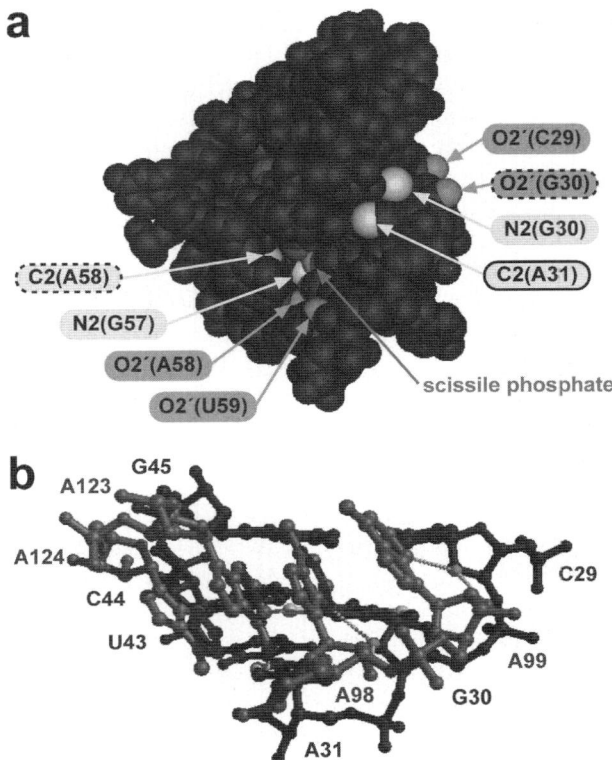

Fig. 6.5 Functional group requirements for P4 domain interaction and active site catalysis determined by NAIM and NAIS. (**a**) Functional group interferences and suppressions within the active site (*left*) and P2.1 (*right*). Depicted is the catalytic core only of the *glmS* ribozyme as described in the legend to Fig. 6.4 and shown from a lower right side angle. The scissile phosphate is colored cyan. (**b**) Interaction of P2.1 with P4 domain adenosines. Shown are four adenosines from the internal loop of the P4 domain (gray) that lie in the minor groove of P2.1 (blue). Nucleotide identities and numbering are depicted. Interferences support hydrogen-bonding interactions (dotted yellow lines) between A123 and G30 and formation of a ribose zipper between C29–G30 and A98–A99 (See figure insert for color reproduction)

6.6   Comprehensive Model of *glmS* Ribozyme Function

Despite the rapid accumulation of a considerable amount of biochemical and bio-physical data pertaining to *glmS* ribozyme structure and coenzyme function, a comprehensive mechanism for GlcN6P-dependent catalysis has not been elaborated. To arrive at such a model of *glmS* ribozyme function, significant insight can be derived from interpretative analysis of pH-reactivity profiles.

An accepted interpretation of hepatitis delta virus (HDV) ribozyme function from pH-reactivity profiles is that the activity of the ribozyme reflects the combined contributions of both the general acid and general base catalysts (Fig. 6.6a; Bevilacqua

2003). For the HDV ribozyme, C75 functions as a general acid catalyst with a shifted pk_a of 5.7, whereas divalent metal ion serves as the general base catalyst (e.g., magnesium ion with a pk_a of 11.4). As pH is decreased or increased, the general acid catalyst becomes more effective at or below its pk_a, and the general base catalyst becomes more effective at or above its pk_a. Importantly, the sole contribution of acid catalysis or combined contribution of each catalyst can be respectively distinguished by reaction in the absence or presence of different divalent metal ions (Nakano et al. 2000). The ribozyme thus exhibits a kinetic profile with an apparent pk_a between the values of the acid and base catalysts, where an intermediate slope of zero reflects the inverse relationship of each catalyst's contribution.

The *glmS* ribozyme exhibits a pH-reactivity profile with an apparent pk_a of 7.8 (McCarthy et al. 2005; Tinsley et al. 2007). By analogy to the HDV ribozyme, if GlcN6P functions primarily as a general acid catalyst for *glmS* ribozyme self-cleavage following its solution pk_a of 8.2, there exists a requirement for a relatively strong general base catalyst with a pk_a of ~10 to achieve the observed kinetic profile (Fig. 6.6b). Based on crystal structure analysis, the N1 of G33 has been proposed to function as the general base, as it is appropriately positioned adjacent to the 2' hydroxyl of A-1 at the scissile phosphodiester linkage (Klein and Ferré-D'Amaré 2006; Cochrane et al. 2007). Although the pk_a of such a nucleobase functionality is consistent with this model, a number of observations contradict this relatively simple interpretation of the pH-reactivity profile. Primarily, the model predicts that such an independently functioning general base catalysts would support measurable activity in the absence of GlcN6P as a general acid catalyst at neutral or basic pH. However, the ribozyme is devoid of activity in the absence of any coenzyme (McCarthy et al. 2005). The apparent pk_a of the ribozyme also does not reflect a pH-averaged value consistent with those of the general acid and proposed general base catalyst. The model additionally predicts that mutation of G33 to adenosine (pk_a 3.9) would support catalysis at near neutral pH, where the functional form of both the general acid and general base would be predominant. However, the G33A mutation has been demonstrated to completely inactivate the *glmS* ribozyme (Cochrane et al. 2007; Klein et al. 2007a). The *glmS* ribozyme therefore defies simple mechanistic explanation based on independently functioning general acid and general base catalysts.

An alternative interpretation of the *glmS* ribozyme pH-reactivity profile is proposed based upon the function of the coenzyme as both a general base and general acid catalyst (Fig. 6.6c). In this model, the kinetic profile of the ribozyme reflects the contributions of GlcN6P operating at its liganded pk_a, which is plausibly equivalent to that of GlcN due to masking of the phosphate group's proximity effect (McCarthy et al. 2005). The ribozyme thus exhibits an apparent pk_a that simply and precisely corresponds to the pk_a of the coenzyme's amine functionality. The shape of the curve also suggests that the coenzyme's function as a general base requisitely precedes its function as a general acid. In other words, maximal activity is provided by deprotonated coenzyme, whose function as a general base reasonably leads to formation of protonated coenzyme and subsequent function as a general acid. Although it has been argued that the apparent role of GlcN6P as a general acid catalyst obviates any invocation of GlcN6P as a general base catalyst (Cochrane et al. 2007),

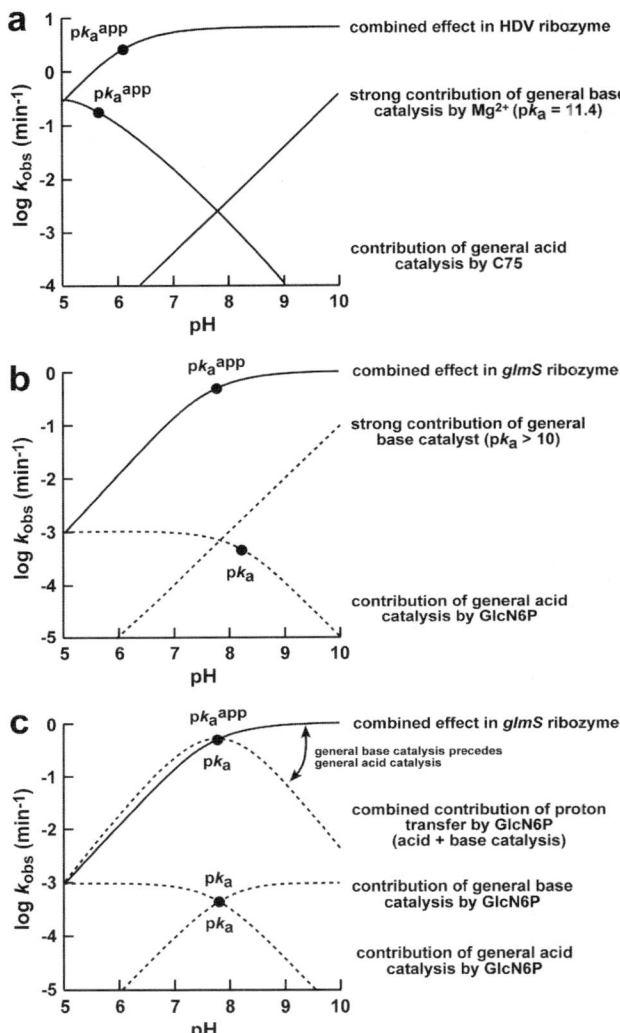

Fig. 6.6 Mechanistic interpretation of pH-reactivity profiles for HDV and *glmS* ribozymes. (**a**) HDV ribozyme pH-reactivity profile. Depicted is the approximate kinetic profile of the HDV ribozyme representing the combined effects of general acid catalysis by C75 determined in the absence of divalent metal ions, and general base catalysis by divalent ions. Solid lines are supported by kinetic analysis (Nakano et al. 2000). (**b**) Simple interpretation of *glmS* ribozyme pH-reactivity profile. By analogy to the HDV ribozyme, the approximate kinetic profile for the *glmS* ribozyme (solid line) would require GlcN6P function solely as a general acid catalyst following its solution pk_a to be combined with a general base catalyst with $pk_a > 10$ (dashed lines). (**c**) Alternative interpretation of *glmS* ribozyme pH-reactivity profile. The approximate kinetic profile for the *glmS* ribozyme (solid line) might represent the combined contributions of general acid and general base catalysis by GlcN6P (dashed lines), in which case the reactivity profile is not diminished above the apparent pk_a of GlcN6P if general base catalysis must precede general acid catalysis

the rationale presented here makes clear the necessity that both general base and general acid catalysis are somehow inherently interdependent. The question remains how the general base catalysis initiated by GlcN6P at a site distal to the 2′ hydroxyl of A-1 can ultimately activate a better-positioned general base catalyst such as G33.

Closer examination of the proximity of functional groups within the active site of the *glmS* ribozyme reveals a plausible mechanism of proton transfer between the coenzyme's amine functionality and the N1 of G33 as the ultimate general base catalyst (Fig. 6.7a). While a proton relay was originally proposed to involve bound water molecules (Klein and Ferré-D'Amaré 2006) which are not observed in other crystal forms of the *glmS* ribozyme (Cochrane et al. 2007; Klein et al. 2007b), we propose that active site nucleotide functional groups proven important for catalysis support a scheme for proton transfer (Fig. 6.7b). The coenzyme's amine group is

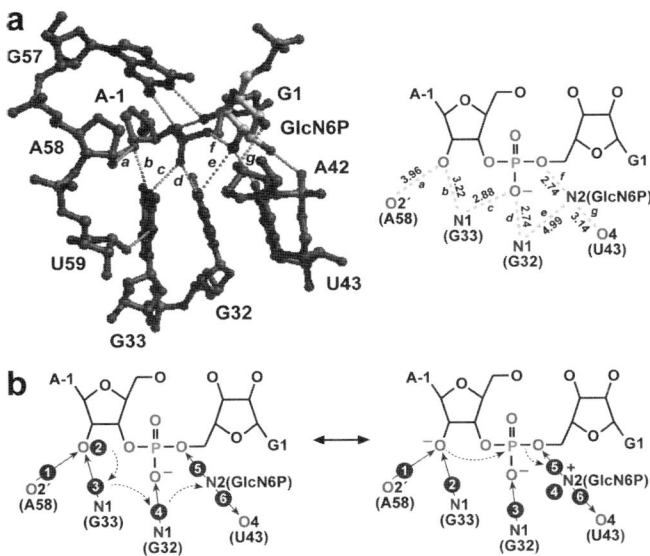

Fig. 6.7 *glmS* ribozyme active site and proposed comprehensive mechanism of action. (**a**) Active site composition. Depicted is a partial structure (*left*) of the catalytic core surrounding the scissile phosphate and including GlcN6P (Cochrane et al. 2007). Nucleotide identities and positions are denoted, where nucleobases for A-1, G1, A42, A58, and U59 are omitted for clarity. Atoms are colored by type with carbon (gray in RNA and white in GlcN6P), oxygen (red), nitrogen (blue), and phosphorus (orange). The methyl group inactivating the 2′ oxygen nucleophile at A-1 is colored green. The diagram at right shows active site functional groups with interatomic distances (dotted yellow lines) given in angstroms. (**b**) Proposed mechanism of action through coordinated proton transfer. The schematic diagram depicts the transfer of active site protons (numbered purple circles) and their hydrogen bonding interactions (purple arrows) between active site functional groups before and after general base catalysis initiated by GlcN6P. In this manner, G33 may be activated to serve as the ultimate general base for deprotonation of the 2′ hydroxyl at A-1 while GlcN6P is concomitantly activated to serve as the general acid catalyst for protonation of the 5′ oxygen at G1 (See figure insert for color reproduction)

specifically hypothesized to initiate a proton relay through the intervening N1 of G32, which contacts a non-bridging phosphate oxygen at the scissile phosphodiester linkage. In this way, the N1 of G33 and the coenzyme's amine are simultaneously and necessarily activated to respectively serve as the ultimate general base for deprotonation of the 2′ hydroxyl of A-1 and the general acid for protonation of the 5′ oxygen leaving group of G1. Importantly, NAIM experiments demonstrate the dependence of *glmS* ribozyme activity on the 2′ hydroxyl groups of U59 and A58 (Jansen et al. 2006), which might respectively influence the pk_a of G33 through interaction at its N3 position, and promote or stabilize formation of the 2′ oxygen nucleophile at the scissile phosphodiester linkage. The potential influence of metal ion in proximity to the N7 of G33 on its pk_a should however not be disregarded. While the role of G32 has not been assessed, the nucleobase identity is strictly conserved in *glmS* ribozymes (Barrick et al. 2004; Link et al. 2006). This comprehensive model appropriately predicts that any perturbation in the chain of events in the proton relay is equally detrimental to ribozyme activity (i.e., general base and general acid catalysis are inherently interdependent), which is consistent with the entirety of available biochemical data.

6.7   Conclusion

Further consideration of available biochemical and biophysical data pertaining to the structure and function of the *glmS* ribozyme reveals that general acid and general base catalysis are inherently interdependent in a coenzyme-dependent active site mechanism of RNA cleavage. The proposed comprehensive mechanistic model wherein the coenzyme, GlcN6P, functions both as the initial general base and consequent general acid catalyst within a proton-relay thus fulfills the apparent biochemical requirements for activity. This analysis in combination with other considerations regarding the effects of coenzyme binding on riboswitch structure and function suggests the development of *glmS* ribozyme agonists as prospective antibiotic compounds must satisfy strict chemical requirement for binding and activity.

References

Barrick JE, Corbino KA, Winkler WC, Nahvi A, Mandal M, Collins J, Lee M, Roth A, Sudarsan N, Jona I, Wickiser JK, Breaker RR (2004) New RNA motifs suggest an expanded scope for riboswitches in bacterial genetic control. Proc Natl Acad Sci U S A 101:6421–6426

Bevilacqua PC (2003) Mechanistic considerations for general acid-base catalysis by RNA: revisiting the mechanism for the hairpin ribozyme. Biochemistry 42:2259–2265

Bevilacqua PC, Yajima R (2006) Nucleobase catalysis in ribozyme mechanism. Curr Opin Chem Biol 10:455–464

Cochrane JC, Lipchock SV, Strobe SA (2007) Structural investigation of the *glmS* ribozyme bound to its catalytic cofactor. Chem Biol 14:97–105

Collins JA, Irnov I, Baker S, Winkler WC (2007) Mechanism of mRNA destabilization by the *glmS* ribozyme. Genes Dev 21:3356–3368

Emilsson GM, Nakamura S, Roth A, Breaker RR (2003) Ribozyme speed limits. RNA 9:907–918

Hampel KJ, Tinsley MM (2006) Evidence for reorganization of the *glmS* ribozyme ligand binding pocket. Biochemistry 45:7861–7871

Irnov I, Kertsburg A, Winkler WC (2006) Genetic control by cis-acting regulatory RNAs in *Bacillus subtilis*: general principles and prospects for discovery. Cold Spring Harb Symp Quant Biol 71:239–249

Jansen JA, McCarthy TJ, Soukup GA, Soukup JK (2006) Backbone and nucleobase contacts to glucosamine-6-phosphate in the *glmS* ribozyme. Nat Struct Mol Biol 13:517–523

Klein DJ, Ferré-D'Amaré AR (2006) Structural basis of *glmS* ribozyme activation by glucosamine-6-phosphate. Science 313:1752–1756

Klein DJ, Been MD, Ferré-D'Amaré AR (2007a) Essential role of an active-site guanine in *glmS* ribozyme catalysis. J Am Chem Soc 129:14858–14859

Klein DJ, Wilkinson SR, Been MD, Ferré-D'Amaré AR (2007b) Requirement of helix P2.2 and nucleotide G1 for positioning the cleavage site and cofactor of the *glmS* ribozyme. J Mol Biol 373:178–189

Li Y, Breaker RR (1999) Kinetics of RNA degradation by specific base catalysis of transesterification involving the 2'-hydroxyl group. J Am Chem Soc 121:5364–5372

Lim J, Grove BC, Roth A, Breaker RR (2006) Characteristics of ligand recognition by a *glmS* self-cleaving ribozyme. Angew Chem Int Ed Engl 45:6689–6693

Link KH, Guo L, and Breaker RR (2006) Examination of the structural and functional versatility of *glmS* ribozymes by using in vitro selection. Nucleic Acids Res 34:4968–4975

McCarthy TJ, Plog MA, Floy SA, Jansen JA, Soukup JK, Soukup GA (2005) Ligand requirements for *glmS* ribozyme self-cleavage. Chem Biol 12:1221–1226

Nakano S, Chadalavada DM, Bevilacqua PC (2000) General acid-base catalysis in the mechanism of the hepatitis delta virus ribozyme. Science 287:1493–1497

Nudler E, Mironov AS (2004) The riboswitch control of bacterial metabolism. Trends Biochem Sci 29:11–17

Roth A, Nahvi A, Lee M, Jona I, Breaker RR (2006) Characteristics of the *glmS* ribozyme suggest only structural roles for divalent metal ions. RNA 12:607–619

Serganov A, Polonskaia A, Phan AT, Breaker RR, Patel DJ (2006) Structural basis for gene regulation by a thiamine pyrophosphate-sensing riboswitch. Nature 441:1167–1171

Sigel RK, Pyle AM (2007) Alternative roles for metal ions in enzyme catalysis and the implications for ribozyme chemistry. Chem Rev 107:97–113

Soukup GA (2006) Core requirements for *glmS* ribozyme self-cleavage reveal a putative pseudoknot structure. Nucleic Acids Res 34:968–975

Soukup GA, Breaker RR (1999) Relationship between internucleotide linkage geometry and the stability of RNA. RNA 5:1308–1325

Thore S, Leibundgut M, Ban N (2006) Structure of the eukaryotic thiamine pyrophosphate riboswitch with its regulatory ligand. Science 312:1208–1211

Tinsley RA, Furchak JR, Walter NG (2007) Trans-acting *glmS* catalytic riboswitch: locked and loaded. RNA 13:468–477

Wilkinson SR, Been MD (2005) A pseudoknot in the 3' non-core region of the *glmS* ribozyme enhances self-cleavage activity. RNA 11:1788–1794

Winkler WC, Breaker RR (2005) Regulation of bacterial gene expression by riboswitches. Annu Rev Microbiol 59:487–517

Winkler WC, Nahvi A, Roth A, Collins JA, Breaker RR (2004) Control of gene expression by a natural metabolite-responsive ribozyme. Nature 428:281–286

Chapter 7
Group I Ribozymes as a Paradigm for RNA Folding and Evolution

Sarah A. Woodson(✉) and Seema Chauhan

Abstract Group I ribozymes are an ancient class of RNA catalysts that serve as a paradigm for the self-assembly of complex structures of non-coding RNA. The diversity of subtypes illustrates the modular character of RNA architecture and the potential for the evolution of new functions. The folding mechanisms of group I ribozymes illustrate the hierarchy of folding transitions and the importance of kinetic partitioning among competing folding pathways. Studies on group I splicing factors demonstrate how proteins facilitate the assembly of splicing complexes by stabilizing tertiary interactions between domains and by ATP-dependent cycles of RNA unfolding.

7.1 Group I Ribozymes as a Paradigm for RNA Folding and Evolution

As discussed throughout this book, RNA is a versatile biomolecule capable of performing a broad range of biological functions. The twin discoveries that the group I intron from *Tetrahymena thermophila* rRNA and the RNA subunit of RNase P were biological catalysts (Cech et al. 1981; Guerrier-Takada et al. 1983) had a profound influence on our understanding of non-coding RNAs in modern cells and the evolution of living systems (Doudna and Cech 2002). First, these discoveries solved the problem of whether enzymatic function or the genetic code appeared first, because RNAs can do both. Second, these discoveries firmly established the idea that non-coding RNAs are active players in the cell's metabolism. This chapter will focus on the structure of group I introns or "ribozymes", and how they have been used to study RNA self-assembly and the evolution of RNA–protein complexes (RNPs).

S.A. Woodson
T.C. Jenkins Department of Biophysics, Johns Hopkins University, 3400 N. Charles St., Baltimore, MD 21218, USA
e-mail: swoodson@jhu.edu.

N.G. Walter et al. (eds.) *Non-Protein Coding RNAs*
doi: 10.1007/978-3-540-70840-7_7, © Springer-Verlag Berlin Heidelberg 2009

145

7.2   Group I Ribozymes: A Theme with Variations

Group I introns are found in the nuclear, mitochondrial, and chloroplast genomes of a diverse collection of organisms (Damberger and Gutell 1994; Lambowitz and Perlman 1990; Michel et al. 1982). Their sporadic presence among phylogenetic lineages is consistent with frequent loss and insertion during evolution (Dujon et al. 1986). All members of this family share a similar structure, and are spliced from their parental RNA via the same two transesterification reactions (reviewed in Cech 1990) (Fig. 7.1). In the first step, the 5′ splice site is cleaved by the nucleophilic attack of the 3′ hydroxyl from a guanosine (exo G) that binds the RNA intermolecularly. In the second step, the 3′ hydroxyl at the end of the 5′ exon attacks the phosphodiester bond at the 3′ splice site, resulting in exon ligation and release of the intron (Fig. 7.1). Although few sequences in group I introns are conserved, nearly all members contain a U·G wobble pair at the 5′ splice site and a G before the 3′ splice site (ωG) (Cech 1990). The 5′ and 3′ exons base pair with an internal guide sequence (IGS), which positions the respective splice sites within the ribozyme active site (Been and Cech 1986; Davies et al. 1982; Suh and Waring 1990).

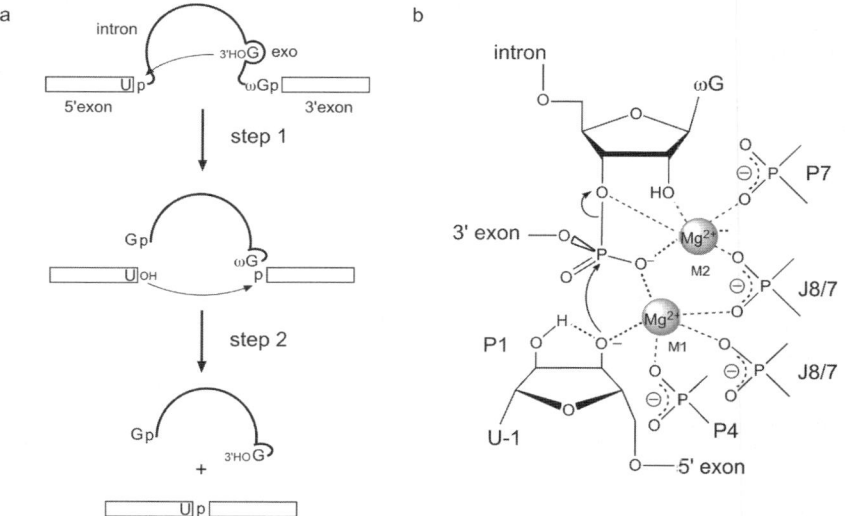

Fig. 7.1 Splicing mechanism of group I introns. (**a**) Splicing requires two phosphodiester transesterification reactions. An intermolecularly bound G (exo G) occupies the G-binding site in step 1; the conserved G and the 3′ end of the intron (ωG) occupies the G-binding site in step 2. Adapted from Cech (1990). (**b**) Coordination of two metal ions at the active site. Adapted from Stahley et al. (2007)

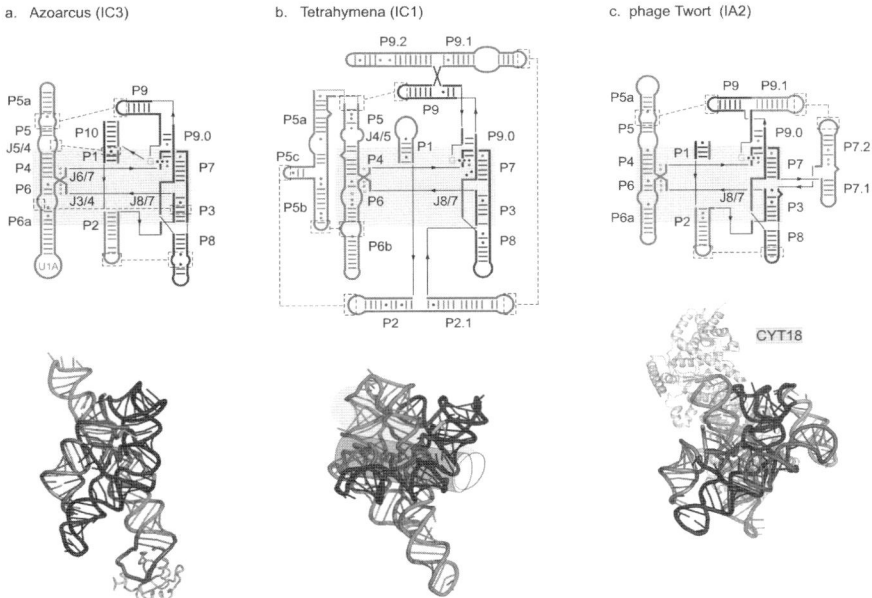

Fig. 7.2 Structures of group I ribozymes. 2D schematics and 3D ribbons are colored by domain as indicated. Red, 5' and 3' exons. (**a**) *Azoarcus* pre-tRNA[ile] ribozyme (*1u6b*; Adams et al, 2004). The gray ribbon is U1A protein, which was used to aid crystallization. (**b**) *Tetrahymena thermophila* LSU rRNA (*1x8w*; Guo et al. 2004). Pink and grey cylinders indicate the predicted position of the P2/P2.1 and P9.1/P9.2 helices (Lehnert et al. 1996), which were not present in the crystal structure. (**c**) Phage Twort orf142 ribozyme complexed with the C-terminal fragment of *N. crassa* CYT18 (*2rkj*; Paukstelis et al. 2008). Figure adapted from Woodson (2005a) (See figure insert for color reproduction)

7.2.1 Tertiary Interactions in the Catalytic Core

The catalytic core, which contains the active site, consists of two major helical domains containing paired (P) regions P4–P6 and P3–P9 (Kim and Cech 1987; Michel and Westhof 1990; Michel et al. 1982) (Fig. 7.2). The P3–P9 domain has many of the active site residues and retains some activity independently of the other domains (Ikawa et al. 2000a). It also contains the G-binding site (Michel et al. 1989), which is formed by an unusual stack of base triples at one end of the helix P7 (Adams et al. 2004a; Guo et al. 2004).

The P4–P6 domain structurally supports the P3–P9 domain (Fig. 7.2) and provides the receptor for the conserved U·G wobble pair in the P1 5' splice site helix (Wang et al. 1993). The minor groove edge of the wobble pair contacts the sheared A58·A87 pair in the joining (J) region 4/5 (Strobel et al. 1998; Szewczak et al. 1998). Ribose 2' hydroxyl groups in the 5' side of the P1 hydrogen bond with conserved residues in the unpaired J8/7 in the P3–P9 domain (Adams et al. 2004b; Pyle et al. 1992; Szewczak et al. 1999). Because the P1 helix lies at the interface between the two major domains, docking of P1 occurs after the core of the ribozyme has folded.

Sequence comparisons, biochemical studies, and crystal structures all show that unpaired segments between the helical domains make the most important contributions to the active site and to the tertiary interactions that hold the catalytic core together (Adams et al. 2004b; Cech et al. 1992; Michel and Westhof 1990). J8/7 alternately contacts the P4–P6, P3–P9, and P1 helices, zigzagging from one domain to the next (Adams et al. 2004a). The other joining regions interact with the double helices of the core in the minor groove, revealing the importance of minor groove motifs for helix packing (Strobel et al. 1998). One of these minor groove motifs are the base triples formed by J3/4 and J6/7 (Michel et al. 1990) that are necessary for association of the P4–P6 and P3–P9 domains (Doudna and Cech 1995). The second minor groove motif is the consecutive type I and type II A-minor contacts (Nissen et al. 2001) between P3 and unpaired adenines in J6/6a (Adams et al. 2004b; Rangan et al. 2003).

7.2.2 Peripheral Helices Fine Tune Stability

Although the tertiary interactions within the catalytic core specify the active site, they are too weak to overcome the unfavorable free energy associated with topological constraints imposed by the P3/P7 pseudoknot and the central triple helix (Brion and Westhof 1997). However, group I ribozymes also encode peripheral tertiary interactions that reinforce the catalytic core. These peripheral helices are modular and flexible in design, and can even be replaced by RNA-binding proteins (Westhof et al. 1996). Their structures, which vary among group I subfamilies, were initially deduced from sequence comparisons (Lehnert et al. 1996; Michel and Westhof 1990). Additional details are now visible in the crystal structures of the ribozymes from *Tetrahymena*, *Azoarcus*, and phage Twort, which each represent different subgroups (reviewed in Woodson 2005a) (Fig. 7.2).

The smallest group I intron from *Azoarcus* pre-tRNA[ile] (subclass IC3) is thought to represent the minimal set of peripheral interactions needed to stabilize the core (Fig. 7.2a) (Reinhold-Hurek and Shub 1992). The helical domains are clamped together by docking of GAAA tetraloops in P2 and P9 with canonical 11-nucleotide receptors in J8/8a and J5/5a, respectively, which can be replaced by related motifs in other members of the IC3 subfamily (Ikawa et al. 1999; Tanner and Cech 1996). The interaction between the tetraloop at the end of P9 and a helical receptor near P5 (J5/5a) is present in nearly all group I ribozymes, and contributes strongly to the stability of the core interactions (Jaeger et al. 1994; Laggerbauer et al. 1994).

In contrast, the peripheral interactions in the *Tetrahymena* ribozyme (subgroup IC1) produce a large, robust tertiary structure, but one that is also prone to misassembly (see Sect. 7.2.2). An extension of P5 (P5abc) packs against the back of P4 and P6, and specifically stabilizes the active conformation of the catalytic core (Fig. 7.2b) (Beaudry and Joyce 1990; Doherty et al. 1999; Johnson et al. 2005). Kissing loop base pairs between P2/P2.1 and P9.1/P.2 create a belt around the exterior of the ribozyme that reinforces helix packing within the core (Lehnert et al. 1996).

The ribozyme from phage *Twort* (subclass IA2) illustrates a different solution to the need for stabilizing the P3/P7 pseudoknot. Members of this subfamily lack P5abc, but contain an insertion within the P3–P9 domain (P7.1–P7.2; Fig. 7.2b) which packs against the minor groove of P3 (Golden et al. 2005; Lehnert et al. 1996). The loop of P7.2 also base pairs with an extension of P9, providing another structural brace to the central fold of the RNA.

7.2.3   Metal Ions in the Active Site

A crystal structure (3.4 Å) of a catalytically active splicing intermediate of the *Azoarcus* intron revealed two Mg2+ ions in the active site, which are almost entirely coordinated by oxygen in the RNA (Stahley and Strobel 2005). The ions bridge the scissile phosphate in geometry remarkably similar to that in DNA polymerases (Fig. 7.1b), as originally proposed by Steitz and Steitz (1993). The requirement for specific coordination of metal ions in the active site of group I ribozymes was first demonstrated by biochemical assays (Grosshans and Cech 1989) and phosphorothioate substitutions in active site residues which made the activity dependent on Mn^{2+} (Piccirilli et al. 1993; Shan et al. 2001; Sjogren et al. 1997; Weinstein et al. 1997). The active site Mg2+ ions bridge all three helical domains of the ribozyme. Thus, the "catalytic" metal ions also maintain the structure of the active site (Rangan and Woodson 2003; Stahley et al. 2007).

In addition to the two metal ions in the active site, many other metal ions are needed to stabilize the folded structure of the RNA. The majority of these ions remain bonded to water and associate non-specifically with the electrostatic field of the RNA (Hermann et al. 1998; Misra and Draper 1998). (For a more detailed discussion, see Chapter 2-DT). Although these non-specific interactions account for most of the thermodynamic stabilization of the folded RNA by metal ions (Draper et al. 2005), site-specific metal ion interactions contribute to the uniqueness of the 3D fold by coordinating atoms within specific tertiary structure motifs, such as a turn in the P5abc subdomain (Basu and Strobel 1999; Cate et al. 1997; Das et al. 2005) or tetraloop receptors (Basu et al. 1998; Stahley et al. 2007).

7.3   Folding of Group I Ribozymes: A Window into RNA Self-Assembly

The common fold of group I introns illustrates how the unusual tertiary interactions required to create an active site from RNA can be reinforced by structural motifs found in many different RNAs. At the same time, studies on the folding pathways of group I ribozymes have contributed fundamental insights into RNA self-assembly and dynamics. Similar concepts have emerged from folding studies on other catalytic RNAs such as RNase P and the hairpin ribozyme (Sosnick and Pan 2003), and thus are likely to apply to many different RNAs.

7.3.1 Hierarchical Model for RNA Folding

The pioneering studies on tRNA folding in the late 1960s and 1970s (reviewed in Crothers 2001) led to a hierarchical model for RNA folding, in which nearest neighbor interactions (base stacking and base pairing) produce the regular elements of secondary structure such as helices, loops, bulges, and junctions (Brion and Westhof 1997; Tinoco and Bustamante 1999). These structural elements create 3D motifs, which subsequently assemble into domains that are stabilized by interhelical tertiary interactions and specific coordination of metal ions (Sosnick and Pan 2003; Treiber and Williamson 2001a; Woodson 2000).

The architectural hierarchy of RNA structure is mirrored by differences in the dynamics and energetics of RNA interactions. RNA secondary structures (-1 to -3 kcal/mol per base pair) are thermodynamically more stable than tertiary structures (reviewed in Burkard et al. 1999; Serra and Turner 1995), and form more quickly (Crothers 2001). Because the tertiary folding greatly increases the local density of negative charge, the tertiary structure is more sensitive than secondary structure to the size, valence, and concentrations of counterions (Woodson 2005b).

The relationships between structural hierarchy, stability, and electrostatics are illustrated by the equilibrium folding pathway of the *Azoarcus* group I ribozyme in Mg2+, for which the two macroscopic tertiary folding transitions are well separated (Rangan et al. 2003) (Fig. 7.3). At low ionic strength, the RNA contains some secondary structure, but little or no tertiary structure (U). Small angle neutron and X-ray scattering studies (SANS and SAXS) showed that the unfolded *Azoarcus* ribozyme has an average radius of gyration (R_g) of 60–65 Å (Chauhan et al. 2005; Perez-Salas et al. 2004). Sub-millimolar Mg2+ (~0.2 mM) neutralizes the phosphate charge and induces the assembly of core helices, resulting in an ensemble of more compact and ordered intermediates (I_c) with an R_g of 31.5 ± 0.5 Å (Perez-Salas et al. 2004). Additional Mg2+ (~2 mM) induces formation of the native (N) tertiary structure ($R_g = 30$ Å), which correlates with the onset of catalytic activity (Rangan et al. 2003). Although Mg2+ is required for organization of the tertiary

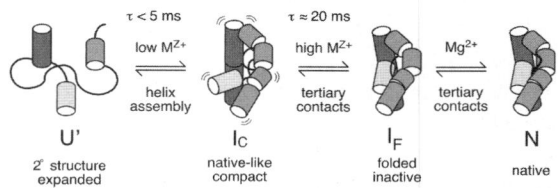

Fig. 7.3 Hierarchical folding of the *Azoarcus* ribozyme. Adapted from Rangan et al. (2003) and Rangan and Woodson (2003)

interactions around the active site, other ions including Na^+ and K^+ stabilize a folded but inactive form (I_F) (Rangan and Woodson 2003).

Although the compact intermediates that are formed at low Mg2+ concentrations (or shorter times) are not active, there is considerable evidence that they contain tertiary interactions. Not only are the intermediates more compact than the unfolded state, but mutations that remove tertiary interactions in the native state also destabilize the intermediates, when collapse transition is monitored by SAXS (Chauhan et al. 2005; Das et al. 2003; Kwok et al. 2005). In the yeast mitochondrial bI5 group I ribozyme, photo-crosslinks confirmed the presence of native-like interactions in compact but non-native intermediates, which form in 5–7 mM Mg2+ and can be detected by a large decrease in the Stokes radius of the RNA (Buchmueller et al. 2000; Webb and Weeks 2001). On the other hand, the compact intermediates are not protected from hydroxyl radical cleavage, demonstrating that the interior of the RNA remains open to solvent (Buchmueller and Weeks 2003; Das et al. 2003; Rangan et al. 2003). This suggests that tertiary interactions in the I state are dynamic (Fig. 7.3).

7.3.2 Folding Intermediates and Dynamics

Temperature-jump relaxation and NMR studies on small hairpins and tRNA showed that RNA secondary structures form more rapidly (10–100 µs) than tertiary structures (10–100 ms) (Cole and Crothers 1972; Craig et al. 1971; Crothers et al. 1974; Lynch and Schimmel 1974). Thus, the hierarchy of RNA structure also extends to the kinetics of RNA folding. Stopped-flow UV spectroscopy and time-resolved small-angle X-ray scattering (SAXS) experiments on RNase P and group I ribozymes showed that the initial collapse transition, which correlates with helix assembly, occurs in 1–10 ms (Fang et al. 1999; Russell et al. 2002b). By contrast, the subsequent search for the native tertiary conformation can take anywhere from 10 ms to several hours.

The variation in these folding times depends on how closely the intermediates resemble the native structure, and thus how much the initial structures must reorganize before reaching the native conformation (see Chapter 2-DT). For example, the catalytic domain of RNase P and the *Azoarcus* ribozyme collapse to native-like intermediates that transform to the native structure in 10–40 ms (Chauhan and Woodson 2008; Fang et al. 1999; Rangan et al. 2003). These RNAs collapse into compact intermediates that are nearly as compact as the native RNA (Chauhan et al. 2005; Fang et al. 2000; Perez-Salas et al. 2004). Biochemical studies suggest that the *Candida* group I ribozyme may behave in a similar way (Xiao et al. 2003; Zhang et al. 2005).

By contrast, classic experiments on the kinetic folding pathway of *Tetrahymena* ribozyme revealed that the P4–P6 domain can fold in 1–2 s, while the P3–P9 domains requires 1 min or longer to fold (Downs and Cech 1996; Sclavi et al. 1998; Zarrinkar and Williamson 1994) (Fig. 7.4). This is because the P3/P7 pseudo-knot

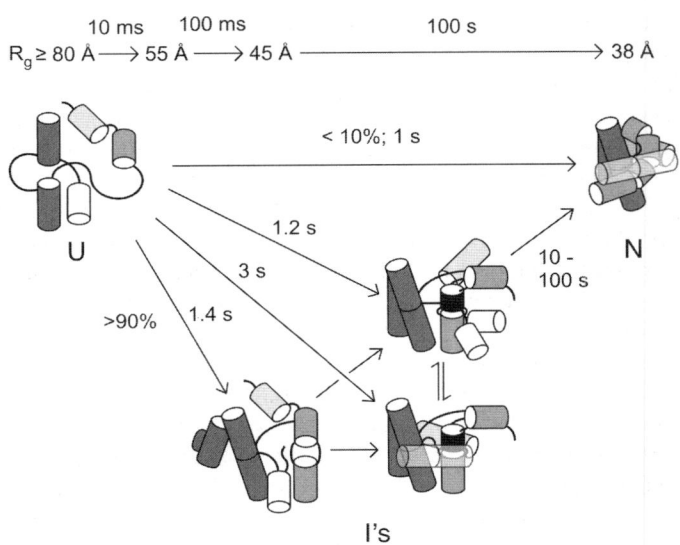

10 ms   100 ms   100 s

$R_g \geq 80$ Å ⟶ 55 Å ⟶ 45 Å ─────────────⟶ 38 Å

< 10%; 1 s

1.2 s

U

3 s

10 - 100 s N

>90% 1.4 s

I's

Fig. 7.4 Kinetic partitioning during folding of the *Tetrahymena* ribozyme. Top, collapse of the unfolded RNA to a series of more compact intermediates was measured by time-resolved SAXS (Russell et al 2002b; Kwok et al. 2005). Bottom, major tertiary folding intermediates detected by time-resolved footprinting (Sclavi et al. 1998; Pan et al. 2000; Laederach et al. 2006). Arrows indicate parallel folding pathways

is replaced by a non-native base pairing (alt P3), which must unfold before the RNA can have a chance to refold into the native structure (Pan and Woodson 1998; Pan et al. 1997). Time-resolved SAXS experiments demonstrated that the initial collapse transition produces intermediates that are at least 20% less compact than the native state (Buchmueller et al. 2000; Fang et al. 2000; Heilman-Miller et al. 2001; Kwok et al. 2005; Russell et al. 2000; Shcherbakova et al. 2004; Swisher et al. 2002; Xiao et al. 2003). Further compaction coincides with additional refolding of the RNA.

Mispairing of P3 and other helices within the pre-rRNA (P1, P2.1, and P9) are further stabilized by tertiary interactions in P5abc, P2, and P2.1/P9.1 (Pan and Woodson 1999; Russell et al. 2002a; Treiber et al. 1998). Perturbations to these interactions by base substitutions or changes in metal ions change the spectrum of intermediates observed in solution (Shcherbakova and Brenowitz 2005; Treiber and Williamson 2001b). Thus, peripheral interactions that stabilize the native state can also lengthen the folding time by prematurely trapping misfolded intermediates.

7.3.3 Kinetic Partitioning

Although more than 90% of the *Tetrahymena* ribozyme becomes kinetically trapped in misfolded intermediates in vitro, a small portion of the RNA folds in a second or less (Pan et al. 1997). Therefore, different molecules in the population can fold along pathways leading directly and rapidly to the native structure or leading through misfolded, kinetically trapped intermediates (Thirumalai and Woodson 1996) (see also Chapter 2-DT). The co-existence of alternative folding pathways for the *Tetrahymena* ribozyme (Fig. 7.4) was shown directly by single molecule fluorescence studies, which detected a fast folding pathway ($1 s^{-1}$) in addition to the slow folding pathways ($1 min^{-1}$) (Zhuang et al. 2000). It is also supported by statistical analyses of time-resolved hydroxyl radical footprinting data (Laederach et al. 2006). Remarkably, a single point mutation in P3 can shift the fraction of rapidly folding RNA to 80%, allowing both domains of the RNA to be protected from hydroxyl radical at $1 s^{-1}$ (Pan et al. 2000).

The overall folding kinetics of group I ribozymes depend on the specificity of the initial collapse and kinetic partitioning, and on the stabilities of the various intermediate structures that can be produced (Thirumalai and Woodson 1996). Since the *Tetrahymena* ribozyme undergoes non-specific collapse, it produces metastable intermediates that refold slowly (1–100 min) to the native state, and in which some tertiary domains are stably folded. By contrast, the *Azoarcus* and *Candida* group I ribozymes fold rapidly, suggesting that the intermediates produced during their initial collapse are close to the native conformation (Rangan et al. 2003; Xiao et al. 2003; Zhang et al. 2005).

There is accumulating evidence that stable RNAs fold by similar mechanisms in cells. The in vivo activity of group I and hairpin ribozymes correlates with the relative stability of the tertiary structure in vitro (Brion et al. 1999; Donahue et al. 2000). Both direct chemical probing in vivo and mutagenesis show that RNAs misfold in cells, albeit to a lesser degree than in the test tube (Nikolcheva and Woodson 1999; Waldsich et al. 2002). The activity and decay rate of the pre-RNA is best explained by kinetic partitioning of transcripts into active and inactive pools (Jackson et al. 2006).

7.3.4 Tertiary Interactions Improve the Specificity of Folding

Although the potential of most group I ribozymes to form more than one secondary structure can trigger partitioning of the RNA population among alternative folding pathways, recent results suggest that the stability of the tertiary interactions is one of the most important determinants of folding specificity. Mutations that disrupt tertiary contacts between helices not only destabilize the compact intermediates (or collapsed states) (Buchmueller and Weeks 2003; Chauhan et al. 2005; Das et al. 2003), but also made base pairing in the core of the *Azoarcus* ribozyme less cooperative (Chauhan and Woodson 2008). Thus, tertiary interactions between helices contribute to the specificity of helix assembly, presumably by biasing the ensemble of base paired states toward native-like conformations.

The wild type ribozyme folds rapidly in Mg2+ at 37°C, with all tertiary contacts nearly saturated within the 5 ms dead-time of the experiment (Chauhan and Woodson 2008; Rangan et al. 2003). By contrast, the mutation in L9 that destabilizes the tertiary structure of the RNA causes half the RNA population to fold rapidly (20 ms) and half to fold very slowly (~100 s). Therefore, by contributing to the specificity of helix assembly in the initial folding transition (from U to I_C), tertiary interactions increase the fraction of the *Azoarcus* ribozyme population that folds directly to the native structure rather than detouring through misfolded states.

7.3.5   Origins of Thermostability

Despite its small size and lack of peripheral helices, the folded structure of the *Azoarcus* ribozyme is very stable, remaining active up to 75°C or in 7.5 M urea (Tanner and Cech 1996). This has been attributed to a high G–C content (71%) and the fact that all of the hairpins are capped by stable GNRA or UNCG tetraloops (Kuo and Piccirilli 2001; Strauss-Soukup and Strobel 2000; Tanner and Cech 1996). Nucleotide swapping experiments between the *Azoarcus* ribozyme and the less stable *Anabaena* group IC3 ribozyme suggested that stronger tertiary interactions in the catalytic core are important for thermostability (Ikawa et al. 2000b). A similar conclusion was reached from selection of a thermostable variant of the *Tetrahymena* ribozyme (Guo and Cech 2002; Guo et al. 2006) and comparisons of the specificity (*S*) domains of mesophilic and thermophilic RNase P ribozymes (Baird et al. 2006).

Surprisingly, the free energy of forming the tertiary interactions in the *I* to *N* transition is only about 2–3 kcal mol^{-1}, based on the Mg2+-dependence of folding (Chauhan and Woodson 2008). However, transient unfolding of the *Azoarcus* ribozyme under physiological conditions results in native-like intermediates that quickly reform the native structure. By contrast, transient unfolding of the *Tetrahymena* ribozyme, which loses activity above 55°C (Guo and Cech 2002), leads to non-native conformations which cannot easily refold (Hopkins and Woodson 2005). The presence of stable non-native intermediates may explain why certain RNAs denature more easily than others, despite the ability to form many favorable tertiary interactions (Fang et al. 2001).

7.4   From Ribozymes to Ribonucleoproteins

Genetic studies uncovered open reading frames (ORFs) that were required for splicing of group I and group II introns in the mitochondria of yeast and other fungi (Faye et al. 1973; Halbreich et al. 1980; Van Ommen et al. 1980). These ORFs, some of which are embedded within the introns themselves, are not splicing enzymes. Rather, they encode proteins that facilitate the RNA-catalyzed splicing reaction by binding and stabilizing the folded RNA (Lambowitz and Perlman 1990). Some proteins facilitate splicing by accelerating the RNA folding reaction. Thus, group I ribozymes and their protein partners provide a window into the evolution of RNPs and the mechanism of their assembly.

7.4.1 RNA Folding Intermediates and Tertiary Structure Capture

One well studied example of a group I splicing factor is the yeast protein CBP2, which is required for splicing of intron bI5 from the cytochrome oxidase *b* (cob) mRNA (McGraw and Tzagoloff 1983). The bI5 RNA can self-splice in vitro in 40 mM $MgCl_2$, but requires CBP2 protein under physiological conditions (5 mM $MgCl_2$) (Gampel and Cech 1991). Footprinting experiments showed that CBP2 binds one face of the P4–P6 helices in the intron, stabilizing the core of the RNA, while additionally contacting the P1/P2 domain containing the 5′ splice site (Shaw and Lewin 1995; Weeks and Cech 1995a). The rate of assembly of the native RNP depended strongly on the conformational state of the RNA, but was independent of CBP2 concentration (Weeks and Cech 1995b). These findings led to the "tertiary structure capture" model, in which unimolecular folding of the RNA is captured and stabilized by interactions with the protein (Weeks and Cech 1996) (Fig. 7.5b).

The probability of trapping the RNA in its native conformation depends on forming a native-like collapsed state prior to binding CBP2 (Buchmueller and Weeks 2003).

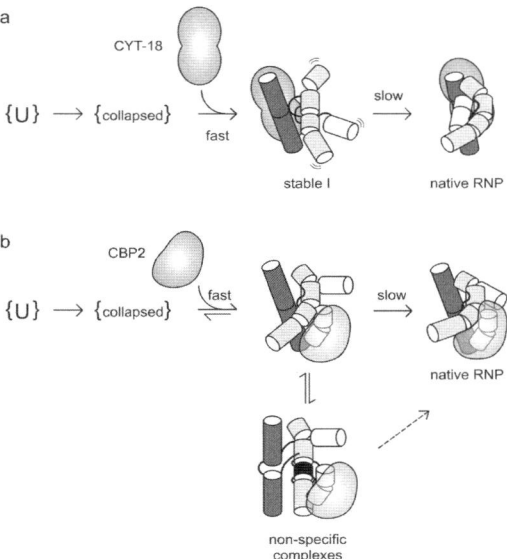

Fig. 7.5 Protein-dependent folding and assembly of splicing RNPs. (**a**) Mitochondrial tRNA synthetase CYT-18 forms a stable complex (K_d ~ 50 pM) with the conserved region of helices P4–P6 (Saldanha et al. 1996). Additional contacts with P3 and P8 form more slowly (~0.5 min⁻¹), leading to the active RNP (Caprara et al. 1996a; Webb et al. 2001a). (**b**) Assembly of splicing complexes with yeast CBP2 occurs via intermediate complexes in which CBP2 rapidly binds the RNA non-specifically and with lower affinity (Bokinsky et al. 2006). Reorganization of non-specific complexes and docking of the P1–P2 domain results in the native complex which is specifically bound by CBP2 (K_d ~ 0.4 nM)

When CBP2 binds the native-like intermediates in 5–7 mM $MgCl_2$, an active complex is formed (Weeks and Cech 1995b). If CBP2 is allowed to bind the unfolded RNA, bI5 is trapped in an unreactive state (Garcia and Weeks 2004). Thus, even in the presence of CBP2, tertiary interactions in the ribozyme core are still critical for assembly.

Single-molecule FRET experiments showed that CBP2 binds the RNA in at least two modes (Bokinsky et al. 2006). Under physiological Mg2+ concentrations, the CBP2 and bI5 RNA form non-specific complexes at near diffusion-controlled rates, which then slowly convert to the specific native complex (Fig. 7.5b). Individual molecules followed distinct trajectories, further supporting the notion that stochastic fluctuations in the RNA partition the complexes among different assembly pathways. Interestingly, CBP2 increases the number of conformational fluctuations (Bokinsky et al. 2006), consistent with observations that CBP2 also increases the refolding rate (Lewin et al. 1995).

7.4.2   Adaptation of Multifunctional Proteins for RNA Stabilization

The concept that group I splicing complexes assemble in several steps is also illustrated by two multifunctional proteins from *Neurospora crassa* and *Aspergillus nidulans* mitochondria. *N. crassa* CYT18 is a mitochondrial tRNA synthetase that is required for the splicing of several mitochondrial group I introns (Collins and Lambowitz 1985). CYT18 binds the P4–P6 helices of group I ribozymes lacking the P5abc extension in the correct orientation (Caprara et al. 1996a, b; Mohr et al. 1992). In turn, this promotes folding of the P3–P9 helices, leading to the active RNP (Caprara et al. 1996a, b). In contrast with CBP2, CYT18 binds the P4–P6 domain with subnanomolar affinity, creating a very stable assembly intermediate (Fig. 7.5a). As a result, the activation energy for reorganization of the P3–P9 domain is high (Webb et al. 2001b).

CYT18 interacts with conserved elements of group I ribozymes and specifically stabilizes core interactions. It can bind a variant of the *Tetrahymena* ribozyme lacking P5abc and compensate for the stabilizing function normally provided by these peripheral helices (Mohr et al. 1994). A crystal structure of CYT18 in complex with the group I intron from phage Twort confirmed that CYT18 contacts conserved residues in the P4–P6 domain of the RNA (Fig. 7.2c) (Paukstelis et al. 2008). Interestingly, two peptide loops from CYT18 reach deep into the junction between P4 and P6, suggesting that the protein not only aligns P4 and P6, but specifically activates the ribozyme core by changing the structure of these helices (Chen et al. 2000).

The I-*Ani*I maturase, which is encoded by an ORF within its cognate intron, facilitates the splicing of a mitochondrial intron in *A. nidulans* and is also a homing DNA endonuclease that restricts intronless alleles of the cob gene (Ho et al. 1997). Like CYT18 and CBP2, I-*Ani*I binds P4–P6 and stabilizes the ribozyme core (Ho and Waring 1999). Chemical probing of the RNA structure showed that the I-*Ani*I encounter complex subsequently promotes docking of the P1–P2 helices (Caprara et al. 2007). Interestingly, docking of P1 increases the affinity of I-*Ani*I for the RNA. Because I-*Ani*I and P1 bind the opposite faces of the ribozyme, this effect must be indirect, possibly via a conformational change in P4 (Bartley et al. 2003; Caprara et al. 2007).

CYT18, I-*Ani*I, and other mitochondrial splicing factors are multi-functional proteins that have been co-opted to facilitate an RNA-catalyzed reaction. Interestingly, splicing activity is usually associated with an additional domain or peptide insertion, allowing these proteins use different surfaces to recognize the intron RNA and their usual substrate (Chatterjee et al. 2003; Hsu et al. 2006; Paukstelis et al. 2008).

7.4.3   DEAD-Box Helicases that Refold Group I Introns

The group I splicing factors described above specifically stabilize the catalytic core, in some cases reaching the active RNP in several stages of induced fit. They do not appear to prevent kinetic partitioning of the RNA population among competing folding pathways, and can even slow down the conformational search for the native structure if the intermediates become too stable. The problem of the kinetics of assembly is solved by a second class of splicing proteins that are members of the "DEAD-box" family of ATP-dependent RNA helicases (Huang et al. 2005; Seraphin et al. 1989). The ATPase activity of DEAD-box proteins controls the affinity of the protein for the RNA, which is coupled to local unwinding of RNA helices (Jankowsky et al. 2001; Yang et al. 2007).

Experiments on the *Tetrahymena* ribozyme demonstrated that the *N. crassa* CYT19 DEAD-box protein was able to stimulate conversion of the non-native altP3 helix to the native P3 pseudo-knot, in the presence of ATP (Mohr et al. 2002). Further experiments showed that, in this heterologous system, CYT19 drives repetitive cycles of RNA unfolding, without discriminating between native and misfolded RNAs (Bhaskaran and Russell 2007; Tijerina et al. 2006). Because the RNA has many more chances to refold, the amount of active ribozyme ultimately increases (Fig. 7.6). During each cycle, the RNA population repartitions among folding pathways leading to the native and misfolded conformations as it would in the absence of CYT19 (Bhaskaran and Russell 2007). Thus, the RNA tertiary interactions

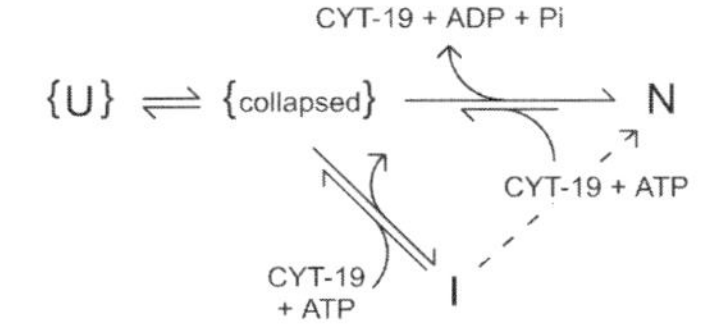

Fig. 7.6 ATP-dependent refolding of RNA by CYT-19 DEAD-box protein. ATP-dependent binding of native or non-native RNA unfolds the RNA. Following ATP hydrolysis and release of CYT19, the RNA goes through another round of folding and kinetic partitioning between misfolded and native states (Bhaskaran and Russell 2007)

continue to dictate the specificity of the folding reaction. In vivo, another important role of DEAD-box ATPases is to ensure degradation of group I ribozymes after splicing (Margossian et al. 1996).

7.5   Evolution of New Ribozyme Functions

The capacity of RNA sequences to form more than one stable structure is not always a liability. This same property may allow new RNA functions to evolve by shuffling of RNA coding segments or even through cumulative point mutations. This principle was demonstrated by Schultes and Bartel, who engineered a sequence intermediate between a ligase ribozyme and the Hepatitis delta virus self-cleaving ribozyme that encoded both activities, albeit inefficiently (Schultes and Bartel 2000).

The group I-like self-cleaving ribozymes (GIR) from slime molds *Naegleria* sp. and *Didimyium iridis* are a natural example of how the group I active site can be rewired to carry out a different chemical reaction (Johansen and Vogt 1994). The GIRs are found inside a normal self-splicing group I intron, and catalyze a self-cleavage reaction that releases the mRNA for the homing endonuclease (Decatur et al. 1995; Einvik et al. 1997). Instead of requiring the usual G-nucleotide, GIR self-cleavage uses the $2'$ hydroxyl from a nearby U as the nucleophile for phosphodiester transesterification, creating a tiny $2'–5'$ lariat at the $5'$ end of the newly released mRNA (Nielsen et al. 2005).

How can the group I active site be modified such that an internal $2'$ hydroxyl becomes the nucleophile, in place of the usual $3'$ hydroxyl? Analysis of sequence conservation among known GIRs and structural modeling showed that a new helix, P15, replaces the P2 helix usually present in typical group I introns (Einvik et al. 1998) (Fig.7.7). As a consequence, J8/7 is shorter, and is proposed to make an entirely new set of tertiary interactions which serve to position the internal U $2'$ hydroxyl in line with the phosphodiester bond to be cleaved (Beckert et al. 2008). Remarkably, this profound change in the topology of the ribozyme core can be explained by shuffling the RNA sequence at only three points in P2 and P8 (Beckert et al. 2008) (Fig. 7.7).

The versatility of the group I framework has also been demonstrated by the evolution of new variants in vitro. In one set of experiments, the P4–P6 domain of the *Tetrahymena* ribozyme was used as a structural scaffold for selection of an RNA ligase active site (Yoshioka et al. 2004). In another set of experiments, the P4–P6 scaffold was replaced by other RNA sequences, which were able to partially support the catalytic activity of the P3–P9 domain (Ohuchi et al. 2002, 2004). Depending on the structural domain, the ribozyme can be made to depend on a protein or a small molecule effector (Atsumi et al. 2003; Kuramitsu et al. 2005). Thus, the modular architecture of group I ribozymes is not only relevant to their assembly and dynamics, but allows useful variations on the theme to emerge through evolution of the sequences.

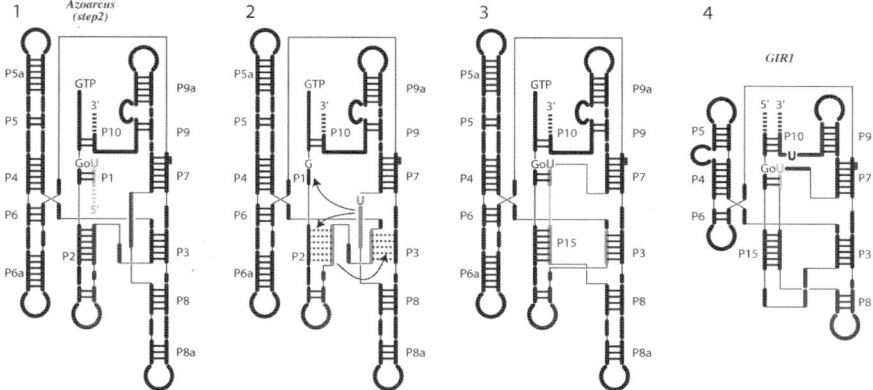

Fig. 7.7 Remodeled core of a group I-like ribozyme (GIR1). A lariat forming GIR1 ribozyme found in *Didymium iridis* (**4**) could have evolved from the smallest self-splicing group I intron (**1**) via a sequence swap in the core (medium grey; **2** and **3**), followed by minimization of peripheral helices (**4**). Reprinted from Beckert et al. (2008) with permission

7.6 Conclusion

The modular architecture and folding dynamics of group I introns reflect themes that are observed in many non-coding RNAs. For example, the substrate binding pockets and active sites in RNase P or the ribosome are located at the interface between helices in the center of the RNA, because this is where complex tertiary interactions are most easily created. On the other hand, active sites are, by their very nature, marginally stable and dynamic. Thus, the structure of the active site must be reinforced by peripheral tertiary interactions, which are less conserved than the active site itself.

The modular structure of the peripheral domains opens up opportunities for diversification of function and for the evolution of RNPs. For example, many helices present in eubacterial rRNAs have been lost in mitochondrial rRNAs; these helices are on the exterior of the ribosome and are apparently compensated by additional proteins (Cavdar Koc et al. 2001; Gutell 1996). Conversely, expansion segments in eukaryotic rRNAs are also located on the exterior of the ribosome, where they may mediate interactions with translation factors (Nilsson et al. 2007).

At the same time, the dichotomy between weak core tertiary interactions and strong peripheral interactions makes group I introns and other large non-coding RNAs vulnerable to misfolding. As the folding mechanisms of additional intron subfamilies are studied, new links between the architecture of RNA and its dynamics are sure to be uncovered. What is certain is that there is more to learn from this ancient family of catalytic RNA.

References

Adams PL, Stahley MR, Kosek AB, Wang J, Strobel SA (2004a) Crystal structure of a self-splicing group I intron with both exons. Nature 430:45–50

Adams PL, Stahley MR, Gill ML, Kosek AB, Wang J, Strobel SA (2004b) Crystal structure of a group I intron splicing intermediate. RNA 10:1867–1887

Atsumi S, Ikawa Y, Shiraishi H, Inoue T (2003) Selections for constituting new RNA-protein interactions in catalytic RNP. Nucleic Acids Res 31:661–669

Baird NJ, Srividya N, Krasilnikov AS, Mondragon A, Sosnick TR, Pan T (2006) Structural basis for altering the stability of homologous RNAs from a mesophilic and a thermophilic bacterium. RNA 12:598–606

Bartley LE, Zhuang X, Das R, Chu S, Herschlag D (2003) Exploration of the transition state for tertiary structure formation between an RNA helix and a large structured RNA. J Mol Biol 328:1011–1026

Basu S, Strobel SA (1999) Thiophilic metal ion rescue of phosphorothioate interference within the Tetrahymena ribozyme P4–P6 domain. RNA 5:1399–1407

Basu S, Rambo RP, Strauss-Soukup J, Cate JH, Ferré d'Amare AR, Strobel SA, Doudna JA (1998) A specific monovalent metal ion integral to the AA platform of the RNA tetraloop receptor. Nat Struct Biol 5:986–992

Beaudry AA, Joyce GF (1990) Minimum secondary structure requirements for catalytic activity of a self-splicing group I intron. Biochemistry 29:6534–6539

Beckert B, Nielsen H, Einvik C, Johansen SD, Westhof E, Masquida B (2008) Molecular modelling of the GIR1 branching ribozyme gives new insight into evolution of structurally related ribozymes. EMBO J 27:667–678

Been MD, Cech TR (1986) One binding site determines sequence specificity of Tetrahymena pre-rRNA self-splicing, trans-splicing, and RNA enzyme activity. Cell 47:207–216

Bhaskaran H, Russell R (2007) Kinetic redistribution of native and misfolded RNAs by a DEAD-box chaperone. Nature 449:1014–1018

Bokinsky G, Nivon LG, Liu S, Chai G, Hong M, Weeks KM, Zhuang X (2006) Two distinct binding modes of a protein cofactor with its target RNA. J Mol Biol 361:771–784

Brion P, Westhof E (1997) Hierarchy and dynamics of RNA folding. Annu Rev Biophys Biomol Struct 26:113–137

Brion P, Schroeder R, Michel F, Westhof E (1999) Influence of specific mutations on the thermal stability of the td group I intron in vitro and on its splicing efficiency in vivo: a comparative study. RNA 5:947–958

Buchmueller KL, Weeks KM (2003) Near native structure in an RNA collapsed state. Biochemistry 42:13869–13878

Buchmueller KL, Webb AE, Richardson DA, Weeks KM (2000) A collapsed, non-native RNA folding state. Nat Struct Biol 7:362–366

Burkard ME, Turner DH, Tinoco IJ (1999) The interactions that shape RNA structure. In: Gesteland RF, Cech TR, Atkins JF (eds.) The RNA World, 2nd edn. Cold Spring Harbor Laboratory Press, Cold Spring Harbor, NY, pp. 233–264

Caprara MG, Mohr G, Lambowitz AM (1996a) A tyrosyl-tRNA synthetase protein induces tertiary folding of the group i intron catalytic core. J Mol Biol 257:512–531

Caprara MG, Lehnert V, Lambowitz AM, Westhof E (1996b) A tyrosyl-tRNA synthetase recognizes a conserved tRNA-like structural motif in the group I intron catalytic core. Cell 87:1135–1145

Caprara MG, Chatterjee P, Solem A, Brady-Passerini KL, Kaspar BJ (2007) An allosteric-feedback mechanism for protein-assisted group I intron splicing. RNA 13:211–222

Cate JH, Hanna RL, Doudna JA (1997) A magnesium ion core at the heart of a ribozyme domain. Nat Struct Biol 4:553–558

Cavdar Koc E, Burkhart W, Blackburn K, Moseley A, Spremulli LL (2001) The small subunit of the mammalian mitochondrial ribosome. Identification of the full complement of ribosomal proteins present. J Biol Chem 276:19363–19374

Cech TR (1990) Self-splicing of group I introns. Annu Rev Biochem 59:543–568

Cech TR, Zaug AJ, Grabowski PJ (1981) In vitro splicing of the ribosomal RNA precursor of Tetrahymena: involvement of a guanosine nucleotide in the excision of the intervening sequence. Cell 27:487–496

Cech TR, Herschlag D, Piccirilli JA, Pyle AM (1992) RNA catalysis by a group I ribozyme. Developing a model for transition state stabilization. J Biol Chem 267:17479–17482

Chatterjee P, Brady KL, Solem A, Ho Y, Caprara MG (2003) Functionally distinct nucleic acid binding sites for a group I intron encoded RNA maturase/DNA homing endonuclease. J Mol Biol 329:239–251

Chauhan S, Woodson SA (2008) Tertiary interactions determine the accuracy of RNA folding. J Am Chem Soc 130:1296–1303

Chauhan S, Caliskan G, Briber RM, Perez-Salas U, Rangan P, Thirumalai D, Woodson SA (2005) RNA tertiary interactions mediate native collapse of a bacterial group I ribozyme. J Mol Biol 353:1199–1209

Chen X, Gutell RR, Lambowitz AM (2000) Function of tyrosyl-tRNA synthetase in splicing group I introns: an induced-fit model for binding to the P4–P6 domain based on analysis of mutations at the junction of the P4–P6 stacked helices. J Mol Biol 301:265–283

Cole PE, Crothers DM (1972) Conformational changes of transfer ribonucleic acid. Relaxation kinetics of the early melting transition of methionine transfer ribonucleic acid (Escherichia coli). Biochemistry 11:4368–4374

Collins RA, Lambowitz AM (1985) RNA splicing in Neurospora mitochondria. Defective splicing of mitochondrial mRNA precursors in the nuclear mutant cyt18–1. J Mol Biol 184:413–428

Craig ME, Crothers DM, Doty P (1971) Relaxation kinetics of dimer formation by self complementary oligonucleotides. J Mol Biol 62:383–401

Crothers DM (2001) RNA conformational dynamics. In: Söll D, Nishimura S, Moore P (eds.) RNA, Elsevier, Oxford, UK, pp. 61–70

Crothers DM, Cole PE, Hilbers CW, Shulman RG (1974) The molecular mechanism of thermal unfolding of Escherichia coli formylmethionine transfer RNA. J Mol Biol 87:63–88

Damberger SH, Gutell RR (1994) A comparative database of group I intron structures. Nucleic Acids Res 22:3508–3510

Das R, Travers KJ, Bai Y, Herschlag D (2005) Determining the Mg2+ stoichiometry for folding an RNA metal ion core. J Am Chem Soc 127:8272–8273

Das R, Kwok LW, Millett IS, Bai Y, Mills TT, Jacob J, Maskel GS, Seifert S, Mochrie SG, Thiyagarajan P, Doniach S, Pollack L, Herschlag D (2003) The fastest global events in RNA folding: electrostatic relaxation and tertiary collapse of the Tetrahymena ribozyme. J Mol Biol 332:311–319

Davies RW, Waring RB, Ray JA, Brown TA, Scazzocchio C (1982) Making ends meet: a model for RNA splicing in fungal mitochondria. Nature 300:719–724

Decatur WA, Einvik C, Johansen S, Vogt VM (1995) Two group I ribozymes with different functions in a nuclear rDNA intron. EMBO J 14:4558–4568

Doherty EA, Herschlag D, Doudna JA (1999) Assembly of an exceptionally stable RNA tertiary interface in a group I ribozyme. Biochemistry 38:2982–2990

Donahue CP, Yadava RS, Nesbitt SM, Fedor MJ (2000) The kinetic mechanism of the hairpin ribozyme in vivo: influence of RNA helix stability on intracellular cleavage kinetics. J Mol Biol 295:693–707

Doudna JA, Cech TR (1995) Self-assembly of a group I intron active site from its component tertiary structural domains. RNA 1:36–45

Doudna JA, Cech TR (2002) The chemical repertoire of natural ribozymes. Nature 418:222–228

Downs WD, Cech TR (1996) Kinetic pathway for folding of the Tetrahymena ribozyme revealed by three UV-inducible crosslinks. RNA 2:718–732

Draper DE, Grilley D, Soto AM (2005) Ions and RNA folding. Annu Rev Biophys Biomol Struct 34:221–243

Dujon B, Colleaux L, Jacquier A, Michel F, Monteilhet C (1986) Mitochondrial introns as mobile genetic elements: the role of intron-encoded proteins. Basic Life Sci 40:5–27

Einvik C, Decatur WA, Embley TM, Vogt VM, Johansen S (1997) Naegleria nucleolar introns contain two group I ribozymes with different functions in RNA splicing and processing. RNA 3:710–720

Einvik C, Nielsen H, Westhof E, Michel F, Johansen S (1998) Group I-like ribozymes with a novel core organization perform obligate sequential hydrolytic cleavages at two processing sites. RNA 4:530–541

Fang XW, Pan T, Sosnick TR (1999) Mg2+-dependent folding of a large ribozyme without kinetic traps. Nat Struct Biol 6:1091–1095

Fang X, Littrell K, Yang XJ, Henderson SJ, Siefert S, Thiyagarajan P, Pan T, Sosnick TR (2000) Mg2+-dependent compaction and folding of yeast tRNAPhe and the catalytic domain of the *B. subtilis* RNase P RNA determined by small-angle X-ray scattering. Biochemistry 39:11107–11113

Fang, XW, Golden, BL, Littrell, K, Shelton, V, Thiyagarajan, P, Pan, T, Sosnick, TR (2001) The thermodynamic origin of the stability of a thermophilic ribozyme. Proc Natl Acad Sci U S A 98:4355–4360

Faye G, Fukuhara H, Grandchamp C, Lazowska J, Michel F, Casey J, Getz GS, Locker J, Rabinowitz M, Bolotin-Fukuhara M, Coen D, Deutsch J, Dujon B, Netter P, Slonimski PP (1973) Mitochondrial nucleic acids in the petite colonie mutants: deletions and repetition of genes. Biochimie 55:779–792

Gampel A, Cech TR (1991) Binding of the CBP2 protein to a yeast mitochondrial group I intron requires the catalytic core of the RNA. Genes Dev 5:1870–1880

Garcia I, Weeks KM (2004) Structural basis for the self-chaperoning function of an RNA collapsed state. Biochemistry 43:15179–15186

Golden BL, Kim H, Chase E (2005) Crystal structure of a phage Twort group I ribozyme-product complex. Nat Struct Mol Biol 12:82–89

Grosshans CA, Cech TR (1989) Metal ion requirements for sequence-specific endoribonuclease activity of the Tetrahymena ribozyme. Biochemistry 28:6888–6894

Guerrier-Takada C, Gardiner K, Marsh T, Pace N, Altman S (1983) The RNA moiety of ribonuclease P is the catalytic subunit of the enzyme. Cell 35:849–857

Guo F, Cech TR (2002) Evolution of Tetrahymena ribozyme mutants with increased structural stability. Nat Struct Biol 9:855–861

Guo F, Gooding AR, Cech TR (2004) Structure of the Tetrahymena ribozyme: base triple sandwich and metal ion at the active site. Mol Cell 16:351–362

Guo F, Gooding AR, Cech TR (2006) Comparison of crystal structure interactions and thermodynamics for stabilizing mutations in the Tetrahymena ribozyme. RNA 12:387–395

Gutell RR (1996) Comparative sequence analysis and the structure of 16S and 23S rRNA. In: Zimmerman RA, Dahlberg AE (eds.) Ribosomal RNA: structure, evolution, processing, and function in protein biosynthesis. CRC Press, Boca Raton, FL, pp. 111–128

Halbreich A, Pajot P, Foucher M, Grandchamp C, Slonimski P (1980) A pathway of cytochrome b mRNA processing in yeast mitochondria: specific splicing steps and an intron-derived circular DNA. Cell 19:321–329

Heilman-Miller SL, Thirumalai D, Woodson SA (2001) Role of counterion condensation in folding of the Tetrahymena ribozyme. I. Equilibrium stabilization by cations. J Mol Biol 306:1157–1166

Hermann T, Auffinger P, Westhof E (1998) Molecular dynamics investigations of hammerhead ribozyme RNA. Eur Biophys J 27:153–165

Ho Y, Waring RB (1999) The maturase encoded by a group I intron from *Aspergillus nidulans* stabilizes RNA tertiary structure and promotes rapid splicing [In Process Citation]. J Mol Biol 292:987–1001

Ho Y, Kim SJ, Waring RB (1997) A protein encoded by a group I intron in Aspergillus nidulans directly assists RNA splicing and is a DNA endonuclease [published erratum appears in Proc Natl Acad Sci U S A 1997 Dec 23;94(26):14976]. Proc Natl Acad Sci USA 94:8994–8999

Hopkins JF, Woodson SA (2005) Molecular beacons as probes of RNA unfolding under native conditions. Nucleic Acids Res 33:5763–5770

Hsu JL, Rho SB, Vannella KM, Martinis SA (2006) Functional divergence of a unique C-terminal domain of leucyl-tRNA synthetase to accommodate its splicing and aminoacylation roles. J Biol Chem 281:23075–23082

Huang HR, Rowe CE, Mohr S, Jiang Y, Lambowitz AM, Perlman PS (2005) The splicing of yeast mitochondrial group I and group II introns requires a DEAD-box protein with RNA chaperone function. Proc Natl Acad Sci USA 102:163–168

Ikawa Y, Naito D, Aono N, Shiraishi H, Inoue T (1999) A conserved motif in group IC3 introns is a new class of GNRA receptor. Nucleic Acids Res 27:1859–1865

Ikawa Y, Shiraishi H, Inoue T (2000a) Minimal catalytic domain of a group I self-splicing intron RNA. Nat Struct Biol 7:1032–1035

Ikawa Y, Naito D, Shiraishi H, Inoue T (2000b) Structure-function relationships of two closely related group IC3 intron ribozymes from *Azoarcus* and *Synechococcus* pre-tRNA. Nucleic Acids Res 28:3269–3277

Jackson, SA, Koduvayur, S, Woodson, SA (2006) Self-splicing of a group I intron reveals partitioning of native and misfolded RNA populations in yeast. RNA 12:2149–2159

Jaeger L, Michel F, Westhof E (1994) Involvement of a GNRA tetraloop in long-range RNA tertiary interactions. J Mol Biol 236:1271–1276

Jankowsky E, Gross CH, Shuman S, Pyle AM (2001) Active disruption of an RNA-protein interaction by a DExH/D RNA helicase. Science 291:121–125

Johansen S, Vogt VM (1994) An intron in the nuclear ribosomal DNA of *Didymium iridis* codes for a group I ribozyme and a novel ribozyme that cooperate in self-splicing. Cell 76:725–734

Johnson TH, Tijerina P, Chadee AB, Herschlag D, Russell R (2005) Structural specificity conferred by a group I RNA peripheral element. Proc Natl Acad Sci USA 102:10176–10181

Kim SH, Cech TR (1987) Three-dimensional model of the active site of the self-splicing rRNA precursor of Tetrahymena. Proc Natl Acad Sci USA 84:8788–8792

Kuo LY, Piccirilli JA (2001) Leaving group stabilization by metal ion coordination and hydrogen bond donation is an evolutionarily conserved feature of group I introns. Biochim Biophys Acta 1522:158–166

Kuramitsu S, Ikawa Y, Inoue T (2005) Rational installation of an allosteric effector on a designed ribozyme. Nucleic Acids Symp Ser (Oxf) 2005(49):349–350

Kwok LW, Shcherbakova I, Lamb JS, Park HY, Andresen K, Smith H, Brenowitz M, Pollack L (2006) Concordant Exploration of the Kinetics of RNA Folding from Global and Local Perspectives. J Mol Biol 355:282–293

Laederach A, Shcherbakova I, Liang MP, Brenowitz M, Altman RB (2006) Local kinetic measures of macromolecular structure reveal partitioning among multiple parallel pathways from the earliest steps in the folding of a large RNA molecule. J Mol Biol 358:1179–1190

Laggerbauer B, Murphy FL, Cech TR (1994) Two major tertiary folding transitions of the Tetrahymena catalytic RNA. EMBO J 13:2669–2676

Lambowitz AM, Perlman PS (1990) Involvement of aminoacyl-tRNA synthetases and other proteins in group I and group II intron splicing. Trends Biochem Sci 15:440–444

Lehnert V, Jaeger L, Michel F, Westhof E (1996) New loop-loop tertiary interactions in self-splicing introns of subgroup IC and ID: a complete 3D model of the *Tetrahymena thermophila* ribozyme. Chem Biol 3:993–1009

Lewin AS, Thomas J, Jr., Tirupati HK (1995) Cotranscriptional splicing of a group I intron is facilitated by the Cbp2 protein. Mol Cell Biol 15:6971–6978

Lynch DC, Schimmel PR (1974) Cooperative binding of magnesium to transfer ribonucleic acid studied by a fluorescent probe. Biochemistry 13:1841–1852

Margossian SP, Li H, Zassenhaus HP, Butow RA (1996) The DExH box protein Suv3p is a component of a yeast mitochondrial 3'- to-5' exoribonuclease that suppresses group I intron toxicity. Cell 84:199–209

McGraw P, Tzagoloff A (1983) Assembly of the mitochondrial membrane system. Characterization of a yeast nuclear gene involved in the processing of the cytochrome b pre-mRNA. J Biol Chem 258:9459–9468

Michel F, Westhof E (1990) Modelling of the three-dimensional architecture of group I catalytic introns based on comparative sequence analysis. J Mol Biol 216:585–610

Michel F, Jacquier A, Dujon B (1982) Comparison of fungal mitochondrial introns reveals extensive homologies in RNA secondary structure. Biochimie 64:867–881

Michel F, Hanna M, Green R, Bartel DP, Szostak JW (1989) The guanosine binding site of the Tetrahymena ribozyme. Nature 342:391–395

Michel F, Ellington AD, Couture S, Szostak JW (1990) Phylogenetic and genetic evidence for base-triples in the catalytic domain of group I introns. Nature 347:578–580

Misra, VK, Draper, DE (1998) On the role of magnesium ions in RNA stability. Biopolymers 48:113–135

Mohr, G, Zhang, A, Gianelos, JA, Belfort, M, Lambowitz, AM (1992) The neurospora CYT-18 protein suppresses defects in the phage T4 td intron by stabilizing the catalytically active structure of the intron core. Cell 69:483–494

Mohr G, Caprara MG, Guo Q, Lambowitz AM (1994) A tyrosyl-tRNA synthetase can function similarly to an RNA structure in the Tetrahymena ribozyme. Nature 370:147–150

Mohr S, Stryker JM, Lambowitz AM (2002) A DEAD-box protein functions as an ATP-dependent RNA chaperone in group I intron splicing. Cell 109:769–779

Nielsen H, Westhof E, Johansen S (2005) An mRNA is capped by a 2′, 5′ lariat catalyzed by a group I-like ribozyme. Science 309:1584–1587

Nikolcheva T, Woodson SA (1999) Facilitation of group I splicing in vivo: misfolding of the Tetrahymena IVS and the role of ribosomal RNA exons. J Mol Biol 292:557–567

Nilsson J, Sengupta J, Gursky R, Nissen P, Frank J (2007) Comparison of fungal 80 S ribosomes by cryo-EM reveals diversity in structure and conformation of rRNA expansion segments. J Mol Biol 369:429–438

Nissen P, Ippolito JA, Ban N, Moore PB, Steitz TA (2001) RNA tertiary interactions in the large ribosomal subunit: the A-minor motif. Proc Natl Acad Sci USA 98:4899–4903

Ohuchi SJ, Ikawa Y, Shiraishi H, Inoue T (2002) Modular engineering of a Group I intron ribozyme. Nucleic Acids Res 30:3473–3480

Ohuchi SJ, Ikawa Y, Shiraishi H, Inoue T (2004) Artificial modules for enhancing rate constants of a Group I intron ribozyme without a P4-P6 core element. J Biol Chem 279:540–546

Pan J, Woodson SA (1998) Folding intermediates of a self-splicing RNA: mispairing of the catalytic core. J Mol Biol 280:597–609

Pan J, Woodson SA (1999) The effect of long-range loop-loop interactions on folding of the Tetrahymena self-splicing RNA. J Mol Biol 294:955–965

Pan J, Thirumalai D, Woodson SA (1997) Folding of RNA involves parallel pathways. J Mol Biol 273:7–13

Pan J, Deras ML, Woodson SA (2000) Fast folding of a ribozyme by stabilizing core interactions: evidence for multiple folding pathways in RNA. J Mol Biol 296:133–144

Paukstelis PJ, Chen JH, Chase E, Lambowitz AM, Golden BL (2008) Structure of a tyrosyl-tRNA synthetase splicing factor bound to a group I intron RNA. Nature 451:94–97

Perez-Salas UA, Rangan P, Krueger S, Briber RM, Thirumalai D, Woodson SA (2004) Compaction of a bacterial group I ribozyme coincides with the assembly of core helices. Biochemistry 43:1746–1753

Piccirilli JA, Vyle JS, Caruthers MH, Cech TR (1993) Metal ion catalysis in the Tetrahymena ribozyme reaction. Nature 361:85–88

Pyle AM, Murphy FL, Cech TR (1992) RNA substrate binding site in the catalytic core of the Tetrahymena ribozyme. Nature 358:123–128

Rangan P, Woodson SA (2003) Structural requirement for Mg2+ binding in the group I intron core. J Mol Biol 329:229–238

Rangan P, Masquida B, Westhof E, Woodson SA (2003) Assembly of core helices and rapid tertiary folding of a small bacterial group I ribozyme. Proc Natl Acad Sci USA 100:1574–1579

Reinhold-Hurek B, Shub DA (1992) Self-splicing introns in tRNA genes of widely divergent bacteria. Nature 357:173–176

Russell R, Millett IS, Doniach S, Herschlag D (2000) Small angle X-ray scattering reveals a compact intermediate in RNA folding. Nat Struct Biol 7:367–370

Russell R, Zhuang X, Babcock HP, Millett IS, Doniach S, Chu S, Herschlag D (2002a) Exploring the folding landscape of a structured RNA. Proc Natl Acad Sci U S A 99:155–160

Russell R, Millett IS, Tate MW, Kwok LW, Nakatani B, Gruner SM, Mochrie SG, Pande V, Doniach S, Herschlag D, Pollack L (2002b) Rapid compaction during RNA folding. Proc Natl Acad Sci U S A 99:4266–4271

Saldanha R, Ellington A, Lambowitz AM (1996) Analysis of the CYT-18 protein binding site at the junction of stacked helices in a group I intron RNA by quantitative binding assays and in vitro selection. J Mol Biol 261:23–42

Schultes EA, Bartel DP (2000) One sequence, two ribozymes: implications for the emergence of new ribozyme folds. Science 289:448–452

Sclavi B, Sullivan M, Chance MR, Brenowitz M, Woodson SA (1998) RNA folding at millisecond intervals by synchrotron hydroxyl radical footprinting. Science 279:1940–1943

Seraphin B, Simon M, Boulet A, Faye G (1989) Mitochondrial splicing requires a protein from a novel helicase family. Nature 337:84–87

Serra MJ, Turner DH (1995) Predicting thermodynamic properties of RNA. Methods Enzymol 259:242–261

Shan S, Kravchuk AV, Piccirilli JA, Herschlag D (2001) Defining the catalytic metal ion interactions in the Tetrahymena ribozyme reaction. Biochemistry 40:5161–5171

Shaw LC, Lewin AS (1995) Protein-induced folding of a group I intron in cytochrome b pre-mRNA. J Biol Chem 270:21552–21562

Shcherbakova I, Brenowitz M (2005) Perturbation of the hierarchical folding of a large RNA by the destabilization of its Scaffold's tertiary structure. J Mol Biol 354:483–496

Shcherbakova I, Gupta S, Chance MR, Brenowitz M (2004) Monovalent ion-mediated folding of the *Tetrahymena thermophila* ribozyme. J Mol Biol 342:1431–1442

Sjogren AS, Pettersson E, Sjoberg BM, Stromberg R (1997) Metal ion interaction with co-substrate in self-splicing of group I introns. Nucleic Acids Res 25:648–653

Sosnick TR, Pan T (2003) RNA folding: models and perspectives. Curr Opin Struct Biol 13:309–316

Stahley MR, Strobel SA (2005) Structural evidence for a two-metal-ion mechanism of group I intron splicing. Science 309:1587–1590

Stahley MR, Adams PL, Wang J, Strobel SA (2007) Structural metals in the group I intron: a ribozyme with a multiple metal ion core. J Mol Biol 372:89–102

Steitz TA, Steitz JA (1993) A general two-metal-ion mechanism for catalytic RNA. Proc Natl Acad Sci U S A 90:6498–6502

Strauss-Soukup JK, Strobel SA (2000) A chemical phylogeny of group I introns based upon interference mapping of a bacterial ribozyme. J Mol Biol 302:339–358

Strobel SA, Ortoleva-Donnelly L, Ryder SP, Cate JH, Moncoeur E (1998) Complementary sets of noncanonical base pairs mediate RNA helix packing in the group I intron active site. Nat Struct Biol 5:60–66

Suh ER, Waring RB (1990) Base pairing between the 3′ exon and an internal guide sequence increases 3′ splice site specificity in the Tetrahymena self-splicing rRNA intron. Mol Cell Biol 10:2960–2965

Swisher JF, Su LJ, Brenowitz M, Anderson VE, Pyle AM (2002) Productive folding to the native state by a group II intron ribozyme. J Mol Biol 315:297–310

Szewczak AA, Ortoleva-Donnelly L, Ryder SP, Moncoeur E, Strobel SA (1998) A minor groove RNA triple helix within the catalytic core of a group I intron. Nat Struct Biol 5:1037–1042

Szewczak AA, Ortoleva-Donnelly L, Zivarts MV, Oyelere AK, Kazantsev AV, Strobel SA (1999) An important base triple anchors the substrate helix recognition surface within the Tetrahymena ribozyme active site. Proc Natl Acad Sci U S A 96:11183–11188

Tanner M, Cech T (1996) Activity and thermostability of the small self-splicing group I intron in the pre-tRNA(Ile) of the purple bacterium Azoarcus. RNA 2:74–83

Thirumalai D, Woodson SA (1996) Kinetics of folding of protein and RNA. Acc Chem Res 29:433–439

Tijerina P, Bhaskaran H, Russell R (2006) Nonspecific binding to structured RNA and preferential unwinding of an exposed helix by the CYT-19 protein, a DEAD-box RNA chaperone. Proc Natl Acad Sci U S A 103:16698–16703

Tinoco IJ, Bustamante C (1999) How RNA folds. J Mol Biol 293:271–261

Treiber DK, Williamson JR (2001a) Beyond kinetic traps in RNA folding. Curr Opin Struct Biol 11:309–314

Treiber DK, Williamson JR (2001b) Concerted kinetic folding of a multidomain ribozyme with a disrupted loop-receptor interaction. J Mol Biol 305:11–21

Treiber DK, Rook MS, Zarrinkar PP, Williamson JR (1998) Kinetic intermediates trapped by native interactions in RNA folding. Science 279:1943–1946

Van Ommen GJ, Boer PH, Groot GS, De Haan M, Roosendaal E, Grivell LA, Haid A, Schweyen RJ (1980) Mutations affecting RNA splicing and the interaction of gene expression of the yeast mitochondrial loci cob and oxi-3. Cell 20:173–183

Waldsich C, Masquida B, Westhof E, Schroeder R (2002) Monitoring intermediate folding states of the td group I intron in vivo. EMBO J 21:5281–5291

Wang JF, Downs WD, Cech TR (1993) Movement of the guide sequence during RNA catalysis by a group I ribozyme. Science 260:504–508

Webb AE, Weeks KM (2001) A collapsed state functions to self-chaperone RNA folding into a native ribonucleoprotein complex. Nat Struct Biol 8:135–140

Webb AE, Rose MA, Westhof E, Weeks KM (2001a) Protein-dependent transition states for ribonucleoprotein assembly. J Mol Biol 309:1087–1100

Webb AE, Rose MA, Westhof E, Weeks KM (2001b) Protein-dependent transition states for ribonucleoprotein assembly. J Mol Biol 309:1087–1100

Weeks KM, Cech TR (1995a) Protein facilitation of group I intron splicing by assembly of the catalytic core and the 5 splice site domain. Cell 82:221–230

Weeks KM, Cech TR (1995b) Efficient protein-facilitated splicing of the yeast mitochondrial bI5 intron. Biochemistry 34:7728–7738

Weeks KM, Cech TR (1996) Assembly of a ribonucleoprotein catalyst by tertiary structure capture. Science 271:345–348

Weinstein LB, Jones BC, Cosstick R, Cech TR (1997) A second catalytic metal ion in group I ribozyme. Nature 388:805–808

Westhof E, Masquida B, Jaeger L (1996) RNA tectonics: towards RNA design. Fold Des 1: R78–88

Woodson SA (2000) Recent insights on RNA folding mechanisms from catalytic RNA. Cell Mol Life Sci 57:796–808

Woodson SA (2005a) Structure and assembly of group I introns. Curr Opin Struct Biol

Woodson SA (2005b) Metal ions and RNA folding: a highly charged topic with a dynamic future. Curr Opin Chem Biol 9:104–109

Xiao M, Leibowitz MJ, Zhang Y (2003) Concerted folding of a Candida ribozyme into the catalytically active structure posterior to a rapid RNA compaction. Nucleic Acids Res 31: 3901–3908

Yang Q, Del Campo M, Lambowitz AM, Jankowsky E (2007) DEAD-box proteins unwind duplexes by local strand separation. Mol Cell 28:253–263

Yoshioka W, Ikawa Y, Jaeger L, Shiraishi H, Inoue T (2004) Generation of a catalytic module on a self-folding RNA. RNA 10:1900–1906

Zarrinkar PP, Williamson JR (1994) Kinetic intermediates in RNA folding. Science 265: 918–924

Zhang L, Xiao M, Lu C, Zhang Y (2005) Fast formation of the P3–P7 pseudoknot: a strategy for efficient folding of the catalytically active ribozyme. RNA 11:59–69

Zhuang X, Bartley LE, Babcock HP, Russell R, Ha T, Herschlag D, Chu S (2000) A single-molecule study of RNA catalysis and folding. Science 288:2048–2051

Chapter 8
Group II Introns and Their Protein Collaborators

Amanda Solem, Nora Zingler, Anna Marie Pyle(✉),
and Jennifer Li-Pook-Than

Abstract Group II introns are an abundant class of autocatalytic introns that excise themselves from precursor mRNAs. Although group II introns are catalytic RNAs, they require the assistance of proteins for efficient splicing in vivo. Proteins that facilitate splicing of organellar group II introns fall into two main categories: intron-encoded maturases and host-encoded proteins. This chapter will focus on the host proteins that group II introns recruited to ensure their function. It will discuss the great diversity of these proteins, define common features, and describe different strategies employed to achieve specificity. Special emphasis will be placed on DEAD-box ATPases, currently the best studied example of host-encoded proteins with a role in group II intron splicing. Since the exact mechanisms by which splicing is facilitated is not known for any of the host proteins, general mechanistic strategies for protein-mediated RNA folding are described and assessed for their potential role in group II intron splicing.

8.1 Introduction to Group II Introns

The splicing of eukaryotic transcripts is typically carried out by a large ribonucleoprotein machine called the spliceosome. However, there are two classes of introns that fold into autocatalytic structures that catalyze their own splicing from precursor RNAs: the group I and group II introns (Lambowitz et al. 1999; Michel and Ferat 1995). Group II introns are very common within the organelles of plants, fungi, protists, and yeast, where they play a major role in pathways for gene expression (Bonen and Vogel 2001; Lehmann and Schmidt 2003). Group II introns are also abundant in diverse bacteria (Ferat and Michel 1993; Martinez-Abarca and Toro 2000). In addition to splicing, many group II introns are mobile, which means

A.M. Pyle
266 Whitney Avenue, Room 334A Bass Building, Yale University, New Haven, CT 06511, USA
e-mail: anna.pyle@yale.edu

N.G. Walter et al. (eds.) *Non-Protein Coding RNAs*   167
doi: 10.1007/978-3-540-70840-7_8, © Springer-Verlag Berlin Heidelberg 2009

that the liberated intron is reactive and it can insert itself, through reverse-splicing, into compatible DNA and RNA targets (Pyle and Lambowitz 2006). By "hopping" and spreading into new genomic locations and hosts, it is believed that group II introns have played a major role in the dispersal of noncoding RNA (including introns) and that they continue to shape the evolution of host genomes (Martin and Koonin 2006; Mattick 1994).

8.1.1 A Ribozyme that Collaborates with Proteins

In order to function and proliferate within new environments, group II introns often require assistance, which they obtain by recruiting or collaborating with proteins (Lambowitz and Zimmerly 2004). Protein recruitment may also be a host adaptation (Lambowitz et al. 1999). In cases where the host derives a selective advantage from the presence of the intron, host proteins might help to "domesticate" the RNA and put it to work in the cell. In all these cases, the intron RNA tends to maintain a functional ribozyme core and the actual chemistry of splicing and reverse-splicing is catalyzed by the RNA itself (Lambowitz and Zimmerly 2004). Recruited proteins are therefore likely to serve structural functions by assisting in the folding or stabilization of active intron structures, or by forming regulatory complexes that link splicing with other metabolic pathways. Earlier reviews on group II intron-associated proteins have focused primarily on maturase proteins, which are encoded rather than recruited by group II introns (Lambowitz and Zimmerly 2004). There is a vast literature on these fascinating proteins, which typically comprise RNA binding motifs, DNA endonuclease motifs, and reverse-transcriptase motifs that are essential for intron mobility (Belfort et al. 2001; Lambowitz and Zimmerly 2004; Matsuura et al. 2001; Pyle et al. 2007). This review will focus on the diversity of host proteins that are recruited and adapted to facilitate group II intron function. It will describe the types of proteins that are harnessed and the growing understanding of their molecular interactions and mechanistic roles in group II intron function.

8.1.2 Group II Intron Architecture and Assembly

A common secondary structure is shared by all group II introns, which are phylogenetically divided into three families (the IIA, IIB and IIC introns) (Toor et al. 2001). The various helical stems can be arranged into six domains that contain motifs important for various aspects of intron assembly and catalysis. The catalytic core and tertiary architecture of the introns are now well-defined and three-dimensional models of IIA and IIB structure have been created (Costa et al. 2000; de Lencastre et al. 2005; de Lencastre and Pyle 2008; Noah and Lambowitz 2003). Moreover, the crystal structure of a IIC intron has recently been solved, elucidating the architecture of the active site of group II introns (Toor et al. 2008). While bacterial and yeast

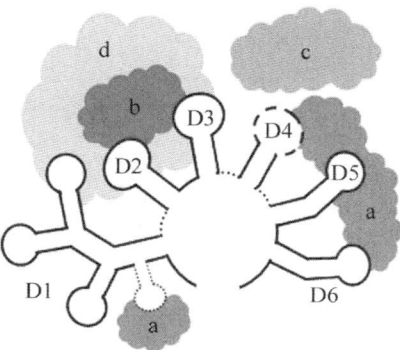

Fig. 8.1 Scheme of the secondary structure of group II introns and their interactions with recruited splicing proteins. D1–D6 represent Domains 1–6, dashed lines in D4 show the position of an optional open reading frame (maturase), dotted lines indicate positions of breaks in trans-spliced introns and dotted lines in D1 specify the position of the inserted stem-loop of the plastid atpF intron. Cloud-shaped figures represent the diverse proteins recruited for splicing: (**a**) proteins that are specific to sites within the intron (CRS1 and Cpn60), (**b**) proteins that are specific to trans-spliced introns (Rat1 and NAP-protein), (**c**) proteins that interact indirectly with the intron and (**c**) proteins that may be part of a "group II intron splicing complex"

group II introns are generally contiguous, many plant introns are split, being encoded on separate pieces of RNA (Bonen 1993, 2008). Indeed, chloroplast introns are often assembled from two and sometimes three sections of RNA that encode the requisite intron domains (Knoop et al. 1997; Perron et al. 2004). It is likely that proteins play a special role in the proper assembly of multi-piece introns. Unlike maturase proteins, which associate with specific regions of the intron (e.g., Domains 1 and 4) (Lambowitz and Zimmerly 2004), recruited proteins can also bind at diverse intronic positions or interact non-specifically with group II introns (Fig. 8.1).

8.1.3 Group II Intron Folding Pathways and Stability

Although maturases and recruited proteins play an important role in the folding pathway and structural stabilization of many group II introns (Perron et al. 2004; Watkins et al. 2007; Zimmerly et al. 1999), there is growing evidence that certain group II intron RNAs can fold autonomously through pathways that are likely to be facilitated by cellular proteins (Fedorova and Zingler 2007). Only one intron has been the subject of detailed kinetic and equilibrium folding analyses; the ai5γ group IIB intron from *S. cerevisiae* mitochondria has been characterized enzymologically, structurally, biophysically and genetically, and it represents a central model system in the study of group II introns (Fedorova and Zingler 2007). This intron, which lacks a maturase, has been shown to fold directly to the native-state through an ordered, stepwise pathway that appears to lack kinetic traps (Fedorova et al. 2007; Su et al. 2003; Swisher et al. 2002).

A diversity of biophysical approaches have demonstrated that the rate-limiting step in ai5γ folding is the slow collapse of intron Domain 1 (D1), which can fold independent of other intron domains (Pyle et al. 2007; Su et al. 2005). Once D1 has folded, catalytic Domains 3, 5 and 6 rapidly dock into respective receptor sites within the D1 scaffold, thereby completing the assembly of the intron (Fedorova et al. 2007; Pyle et al. 2007). D1 collapse is mediated by a tiny RNA junction motif that is located in the center of this extended domain (Waldsich and Pyle 2007; Waldsich and Pyle 2008). Until this folding control element adopts the correct conformation, the long-range tertiary interactions within D1 cannot form, and the intron maintains an extended conformation (Waldsich and Pyle 2007, 2008). If the ai5γ pathway is general, then the stabilization of early folding intermediates is the most important factor in promoting faithful assembly of group II introns. By extension, associated proteins are likely to play a vital role in stabilizing obligate folding intermediates and perhaps the native state itself (Fedorova et al. 2007; Pyle et al. 2007). An important caveat is that folding studies on the ai5γ intron have been conducted with ribozyme constructs that lack peripheral structures and exons, and therefore the folding may represent an idealized scenario. Nonetheless, folding studies on ai5γ ribozymes have provided new paradigms for understanding the folding of large, multidomain RNA molecules.

8.1.4   Introduction to the Protein Collaborators

For many introns, the most important protein partner is the intron-encoded maturase protein, which co-evolves with its cognate intron and should be considered an integral component of any mobile intron. However, it is becoming clear that most, if not all introns also rely on the action of host proteins that have been recruited or adapted from other metabolic functions. These latter proteins, and their cooperative interactions with the intron invader, are the focus of this review. For purposes of the forthcoming discussion we will first give an overview of the variety of proteins described so far and then focus on the example of the well characterized DEAD-box proteins.

8.2   A Kaleidoscope of Recruited Proteins in Higher Eukaryotes

Recent studies have uncovered a startling diversity of proteins that have been recruited to promote efficient splicing of group II introns. These frequently maturase-less group II introns have co-evolved with their hosts and require compensatory nuclear-encoded splicing proteins that are targeted to their respective organelles. The appropriated proteins have diverse functions, and in some cases have retained features of their progenitors. Such proteins include a pseudouridine synthetase (Perron et al. 1999), a peptidyl-tRNA hydrolase (Jenkins and Barkan 2001), a ribonuclease (Watkins et al. 2007), a heat shock chaperonin (Balczun et al. 2006), an ion transporter (Weghuber et al. 2006) and nucleosome assembly proteins (Glanz et al. 2006).

Some of these proteins can indirectly affect splicing by altering the cellular environment. For example, the yeast mitochondrial membrane Mg^{2+} transporter protein, MRS2 (*m*itochondrial *RNA* *s*plicing) regulates the concentration of magnesium in mitochondria (Weghuber et al. 2006; Wiesenberger et al. 1992). Indeed, mutants of MRS2 attenuate the efficiency of group II intron splicing (Gregan et al. 2001). A complement to this yeast Mg^{2+} influx protein was also found in plants (Schock et al. 2000). Thus proteins can have profound but indirect effects on group II introns.

Only a handful of proteins with a direct role in splicing have been described. Many have only recently been identified and implicated in group II intron function. Their mechanistic behavior is not yet fully understood, but key similarities between these diverse proteins are observed. The intrinsic characteristics of proteins that facilitate organellar group II intron splicing are that they contain RNA recognition and/or RNA binding motifs, and in some cases, they have protein–protein interaction motifs.

8.2.1 RNA Recognition Motifs of Co-opted Proteins

Proteins that facilitate group II intron splicing generally bind RNA, but they do not use a single platform for this purpose. For instance, the plant *c*hloroplast *RNA* *s*plicing *2* protein, CRS2 shares similar domains with peptidyl-tRNA hydrolases that cleave the ester bond between tRNAs and the emerging translational peptide (Jenkins and Barkan 2001). Crystallographic studies of CRS2 revealed specialized features, despite the high structural conservation with bacterial peptidyl-tRNA hydrolases (Fig. 8.2a). For instance, CRS2 has a unique hydrophobic patch that contains residues shown to be important in binding its protein cofactors (CAF1 and 2; *c*hloroplast-*a*ssociated *f*actors). Also, the basic region that is believed to interact with RNA in peptidyl-tRNA hydrolases is expanded in CRS2 and is hypothesized to associate with group II introns (Ostheimer et al. 2005) (Fig. 8.2a).

Other proteins involved in group II intron splicing contain motifs or domains found in many RNA binding proteins. One ancient RNA binding motif that is present in several plant chloroplast proteins (CRS1, CAF1 and CAF2) and which plays a role in group II intron splicing, is called the CRM domain (*c*hloroplast *RNA* splicing and ribosome *m*aturation). As the name suggests, CRMs are involved in ribosome assembly and are analogous to the YhbY protein that is found in archea and eubacteria (Barkan et al. 2007). The crystal structure of YhbY (Fig. 8.2b), containing a GxxG motif and α–β–α–β–α–β–β motif similar to the translation initiation factor (IF3C), shows a compact structure with a rich basic surface that is implicated in the binding of 16S rRNA. Analogously, this basic surface of the CRM domain could also be used to bind group II introns.

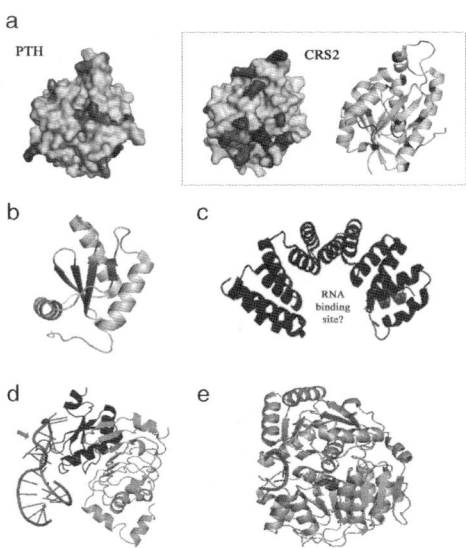

Fig. 8.2 Structures of selected proteins associated with RNA processing. (**a**) Electrostatic surface representations of peptidyl-tRNA hydrolases (PTH) and CRS2 (also in ribbon form, right). Basic residues (blue) are sites for tRNA and prospective group II intron binding, respectively (Ostheimer et al. 2005; Schmitt et al. 1997). (**b**) Crystal structure of an ancient RNA domain (YhbY) homologous with CRS1. The β sheet face (blue), containing conserved basic residues, and the GxxG motif (green), are both implicated in nucleic acid recognition (Ostheimer et al. 2002). (**c**) Model of 6 PPR motifs based on the crystal structure of a closely related TPR protein (Kim et al. 2006; Tavares-Carreon et al. 2008). (**d**) Spliceosomal RRM protein U2B" (blue) interacting with the hairpin region of U2 snRNA (red); shown with arrow. This association occurs only when U2B" is interacting with U2A', a leucine-rich protein (green) (Maris et al. 2005; Price et al. 1998). (**e**) Crystal structure of the DEAD-box protein Vasa, showing the DEAD-box motif (blue) and RNA binding sites (green) in the helicase domain (Sengoku et al. 2006). In (**d**) and (**e**), RNA strands are shown in red. Figures were adapted from the references indicated and generated using PyMol from protein data bank accession numbers 2PTH, 1RYB, 1LN4, 2FI7 and IA9N, respectively (See figure insert for color reproduction)

A novel RNA binding motif that was found to interact with organellar mitochondrial transcripts is the *p*entatrico*p*eptide *r*epeat family (PPR). These polypeptides contain tandem repeats of a 35-amino acid motif and, like TPRs (tricopeptides), are thought to form a solenoid structure that contains a hydrophilic groove for interaction with RNA (Fig. 8.2c) (Saha et al. 2007; Small and Peeters 2000). PPR proteins have been implicated in a wide range of RNA metabolic processes including translation, RNA stability and RNA editing (Kotera et al. 2005). Of the ~400 PPR proteins encoded in the *Arabidopsis* plant nucleus, two thirds are predicted to target the mitochondria or chloroplast (Geddy and Brown 2007; Lurin et al. 2004). A characteristic PPR (OTP43) was recently found to take part in the trans-splicing of a plant mitochondrial intron (de Longevialle et al. 2007). Another PPR (PPR4) participates in trans-splicing in a plant plastid, and is interesting because it also contains an *RNA r*ecognition *m*otif (RRM) (Schmitz-Linneweber et al. 2006). RRM motifs are important features of many group II-associated proteins. The RRM contains a well-defined fold and a consensus sequence comprised of aromatic and charged residues that interact with RNA. These proteins are found in both eukaryotes and prokaryotes, and in the former these RRMs are intrinsic components of the spliceosomal machinery (Fig. 8.2d). Many RRM proteins also promote protein–protein interactions (Maris et al. 2005).

8.2.2 Intron-Specific Proteins and General Splicing Factors

Proteins that are recruited for organellar splicing can be split into those that interact with specific regions of the RNA and those that are involved in a generalized splicing mechanism.

A wide range of proteins have evolved to associate with specific sites within contemporary group II introns. The aforementioned CRS1 protein binds an inserted stem-loop region that is found only in D1 of the atpF intron of plant chloroplasts (Ostersetzer et al. 2005). Other proteins interact with trans-spliced introns, such as those found in *Chlamydomonas* plastids. A unique example of a tripartite trans-splicing intron is psaA intron 1 which consists of its 5′ intronic region, a middle portion (tscA) and its 3′ intronic end on three separated transcripts (Goldschmidt-Clermont et al. 1991). This intriguing tripartite structure has often been compared to snRNPs, and has contributed to the notion that group II introns share a common ancestor with nuclear spliceosomal introns. Interestingly, two very diverse proteins, a NAP-like protein (*n*ucleosome *a*ssembly *p*rotein) and the Rat1 protein (NAD$^+$-binding domain of a poly(ADP-ribose) polymerase) bind specifically to the tscA portion of this trans-intron (Balczun et al. 2005; Glanz et al. 2006).

There is considerable evidence that plant chloroplasts contain relatively large spliceosome-like complexes that are built around group II introns rather than snRNAs. However, the behavior of these complexes is just beginning to be characterized (Watkins et al. 2007). A general plant chloroplast "group II intron splicing complex" may involve different combinations of CRS2, CAF1 and CAF2 aiding in the splicing of multiple plastid group II introns (Jenkins et al. 1997; Ostheimer et al. 2003). Interestingly, experiments suggest that CRS2-CAF complexes form splicing RNPs and that CAF1 and CAF2 confer specificity to overlapping sets of group II introns (Ostheimer et al. 2006). Recent experiments have shown that RNC1 is also involved in splicing of a subset of the introns that interact with CRS2-CAF complexes as well as some group IIA introns that are CRS2 independent. RNC1 is a ribonucleaseIII-derived protein that has lost its endonuclease activity, but has retained its capacity to bind RNA (Watkins et al. 2007). Thus we are just beginning to understand the role of proteins in large complexes involved in group II intron splicing in plant chloroplasts.

In an analogous system, the 14 nuclear genes implicated in *Chlamydomonas* intron splicing may also participate in an algal plastid "group II intron splicing complex." Among these components is a novel membrane-associated protein,

Raa1, characterized by two separate domains associated with the splicing of two distinct trans-introns (Merendino et al. 2006). Cpn60, which resembles a heat-shock chaperonin, has affinity to regions within Domains 4–6 in two different group II introns, and is thus thought to behave as a general splicing factor (Balczun et al. 2006). It is also similar to the bacterial GroEL family of chaperonins that facilitates protein folding. This emphasizes again the diversity of proteins implicated in group II intron splicing.

Overall, these co-opted proteins have elastic features that allow for a rapid co-evolution with group II introns. Although many of these introns face considerable barriers to proper folding and splicing activity, it is important to note that there is no evidence that proteins are performing a catalytic role in group II intron splicing. In each of these cases the novel RNP complexes that are being formed are not yet fully elucidated, but are likely to influence the folding and/or stabilization of intron structure, as discussed in the next section.

8.3 DEAD-Box Proteins Involved in Group II Intron Splicing

8.3.1 General Characteristics of DEAD-Box ATPase Proteins

A distinctive subgroup of ATPase proteins plays an important role in the general splicing of group II introns. These proteins, which include Mss116 from yeast and CYT-19 from *Neurospora*, are DEAD-box proteins (Cordin et al. 2006). Members of the DEAD-box family are ligand-regulated RNA-binding proteins that are involved in every aspect of RNA metabolism. These proteins cleave ATP in an RNA-stimulated manner and they are named after one of the conserved amino acid motifs that are involved in ATP hydrolysis (the DEAD sequence in Motif II) (Fig. 8.2e). Several family members have been shown to act as helicases, i.e. they separate strands of RNA duplexes in an ATP-dependent manner (Cordin et al. 2006; Pyle 2008). However, many DEAD-box proteins have no measured helicase activity in vitro. At least one member has no need for helicase activity to perform its function in vivo as it acts as a clamp (Shibuya et al. 2004). Despite a small number of well-studied examples, at this time the precise role and mechanism of most DEAD-box proteins in the cell is unclear.

8.3.2 Biological Roles of Mss116, Cyt-19, and Ded1

For Mss116, the primary biological role is well-defined. After being identified in a genetic screen in *S. cerevisiae* (Seraphin et al. 1989; Tzagoloff et al. 1975), Mss116 was shown to be important for splicing of all mitochondrial group I and group II introns in vivo (Huang et al. 2005). CYT-19 was found through a genetic screen for genes that have an effect on group I intron splicing in *N. crassa* (Bertrand et al. 1982) and can functionally replace Mss116 in vivo (Huang et al. 2005). Unlike other proteins mentioned in previous sections of this chapter, CYT-19- and Mss116-mediated splicing have also been studied in vitro: these proteins are able to promote self-splicing of introns under near-physiological conditions, requiring only ATP as a co-factor (Mohr et al. 2002, 2006; Solem et al. 2006). Another protein from the same family, Ded1, has also been shown to facilitate group II intron splicing in vitro (Halls et al. 2007; Solem et al. 2006). This *S. cerevisiae* protein has been implicated in translation, nuclear splicing, and ribosome assembly (Chuang et al. 1997). Curiously, however, as there is no evidence that Ded1 is localized to the mitochondrion, it is unlikely to encounter self-splicing introns in vivo. Therefore, this subgroup of DEAD-box proteins might not have specifically evolved for splicing, but could have a broader function in RNA folding or structural remodeling. The in vitro self-splicing system thus represents a valuable model system to study the general mechanism by which these proteins facilitate folding of large RNAs. However, despite a growing body of information on protein-facilitated splicing, the mechanisms of these proteins are still not completely elucidated. In the following sections, we will first list the general mechanisms by which a protein can promote proper folding of a large RNA molecule. Subsequently, we will discuss which of these activities these proteins might employ.

8.3.3   General Mechanisms by Which Proteins Can Facilitate RNA Folding

RNA and proteins exhibit several differences in their folding behavior. Since most proteins are able to fold to their native conformation, protein chaperones are mainly required to prevent aggregation and protect nascent strands from misfolding (Hartl and Hayer-Hartl 2002). RNA, on the other hand, can form many alternative stable structures and base pairings, and therefore has a propensity to form misfolded species (kinetic traps) (Sosnick and Pan 2003). These misfolded conformations can be almost as stable as the native state and resolve very slowly.

Proteins can use several mechanisms to promote folding of large RNAs. First, a protein could facilitate collapse of a large RNA by relieving unfavorable charge–charge interactions between sections of RNA backbone that occur when the RNA compacts (Fig. 8.3a). Next, a protein could prevent misfolded structures from forming, or it could resolve misfolded structures by recognizing and unfolding misfolded RNAs. Alternatively, the protein could non-specifically bind and unfold both native and misfolded RNAs, thus allowing the RNA a new chance to fold. These mechanisms are considered chaperone activities (Fig. 8.3b) (Herschlag 1995). Interestingly most protein chaperones are ATP-dependent (Hartl and Hayer-Hartl 2002), while the majority of identified RNA chaperones are ATP-independent (Russell 2008). Finally, the protein could also stabilize the correctly folded RNA in two different ways. It could bind the substructure of a large RNA to form a platform for proper RNA folding (scaffolding (Chen et al. 2000)) (Fig. 8.3c) or bind and stabilize the correctly folded form of a large RNA (tertiary-structure capture

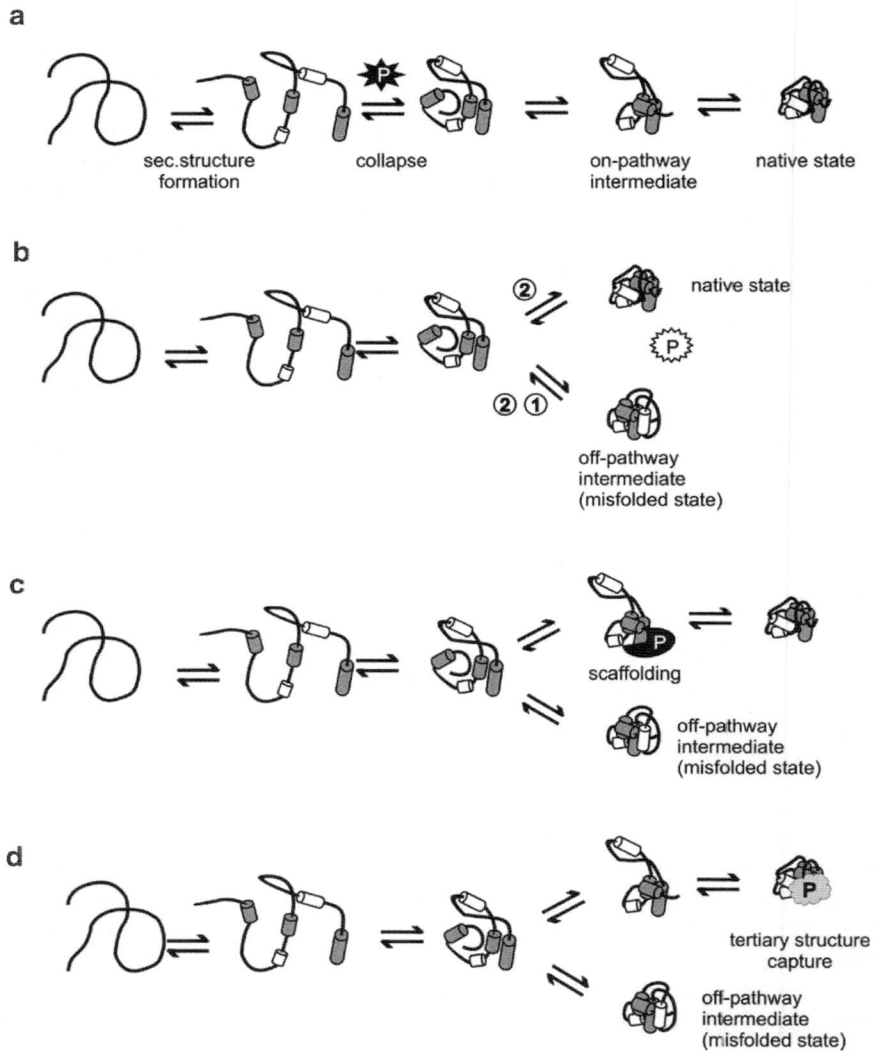

Fig. 8.3 Protein-facilitated RNA folding. (**a**) A protein (P) can facilitate collapse of the RNA. (**b**) A protein (P) can promote formation of the native state by either specifically recognizing and unfolding a misfolded state (1) or disrupting both misfolded and native states of the RNA (2) and allowing the RNA a new chance to fold. (**c**) A protein (P) can promote formation of the native state by stabilizing an important substructure of the RNA. (**d**) Alternatively, a protein (P) can recognize and stabilize the native state

(Weeks and Cech 1996)) (Fig. 8.3d). Thus there are at least three major ways by which proper folding of an RNA can be promoted: by facilitating collapse, preventing or resolving misfolded structures, or by stabilizing correctly folded structures.

8.3.4 Possible Mechanisms by Which DEAD-Box Proteins Promote Splicing

Currently there are several plausible ways in which DEAD-box proteins could be acting on the group II intron ai5γ. The first mechanism proposed was resolution of misfolded structures (kinetic traps) through helicase activity (Seraphin et al. 1989). This is a straightforward explanation, but it is somewhat controversial. For instance, experiments studying folding of the intron core do not indicate the presence of kinetic traps (Fedorova et al. 2007; Su et al. 2003; Swisher et al. 2002), although they could exist in the context of the full-length intron. Also, it has not been conclusively proven that helicase activity is indeed responsible for the splicing activity (Del Campo et al. 2007). It is clear, however, that ATP is required for the protein to facilitate splicing of the intron RNA (Halls et al. 2007; Mohr et al. 2006; Solem et al. 2006). A chaperone function of those proteins is supported by a recent paper from the Russell lab on the action of CYT-19 on the well-characterized *Tetrahymena* group I intron (Bhaskaran and Russell 2007). It showed that CYT-19 partially unfolds both native and misfolded forms of the RNA and allows the intermediates a new chance to refold. The protein does not specifically recognize misfolded RNAs, but has more difficulty in unfolding more stable RNA structures. Therefore CYT-19 is able to push the equilibrium towards the native state if this state is sufficiently stable compared to the misfolded state. As the Russell lab states in their paper, it is logical that a protein would have difficulty distinguishing between native and misfolded RNAs which have similar structures and electrostatics. Most protein chaperones recognize nascent and misfolded proteins by exposed hydrophobic patches (Hartl and Hayer-Hartl 2002); a similar general hallmark of misfolded RNAs does not exist.

Despite the enticing simplicity of the chaperone theory, it is important to keep in mind that other mechanisms are also possible. It has also been proposed that DEAD-box proteins could stabilize a correctly folded structure (Del Campo et al. 2007; Solem et al. 2006). As Mss116 and CYT-19 act on many different RNAs (Huang et al. 2005), it is unlikely that they are specifically recognizing one intron; however, it is possible that this class of proteins binds and stabilizes a common structural feature of RNA such as stacked helices or a stem-loop and thus appears to act non-specifically. As ATP is required for splicing, it is possible that ATP binding or hydrolysis causes a necessary conformational change in the protein, as in the case of the protein chaperone Hsp70 (Hartl and Hayer-Hartl 2002). Alternatively, ATP hydrolysis could act as a switch to alter RNA-binding affinity, for instance, allowing the protein to bind an RNA substructure and then release it to allow other regions of the RNA to fold in that area.

Most attempts to understand the mechanism of these proteins have focused on the RNA binding and helicase activities conferred by the conserved helicase domains. However, Mss116, CYT-19 and Ded1 all have non-conserved C-terminal domains that contain an abundance of positively charged residues (Halls et al. 2007; Solem et al. 2006). For CYT-19, experiments suggest that the C-terminus contributes to

RNA binding (Grohman et al. 2007). In fact, the positive residues in the C-terminus may contribute to RNA binding in all of these proteins. In addition, it is possible that these charged residues could facilitate collapse of large RNAs by relieving unfavorable charge–charge interactions.

It is interesting to note that when activities of Mss116 and CYT-19 are compared in different assay systems, very different results can be observed. For example, CYT-19 promotes reverse splicing of the group II intron bI1 in vitro, but Mss116 does not (Halls et al. 2007; Mohr et al. 2006). In contrast, Mss116 is able to stimulate maturase-dependent splicing of a group I intron in the presence of ADP or ATP, while CYT-19 can only perform this function with ATP (Halls et al. 2007; Mohr et al. 2006). This indicates that these proteins are not just mono-functional, but probably use diverse mechanisms. By fine-tuning the contribution of each mechanism to the overall activity, the members of the DEAD-box family may be able to adapt to new niches in the biology of the cell.

8.4 Conclusion

The wide range of host proteins that are implicated in group II intron splicing can employ diverse mechanistic strategies for facilitation of intron function. Future work will expand our understanding of how these proteins are indeed allowing group II introns to fold and splice efficiently in vivo. Studies on large group II intron splicing complexes may even reveal new insights into the evolution of spliceosomal splicing. At this time we can only appreciate the intricate co-evolution between host and organellar systems and the opportunistic nature of the introns to recruit host proteins that allow them to persist in dynamic cellular environments.

References

Balczun C, Bunse A, Schwarz C, Piotrowski M, Kück U (2006) Chloroplast heat shock protein Cpn60 from *Chlamydomonas reinhardtii* exhibits a novel function as a group II intron-specific RNA-binding protein. FEBS Lett 580:4527–4532

Balczun C, Bunse A, Hahn D, Bennoun P, Nickelsen J, Kück U (2005) Two adjacent nuclear genes are required for functional complementation of a chloroplast trans-splicing mutant from *Chlamydomonas reinhardtii*. Plant J 43:636–648

Barkan A, Klipcan L, Ostersetzer O, Kawamura T, Asakura Y, Watkins KP (2007) The CRM domain: an RNA binding module derived from an ancient ribosome-associated protein. RNA 13:55–64

Belfort M, Derbyshire V, Parker MM, Cousineau B, Lambowitz AM (2001) Mobile introns: pathways and proteins. In: NL Craig, R Gragie, M Gellert, AM Lambowitz (eds.) Mobile DNA II. ASM Press, Washington, DC, pp. 761–782

Bertrand H, Bridge P, Collins RA, Garriga G, Lambowitz AM (1982) RNA splicing in Neurospora mitochondria. Characterization of new nuclear mutants with defects in splicing the mitochondrial large rRNA. Cell 29:517–526

Bhaskaran H, Russell R (2007) Kinetic redistribution of native and misfolded RNAs by a DEAD-box chaperone. Nature 449:1014–1018

Bonen L (1993) Trans-splicing of pre-mRNA in plants, animals, and protists. FASEB J 7:40–46

Bonen L (2008) Cis- and trans-splicing of group II introns in plant mitochondria. Mitochondrion 8:26–34

Bonen L, Vogel J (2001) The ins and outs of group II introns. Trends Genet 17:322–331

Chen X, Gutell RR, Lambowitz AM (2000) Function of tyrosyl-tRNA synthetase in splicing group I introns: an induced-fit model for binding to the P4–P6 domain based on analysis of mutations at the junction of the P4-P6 stacked helices. J Mol Biol 301:265–283

Chuang RY, Weaver PL, Liu Z, Chang TH (1997) Requirement of the DEAD-Box protein Ded1p for messenger RNA translation. Science 275:1468–1471

Cordin O, Banroques J, Tanner NK, Linder P (2006) The DEAD-box protein family of RNA helicases. Gene 367:17–37

Costa M, Michel F, Westhof E (2000) A three-dimensional perspective on exon binding by a group II self-splicing intron. EMBO J 19:5007–5018

de Lencastre A, Pyle AM (2008) Three essential and conserved regions of the group II intron are proximal to the 5′-splice site. RNA 14:11–24

de Lencastre A, Hamill S, Pyle AM (2005) A single active-site region for a group II intron. Nat Struct Mol Biol 12:626–627

de Longevialle AF, Meyer EH, Andres C, Taylor NL, Lurin C, Millar AH, Small ID (2007) The pentatricopeptide repeat gene OTP43 is required for trans-splicing of the mitochondrial nad1 Intron 1 in *Arabidopsis thaliana*. Plant Cell 19:3256–3265

Del Campo M, Tijerina P, Bhaskaran H, Mohr S, Yang Q, Jankowsky E, Russell R, Lambowitz AM (2007) Do DEAD-box proteins promote group II intron splicing without unwinding RNA? Mol Cell 28:159–166

Fedorova O, Zingler N (2007) Group II introns: structure, folding and splicing mechanism. Biol Chem 388:665–678

Fedorova O, Waldsich C, Pyle AM (2007) Group II intron folding under near-physiological conditions: collapsing to the near-native state. J Mol Biol 366:1099–1114

Ferat JL, Michel F (1993) Group II self-splicing introns in bacteria. Nature 364:358–361

Geddy R, Brown GG (2007) Genes encoding pentatricopeptide repeat (PPR) proteins are not conserved in location in plant genomes and may be subject to diversifying selection. BMC Genomics 8:130

Glanz S, Bunse A, Wimbert A, Balczun C, Kück U (2006) A nucleosome assembly protein-like polypeptide binds to chloroplast group II intron RNA in *Chlamydomonas reinhardtii*. Nucleic Acids Res 34:5337–5351

Goldschmidt-Clermont M, Choquet Y, Girard-Bascou J, Michel F, Schirmer-Rahire M, Rochaix JD (1991) A small chloroplast RNA may be required for trans-splicing in *Chlamydomonas reinhardtii*. Cell 65:135–143

Gregan J, Kolisek M, Schweyen RJ (2001) Mitochondrial Mg2+ homeostasis is critical for group II intron splicing in vivo. Genes Dev 15:2229–2237

Grohman JK, Del Campo M, Bhaskaran H, Tijerina P, Lambowitz AM, Russell R (2007) Probing the mechanisms of DEAD-box proteins as general RNA chaperones: the C-terminal domain of CYT-19 mediates general recognition of RNA. Biochemistry 46:3013–3022

Halls C, Mohr S, Del Campo M, Yang Q, Jankowsky E, Lambowitz AM (2007) Involvement of DEAD-box proteins in group I and group II intron splicing. Biochemical characterization of Mss116p, ATP hydrolysis-dependent and -independent mechanisms, and general RNA chaperone activity. J Mol Biol 365:835–855

Hartl FU, Hayer-Hartl M (2002) Molecular chaperones in the cytosol: from nascent chain to folded protein. Science 295:1852–1858

Herschlag D (1995) RNA chaperones and the RNA folding problem. J Biol Chem 270: 20871–20874

Huang HR, Rowe CE, Mohr S, Jiang Y, Lambowitz AM, Perlman PS (2005) The splicing of yeast mitochondrial group I and group II introns requires a DEAD-box protein with RNA chaperone function. Proc Natl Acad Sci U S A 102:163–168

Jenkins BD, Barkan A (2001) Recruitment of a peptidyl-tRNA hydrolase as a facilitator of group II intron splicing in chloroplasts. EMBO J 20:872–879

Jenkins BD, Kulhanek DJ, Barkan A (1997) Nuclear mutations that block group II RNA splicing in maize chloroplasts reveal several intron classes with distinct requirements for splicing factors. Plant Cell 9:283–296

Kim K, Oh J, Han D, Kim EE, Lee B, Kim Y (2006) Crystal structure of PilF: functional implication in the type 4 pilus biogenesis in *Pseudomonas aeruginosa*. Biochem Biophys Res Commun 340:1028–1038

Knoop V, Altwasser M, Brennicke A (1997) A tripartite group II intron in mitochondria of an angiosperm plant. Mol Gen Genet 255:269–276

Kotera E, Tasaka M, Shikanai T (2005) A pentatricopeptide repeat protein is essential for RNA editing in chloroplasts. Nature 433:326–330

Lambowitz AM, Zimmerly S (2004) Mobile group II introns. Annu Rev Genet 38:1–35

Lambowitz AM, Caprara MG, Zimmerly S, Perlman PS (1999) Group I and group II ribozymes as RNPs: clues to the past and guides to the future. In: RF Gesteland, TR Cech, JF Atkins (eds.) The RNA World. Cold Spring Harbor Laboratory Press, Cold Spring Harbor, New York, pp. 451–485

Lehmann K, Schmidt U (2003) Group II introns: structure and catalytic versatility of large natural ribozymes. Crit Rev Biochem Mol Biol 38:249–303

Lurin C, Andres C, Aubourg S, Bellaoui M, Bitton F, Bruyere C, Caboche M, Debast C, Gualberto J, Hoffmann B, Lecharny A, Le Ret M, Martin-Magniette ML, Mireau H, Peeters N, Renou JP, Szurek B, Taconnat L, Small I (2004) Genome-wide analysis of *Arabidopsis* pentatricopeptide repeat proteins reveals their essential role in organelle biogenesis. Plant Cell 16: 2089–2103

Maris C, Dominguez C, Allain FH (2005) The RNA recognition motif, a plastic RNA-binding platform to regulate post-transcriptional gene expression. FEBS J 272:2118–2131

Martin W, Koonin EV (2006) Introns and the origin of nucleus-cytosol compartmentalization. Nature 440:41–45

Martinez-Abarca F, Toro N (2000) Group II introns in the bacterial world. Mol Microbiol 38: 917–926

Matsuura M, Noah JW, Lambowitz AM (2001) Mechanism of maturase-promoted group II intron splicing. EMBO J 20:7259–7270

Mattick JS (1994) Introns: evolution and function. Curr Opin Genet Dev 4:823–831

Merendino L, Perron K, Rahire M, Howald I, Rochaix JD, Goldschmidt-Clermont M (2006) A novel multifunctional factor involved in trans-splicing of chloroplast introns in *Chlamydomonas*. Nucleic Acids Res 34:262–274

Michel F, Ferat JL (1995) Structure and activities of group II introns. Annu Rev Biochem 64:435–461

Mohr S, Stryker JM, Lambowitz AM (2002) A DEAD-box protein functions as an ATP-dependent RNA chaperone in group I intron splicing. Cell 109:769–779

Mohr S, Matsuura M, Perlman PS, Lambowitz AM (2006) A DEAD-box protein alone promotes group II intron splicing and reverse splicing by acting as an RNA chaperone. Proc Natl Acad Sci U S A 103:3569–3574

Noah JW, Lambowitz AM (2003) Effects of maturase binding and Mg^{2-} concentration on group II intron RNA folding investigated by UV cross-linking. Biochemistry 42:12466–12480

Ostersetzer O, Cooke AM, Watkins KP, Barkan A (2005) CRS1, a chloroplast group II intron splicing factor, promotes intron folding through specific interactions with two intron domains. Plant Cell 17:241–255

Ostheimer GJ, Barkan A, Matthews BW (2002) Crystal structure of *E. coli* YhbY: a representative of a novel class of RNA binding proteins. Structure 10:1593–1601

Ostheimer GJ, Williams-Carrier R, Belcher S, Osborne E, Gierke J, Barkan A (2003) Group II intron splicing factors derived by diversification of an ancient RNA-binding domain. EMBO J 22:3919–3929

Ostheimer GJ, Hadjivassiliou H, Kloer DP, Barkan A, Matthews BW (2005) Structural analysis of the group II intron splicing factor CRS2 yields insights into its protein and RNA interaction surfaces. J Mol Biol 345:51–68

Ostheimer GJ, Rojas M, Hadjivassiliou H, Barkan A (2006) Formation of the CRS2-CAF2 group II intron splicing complex is mediated by a 22-amino acid motif in the COOH-terminal region of CAF2. J Biol Chem 281:4732–4738

Perron K, Goldschmidt-Clermont M, Rochaix JD (1999) A factor related to pseudouridine synthases is required for chloroplast group II intron trans-splicing in *Chlamydomonas reinhardtii*. EMBO J 18:6481–6490

Perron K, Goldschmidt-Clermont M, Rochaix JD (2004) A multiprotein complex involved in chloroplast group II intron splicing. RNA 10:704–711

Price SR, Evans PR, Nagai K (1998) Crystal structure of the spliceosomal U2B″-U2A′ protein complex bound to a fragment of U2 small nuclear RNA. Nature 394:645–650

Pyle AM (2008) Translocation and unwinding mechanisms of RNA and DNA helicases. Annu Rev Biophys 37:317–336

Pyle AM, Lambowitz AM (2006) Group II introns: ribozymes that splice RNA and invade DNA. In: RF Gesteland, TR Cech, JF Atkins (eds.) The RNA World. Cold Spring Harbor Laboratory Press, Cold Spring Harbor, New York, pp. 469–506

Pyle AM, Fedorova O, Waldsich C (2007) Folding of group II introns: a model system for large, multidomain RNAs? Trends Biochem Sci 32:138–145

Russell R (2008) RNA misfolding and the action of chaperones. Front Biosci 13:1–20

Saha D, Prasad AM, Srinivasan R (2007) Pentatricopeptide repeat proteins and their emerging roles in plants. Plant Physiol Biochem 45:521–534

Schmitt E, Mechulam Y, Fromant M, Plateau P, Blanquet S (1997) Crystal structure at 1.2 A resolution and active site mapping of *Escherichia coli* peptidyl-tRNA hydrolase. EMBO J 16:4760–4769

Schmitz-Linneweber C, Williams-Carrier RE, Williams-Voelker PM, Kroeger TS, Vichas A, Barkan A (2006) A pentatricopeptide repeat protein facilitates the trans-splicing of the maize chloroplast rps12 pre-mRNA. Plant Cell 18:2650–2663

Schock I, Gregan J, Steinhauser S, Schweyen R, Brennicke A, Knoop V (2000) A member of a novel *Arabidopsis thaliana* gene family of candidate Mg^{2+} ion transporters complements a yeast mitochondrial group II intron-splicing mutant. Plant J 24:489–501

Sengoku T, Nureki O, Nakamura A, Kobayashi S, Yokoyama S (2006) Structural basis for RNA unwinding by the DEAD-box protein *Drosophila* Vasa. Cell 125:287–300

Seraphin B, Simon M, Boulet A, Faye G (1989) Mitochondrial splicing requires a protein from a novel helicase family. Nature 337:84–87

Shibuya T, Tange TO, Sonenberg N, Moore MJ (2004) eIF4AIII binds spliced mRNA in the exon junction complex and is essential for nonsense-mediated decay. Nat Struct Mol Biol 11: 346–351

Small ID, Peeters N (2000) The PPR motif – a TPR-related motif prevalent in plant organellar proteins. Trends Biochem Sci 25:46–47

Solem A, Zingler N, Pyle AM (2006) A DEAD protein that activates intron self-splicing without unwinding RNA. Mol Cell 24:611–617

Sosnick TR, Pan T (2003) RNA folding: models and perspectives. Curr Opin Struct Biol 13: 309–316

Su LJ, Brenowitz M, Pyle AM (2003) An alternative route for the folding of large RNAs: apparent two-state folding by a group II intron ribozyme. J Mol Biol 334:639–652

Su LJ, Waldsich C, Pyle AM (2005) An obligate intermediate along the slow folding pathway of a group II intron ribozyme. Nucleic Acids Res 33:6674–6687

Swisher JF, Su LJ, Brenowitz M, Anderson VE, Pyle AM (2002) Productive folding to the native state by a group II intron ribozyme. J Mol Biol 315:297–310

Tavares-Carreon F, Camacho-Villasana Y, Zamudio-Ochoa A, Shingu-Vazquez M, Torres-Larios A, Perez-Martinez X (2008) The pentatricopeptide repeats present in Pet309 are necessary for translation but not for stability of the mitochondrial COX1 mRNA in yeast. J Biol Chem 283:1472–1479

Toor N, Hausner G, Zimmerly S (2001) Coevolution of group II intron RNA structures with their intron-encoded reverse transcriptases. RNA 7:1142–1152

Toor N, Keating KS, Taylor SD, Pyle AM (2008) Crystal structure of a self-spliced group II intron. Science 320:77–82

Tzagoloff A, Akai A, Needleman RB (1975) Assembly of the mitochondrial membrane system. Characterization of nuclear mutants of *Saccharomyces cerevisiae* with defects in mitochondrial ATPase and respiratory enzymes. J Biol Chem 250:8228–8235

Waldsich C, Pyle AM (2007) A folding control element for tertiary collapse of a group II intron ribozyme. Nat Struct Mol Biol 14:37–44

Waldsich C, Pyle AM (2008) A kinetic intermediate that regulates proper folding of a group II intron RNA. J Mol Biol 375:572–580

Watkins KP, Kroeger TS, Cooke AM, Williams-Carrier RE, Friso G, Belcher SE, van Wijk KJ, Barkan A (2007) A ribonuclease III domain protein functions in group II intron splicing in maize chloroplasts. Plant Cell 19:2606–2623

Weeks KM, Cech TR (1996) Assembly of a ribonucleoprotein catalyst by tertiary structure capture. Science 271:345–348

Weghuber J, Dieterich F, Froschauer EM, Svidova S, Schweyen RJ (2006) Mutational analysis of functional domains in Mrs2p, the mitochondrial Mg^{2+} channel protein of *Saccharomyces cerevisiae*. FEBS J 273:1198–1209

Wiesenberger G, Waldherr M, Schweyen RJ (1992) The nuclear gene MRS2 is essential for the excision of group II introns from yeast mitochondrial transcripts in vivo. J Biol Chem 267:6963–6969

Zimmerly S, Moran JV, Perlman PS, Lambowitz AM (1999) Group II intron reverse transcriptase in yeast mitochondria. Stabilization and regulation of reverse transcriptase activity by the intron RNA. J Mol Biol 289:473–490

Chapter 9
Understanding the Role of Metal Ions in RNA Folding and Function: Lessons from RNase P, a Ribonucleoprotein Enzyme

Michael E. Harris(✉) and Eric L. Christian

Abstract There is a large and rapidly growing literature relating RNA function to metal ion identity and concentration; however, due to the complexity and large number of interactions it remains a significant experimental challenge to tie the interactions of individual ions to specific aspects of RNA function. Investigation of the ribonculeoprotein enzyme RNase P function has assisted in defining characteristics of RNA–metal ion interactions and provided a useful model system for understanding RNA catalysis and ribonucleoprotein assembly. The goal of this chapter is to review progress in understanding the physical basis of functional metal ion interactions with P RNA and relate this progress to the development of our understanding of RNA metal ion interactions in general. The research results reviewed here encompass: (1) Determination of the contribution of divalent metal ion binding to specific aspects of enzyme function, (2) Identification of individual metal ion binding sites in P RNA and their contribution to function, and (3) The effect of protein binding on RNA–metal ion affinity.

9.1   Introduction: The Fundamental Nature of RNA–Metal Ion Interactions

RNAs must bind a multitude of positively charged ions in order to function. Understanding how strongly ions bind, and quantifying their contribution to biological function is therefore of intense interest (e.g. Draper et al. 2005; Sigel and Pyle 2007). Currently, there is a large and rapidly growing body of experimental literature relating RNA function to metal ion identity and concentration (e.g. Bai et al. 2007; Draper 2004; Pyle 2002). High resolution structures of RNA–metal ion complexes are now abundant (see e.g. Stefan et al. 2006)[1]. However, due to the

---

M.E. Harris

Center for RNA Molecular Biology, Department of Biochemistry, CWRU - School of Medicine, Cleveland, OH 44106, USA

e-mail: meh2@cwru.edu

[1] http://merna.lbl.gov/

N.G. Walter et al. (eds.) *Non-Protein Coding RNAs*

doi: 10.1007/978-3-540-70840-7_9, © Springer-Verlag Berlin Heidelberg 2009

complexity and large number of interactions it remains a significant experimental challenge to relate the interactions of individual ions with specific aspects of RNA function. The limitations and difficulties in application of current biochemical approaches are significant barriers to achieving a complete understanding of RNA–metal ion interactions, necessitating the development of new methods to characterize the atomic contacts that underlie ion association in solution.

Detailed analysis of a growing number of model systems provide insight into the modes of ion interactions with RNA as developed in several recent reviews on RNA–metal ion interactions and RNA catalysis (Fedor and Williamson 2005; Sigel and Pyle 2007). There is structural, biochemical and biophysical evidence that divalent metal ions bind site-specifically, and associate electro-statically with the negatively charged phosphodiester backbone (Draper et al. 2005; Misra and Draper 2002; Rueda et al. 2003; Stahley et al. 2007; Stahley and Strobel 2005; Wilson and Lilley 2002). Site-binding typically occurs at internal helix bulges and complex helix junctions where it can assist in organizing non-Watson–Crick structure important for function. For ribozymes that catalyze phosphoryl transfer, active site ions bound to the reactive phosphate and nucleophile can provide significant catalysis (Anderson et al. 2006; DeRose 2003; Lonnberg and Lonnberg 2005; Sigel and Pyle 2007). As ion binding often results in formation of both local and global structure, the interactions of individual ions can be strongly coupled. Thus, interactions distant from the active site nonetheless contribute significantly to catalysis by organizing catalytic centers, or positioning the substrate. Finally, most RNAs operate as ribonucleoproteins and the binding of specific proteins can significantly modulate the apparent affinity of metal ions that bind to the functional RNA (Batey and Williamson 1998; Buck et al. 2005a; Caprara et al. 2001).

Investigation of the function of the ribonculeoprotein enzyme RNase P has assisted in defining many of these characteristics of RNA–metal ion interactions and has provided a useful model system for understanding RNA catalysis and ribonucleoprotein assembly (Altman 2007; Kazantsev and Pace 2006; Smith et al. 2007). The goal of this chapter is to review progress in understanding the physical basis of functional metal ion interactions with P RNA and to relate this progress to the development of our understanding of RNA metal ion interactions in general. In the following sections we will review our research results and that of others as they relate to: (1) Determination of the contribution of divalent metal ion binding to specific aspects of enzyme function, (2) Identification of individual metal ion binding sites in P RNA and their contribution to function, and (3) The effect of C5 protein on RNA–metal ion affinity.

RNase P was originally identified by Altman and Robertson as an essential enzyme in tRNA processing responsible for generating the mature 5' end of tRNA via endo-nucleolytic cleavage of pre-tRNAs (the P stands for '*Processing*') (Fig. 9.1) (Robertson et al. 1972; Stark et al. 1978). Ultimately, it was revealed that RNase P is a ribonucleoprotein enzyme composed of a large (*ca.* 400 nucleotide) RNA subunit (P RNA) and a smaller protein subunit (100 amino acids) termed C5 in *E. coli* (Gardiner and Pace 1980; Kole et al. 1980). The P RNA is a ribozyme able to bind

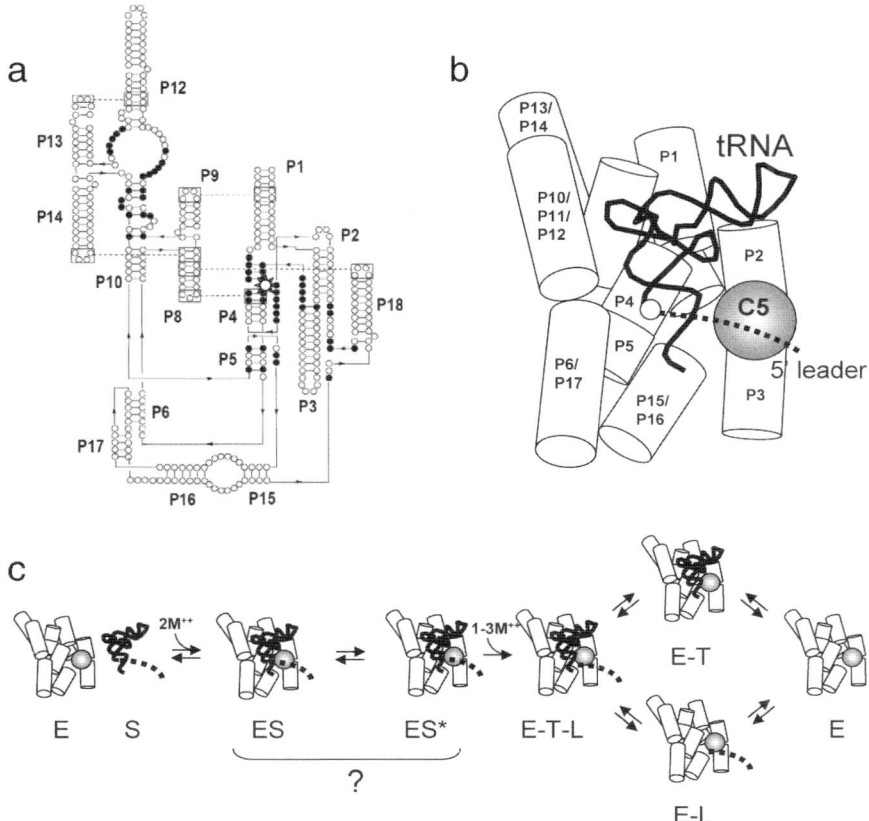

Fig. 9.1 (**a**) Secondary structure of P RNA from *E. coli*. Nucleotide positions are depicted as dots. The secondary structure is arranged according to the three dimensional structure derived from solution probing and crystallographic data. Base pairs are indicated by lines, connections between adjacent nucleotides are shown as arrows in cases where such connections are disrupted by the two dimensional projection. The helices are given the designation P for paired regions and are numbered from the 5′ end (P1, P2, P3, *etc.*). Loop-helix interactions are depicted by boxes connected by dashed lines. The locations of the U69Δ mutation is indicated by a star (see text and Fig. 9.3). (**b**) Cartoon depiction of the three dimensional structure of the RNase P-pre-tRNA complex. Helical elements of P RNA are shown as cylinders of proportional length and labeled according to part A. The tRNA phosphodiester backbone is shown as a black ribbon and the 5′ leader sequence is shown as a dashed line. The location of the cleavage site on pre-tRNA is indicated by a white sphere. The C5 protein is depicted as a gray sphere that is proportional to its size relative to P RNA. (**c**) Minimal kinetic scheme for the cleavage or pre-tRNA by the RNase P holoenzyme. The RNA and protein subunits of RNase P (E) and pre-tRNA (S) are depicted as in part B. The scheme shows formation of an initial ES complex that is proposed to isomerize to form a catalytically competent complex (ES*). Catalysis yields the enzyme product complex (E–T–L) from which the leader (L) and tRNA (T) dissociate to regenerate the free enzyme. As described in the text, metal ions in addition to those required for folding are thought to bind to form the ES complex and the transition state

to pre-tRNA, and catalyze phosphodiester hydrolysis at the correct position (Guerrier-Takada and Altman 1984a, b). In contrast to the catalytic role of the RNA subunit, the RNase P protein subunit acts primarily to enhance substrate binding affinity and stabilize P RNA structure (Buck et al. 2005a; Crary et al. 1998; Sun et al. 2006). The essential role of metal ion interactions for folding, substrate recognition and catalysis by RNase P was recognized early on, and to date there has been significant effort in identifying functionally important sites of metal ion binding by both biochemical and structural means. Several key sites of metal ion association are known and recent results provide information on how their association is likely to contribute to enzyme function (Harris and Christian 2003; Kazantsev and Pace 2006). Important progress has also been made in characterizing active site metal ion interactions in P RNA as well as the role of the RNase P protein subunit in modulating metal ion affinity (Hsieh et al. 2004; Smith et al. 2007).

9.2 Parsing Out the Metal Ion Requirements of P RNA for Folding, Substrate Binding and Catalysis

As indicated above, biophysical and theoretical studies of RNA–metal ion interactions lead to the perspective that cations such as magnesium bind to large RNAs like P RNA in two chief modes- referred to by Draper as *site bound* and *diffuse* interactions (Draper 2004; Draper et al. 2005). Association of magnesium by site-specific binding involves the outer hydration sphere of the ion making hydrogen bonding contacts to RNA functional groups. In this mode, one or more of the metal ion bound waters can be displaced and the site-bound magnesium can make direct (inner-sphere) contact with an RNA functional group. Metal ions typically form contacts with non-bridging phosphate oxygen due to the presence of lone-pair electrons and their relative chemical hardness. In contrast, diffuse metal ion association occurs via weak electrostatic interactions of ions with the negatively charged phosphodiester backbone. These weak interactions nonetheless make a significant thermodynamic contribution to RNA structure stability, and therefore biological function due to the large number of ions bound. Diffuse divalent metal ion interactions make a large contribution to folding of P RNA (Baird et al. 2007). Higher concentrations of monovalent ions increase substrate binding affinity to the P RNA subunit alone, it is thought, due to the reduction of electrostatic repulsion between ribozyme and substrate RNAs. Interestingly, cross linking between the pre-tRNA substrate and P RNA can be detected at high monovalent ion concentrations in the absence of added divalent metal ions (Smith et al. 1992). Nonetheless, the interactions of site bound divalent metals are clearly required for folding of P RNA into its native, functional geometry, for high affinity substrate binding, and current models of catalytic mechanism evoke direct coordination with the reactive substrate phosphate (e.g. see Kazantsev and Pace 2006).

The metal ion requirement for folding of RNase P has been extensively studied by Pan and Soznik and expertly reviewed elsewhere (Baird et al. 2007). Briefly, like most large RNAs, the folding of RNase P occurs in two phases, with an initial but nonspecific packing of RNA at low ionic strength (approximately 0.1 M) where predominantly diffuse metal ion interactions in the form of an "ion cloud" act to screen electrostatic repulsion of the negatively charged phosphodiester backbone. The second phase which occurs at higher ionic strength is characterized by the formation of increasingly ordered RNA structure and a small number of specific divalent metal ion interactions. Magnesium dependent folding of P RNA is cooperative with a Hill coefficient of about 3 and a transition midpoint of 2–3 mM in the presence of 0.1 M monovalent ion. Additional ions that support substrate binding and catalysis have apparent dissociation constants in the 10–50 mM range (see below). Thus, like most large RNAs, incubation with millimolar concentrations of divalent metal ions at near physiological ionic strength, permits folding to the native state. Nonetheless, optimal substrate binding affinity and catalytic rate require the binding of additional divalent ions.

Indeed, early studies of RNase P by Pace and by Altman recognized the dominant influence of divalent metal ion concentration and identity, on P RNA binding and catalysis, and demonstrated the ability of the RNase P protein subunit to alter metal ion requirements for function (Gardiner et al. 1985; Guerrier-Takada et al. 1986). Initial reports using multiple turnover assays with radio-labeled substrates showed that P RNA alone is active in high (M) monovalent ion concentrations as long as divalent ions were present. Binding of the RNase P protein subunit resulted in higher activity at lower (0.1 M) monovalent ion concentrations, and also decreased the concentrations of magnesium ions required for optimal activity.

An appreciation of the contributions of magnesium ion binding to individual steps in the reaction pathway necessarily progressed alongside the development of an increasingly detailed understanding of the kinetic mechanism of the P RNA reaction (e.g. see Kurz and Fierke 2000, 2002; Sun and Harris 2007). Kinetic, thermodynamic and structure-function studies to date support a minimal kinetic scheme in which the enzyme binds conserved tRNA residues as summarized in Fig. 9.1c. The affinity of the holoenzyme (E) for pre-tRNA (S) at 0.1 M ionic strength and 5–10 mM $MgCl_2$ is high with the observed dissociation constant (K_d) in the nano molar range. Formation of contacts at the cleavage site is likely to be linked to a conformational change (ES $\rightarrow$ ES*) that is necessary for catalysis. Substrate cleavage by phosphodiester bond hydrolysis is then catalyzed by the P RNA active site. Subsequent product dissociation, where it has been examined, appears ordered with the dissociation of the leader (L in Fig. 9.1c) preceding that of the mature tRNA (T).

Evidence of a two step binding mechanism arises from the observation of non-additivity in the effect of mutations that disrupt interactions between P RNA and conserved tRNA nucleotides adjacent to the cleavage site that are required for correct cleavage (Loria and Pan 1998; Zahler et al. 2005). These interactions include recognition of nucleotides R(73)C(74)C(75) at the tRNA 3' end and nucleotide U(−1) in the pre-tRNA leader which interact with P15 and J5/15 elements of P RNA, respectively (Fig. 9.2). The conserved G(+1)-C(72) base pair at the base of the acceptor stem is also an important recognition element; however, the precise

Fig. 9.2 Substrate structures recognized by RNase P. The conserved sequences flanking the RNase P cleavage site are shown in the context of the three-dimensional structure model (*top*) and in secondary structure projection (*below*). Nucleotides contacts on pre-tRNA that are discussed in the text are shown as capital letters. The location of the RNase P cleavage site is shown by an arrow. The P RNA and C5 protein subunits are depicted as in Fig. 9.1

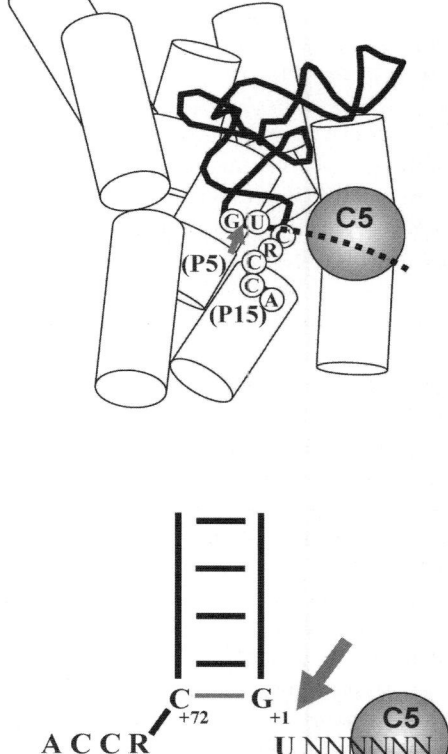

interaction with P RNA is not yet known (e.g. Kikovska et al. 2005, 2006). Additional interactions include H-bonding between 2′ hydroxyl groups (2′OH) in the T-stem and conserved adenosines in J10/11 as well as 2′OH residues in the acceptor stem positioned near the conserved P4 helix of P RNA (Christian et al. 2006; Cuzic and Hartmann 2007; Loria and Pan 1997).

Substrate binding by P RNA at physiological ionic strength (0.1 M) is relatively weak (K_d = micromolar); however, as indicated above, binding by the RNase P holoenzyme under these conditions is much more stable with dramatically slower dissociation rate resulting in dissociation constants that are 1–5 nM (Buck et al. 2005a; Kurz et al. 1998; Sun et al. 2006). Binding is generally measured by photo cross linking to trap the ES complex (Beebe and Fierke 1994; Smith et al. 1992), by separation of free and bound substrate by gel-shift analysis (Hardt et al. 1993), or by gel-filtration spin column (Beebe and Fierke 1994). To isolate binding from cleavage two basic strategies have been used to slow the cleavage rate sufficiently and allow the accumulation of enzyme–substrate complexes. The first approach, demonstrated by Smith and Pace, involves substitution of Mg^{2+} with Ca^{2+} which slows the cleavage rate by > 1,000-fold but still supports high affinity substrate binding (Smith et al. 1992). The second approach is to include a $2'$-deoxynucleotide modification at the cleavage site, which also slows catalysis by several orders of magnitude (Loria and Pan 1999; Smith and Pace 1993). Analysis of the dependence of binding on divalent metal ions by gel-shift or analysis of dissociation rates shows cooperative dependence on ion concentration. Based on these data, substrate binding involves the uptake of two ions that bind to the enzyme substrate complex which increase the substrate affinity 10^3–10^5 fold (Beebe et al. 1996; Hardt et al. 1993). As discussed in more detail below, one of these sites is likely to involve interactions with the reactive phosphate, while the second site may reflect ion binding associated with positioning of the base of the pre-tRNA acceptor stem by the conserved core of P RNA.

To assess the rate of catalysis, single turnover catalytic assays performed under saturating enzyme concentrations allow measurement of the apparent rate constant for hydrolysis (k_{chem}) (Beebe and Fierke 1994; Fierke and Hammes 1995). Using an optimal model substrate under identical reaction conditions there is a less than tenfold difference in k_{chem} for the holoenzyme compared to P RNA alone (Crary et al. 1998; Sun et al. 2006). These data demonstrate that, the RNA subunit does the heavy-lifting for transition state stabilization. At moderate ionic strength (0.1 M) dependence of k_{chem} on magnesium ion concentration for both P RNA and the RNase P holoenzyme is non-cooperative (Crary et al. 1998; Sun et al. 2006) with apparent dissociation constants of 30–40 mM. Interestingly, data for the RNA alone under higher ionic strength, using either single turnover kinetics (Oh et al. 1998) or multiple turnover kinetics with a slow-cleaving substrate (Smith and Pace 1993) to isolate the metal ion dependence of k_{chem}, show much higher cooperativity with Hill values of 2–3 (e.g. see Fig. 9.3). The basis for these differences is not known but is likely to arise from differential effects of competing monovalent ions on divalent metal ion binding sites. This data falls short of demonstrating that P RNA is a *bona fide* metalloenzyme, but it supports the conclusion that at least one divalent metal ion binding site is detected kinetically as essential for transition state stabilization.

A simple approach to test for residues and functional groups in RNAs important for metal ion inter actions, is to embed site specific mutations and modifications in the enzyme or substrate and compare the Mg^{2+} ion dependence of binding affinity or the single turnover rate constant. The detection of effects on k_{chem} for example, provides evidence that the functional group is linked to functional metal ion interactions. An example is shown in Fig. 9.3 in which deletion of the universally conserved bulged U residue in P4 within the catalytic core results in a significant change in the Mg^{2+} dependence of the catalytic step (Kaye et al. 2002b). As described below there

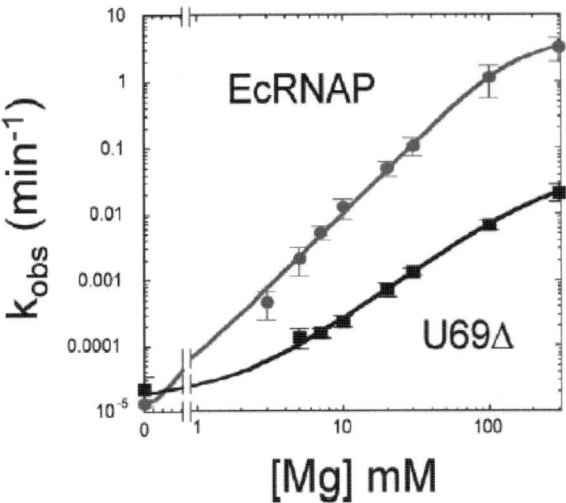

Fig. 9.3 Exemplary effects of mutations on the Mg^{2+} ion dependence of the catalytic step of the RNase P reaction. In this example deletion of a universally conserved bulged uridine (U69Δ) residue in the catalytic core of P RNA is tested (see Fig. 9.1). The effect of increasing Mg^{2+} ion concentration on the observed single turnover rate constant (k_{obs}) for the native P RNA (EcRNAP) and for the U69Δ ribozyme is shown. The data is fit to a cooperative binding model as described in Kaye et al. (2002b)

are several lines of evidence including these observations that support a role for P4 in positioning metal ions. As shown in the figure, this U69 deletion (U69Δ) results in the requirement of higher Mg^{2+} ion concentrations to achieve the maximal catalytic rate constant. Fitting these data to the Hill equation is standard procedure for quantitative comparisons of the native and mutant enzymes which in this case demonstrates that the mutations decreases both the apparent affinity and cooperativity of Mg^{2+} ions contributing to catalysis. Bulk Mg^{2+} titration experiments like these are necessary to define the metal ion requirements for different enzymatic steps; and additional kinetic dissection will be very revealing, especially as efforts to define the nature of conformational changes have indicated indirectly this far. However, these data alone cannot provide information concerning the site of ion binding, nor the manner in which binding is linked to enzyme function.

9.3   Initial Studies of Metal Binding Sites in P RNA by Metal Ion Induced Cleavage and NAIM

To define the site of site-specific metal ion binding in P RNA, several biochemical approaches were considered including metal ion induced cleavage and modification interference. Both techniques have been successful in localizing regions or even individual functional groups involved in metal ion association. However, both approaches have inherent limitations due to differences in the binding sites of different metal ions used for cleavage, or, in the case of modification interference, the limitation of examining only those modifications that can be incorporated into RNA via *in vitro* transcription and the possibility of indirect effects on metal ion binding. Nonetheless, these data significantly narrow the regions of P RNA associated with metal ions and provide strong evidence for cooperative binding of site-bound ions in the conserved core of P RNA.

It is established that intermolecular cleavage of RNA can be catalyzed by both acid and base by attacking the adjacent $2'OH$ (Oivanen et al. 1998). Thus, metal ions with pK_as sufficiently close to neutrality (e.g. pH 8 or 9 vs. pH 11) can be used as a source of a general base, if metal ion binding occurs proximal to the phosphodiester backbone (Brown et al. 1983; Kuusela and Lonnberg 1996). Phosphoryl transfer occurs through a S_N2 mechanism necessitating inline nucleophilic attack and leaving group displacement; thus if the geometry of the phosphodiester backbone is permissible, the associated metal ion can catalyze cleavage of phosphodiester bonds in its vicinity. RNA helices restrict the conformation of the backbone and are generally resistant to metal ion cleavage; however, sites of ion association in regions of non-Watson–Crick structure can be cleaved quite efficiently (e.g. Soukup and Breaker 1999).

Pb^{2+} cleaves P RNA in several discrete locations with most efficient cleavage at sites in P15 in the catalytic domain, a region demonstrably involved in enzyme–substrate contacts with the pre-tRNA $3'$ end (Brannvall et al. 2001; Ciesiolka et al. 1994; Zito et al. 1993) (Fig. 9.4). The phylogenetic conservation of these cleavage sites argues that they reflect common elements of P RNA structure and thus may be functionally relevant. Indeed, as described in more detail below, additional structural and functional data support functional metal ion binding in P15. The relevance of additional sites of Pb^{3+} cleavage to functional Mg^{2+} interactions is more tenuous; however, these cleavages can nonetheless be diagnostic for formation of native structure and detection of structure perturbation due to mutation.

The cleavage pattern from Tb^{3+} is generally considered more diagnostic of the behavior of Mg^{2+} in RNA (e.g. Hargittai and Musier-Forsyth 2000; Sigel et al. 2000) than Pb^{2+} because of its similarity to Mg^{2+} in ionic radius (0.72 Å, 0.92 Å, 1.19 Å, for Mg^{2+}, Tb^{3+}, Pb^{2+}, respectively) (Shannon 1976), and preference for the coordination of oxygen ligands (Nieboer 1975). Nevertheless, Tb^{3+} cleavage is generally consistent with the results of Pb^{2+} reactivity, but identifies a larger number of positions in RNase P where ions appear to congregate (Kaye et al. 2002b). Addition of increasing concentration of Mg^{2+} competes for Tb^{3+} binding and quantitative analysis reveals that different Tb^{3+} cleavage sites have differential sensitivity to competition by Mg^{2+}. One simple interpretation of such results is that the Tb^{3+} cleavage sites, most sensitive to Mg^{2+} competition, represent regions of the RNA structure that both ions bind, and thus, information about *bona fide* Mg^{2+} binding sites is obtained. However, consideration of the interplay between site-bound and diffuse interactions suggests that weak site bound interactions can be completed by simply increasing ionic strength. Additionally, different metal ions have different

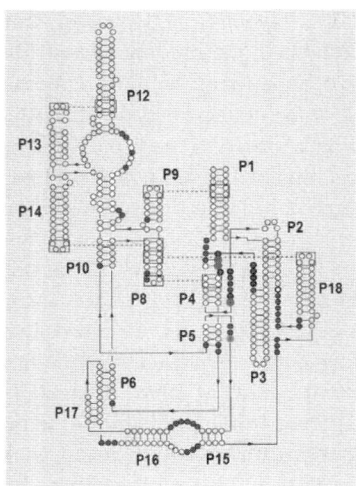 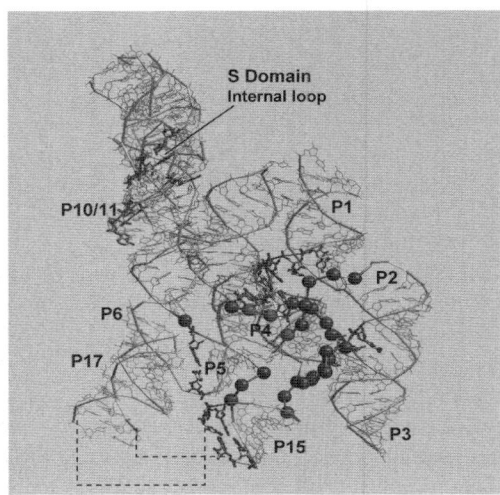

Fig. 9.4 Sites of metal ion dependent cleavage and phosphorothioate interference indicate residues involved in Mg^{2+} interactions. The secondary structure of *E. coli* P RNA is shown on the left and the three-dimensional structure of *T. thermophilus*, both are Type A P RNAs. The sites of Tb^{3+} cleavage of *E. coli* P RNA are shown on the secondary structure diagram as blue circles. Blue nucleotides in the three-dimensional structure are the homologous residues in the *T. thermophilus* structure. Similarly, the sites of strong phosphorothioate interference observed in both *E. coli* and *B. subtillis* P RNAs are indicated by red spheres (See figure insert for color reproduction)

coordination preferences and interact with different geometries even while binding at the same general/local site in the RNA structure, as observed for different ions in the crystal structures of tRNA (Shi and Moore 2000) and SRP RNA (Batey and Doudna 2002).

Interpretation of metal cleavage results in P RNA is further complicated by the fact that different Tb^{3+} or Pb^{2+} cleavage sites may have differential sensitivity to ionic strength. In addition, Mg^{2+} binding nearby could alter the RNA structure at a distance, to make the in-line attack geometry for necessary metal ion induced cleavage less favorable. Further, localization of ions determined by cleavage does not necessarily imply specific functional relevance, and tight binding is not necessarily a prerequisite for functional importance either. Nonetheless, as revealed in other systems, the sites of metal ion cleavage correlate well with those of high negative electrostatic potential (Sigel and Pyle 2007). Indeed, elucidation of crystal structures of P RNA show that Tb^{3+} cleavage sites are clearly localized to the folded catalytic core of universally conserved sequence, where close approach of the phosphodiester will necessarily create regions of high electrostatic potential (Fig. 9.4).

A more incisive method for finding sites of direct metal coordination that contribute to function, albeit with its own limitations, is phosphorothioate (PS) modification interference (Christian 2006; Vortler and Eckstein 2000). This approach relies on the principle that Mg^{2+}, as a hard metal ion, prefers hard ligands such as the non-bridging phosphate oxygen in nucleic acids. Replacing one of these atoms with a softer atom, such as sulfur, dramatically weakens Mg^{2+} ion affinity for the resultant phosphorothioate (Cohn and Hu 1978). If such a site is involved in a functionally important metal ion interaction, the modified molecule will have reduced activity due to disruption of the metal contact. Inclusion of a softer metal ion (such as Mn^{2+} or Cd^{2+}) that supports enzyme activity, and t more readily coordinates with soft ligands like sulfur, can rescue ribozyme activity if the site of disruptive PS modification was one of direct metal ion coordination (Piccirilli et al. 1993). Such a "thiophilic" metal ion rescue of a PS modification is taken as strong evidence that a genuine functional metal ion binding site is being interrogated. These kinds of experiments can be done in a site specific modification fashion and by modification interference in which all positions in the RNA are scanned simultaneously.

The survey of sensitive positions the modification-interference approach involves generating a population of RNAs randomly substituted at a level of *ca.* one PS modification per molecule. Partial reaction allows productive molecules to be converted into product while molecules rendered non-functional due to modification remain in the precursor population. End-labeling of the RNA followed by cleavage at the site of phosphorothioate modification with iodine, allows the sites of deleterious modification to be identified by comparison of the cleavage patterns of the precursor and product populations on sequencing gels (for a more full discussion see Ryder et al. 2000). While this approach is rapid and highly accurate, it is limited by two factors. First, the PS modification renders the phosphorothioate chiral and only the Rp position can be assayed as this is the only isomer that RNA polymerases will incorporate. Additionally, the methodology requires that functional and non-functional RNAs be separated, and therefore the intrinsic intermolecular activity of P RNA is a limitation in this regard. Nonetheless, this approach was highly successful in the analysis of functionally important metal ion contacts in the Group I intron ribozyme (Christian and Yarus 1993) as well as other RNAs,. Subsequent high resolution biochemical and structural studies have confirmed its accuracy (Stahley and Strobel 2006).

For P RNA, tethered ribozyme–substrate complexes in which the substrate sequences are appended to circularly permuted ribozymes that react with specificity and kinetics similar to the native ribozyme were engineered in order to permit active and inactive population to be purified (Frank et al. 1994). Attachment sites were based on intermolecular cross linking results that helped define the pre-tRNA binding interface, and application of these reagents in interference and selection studies has been highly successful. Initial application in PS interference demonstrated three strong interference sites. These were located in universally conserved sequence in the catalytic domain of the ribozyme, centered in and adjacent to P4 (Harris and Pace 1995) (Fig. 9.4). In the presence of Mn^{2+}, one of the sites was partially rescued. Additional selection experiments using the tethered ribozyme–substrate reagents for altered metal ion specificity yielded a point mutant that more readily accepts Ca^{2+} as an activating metal ion (Frank and Pace 1997). While these studies were consistent with P4 being a true functionally important metal binding site, the evidence is still qualified by several factors. The most important issue is that the sensitivity to phopsphorothioate modification in P4 could disrupt folding, substrate binding or active site metal ion binding as discussed in more detail below.

A powerful variation to this experiment is to couple the phosphorothioate tag to nucleobase analogs. This analysis, termed nucleotide analog interference mapping (NAIM), allows the identification of additional RNA functional groups that when modified or deleted are deleterious (Ryder et al. 2000). Using the tethered P RNA-tRNA constructs in this capacity identified a collection of additional functional groups in the catalytic domain that are clearly important for function (Kaye et al. 2002a) (Fig. 9.4). Analysis at lower Mg^{2+} concentrations should increase sensitivity to modification of functional groups linked to metal ion binding, if selection conditions are altered to allow for slower overall reaction rate. In this manner many RNA residues with functional groups with thermodynamic contributions linked to metal ion binding have been identified. The PS modifications identified in P4 are the least sensitive to increasing Mg^{2+} concentration and thus appear to cause by far the largest thermodynamic defect in P RNA activity. However, the interference of most nucleobase modifications can be 'rescued' simply by raising the Mg^{2+} ion concentration in the selection reaction. This phenomenon illustrates a fundamental principle of RNA structure. That is, the thermodynamic contribution of an individual interaction depends on its coupling with other interactions that occur in the same folded state. To explain the context dependent contribution of macromolecular interaction in enzymes interactions, Herschlag and colleagues have used the analogy of the differential effect of removing a support beam in a well built, or poorly built house (Kraut et al. 2003). Similarly, the suppression of interference by functional group modification at higher Mg^{2+} concentrations appears to be due to the increase in metal ion interactions that stabilize folding and/or substrate binding. Thus, this data identified functional groups involved in Mg^{2+} dependent structure, but not necessarily directly linked to Mg^{2+} ion site specific binding *per se*. It is this realization that drives the analysis of site-specific modifications directed at isolating the binding of individual ions and linking them more directly to enzyme function.

Such site-specifically modified molecules are generated by oligonucleotide directed ligation of RNA fragments that contain modifications generated by solid-phase synthesis (Moore and Sharp 1992). This approach has been successful in developing both *E. coli* and *B. subtilis* P RNAs in which PS modifications in P4 and elsewhere are embedded (Christian et al. 2000, 2002a; Crary et al. 2002). The analyses of the effects of such "atomic mutations" on reaction kinetics confirmed the PS results from modification interference, and allowed the analysis of Sp positions in P4 as well. Generation of the native ribozyme allowed Fierke and colleagues to assess the effects of P4 PS modification on substrate binding, which was determined to be correspondingly small. Thus, the primary effect of these modifications is on transition state stabilization. Quantitative analysis of the PS rescue by thiophilic ions has been pioneered by Piccirilli and Herschlag and has yielded detailed insights into *Tetrahymena* Group I (GI) intron ribozyme active site metal ion interactions (Shan et al. 1999). Briefly, information on the thermodynamics of binding of the rescuing ion is obtained from quantitative analysis of the ion concentration dependence of rate constant for reaction of the modified substrate/enzyme. We adapted this approach to the tethered P RNA-tRNA constructs described above and found cooperativity of rescue at the Rp oxygen of G68 consistent with two or more ions coordinated with this position as well as a single ion with the Sp phosphate oxygen at this site (Christian et al. 2002a). Such bi-dentate metal ion interactions have been observed in the metal ion core of the P4–P6 domain of GI intron as well as in 5S rRNA (Cate et al. 1997; Correll et al. 1997). The presence of such a poly-nuclear metal binding site within the catalytic core that contributes to transition state stabilization provides an excellent candidate for a binding site for catalytic metal ions .P4 and associated interferences in the catalytic domain are the most attractive residues in this regard.

9.4 Contribution of Metal Ions to Enzyme–Substrate Interactions and Interpreting Their Links to Enzyme Specificity

A second metal ion binding site that has received attention is located in P15 where the 3′ terminal RCCA sequence of pre-tRNA pairs with a conserved UGG sequence (e.g. see Busch et al. 2000; Kirsebom and Svard 1994; Oh and Pace 1994). As introduced above, this is one of the most prominently cleaved element by metal ions, highlighting it as a region of high negative electrostatic potential (Brannvall et al. 2001; Ciesiolka et al. 1994; Zito et al. 1993). Indeed, probing this element in isolation mimics metal cleavage seen in larger RNA; however, site-specific PS substitutions in P15 give relatively small inhibitory effects inconsistent with a role in positioning active site metal ions (Kufel and Kirsebom 1998). However, P15 is clearly proximal to the reactive phosphate since it serves as the binding site for the substrate 3′ CCA sequence. When single turnover kinetics is used to isolate the chemical step, the deletion of the C74–G292 pair between tRNA and P RNA in P15 reduces the single turnover cleavage rate constant by as much as 60-fold. Elevated Mg^{2+} concentration suppresses the mutation effect, providing evidence that interactions in this region are not absolutely necessary for the catalytic mechanism (Oh et al. 1998), but they promote the binding of metal ions important for optimal substrate cleavage through positioning effects. Additional mutational studies show that the role of the G293–C74 interaction is essentially confined to Watson–Crick base-pairing, with no indication of crucial tertiary contacts involving this base-pair (Busch et al. 2000).

A series of studies have analyzed the degree of mis-cleavage due to disruption of the 3′ RCC contact with P15 showing that modifications at these sites influence specificity as a function of metal ion concentration and identity (Brannvall and Kirsebom 1999; Brannvall et al. 2003, 2004). Structure function studies in which mis-cleavage and multiple turnover kinetics were monitored show that the identity of the + 73/294 base pair (bp) influences catalytic efficiency. Additionally, 2′-deoxy or 2′-deoxy-N7-deaza substitutions at substrate nucleotide G72 result in an increase in the concentration of Mg^{2+} required to suppress mis-cleavage induced by the presence of Mn^{2+} (Kikovska et al. 2006). Thus, the identity and orientation of this pair, as well as to some extent the 2′OH and N7 of G73 in the substrate are considered to contribute to catalysis, by affecting Mg^{2+} at the cleavage site. These studies provide important information on the substrate functional groups that form interactions in the enzyme–substrate complex. Additionally they are an important demonstration of the complex dependent effects of variations in substrate structure on RNase P specificity and provide information on the potential mechanisms of cleavage site choice.

However, despite claims in the literature on RNase P the results from such experiments are not readily interpretable in terms of identifying functional groups that serve as ligands for divalent metal ions. The reason for this ambiguity is that the fraction of substrate miscleaved depends both on the intrinsic rates, as well as on the binding affinities of the correct and miscleavaged ES complexes. Thus, there is a fundamental limitation to interpreting observed rates of cleavage and miscleavage in terms of reflecting interactions that are important for transition state stabilization. Figure 9.5 shows a simple reaction mechanism for formation of a correct miscleavage complex. In such a model the increase in the cleavage products P_C (correct cleavage) and P_{MC} (miscleavage) are described as,

$$P_c = F_c(1-e^{-k(obs)t})$$

(1)

and

$$P_{MC} = (1-F_c)(1-e^{-k(obs)t})$$

(2)

where P_c and P_{MC} are the fractions of substrate that are cleaved at the correct and miscleavage sites, respectively; F_c is the fraction of substrate cleavage at the P_C site at infinite time; $k_{(obs)}$ is the observed rate constant, and t is time. For cleavage at the P_C and P_{MC} sites at saturating enzyme concentration,

$$K_{(obs)} = K_c k_c + K_{mc} k_{mc}/(K_c + K_{mc})$$

(3)

Where K_c, K_{mc}, k_c and k_{mc} are equilibrium substrate association and rate constants shown in Fig. 9.4a. Thus, if cleavage is rate limiting, the fraction of substrate cleaved at the correct site (that is, F_c), which is the parameter measured in studies of the condition dependence of fidelity for RNase P, is defined as,

$$F_c = K_c k_c / (K_c k_c + K_{mc} k_{mc})$$

(4)

Thus, the fraction of substrate that is correctly cleaved depends not only on the intrinsic rates of reaction of the ES_c and ES_{mc} complexes, but also on the partitioning of the substrate into these complexes as defined by the relative magnitudes of K_c and K_{mc}. Accordingly, interpretation of F_c solely in terms of changes in active site interactions that influence k_c is only one potential mechanistic interpretation of the experiment. An increase in the observed fraction of substrate miscleaved (a decrease in F_c, that is) can be due to an increase in intrinsic rate of miscleavage

Fig. 9.5 (**a**) Minimal scheme for substrate miscleavage. The RNase P enzyme (E) binds pre-tRNA (S) to form complexes with either the correct phosphodiester bond positioned in the active site (E–S$_c$) or an adjacent phosphodiester bond (E–S$_m$) described by equilibrium constants K_c and K_{mc}, respectively. The complexes undergo catalysis at intrinsic rates for reaction of the complexes designated k_c and k_{mc} for reaction of the E–S$_c$ and E–S$_m$ complexes, respectively. (**b**) Dependence of the fraction of substrate that undergoes miscleavage (F$_c$) as a function of Mg^{2+} concentration for P RNA cleavage (P RNA and A248U) of two modified pre-tRNAs (dU(−1) and dA(−1)). The data are fit to (6) as described in the text

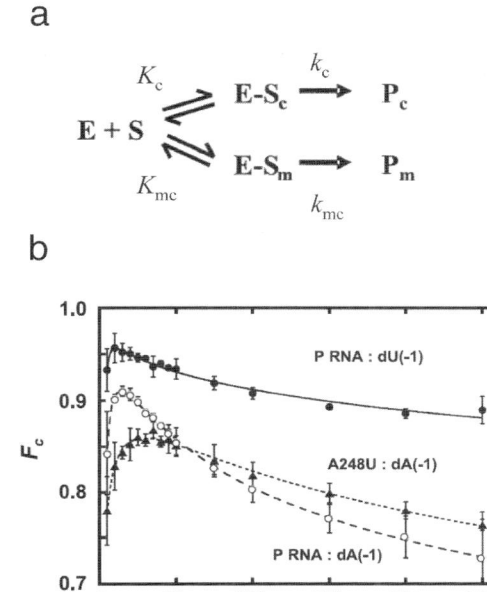

(k_{mc}), a decrease in the intrinsic rate of correct cleavage (k_c), an increase in the binding affinity of the miscleavage complex (K_{mc}), or a decrease in the affinity of the correct cleavage complex (K_c), or any combination of the above.

As indicated above, metal ion identity and concentration can influence degree of miscleavage observed, and this result has been interpreted as providing some kind of information on active site interactions that are important for catalysis, that is, the first order rate constant for cleavage at the correct site (k_c). Yet, here again there is a fundamental limitation. Considering (4), above, and the dependence of catalysis on Mg^{2+} ion concentration it can be shown that,

$$F_c = [1 + (K_{mc}/k_c)([Mg^{2+}](K_{Mg,c} + [Mg^{2+}]/(K_{Mg,mc} + [Mg^{2+}]))]^{-1} \qquad (5)$$

Where $K_{Mg,c}$ and $K_{Mg,mc}$ are the Mg^{2+} concentrations required to achieve half-maximal cleavage rates at the C$_0$ and M$_{-1}$ sites, respectively. The crucial point is that while the ratio of k_{mc} and k_c can be determined, these two variables are not independent. Thus, in addition to the potential for indirect effects of metal ion binding on catalysis, it is fundamentally not possible to interpret changes in F$_c$ to changes in metal ion concentration in terms of active site interactions involved in metal ion utilization in either the correct or incorrect cleavage complexes. Similarly, rescue of miscleavage by increasing pH has been interpreted as evidence for a titratable group necessary for cleavage at the correct site. However, as the discussion above regarding metal ion dependence illustrates, such a result can also be obtained if there is simply a difference in the rate limiting steps of the two pathways such that the correct cleavage pathway is more dependent on the intrinsic, pH dependent catalytic rate.

The structure-function data to date on the role of P15 in substrate recognition and catalysis has provided important insights into the RNA–RNA interactions that occur between P RNA and tRNA in this region. Structure-modeling of this interaction as well as determination of the structure of this element in isolation demonstrates a complex internal bulge that forms a key binding pocket for the tRNA 3′ end. Due to the distorted major and minor grooves of P15 resulting from non-Watson–Crick pairing in this helix, and the close approach of the substrate and enzyme phosphodiester backbones in this interaction, it is not surprising that divalent metal ions play an important role in the folding and function of this important site of substrate recognition. Clearly, RNA–RNA and RNA–metal ion interactions in P15 can influence enzyme specificity and catalytic rate indirectly via some form of "cross talk". However, one important additional lesson that can be drawn is that while it is compelling to interpret metal ion dependent changes in specificity in terms of active site interactions, it is inadvisable to do so if the intrinsic binding affinities and catalytic rates for the correct and miscleavage complexes cannot be measured, as is generally the case.

9.5 Does the P4 Metal Binding Site Play a Direct Or Indirect Role in P RNA Catalysis?

The demonstrable importance of metal ion binding sites in P4 raises the question of specific role(s) of these ions in P RNA function. Indeed, the issue of attributing specific functional roles to individual metal ion interactions and quantifying their thermodynamic contribution to enzyme function is a major question in the field of RNA structure and function. As illustrated above, interpretation of active site interactions from biochemical data can be tenuous and in the case of P4 both direct and indirect contributions to catalysis have been hypothesized Although current models favor an indirect role, the issue of the role of P4 metal ions in catalytic function is not entirely settled. The identification of metal binding in P4 and the difficulties inherent in analysis of the functional role of these interactions is particularly illustrative of the general considerations in relating metal ion interactions with specific functional roles.

Formally, important Mg^{2+} coordination interactions in P4 could function in a number of ways to contribute to the apparent degree of transition state stabilization (Anderson et al. 2006; Cowan 1998; Sigel and Pyle 2007). First, they could interact directly with the reactive phosphate, in which case they would be considered *bona fide* 'active site' metal ion interactions. Such a role could include metal ions that act by electrostatic catalysis or by in proton transfer as well. Second, many enzymes also use electrostatic effects to *indirectly* activate functional groups in an active site for example by modulating the reactivity of active site residues. Third, ion binding in P4 could alter the charge distribution of the transition state by long-range electrostatic effects on solvation, counterion accessibility, or local geometry. It is difficult to distinguish between these possibilities but under favorable circumstances evidence for one or another mode can be inferred from incisive biochemical data as described in more detail below.

A direct role for P4 in coordinating active site metal ions is inferred from the observation that the largest phosphorothioate effects (2–3 orders-of-magnitude) are restricted to it (and adjacent J2–4), and that such modifications reduce catalysis without changing the rate limiting step (Christian et al. 2000, 2002a; Crary et al. 2002). However, this result alone does not imply a direct coordination of an active site metal ion, since the metal ion that is disrupted could influence structure necessary for binding a catalytic metal ion or for correct geometry of the active site. Nevertheless, such a change would have to be quite subtle since even mutations in P4 do not significantly alter P RNA folding as indicated by comparison of the Tb^{3+} cleavage pattern of the mutant and wild type RNAs (Kaye et al. 2002b). Alternatively, a multi-step binding mechanism which has been implicated by kinetic data could permit effects of PS modification to be expressed on a conformational change step that is not monitored in the binding assay, but nonetheless contributes to the observed catalytic rate (Fig. 9.1c). For example, if the substrate binds in an open complex but reacts from a closed one that has a new interaction with P4 that also forms in the transition state, then destabilization of the closed complex will not affect the apparent binding affinity. However, since these interactions are formed in the transition state, they will also be destabilized relative to the undocked ground state. In this case the free energy difference between the free and bound substrate is unaltered while the difference between the ground state and transition state increases resulting in an effect on k(obs) but not K_d(app). Thus, a more detailed understanding of the P RNA binding reaction and the involvement of conformational changes is necessary to test such a mechanism.

Structure function studies clearly demonstrate a linkage between P4 sequence and structure and the positioning of metal ions that are important for catalysis. Mutations in P4 that alter the position of the bulged U clearly disrupt catalysis and decrease the affinity of metal ions essential for catalysis (Kaye et al. 2002b)., This result is consistent with positioning important metal ions, but falls short of indicating a role in positioning "catalytic" metal ions. Additional evidence cited in favor of a direct role in catalytic function comes from in vitro selection experiments which generated ribozymes that can function more efficiently with Ca^{2+} as a metal ion (Frank and Pace 1997). Specifically, a single C to U mutation in P4 was found to largely increase the Ca^{2+} rate with little effect on the Mg^{2+} dependent reaction. This result suggests that the geometry of the ribozyme is altered such that a metal ion essential for transition state stabilization has been altered to more readily accept Ca^{2+}. The observation of a higher catalytic rate constant for the mutant in the presence of Ca^{2+} could be due to an increased affinity of Ca^{2+} as a catalytic metal. However, the negative effect of the mutation on the Mg^{2+} dependent rate suggests that any change in specificity involves only accommodation of Ca^{2+} and not a change in specificity from Mg^{2+} to Ca^{2+}. Consideration of alternative hypotheses suggest that several mechanisms could give rise to this altered metal ion specificity without evoking a direct role for P4 in coordinating active site metal ions. Most notably, the mutation could simply weaken an inhibitory Ca^{2+} binding site that binds away from the active site and thus does not affect active site metal ion interactions at all. Additionally, although the mutant ribozyme retains pH sensitivity similar to the wild type ribozyme and thus does not appear to have a different rate limiting step, as discussed above, a difference in kinetic mechanisms of the mutant and wild type ribozymes could also give rise to enhanced reactivity in the presence of Ca^{2+}.

A potentially more direct analysis of the role of P4 in catalysis is suggested by the ability to monitor active site metal ion interactions using quantitative PS rescue analysis. Hartmann and colleagues showed that an Rp phosphorothioate modification at the scissile phosphate inhibits catalysis but has little effect on ground state binding affinity (Warnecke et al. 1996, 1999). Replacement of Mg^{2+} by Cd^{2+} in the reaction rescues catalysis providing evidence for direct coordination of metal ions to the reactive phosphate. Indeed, this kind of data is the best evidence yet that P RNA is a real metalloenzyme. This model for direct active site metal interaction with the reactive phosphate is refined by quantitative analysis of the concentration dependence of Cd^{2+} rescue in a background of constant Mg^{2+} (Christian et al. 2006; Sun and Harris 2007). Under such conditions the thermodynamic signature of the rescuing active site metal ions can be assessed and data demonstrates that two metal ions bind with the Rp non-bridging oxygen of the substrate with an apparent affinity of 10–30 mM. Combining mutagenesis and quantitative PS rescue approach allowed us to question whether these mutations have a specific effect on the affinity of rescuing Cd^{2+} ions and, by extension, that of the native active site Mg^{2+} interactions. We found that deletion or repositioning of the bulged U significantly reduces the affinity of the rescuing Cd^{2+} ions and reduces the apparent cooperativity.

Comparison of the pattern of Tb^{3+} cleavage patterns of the native and mutant P RNAs shows that changes in P4 bulged structure have relatively small, but detectable changes in J3/4 that appears to organize tertiary structure in the catalytic domain (Kaye et al. 2002b). Additionally, intermolecular cross linking studies show that the substrate binds in a different ground state conformation when P4 structure is altered (Christian et al. 2006). Importantly, in both the wild-type and mutant ribozymes, the metal ion binding sites in P4, are located several nucleotides distant from the reactive phosphate. In addition, a model of the P RNA–tRNA complex based on the extensive intermolecular cross linking data available positions the reactive phosphate at a distance from P4 but requires that the tRNA acceptor stem come very close to the metal binding sites in P4 (Kazantsev and Pace 2006; Niranjanakumari et al. 2007). This model is consistent with the cross linking of the P4 U bulge that positions P4 near the acceptor stem nucleotide +5 instead of within coordination of the reactive phosphate 5′ to nucleotide +1, although this constraint was not included in the modeling. Thus, the positioning of the substrate in the ground state, the demonstrable disruption in local structure, and repositioning of the substrate in the ground state resulting from P4 mutation, is more consistent with an indirect role for P4 in positioning metal ions rather than a direct role in providing coordination ligands.

Provided that the cross linking results yield some insight into the structure in the transition state, the data together would support a model in which the P4 metal binding site is peripheral to the active site, but nonetheless positions the substrate, potentially via a salt bridge in the active site. It is worth noting that the hammer-head ribozyme was first characterized structurally in an abbreviated form which necessitated a large conformational change to assemble the active site. The P RNA structure has been solved in the absence of substrate, and it could also reflect a conformation that must undergo rearrangement in order to assemble the active site. Accordingly, the fact that a significant conformational rearrangement repositions P4 more proximal to the pre-tRNA cleavage site in the transition state cannot be excluded.

9.6   Application of Kinetic Isotope Effects to Probe Mechanism and Active Site Interactions in P RNA Catalysis

The number and complexity of metal ion interactions in large RNAs like P RNA make it difficult to characterize biochemical individual sites. Bulk titration experiments are generally impossible to interpret in terms of specific individual contacts, and structure-functional approaches in general suffer from the ambiguity that it is difficult to distinguish between the direct and indirect effects of structure perturbation. High resolution structure determination can also yield inaccuracies due to trapping of non-native conformations. They are by definition static models that may reflect ground states that must rearrange to form the transition one in which the true catalytic interactions are formed. These considerations necessitate exploration of new means of probing native Mg^{2+} contacts. For P RNA catalysis, like all ribozymes, a complete understanding of its function must necessarily include the interactions that are present in the transition state.

A powerful method for defining the changes in bonding that occur in the transitions state is he analysis of kinetic isotope effects (KIE) (Cassano et al. 2004b; Northrop 2001). Practically, KIE experiments monitor the effect on reaction rate of substituting a heavier, stable isotope for one of the reacting atoms of the substrate (^{18}O in place of ^{16}O for one of the phosphate oxygens, for example). Conceptually, such measurements reveal the changes in the bonding environment that occur at a particular substrate atom going from the ground state to the transition state (Cleland 1995). If that change in bonding environment involves the direct coordination of a divalent metal ion such as Mg^{2+}, this interaction must necessarily be reflected in the effect on reaction rate caused by changing the atomic mass of that atom to a heavier isotope. The reason for this is that chemical reactions are dependent on the vibrational characteristics of the reacting atoms. Simply put, the stiffening of the bonding environment of a particular atom due to metal ion coordination will result in a slight advantage for a heavier isotope. There is a *ca.* 4% enrichment of $H^{18}O-$ in the water molecules coordinated with Mg^{2+} ions in solution, due to equilibrium isotope effects on metal coordination and deprotonation (Hunt and Taube 1959; Taube 1954). Therefore, if Mg^{2+} is coordinating the nucleophile in a phosphodiester bond hydrolysis reaction, there should be an enrichment in nucleophiles that are [18]O versus [16]O. Expressed as a ratio of the equilibrium constants for the ^{18}O and ^{16}O atoms, the equilibrium isotope effect of 1.04 is obtained ($^{16}K/^{18}K$). Thus, there is a significant 'equilibrium' isotope effect on metal ion coordination. This effect will also contribute to a difference in the rate of reaction (giving rise to a "kinetic" isotope effect) of the two different isotopes in a hydrolysis reaction in which the nucleophile is coordinated to a Mg^{2+} ion. This enrichment results in a small, but predictable enrichment of [18]O in the hydrolyzed ester product.

This principle in action is demonstrated by comparing the nucleophile isotope effects for hydrolysis of model diesters in solution in the presence and absence of Mg^{2+} (Cassano et al. 2004a). Alkaline hydrolysis of an alkyl *p*-nitrophenol phosphate diester has an $^{18}k_{(nuc)}$ of 1.07 (Cassano et al. 2002). Mg^{2+} provides significant catalysis of phosphodiesters in solution and chemical kinetic data support a model in which Mg^{2+} coordinates directly with the hydroxide ion nucleophile. This direct coordination is reflected in a lower $^{18}k_{(nuc)}$ for the Mg^{2+} catalyzed reaction compared with catalysis by base alone. For the Mg^{2+} catalyzed reaction a value of 1.04 is obtained that is consistent with a stiffening of the bonding environment by equilibrium coordination of the nucleophile by the catalytic Mg^{2+} ion (Fig. 9.6). For RNase P catalysis an $^{18}k_{(nuc)}$ of 1.04 is obtained, similar to the magnitude of Mg^{2+} catalyzed phosphodiester hydrolysis in solution. Thus, comparison of $^{18}k_{(nuc)}$ determined for the RNase P catalyze reaction with that for alkaline hydrolysis in solution provides evidence that active site interactions act to stiffen the nucleophile bonding environment in the transition state. The similarity of the values for RNase P catalysis and Mg^{2+} catalysis in solution is consistent with direct coordination of the nucleophile by active site metal ions. Such a result is parsimonious with the proposed mechanism and provides some of the best evidence to date of a metalloenzyme active site for this ribozyme.

9.7 The Role P Protein in Modulating P RNA–Metal Ion Interactions

Much has been learned by analyzing the enzymology and metal ion binding properties of P RNA alone; yet, these studies are typically performed at very high monovalent (1 M) or divalent (100 mM) ion concentrations in order to promote high affinity substrate binding. Such conditions place constraints on analyses of individual ion binding interactions and could obscure functionally relevant interactions. Fundamentally, the RNA subunit functions in concert with its cognate protein cofactor *in vivo* and a complete understanding of the biology of this enzyme necessitates

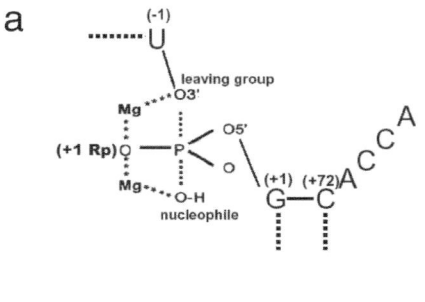

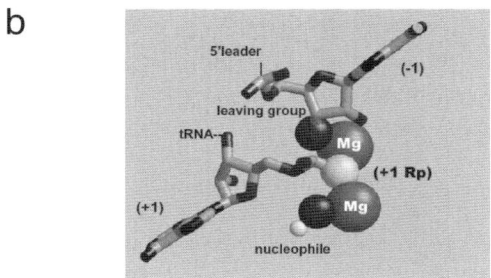

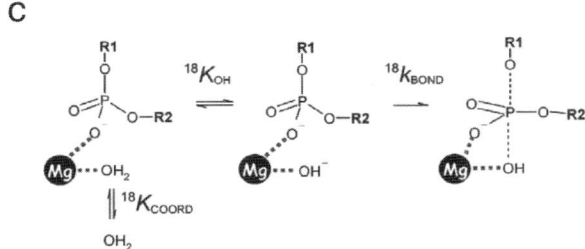

Fig. 9.6 Involvement of metal ions in P RNA catalysis. (**a**) General two metal ion mechanism of P RNA catalyzed hydrolysis of pre-tRNA. The hydrolysis reaction occurs between nucleotides G(+1) and U(−1) in pre-tRNA. The reaction involves nucleophilic attack by water (hydroxide ion) and the leaving group is the 3′O of the U(−1) nucleotide yielding a 5′ phosphate terminus of tRNA and a 3′OH at the 3′ terminus of the 5′ leader sequence product. Kinetic data is most consistent with a minimum of two metal ions that both coordinate to the pro-Rp phosphate oxygen. Proposed catalytic roles for these ions involve direct coordination of the nucleophile to assist in positioning, and increase in acidity favoring formation of the more nucleophilic hydroxide ion. Additionally, metal ion coordination with the leaving group could provide catalysis by offsetting the unfavorable formation of a negative charge at this position in the transition state. (**b**) Three dimensional models of the two metal ion mechanisms indicated in part A. The geometry is based on the position of metal ions and nucleotide organization found in the *Tetrahymena* GI intron active site. (**c**) Microscopic steps involved in nucleophilic activation based on analysis of nucleophile kinetic isotope effects. As described in the text the bonding environment of the nucleophilic water will be influenced by metal ion coordination ($^{18}K_{COORD}$), by deprotonation ($^{18}K_{OH}$) as well as by the formation of the new bond to phosphorus ($^{18}k_{BOND}$). Changes in $^{18}K_{COORD}$ and $^{18}K_{OH}$ will necessarily influence the observed isotope effect providing a means to test the involvement of metal ions in nucleophilic activation

comparative analysis of the ribonucleoprotein form of the enzyme. Accordingly, there has been renewed interest in the function of the holoenzyme and as protocols for the stoichiometric assembly of the holoenzyme are now standard (Buck et al. 2005a; Crary et al. 1998; Sun et al. 2006), new insights have been gained demonstrating that the binding of this small (*ca.* 90 amino acid) basic protein to P RNA has dramatic effects on RNase P function, including the metal ion requirements for folding, substrate binding, and catalysis.

The homologous structures of three different bacterial RNase P proteins show that they all adopt an α–β sandwich fold that is structurally homologous to other RNA binding proteins, including the ribosomal protein S5 and elongation factor G (Kazantsev et al. 2003; Spitzfaden et al. 2000; Stams et al. 1998). Three regions of the RNase P protein are likely to interact with RNA: (1) A left-handed β–α–β connection that contains a highly conserved RNR motif, (2) A central cleft formed by four anti-parallel β-strands, and (3) An α-helix and a cluster of polar residues-termed the 'metal binding loop'- that bind two zinc ions observed in the crystal structure of *B. subtilis* protein that could make bridging interactions with RNA functional groups. Spectroscopic as well as solution chemical probing under metal ion conditions that are saturating for P RNA folding provides evidence that the protein does not dramatically alter the structure of P RNA (Guo et al. 2006; Loria and Pan 2001; Rox et al. 2002). However, CD analysis of both RNase P RNA and protein as well as the study of intrinsic tryptophan fluorescence of P protein show that both RNA and protein subunits undergo conformational changes on assembly (Guo et al. 2006). From these and other studies P protein is now recognized as an intrinsically unstructured molecule that undergoes RNA-dependent folding during assembly. Conformational changes in the RNA subunit, under ion conditions saturating for folding, are more subtle and restricted to the locality of the protein binding site in the catalytic domain.

The first detailed comparative studies of the enzymology of P RNA and the RNase P holoenzyme from *B. subtilis* demonstrated that the protein subunit significantly enhances the affinity of pre-tRNA by interacting directly with the 5' leader sequence (Crary et al. 1998). For the *E. coli* enzyme comparative studies showed that the protein also enhances the binding of the tRNA product (Buck et al. 2005a); however, there are differential effects on product binding affinity resulting in differences in holoenzyme affinity for different tRNAs (Sun et al. 2006). In contrast, the effects of protein binding results in uniform affinity for different pre-tRNA substrates. Fierke and colleagues examined the *B. subtilis* enzyme by a combination of single turnover kinetics and equilibrium analysis and showed that the binding of P protein increases the affinity of *ca.* four metal ion binding sites that stabilize pre-tRNA binding (Beebe et al. 1996). However, direct measurement of total metal ion binding to P RNA provides evidence that the protein component does not alter the number or apparent affinity of magnesium ions that are either diffusely associated with the RNase P RNA polyanion or are specifically involved in the binding of mature tRNA. By comparing results obtained with pre-tRNAs having different leader sequence lengths, Fierke and colleagues provided evidence that this stabilizing effect is coupled to the P protein/5'-leader contact in the RNase P holoenzyme-pre-tRNA complex. These results suggest that the protein component enhances the magnesium affinity of the RNase P–pre-tRNA complex indirectly by binding and positioning pre-tRNA.

Indeed, analysis of the effects of protein binding on P RNA mediated catalysis also support an important, but indirect role for P protein on binding metal ions for transition state stabilization. Structure-function studies demonstrate that the RNA subunit interacts directly with a series of functional groups at the cleavage site (Christian et al. 2002b; Kirsebom 2007). Although only a subset of tRNAs has been tested in any organism, a limited survey of consensus substrates shows that the P protein has only moderate (*ca.* tenfold) effect on the rate of catalysis (Buck et al. 2005a; Crary et al. 1998; Sun et al. 2006). However, for substrates lacking consensus recognition sequences this apparent effect can be up to 1,000-fold (Sun et al. 2006). These results raise questions of the P protein subunit contribution directly to the catalytic step for these RNAs and whether such large apparent contributions to catalytic rate are related to effects on binding of divalent metal ions essential (either directly or indirectly) for transition state stabilization. In order to address these questions we employed quantitative analyses of the Cd^{2+}-dependent rescue of a cleavage site PS modification to monitor the specific affinity of active site metal ions (Sun and Harris 2007). This basic approach was developed for analysis of hammerhead ribozyme and GI intron ribozyme catalytic function as reviewed elsewhere (Christian 2006). High resolution structures of both of these ribozymes have been determined that contain the interactions demonstrated by biochemical studies, thus providing strong support for the proposed active site interactions in these systems. Kinetic analysis of the RNase P reaction show that a pre-tRNA cleavage site Rp-PS modification makes catalysis dependent on the binding of the two Cd^{2+} ions coordinated with the incorporated sulfur atom. Quantitative analysis of the concentration dependence of the Cd^{2+} rescue of the observed catalytic rate constant ascertains the thermodynamics of the binding of these metal ions to the E–S complex (Fig. 9.7). Comparing the results obtained under identical conditions for the *E. coli* P RNA subunit and the reconstituted holoenzyme showed that protein binding does not change the rate limiting step of the reaction as observed for the *B. subtilis* enzyme. Additionally, these results provided solid quantitative confirmation of earlier results that suggested two catalytic metal ions coordinated with the reactive phosphate of pre-tRNA. Most importantly it was noted that, the P protein binding with the RNA subunit clearly increased the apparent affinity of the rescuing metal ions without altering cooperativity or the number of metal ions involved in formation of the E–S complex. Furthermore comparative studies of the Mg^{2+} dependence of the rates of cleavage for canonical and non-canonical pre-tRNAs, also reveal large effects of protein binding on increasing apparent metal ion binding essential for transition state stabilization.

The findings above raise the question of how the protein has differential effects on catalysis, resulting in essential uniformity in catalytic rate; and, how it influences metal ion binding, the key issue being whether it does so directly by contributing

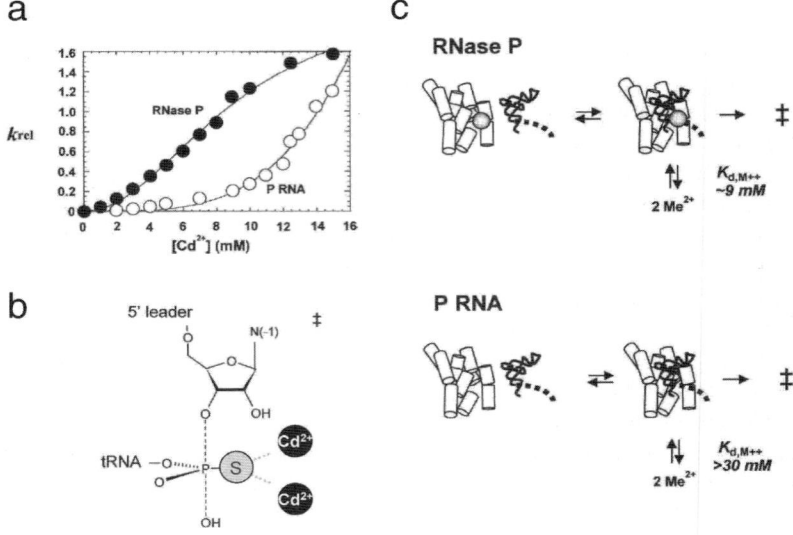

Fig. 9.7 (**a**) Metal ion concentration dependence of the rescue of a substrate cleavage site phosphorothioate modification provides a means for monitoring active site metal ion affinity. The dependence of k_{rel} (the ratio of rate constants obtained using the unmodified and PS-modified substrates) on Cd^{2+} concentration is shown for the reaction of the RNase P holoenzyme (filled circles) and for P RNA alone under identical conditions (open circles). Data is fit to a cooperative binding equation to extract the Hill constant and the apparent affinities of the two different enzyme forms. (**b**) Schematic representation of the coordination of two rescuing Cd^{2+} ions with the site specific PS modification at the reactive phosphate of the pre-tRNA substrate in the transition state. (**c**) Mechanistic interpretation of data leads to a model in which two active site metal ions bind cooperatively to the E–S complex. The affinities of these ions is increased in the RNase P holoenzyme resulting in a larger rate constant for pre-tRNA cleavage under conditions of sub saturating metal ion

functional groups that coordinate metal ions, or indirectly, by influencing P RNA structure or kinetic mechanism. Although structures of P RNA, protein and substrate are available, structure of component complexes are not. However, photo cross linking and high resolution chemical protection data have served as the basis for models of the RNase P holoenzyme (Buck et al. 2005b; Niranjanakumari et al. 2007). Chemical probing and intermolecular photo cross-linking demonstrate that important regions of the P protein are proximal to functionally important regions of P RNA and pre-tRNA in solution (Biswas et al. 2000; Niranjanakumari et al. 2007; Rox et al. 2002). The metal binding loop and N-terminus of the P protein are near the P3 stem-loop of P RNA. Additionally, these models place the conserved RNR motif close to the metal binding sites in helix P4. Cross-linking and affinity cleavage studies indicate that the central cleft of the P protein is vital to the recognition of pre-tRNA substrates and it is proposed to interact with the 5′ leader of the pre-tRNA substrates. Thus, there is ample evidence that the protein subunit is "at the scene of the crime" and could directly bind catalytic metal ions. However, no direct evidence exists yet to support such an intimate relationship between RNA and protein in the RNase P active site. Additionally, no direct role can be attributed as the protein only increases the catalytic rate by <10-fold for some substrates.

The long-standing hypothesis that a conformational change is necessary to position the pre-tRNA substrate for catalysis suggests a potential mechanism by which the P protein subunit could have differential effects on the apparent rate of catalysis for different substrates, and increase the apparent affinity of catalytic metal ions as well. In such a scenario the protein could contribute to a substrate docking step, after initial substrate binding, in which the binding sites for active site metal ions are formed. If catalysis is slow relative to equilibration between the docked and undocked forms, the pH dependence will be the same for the P RNA and holoenzyme reactions since phosphodiester bond cleavage is the rate limiting in both cases (Loria and Pan 1999). Since in this model, docking results in formation of metal ion binding sites associated with catalysis, shifting the docked state equilibrium should result in a decrease in the concentration of Mg^{2+} required to reach saturation, This is consistent with previous studies (Crary et al. 1998; Sun and Harris 2007). As discussed above, studies by Fierke and colleagues showed that shortening the leader sequence of a model pre-tRNA substrate decreased the apparent affinity of metal ion interactions necessary for cleavage (Kurz and Fierke 2002). Thus, such a substrate docking step could be linked to protein interactions with the leader sequence. This hypothetical mechanism has several attractive features in that it resolves the differential effect of the protein on catalysis for different substrates as well as the effect of the protein on active site metal ion affinity without invoking a novel functional property for the protein subunit. Importantly, this view provides a testable hypothesis and several groups are converging on assays for the proposed docking step and defining the structural changes that occur contemporaneously. Therefore, continued careful comparative analysis of the kinetic mechanisms of the RNA alone and holoenzyme reactions will either resolve such models or provide the information necessary for the development of an alternative and more detailed mechanistic view.

9.8 Future Directions

The past few years have seen an enormous increase in our understanding of the structure and functional properties of the RNase P ribonucleoprotein, and each advance has at some level encompassed an increased understanding of the central role that metal ion interactions play. The general divalent metal ion requirements for folding, substrate binding and catalysis have been defined and in each case the uptake of one or more metal ions is a necessary component. It is clear that the complexity in the RNase P reaction mechanism needs to be discovered, and it is expected that additional important details will be revealed. In this regard, application of rapid kinetic methods to fluorescence detection will be an important component of RNase P research as it has most other enzymes. As new detail about the reaction pathway and assays for analyzing individual steps come to light it will be important to understand their metal ion requirements.

It is more problematic to define the sites of metal ion interaction by biochemical methods than the metal ion concentration dependence on function. In this case, increasingly high resolutions structures from X-ray crystallography and NMR is likely to provide the key to identifying the best candidates for functional metal ion interactions. It cannot be overlooked however, that these advances, should they come as expected, must necessarily be coupled to biochemical assays for testing the functional relevance of the sites that are observed. The lessons that have been learned from analysis of metal ion interactions in other ribozymes including the hammerhead and GI intron systems, tell us that the most powerful perspective will come from coupling high resolution structure determination and incisive biochemical probing.

Similarly, there are likely to be additional surprises in store regarding the structure and function of the RNase P protein subunit. It is an open question whether the protein contributes residues to the RNase P active site, and although it can be argued that the protein is unlikely to participate in a hybrid RNA–protein active site, this possibility cannot be entirely ruled out based on the available data. As summarized above, an attractive model for the observed effects of protein binding on catalysis and active site metal ion binding can be hypothesized. However, one of the most useful aspects of such a model is that it is sufficiently detailed that it can, and may very well be, proved utterly incorrect. It is anticipated that structural biology will provide a detailed (but static) model of the P RNA–P protein complex, and it will be the job of biochemistry to integrate that information into experiments to test its role in P RNA substrate recognition and catalysis.

References

Altman S (2007) A view of RNase P. Mol Biosyst 3:604–607

Anderson VE, Ruszczycky MW, Harris ME (2006) Activation of oxygen nucleophiles in enzyme catalysis. Chem Rev 106:3236–3251

Bai Y, Greenfeld M, Travers KJ, Chu VB, Lipfert J, Doniach S, Herschlag D (2007) Quantitative and comprehensive decomposition of the ion atmosphere around nucleic acids. J Am Chem Soc 129:14981–14988

Baird NJ, Fang XW, Srividya N, Pan T, Sosnick TR (2007) Folding of a universal ribozyme: the ribonuclease P RNA. Q Rev Biophys 40:113–161

Batey RT, Doudna JA (2002) Structural and energetic analysis of metal ions essential to SRP signal recognition domain assembly. Biochemistry 41:11703–11710

Batey RT, Williamson JR (1998) Effects of polyvalent cations on the folding of an rRNA three- way junction and binding of ribosomal protein S15. RNA 4:984–997

Beebe JA, Fierke CA (1994) A kinetic mechanism for cleavage of precursor tRNA(Asp) catalyzed by the RNA component of *Bacillus subtilis* ribonuclease P. Biochemistry 33:10294–10304

Beebe JA, Kurz JC, Fierke CA (1996) Magnesium ions are required by *Bacillus subtilis* ribonuclease P RNA for both binding and cleaving precursor tRNAAsp. Biochemistry 35:10493–10505

Biswas R, Ledman DW, Fox RO, Altman S, Gopalan V (2000) Mapping RNA-protein interactions in ribonuclease P from *Escherichia coli* using disulfide-linked EDTA-Fe. J Mol Biol 296:19–31

Brannvall M, Kirsebom LA (1999) Manganese ions induce miscleavage in the *Escherichia coli* RNase P RNA- catalyzed reaction. J Mol Biol 292:53–63

Brannvall M, Mikkelsen NE, Kirsebom LA (2001) Monitoring the structure of *Escherichia coli* RNase P RNA in the presence of various divalent metal ions. Nucleic Acids Res 29:1426–1432

Brannvall M, Pettersson BM, Kirsebom LA (2003) Importance of the +73/294 interaction in *Escherichia coli* RNase P RNA substrate complexes for cleavage and metal ion coordination. J Mol Biol 325:697–709

Brannvall M, Kikovska E, Kirsebom LA (2004) Cross talk between the +73/294 interaction and the cleavage site in RNase P RNA mediated cleavage. Nucleic Acids Res 32:5418–5429

Brown RS, Hingerty BE, Dewan JC, Klug A (1983) Pb(II)-catalysed cleavage of the sugar-phosphate backbone of yeast tRNAPhe–implications for lead toxicity and self-splicing RNA. Nature 303:543–546

Buck AH, Dalby AB, Poole AW, Kazantsev AV, Pace NR (2005a) Protein activation of a ribozyme: the role of bacterial RNase P protein. EMBO J 24:3360–3368

Buck AH, Kazantsev AV, Dalby AB, Pace NR (2005b) Structural perspective on the activation of RNAse P RNA by protein. Nat Struct Mol Biol 12:958–964

Busch S, Kirsebom LA, Notbohm H, Hartmann RK (2000) Differential role of the intermolecular base-pairs G292-C(75) and G293- C(74) in the reaction catalyzed by *Escherichia coli* RNase P RNA. J Mol Biol 299:941–951

Caprara MG, Myers CA, Lambowitz AM (2001) Interaction of the *Neurospora crassa* mitochondrial tyrosyl-tRNA synthetase (CYT-18 protein) with the group I intron P4-P6 domain. Thermodynamic analysis and the role of metal ions. J Mol Biol 308:165–190

Cassano AG, Anderson VE, Harris ME (2002) Evidence for direct attack by hydroxide in phosphodiester hydrolysis. J Am Chem Soc 124:10964–10965

Cassano AG, Anderson VE, Harris ME (2004a) Analysis of solvent nucleophile isotope effects: evidence for concerted mechanisms and nucleophilic activation by metal coordination in nonenzymatic and ribozyme-catalyzed phosphodiester hydrolysis. Biochemistry 43:10547–10559

Cassano AG, Anderson VE, Harris ME (2004b) Understanding the transition states of phosphodiester bond cleavage: insights from heavy atom isotope effects. Biopolymers 73:110–129

Cate JH, Hanna RL, Doudna JA (1997) A magnesium ion core at the heart of a ribozyme domain. Nat Struct Biol 4:553–558

Christian E (2006). Identification and characterization of metal ion binding to RNA by thiophylic metal ion rescue. In: Hartmann K, Binderief A, Schon A, Westhof E (eds.) Handbook of RNA biochemistry. Wiley-VCH, New York. pp. 319–341

Christian EL, Yarus M (1993) Metal coordination sites that contribute to structure and catalysis in the group I intron from Tetrahymena. Biochemistry 32:4475–4480

Christian EL, Kaye NM, Harris ME (2000) Helix P4 is a divalent metal ion binding site in the conserved core of the ribonuclease P ribozyme. RNA 6:511–519

Christian EL, Kaye NM, Harris ME (2002a) Evidence for a polynuclear metal ion binding site in the catalytic domain of ribonuclease P RNA. EMBO J 21:2253–2262

Christian EL, Zahler NH, Kaye NM, Harris ME (2002b) Analysis of substrate recognition by the ribonucleoprotein endonuclease RNase P. Methods 28:307–322

Christian EL, Smith KM, Perera N, Harris ME (2006) The P4 metal binding site in RNase P RNA affects active site metal affinity through substrate positioning. RNA 12:1463–1467

Ciesiolka J, Hardt WD, Schlegl J, Erdmann VA, Hartmann RK (1994) Lead-ion-induced cleavage of RNase P RNA. Eur J Biochem 219:49–56

Cleland WW (1995) Isotope effects: determination of enzyme transition state structure. Methods Enzymol 249:341–373

Cohn M, Hu A (1978) Isotopic (18O) shift in 31P nuclear magnetic resonance applied to a study of enzyme-catalyzed phosphate–phosphate exchange and phosphate (oxygen)–water exchange reactions. Proc Natl Acad Sci U S A 75:200–203

Correll CC, Freeborn B, Moore PB, Steitz TA (1997) Metals, motifs, and recognition in the crystal structure of a 5S rRNA domain. Cell 91:705–712

Cowan JA (1998) Metal activation of enzymes in nucleic acid biochemistry. Chem Rev 98: 1067–1088

Crary SM, Niranjanakumari S, Fierke CA (1998) The protein component of *Bacillus subtilis* ribonuclease P increases catalytic efficiency by enhancing interactions with the 5´ leader sequence of pre-tRNAAsp. Biochemistry 37:9409–9416

Crary SM, Kurz JC, Fierke CA (2002) Specific phosphorothioate substitutions probe the active site of *Bacillus subtilis* ribonuclease P. RNA 8:933–947

Cuzic S, Hartmann RK (2007) A 2´-methyl or 2´-methylene group at G+ 1 in precursor tRNA interferes with Mg2+ binding at the enzyme-substrate interface in E-S complexes of *E. coli* RNase P. Biol Chem 388:717–726

DeRose VJ (2003) Metal ion binding to catalytic RNA molecules. Curr Opin Struct Biol 13:317–324

Draper DE (2004) A guide to ions and RNA structure. RNA 10:335–343

Draper DE, Grilley D, Soto AM (2005) Ions and RNA folding. Annu Rev Biophys Biomol Struct 34:221–243

Fedor MJ, Williamson JR (2005) The catalytic diversity of RNAs. Nat Rev Mol Cell Biol 6:399–412

Fierke CA, Hammes GG (1995) Transient kinetic approaches to enzyme mechanisms. Methods Enzymol 249:3–37

Frank DN, Pace NR (1997) In vitro selection for altered divalent metal specificity in the RNase P RNA. Proc Natl Acad Sci U S A 94:14355–14360

Frank DN, Harris ME, Pace NR (1994) Rational design of self-cleaving pre-tRNA-ribonuclease P RNA conjugates. Biochemistry 33:10800–10808

Gardiner K, Pace NR (1980) RNase P of *Bacillus subtilis* has a RNA component. J Biol Chem 255:7507–7509

Gardiner KJ, Marsh TL, Pace NR (1985) Ion dependence of the *Bacillus subtilis* RNase P reaction. J Biol Chem 260:5415–5419

Guerrier-Takada C, Altman S (1984a) Catalytic activity of an RNA molecule prepared by transcription in vitro. Science 223:285–286

Guerrier-Takada C, Altman S (1984b) Structure in solution of M1 RNA, the catalytic subunit of ribonuclease P from *Escherichia coli*. Biochemistry 23:6327–6334

Guerrier-Takada C, Haydock K, Allen L, Altman S (1986) Metal ion requirements and other aspects of the reaction catalyzed by M1 RNA, the RNA subunit of ribonuclease P from *Escherichia coli*. Biochemistry. 25:1509–1515

Guo X, Campbell FE, Sun L, Christian EL, Anderson VE, Harris ME (2006) RNA-dependent folding and stabilization of C5 protein during assembly of the *E. coli* RNase P holoenzyme. J Mol Biol 360:190–203

Hardt WD, Schlegl J, Erdmann VA, Hartmann RK (1993) Gel retardation analysis of *E. coli* M1 RNA-tRNA complexes. Nucleic Acids Res. 21:3521–3527

Hargittai MR, Musier-Forsyth K (2000) Use of terbium as a probe of tRNA tertiary structure and folding. RNA 6:1672–1680

Harris ME, Christian EL (2003) Recent insights into the structure and function of the ribonucleoprotein enzyme ribonuclease P. Curr Opin Struct Biol 13:325–333

Harris ME, Pace NR (1995) Identification of phosphates involved in catalysis by the ribozyme RNase P RNA. RNA 1:210–218

Hsieh J, Andrews AJ, Fierke CA (2004) Roles of protein subunits in RNA-protein complexes: lessons from ribonuclease P. Biopolymers 73:79–89

Hunt HR, Taube H (1959) The relative acidities of H_2O^{18} and H_2O^{16} coordinated to a tripositive ion. J Phys Chem 63:124–125

Kaye NM, Christian EL, Harris ME (2002a) NAIM and site-specific functional group modification analysis of RNase P RNA: magnesium dependent structure within the conserved P1-P4 multihelix junction contributes to catalysis. Biochemistry 41:4533–4545

Kaye NM, Zahler NH, Christian EL, Harris ME (2002b) Conservation of helical structure contributes to functional metal ion interactions in the catalytic domain of ribonuclease P RNA. J Mol Biol 324:429–442

Kazantsev AV, Pace NR (2006) Bacterial RNase P: a new view of an ancient enzyme. Nat Rev Microbiol 4:729–740

Kazantsev AV, Krivenko AA, Harrington DJ, Carter RJ, Holbrook SR, Adams PD, Pace NR (2003) High-resolution structure of RNase P protein from *Thermotoga maritima*. Proc Natl Acad Sci U S A 100:7497–7502

Kikovska E, Brannvall M, Kufel J, Kirsebom LA (2005) Substrate discrimination in RNase P RNA-mediated cleavage: importance of the structural environment of the RNase P cleavage site. Nucleic Acids Res 33:2012–2021

Kikovska E, Brannvall M, Kirsebom LA (2006) The exocyclic amine at the RNase P cleavage site contributes to substrate binding and catalysis. J Mol Biol 359:572–584

Kirsebom LA (2007) RNase P RNA mediated cleavage: substrate recognition and catalysis. Biochimie 89:1183–1194

Kirsebom LA, Svard SG (1994) Base pairing between *Escherichia coli* RNase P RNA and its substrate. EMBO J 13:4870–4876

Kole R, Baer MF, Stark BC, Altman S (1980) *E. coli* RNAase P has a required RNA component. Cell 19:881–887

Kraut DA, Carroll KS, Herschlag D (2003) Challenges in enzyme mechanism and energetics. Annu Rev Biochem 72:517–571

Kufel J, Kirsebom LA (1998) The P15-loop of *Escherichia coli* RNase P RNA is an autonomous divalent metal ion binding domain. RNA 4:777–788

Kurz JC, Fierke CA (2000) Ribonuclease P: a ribonucleoprotein enzyme. Curr Opin Chem Biol 4:553–558

Kurz JC, Fierke CA (2002) The affinity of magnesium binding sites in the *Bacillus subtilis* RNase P x pre-tRNA complex is enhanced by the protein subunit. Biochemistry 41:9545–9558

Kurz JC, Niranjanakumari S, Fierke CA (1998) Protein component of *Bacillus subtilis* RNase P specifically enhances the affinity for precursor-tRNAAsp. Biochemistry 37:2393–2400

Kuusela S, Lonnberg H (1996) Effect of metal ions on the hydrolytic reactions of nucleosides and their phosphoesters. Met Ions Biol Syst 32:271–300

Lonnberg T, Lonnberg H (2005) Chemical models for ribozyme action. Curr Opin Chem Biol 9:665–673

Loria A, Pan T (1997) Recognition of the T stem-loop of a pre-tRNA substrate by the ribozyme from *Bacillus subtilis* ribonuclease P. Biochemistry 36:6317–6325

Loria A, Pan T (1998) Recognition of the 5′ leader and the acceptor stem of a pre-tRNA substrate by the ribozyme from *Bacillus subtilis* RNase P. Biochemistry 37:10126–10133

Loria A, Pan T (1999) The cleavage step of ribonuclease P catalysis is determined by ribozyme-substrate interactions both distal and proximal to the cleavage site. Biochemistry 38:8612–8620

Loria A, Pan T (2001) Modular construction for function of a ribonucleoprotein enzyme: the catalytic domain of *Bacillus subtilis* RNase P complexed with *B. subtilis* RNase P protein. Nucleic Acids Res 29:1892–1897

Misra VK, Draper DE (2002) The linkage between magnesium binding and RNA folding. J Mol Biol 317:507–521

Moore MJ, Sharp PA (1992) Site-specific modification of pre-mRNA: the 2′-hydroxyl groups at the splice sites. Science 256:992–997

Nieboer E (1975). The lanthanide ions as structural probes in biological and model systems. In: Rare earths, F.H. Spedding & A.H. Daane, Eds. Wiley, New York pp. 1–47

Niranjanakumari S, Day-Storms JJ, Ahmed M, Hsieh J, Zahler NH, Venters RA, Fierke CA, Getz MM, Andrews AJ, Al-Hashimi HM (2007) Probing the architecture of the *B. subtilis* RNase P holoenzyme active site by cross-linking and affinity cleavage. RNA 13:521–535

Northrop DB (2001) Uses of isotope effects in the study of enzymes. Methods 24:117–124

Oh BK, Pace NR (1994) Interaction of the 3′-end of tRNA with ribonuclease P RNA. Nucleic Acids Research. 22:4087–4094

Oh BK, Frank DN, Pace NR (1998) Participation of the 3′-CCA of tRNA in the binding of catalytic Mg2+ ions by ribonuclease P. Biochemistry 37:7277–7283

Oivanen M, Kuusela S, Lonnberg H (1998) Kinetics and mechanism for the cleavage and isomerization of phosphodiester bonds of RNA by Bronsted acids and bases. Chem Rev 98:961–990

Piccirilli JA, Vyle JS, Caruthers MH, Cech TR (1993) Metal ion catalysis in the Tetrahymena ribozyme reaction. Nature 361:85–88

Pyle AM (2002) Metal ions in the structure and function of RNA. J Biol Inorg Chem 7:679–690

Robertson HD, Altman S, Smith JD (1972) Purification and properties of a specific *Escherichia coli* ribonuclease which cleaves a tyrosine transfer ribonucleic acid presursor. J Biol Chem 247:5243–5251

Rox C, Feltens R, Pfeiffer T, Hartmann RK (2002) Potential contact sites between the protein and RNA subunit in the *Bacillus subtilis* RNase P holoenzyme. J Mol Biol 315:551–560

Rueda D, Wick K, McDowell SE, Walter NG (2003) Diffusely bound Mg2+ ions slightly reorient stems I and II of the hammerhead ribozyme to increase the probability of formation of the catalytic core. Biochemistry 42:9924–9936

Ryder SP, Ortoleva-Donnelly L, Kosek AB, Strobel SA (2000) Chemical probing of RNA by nucleotide analog interference mapping. Methods Enzymol 317:92–109

Shan S, Yoshida A, Sun S, Piccirilli JA, Herschlag D (1999) Three metal ions at the active site of the Tetrahymena group I ribozyme. Proc Natl Acad Sci U S A 96:12299–12304

Shannon R (1976) Revised effective ionic radii and systematic studies of interatomic distances in halides and chalcogenides. Acta Crystallogr A 32:751–767

Shi H, Moore PB (2000) The crystal structure of yeast phenylalanine tRNA at 1.93 A resolution: a classic structure revisited. RNA 6:1091–1105

Sigel RK, Pyle AM (2007) Alternative roles for metal ions in enzyme catalysis and the implications for ribozyme chemistry. Chem Rev 107:97–113

Sigel RK, Vaidya A, Pyle AM (2000) Metal ion binding sites in a group II intron core. Nat Struct Biol 7:1111–1116

Smith D, Pace NR (1993) Multiple magnesium ions in the ribonuclease P reaction mechanism. Biochemistry. 32:5273–5281

Smith D, Burgin AB, Haas ES, Pace NR (1992) Influence of metal ions on the ribonuclease P reaction. Distinguishing substrate binding from catalysis. J Biol Chem 267:2429–2436

Smith JK, Hsieh J, Fierke CA (2007) Importance of RNA-protein interactions in bacterial ribonuclease P structure and catalysis. Biopolymers 87:329–338

Soukup GA, Breaker RR (1999) Relationship between internucleotide linkage geometry and the stability of RNA. RNA 5:1308–1325

Spitzfaden C, Nicholson N, Jones JJ, Guth S, Lehr R, Prescott CD, Hegg LA, Eggleston DS (2000) The structure of ribonuclease P protein from *Staphylococcus aureus* reveals a unique binding site for single-stranded RNA. J Mol Biol 295:105–115

Stahley MR, Strobel SA (2005) Structural evidence for a two-metal-ion mechanism of group I intron splicing. Science 309:1587–1590

Stahley MR, Strobel SA (2006) RNA splicing: group I intron crystal structures reveal the basis of splice site selection and metal ion catalysis. Curr Opin Struct Biol 16:319–326

Stahley MR, Adams PL, Wang J, Strobel SA (2007) Structural metals in the group I intron: a ribozyme with a multiple metal ion core. J Mol Biol 372:89–102

Stams T, Niranjanakumari S, Fierke CA, Christianson DW (1998) Ribonuclease P protein structure: evolutionary origins in the translational apparatus. Science 280:752–755

Stark BC, Kole R, Bowman EJ, Altman S (1978) Ribonuclease P: an enzyme with an essential RNA component. Proc Natl Acad Sci U S A 75:3717–3721

Stefan LR, Zhang R, Levitan AG, Hendrix DK, Brenner SE, Holbrook SR (2006) MeRNA: a database of metal ion binding sites in RNA structures. Nucleic Acids Res 34:D131–D134

Sun L, Harris ME (2007) Evidence that binding of C5 protein to P RNA enhances ribozyme catalysis by influencing active site metal ion affinity. RNA 13:1505–1515

Sun L, Campbell FE, Zahler NH, Harris ME (2006) Evidence that substrate-specific effects of C5 protein lead to uniformity in binding and catalysis by RNase P. EMBO J 25:3998–4007

Taube H (1954) Use of oxygen-isotope effects in the study of hydration of ions. J Phys Chem 58:523–528

Vortler LC, Eckstein F (2000) Phosphorothioate modification of RNA for stereochemical and interference analyses. Methods Enzymol 317:74–91

Warnecke JM, Furste JP, Hardt WD, Erdmann VA, Hartmann RK (1996) Ribonuclease P (RNase P) RNA is converted to a Cd(2+)-ribozyme by a single Rp-phosphorothioate modification in the precursor tRNA at the RNase P cleavage site. Proc Natl Acad Sci U S A 93:8924–8928

Warnecke JM, Held R, Busch S, Hartmann RK (1999) Role of metal ions in the hydrolysis reaction catalyzed by RNase P RNA from *Bacillus subtilis*. J Mol Biol 290:433–445

Wilson TJ, Lilley DM (2002) Metal ion binding and the folding of the hairpin ribozyme. RNA 8:587–600

Zahler NH, Sun L, Christian EL, Harris ME (2005) The pre-tRNA nucleotide base and 2'-hydroxyl at N(-1) contribute to fidelity in tRNA processing by RNase P. J Mol Biol 345:969–985

Zito K, Huttenhofer A, Pace NR (1993) Lead-catalyzed cleavage of ribonuclease P RNA as a probe for integrity of tertiary structure. Nucleic Acids Res 21:5916–5920

Chapter 10
Beyond Crystallography: Investigating the Conformational Dynamics of the Purine Riboswitch

Colby D. Stoddard and Robert T. Batey(⊠)

Abstract Riboswitches are structured elements located in the 5′-untranslated regions of numerous bacterial mRNAs that serve to regulate gene expression via their ability to specifically bind metabolites. The purine riboswitch ligand-binding domain has emerged as an important model system for investigating the relationship between RNA structure and function. Directed by NMR and crystallographically generated structures of this RNA, a variety of biophysical and biochemical techniques have been utilized to understand its dynamic nature. In this review, we describe these various approaches and what they reveal about the purine riboswitch.

Abbreviations *2AP* 2-aminopurine; *FRET* fluorescence resonance energy transfer; *J* joining region; *L* loop; *NMIA* *N*-methylisatoic anhydride; *SHAPE* selective 2′-hydroxyl acylation analyzed by primer extension; *smFRET* single molecule fluorescence resonance energy transfer; *TPP* thiamine pyrophosphate; *UTR* untranslated region

10.1 Introduction

X-ray crystallography has the power to yield detailed information about the structure and function of biological molecules at the atomic level. However, it is common for a crystal structure to leave unresolved questions relating to the fact that a structural model based upon diffraction data is a static representation of an otherwise dynamic molecule. This is an accepted consequence of the requirement that the molecule be organized within the context of a crystalline lattice. Consequently, disordered regions of the macromolecule such as mobile loops will not be observed in the resulting electron density map. Since dynamic regions of proteins and RNA are often intimately linked to function, even if a structure is obtained, information about the most interesting aspects of the molecule may not be present within the model.

R.T. Batey
Department of Chemistry and Biochemistry, Campus Box 215, University of Colorado-Boulder, Boulder, CO 80309, USA
e-mail: Robert.Batey@colorado.edu

N.G. Walter et al. (eds.) *Non-Protein Coding RNAs*
doi: 10.1007/978-3-540-70840-7_10, © Springer-Verlag Berlin Heidelberg 2009

Integration of other biophysical techniques into a structural study is required to present a complete picture of the RNA of interest and its biological activity. This is primarily rooted in the observation that RNA-mediated processes often involve significant conformational changes e.g. many RNAs undergo an induced fit mechanism when interacting with proteins, other nucleic acids, or small molecules (Williamson 2000; Leulliot and Varani 2001). In this mode of recognition, the global and/or local conformation of the RNA in the unbound state is significantly different from the bound state. Recently, a host of ligand-binding mRNAs, termed riboswitches, that directly couple metabolite binding to genetic regulation through conformational changes have been discovered (reviewed in (Winkler and Breaker 2005)), making them an ideal system to study this relationship. The goal of this review is to (1) discuss the questions arising from the crystal structure of a purine-binding riboswitch and (2) highlight the biophysical approaches used to address these issues.

10.1.1   *Genetic Regulation by Riboswitches*

Riboswitches are *cis*-acting regulatory elements most commonly found in the 5′-untranslated region (UTR) of bacterial mRNAs (Winkler and Breaker 2005). These RNAs contain highly structured regions that bind a specific metabolite to activate or repress the expression of downstream genes. Currently, at least twelve classes of riboswitches have been identified that recognize diverse biological compounds including purine nucleobases (Mandal et al. 2003; Mandal and Breaker 2004), amino acids (Rodionov et al. 2003; Sudarsan et al. 2003; Mandal et al. 2004), cofactors (Grundy and Henkin 1998; Miranda-Rios et al. 2001; Mironov et al. 2002; Nahvi et al. 2002; Winkler et al. 2002), a sugar (Winkler et al. 2004), and metal ions (Cromie et al. 2006; Dann et al. 2007). In addition, a number of "orphan" elements have been recognized in bacterial mRNAs that appear to have the hallmarks of a riboswitch, but for which the effector ligand has not yet been identified (Barrick et al. 2004; Weinberg et al. 2007). In some bacteria, these mRNA elements regulate expression of >4% of all genes (*B. subtilis* has at least 29 riboswitches) underscoring their fundamental importance in cellular metabolism (Irnov et al. 2006).

With the exception of the *glmS* riboswitch-ribozyme (Winkler et al. 2004), genetic regulation by all bacterial riboswitches is determined at the level of transcriptional termination or translation initiation. These elements exert control over gene expression via the interplay of two distinct domains: a ligand-binding *aptamer domain* and a downstream regulatory domain termed the *expression platform* (reviewed in (Winkler and Breaker 2005)). Binding of the ligand drives structural changes within the aptamer domain that in turn directs the formation of one of two mutually exclusive secondary structural elements in the expression platform. In transcriptional regulators, these stem-loop structures, called the terminator and antiterminator, cause RNA polymerase to either abort or proceed with transcription, respectively. Switching is achieved within the expression platform through a shared sequence element, the *switching sequence*, that is required for helix formation at the

3'-end of the aptamer domain or the 5'-end of the antiterminator (reviewed in (Stoddard and Batey 2006)). This mutually exclusive switching is at the heart of the linkage between ligand binding and gene expression; and thus riboswitch function must be understood from the perspective of conformational changes and dynamics. More recently, the aptamer domain of the TPP riboswitch has been found in plant and fungal mRNAs (Sudarsan et al. 2003) where it functions to influence intron splicing or processing of the 3'-terminus (Cheah et al. 2007; Wachter et al. 2007), further expanding the scope of how ligand binding can be coupled to various mRNA-related processes.

10.1.2   The Purine Riboswitch

Of all the riboswitches, purine responsive mRNAs have been the most extensively studied using a wide variety of biophysical and structural techniques. This is due to a combination of favorable features including high affinity for cognate ligands, small overall size of the aptamer domain, and the ability to discriminate between chemically similar compounds (Mandal et al. 2003; Mandal and Breaker 2004). The secondary structure of the aptamer domain, as deduced by a phylogenetic analysis of currently over 120 sequences (Griffiths-Jones et al. 2005), is comprised of three paired regions (P1, P2, and P3) positioned around a three-way junction of joining regions (J1/2, J2/3, and J3/1) (Mandal et al. 2003; Mandal and Breaker 2004). In addition, the paired regions P2 and P3 are capped by terminal loop motifs (L2 and L3) that, along with the three-way junction, contain the majority of highly conserved nucleotides. One of the most conserved features is a pyrimidine residue in J3/1 that is always a cytidine in guanine-responsive RNAs and a uridine in adenine-responsive RNAs, implying that this nucleotide forms a Watson–Crick pair with the purine nucleobase (Mandal et al. 2003; Mandal and Breaker 2004).

Crystallographic studies of the purine riboswitch aptamer domain complexed to guanine, hypoxanthine, or adenine reveal a complex and compact architecture (Fig. 10.1a) (Batey et al. 2004; Serganov et al. 2004). Terminal loops L2 and L3 form a series of base interactions, including two quartets, to bring the P2 and P3 helices into a parallel arrangement. The three-way junction directly interacts with the ligand to form an intricate network of hydrogen bonding interactions using highly conserved nucleotides (Batey et al. 2004; Serganov et al. 2004; Noeske et al. 2005). The purine nucleobase is found at the center of the junction directly contacting a set of four pyrimidine residues (Fig. 10.1b) that almost completely encapsulate it. This indicates ligand binding must be accompanied by a folding event. Strikingly, the primary means of interaction is through hydrogen bonds as there is very little base stacking with the RNA (Gilbert and Batey 2005). While these structures provide concrete insights into how ligand recognition occurs, they do not reveal how the RNA folds around the ligand or how binding directs the switching mechanism that drives gene regulation. To shed light on these processes, it is necessary to employ other experimental techniques.

Fig. 10.1 (**a**) Three-dimensional structure of the purine riboswitch bound to hypoxanthine (PDB 1U8D) (Batey et al. 2004). Paired regions are indicated in white while nucleotides involved in tertiary contacts are dark. Loop (L) and junction (J) regions are indicated along with sites for FRET donor and acceptor attachment (☆) in a smFRET study by Lafontaine (Lemay et al. 2006). The specificity pyrimidine (C74) forms a Watson–Crick basepairing interaction with hypoxanthine (black spheres). (**b**) Hypoxanthine (HX)-binding illustrates that specificity is achieved through extensive hydrogen bonding within the core of the three-way junction that allows the RNA to sample all aspects of a ligand

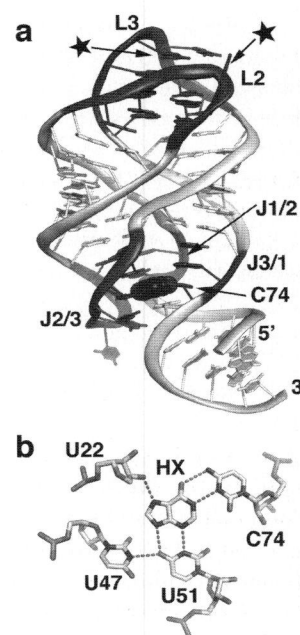

10.2   Global Folding of the Aptamer Domain

One of the first events that happens during the life of the riboswitch is that the aptamer domain productively folds into a conformation capable of efficiently recognizing the effector ligand. A feature of riboswitch aptamer domains that distinguishes them from many of the well-characterized in vitro selected aptamers is they almost always contain tertiary structural elements spatially distant from the ligand binding site that facilitate formation of the biologically active global architecture of the RNA (Gilbert and Batey 2005). For instance, while the L2–L3 interaction of the purine riboswitch (Fig. 10.1a) does not form direct contacts with ligand, it is an essential feature for function under physiological conditions (Mandal et al. 2003; Batey et al. 2004; Serganov et al. 2004). Other riboswitches contain common structural motifs such as kink-turns and pseudoknots (Winkler et al. 2001; McDaniel et al. 2005; Blouin and Lafontaine 2007) that also create complex tertiary architectures surrounding the binding pocket. Some of the first questions addressed following the determination of the crystal structure of the purine riboswitch involved the relationship between tertiary structure formation and ligand binding.

Chemical probing studies indicated that the L2–L3 interaction forms in a ligand-independent fashion for both the guanine and adenine binding riboswitches (Mandal et al. 2003; Mandal and Breaker 2004). To directly observe formation of tertiary structure, Lafontaine and coworkers employed a single molecule fluorescence resonance energy transfer (smFRET) approach. Dynamics were monitored by attaching a fluorescein-Cy3 FRET pair at two uridine residues at the tips of L2 and L3 in the *B. subtilis pbuE* adenine riboswitch (Lemay et al. 2006). These positions are ideal due to their solvent accessibility and low conservation (positions are marked by stars in Fig. 10.1a). In the absence of adenine, ensemble measurements of FRET efficiency as a function of magnesium clearly showed that this tertiary interaction forms in the absence of ligand with a $[Mg^{2+}]_{1/2} \approx 50 \mu M$, serving to globally organize the RNA as observed in the crystal structure. The low magnesium requirement for formation of the L2–L3 interaction is somewhat surprising in light of the number of multivalent ions observed in crystal structures (Batey et al. 2004; Serganov et al. 2004), but is independently supported by NMR studies (Noeske et al. 2007).

The single molecule approach allows for observation of the dynamic opening and closing of the loop–loop interaction. Under roughly physiological magnesium ion concentrations (0.5 mM Mg^{2+}), rates of docking (k_{fold}) and undocking (k_{unfold}) of the two loops were measured to be $0.76 s^{-1}$ and $1.3 s^{-1}$, respectively (Lemay et al. 2006). Thus, the globally folded form of the RNA is in a dynamic equilibrium with the unfolded state, despite the extensive network of base–base and base–backbone interactions observed in the crystal structure (Batey et al. 2004; Serganov et al. 2004). Addition of 50 μM adenine, a saturating concentration that ensures all of the RNA is bound, has a twofold effect on both the folding and unfolding rates ($1.5 s^{-1}$ and $0.84 s^{-1}$, respectively) indicating a small but distinct influence on the dynamics of the L2–L3 interaction. These data are supported by NMR studies from the Schwalbe group that detect a small change in the L2–L3 region upon purine binding (Buck et al. 2007; Noeske et al. 2007). This stabilizing influence on the tertiary structure becomes markedly more pronounced in the complete absence of magnesium.

Another powerful aspect of the single molecule approach is the ability to observe intermediate states not easily accessible by ensemble techniques. In the case of the adenine riboswitch, smFRET reveals a discrete folding intermediate at low magnesium ion concentrations (>50 μM); as divalent ion concentrations increase, the unfolded to intermediate transition becomes too fast to observe (Lemay et al. 2006). Detailed analysis of these data further suggests that this intermediate is an obligate step in the folding process. Collectively, these studies reveal a small but clear interplay between binding of adenine to the three-way junction and the structure of the L2–L3 interaction. This relationship has been proposed to be a consequence of a closing U·U pair in the P2 helix (Noeske et al. 2007a; Noeske et al. 2007b) that destabilizes the L2–L3 interaction. In contrast, the *B. subtilis xpt-pbuX* guanine riboswitch, which has stabilizing G–C pairs closing L2, appears to stably form the L2–L3 interaction in the absence of ligand and magnesium (Noeske et al. 2007). As formation of L2–L3 is essential for purine riboswitch function, it is likely that the stability of this interaction influences the effective intracellular purine concentration that regulates a particular mRNA. In this fashion, individual genes or operons regulated by purine riboswitch in the same organism can be tuned to respond to different intracellular effector concentrations.

10.3 Ligand-Dependent Conformational Changes in the Aptamer Domain

A second set of questions arising from the structural analysis of the purine riboswitch relates to how the ligand gains entry to the binding site at the center of the three-way junction. Complete burial of the ligand clearly implies that at least one of the joining (J) strands must be flexible in the unbound state (Batey et al. 2004). To probe local conformational changes in the RNA that accompany ligand binding, as opposed to the global changes in the RNA imparted by formation of the L2–L3 interaction, two approaches have been used: chemical footprinting and site-specific 2-aminopurine (2AP) labeling.

10.3.1 Footprinting Analysis of Ligand Binding

The technical simplicity of chemical probing makes it one of the most commonly used methods to investigate aspects of an RNA's secondary and tertiary structure. A variety of probing reagents are used including nucleases, dimethyl sulfate (DMS), the Fenton reaction (Fe^{II}-EDTA), and autohydrolysis (commonly referred to as in-line probing) to build a working model of the RNA (Ehresmann et al. 1987). However, idiosyncrasies inherent to each of these traditional tools have motivated the development of novel methods and reagents that act as more general and flexible probes of RNA structure. In particular, in-line probing (Nahvi et al. 2002) and SHAPE chemistry (Wilkinson et al. 2006) are becoming increasingly popular methods.

Ligand-dependent structural alterations of the purine riboswitch, as well as many other riboswitches, have been mostly investigated by in-line probing. This technique, pioneered by Breaker and colleagues, exploits the natural tendency of loosely structured regions of RNA to undergo backbone scission in the presence of magnesium ions when the 2'-hydroxyl group makes an in-line attack on the adjacent phosphodiester linkage (Nahvi et al. 2002). In the purine riboswitch aptamer domain, regions within the three-way junction are particularly susceptible to scission and are interpreted to be the conformationally dynamic regions of the RNA (Mandal et al. 2003). Addition of the appropriate ligand results in complete protection of the junction implicating this region of the RNA as the region most responsive to binding. Strikingly, the only two sites of strand scission in the bound form correspond to the two nucleotides that are flipped out into solution in the crystal structure, indicating that this technique is indeed an accurate reporter of structure. While this is a simple method, its use may be limited by the requirement of elevated concentrations of Mg^{2+} (20 mM) and long reaction times (24–48 h).

A powerful probing technique that has emerged from the Weeks laboratory is *S*elective 2'-*H*ydroxyl *A*cylation analyzed by *P*rimer *E*xtension (SHAPE)

(Wilkinson et al. 2006). This method exploits the fact that non-bridging phosphate oxygens influence the local pK_a of their adjacent 2'-hydroxyl group. Sugars in the C3'-*endo* configuration (as in A-form helices) place the 2'-hydroxyl group close to the backbone, thereby increasing its local pK_a, disfavoring formation of the reactive alkoxide via deprotonation. Conversely, sugars in the C2'-*endo* configuration place the 2'-hydroxyl group further away, lowering their pK_a relative to C3'-*endo* sugars. This phenomenon results in a conformation-dependent reactivity towards N-methylisatoic anhydride (NMIA) such that sugars in flexible regions of the RNA that access the C2'-*endo* conformation are acylated, which is subsequently detected as a reverse transcription stop in a sequencing gel. Thus, SHAPE accurately reports backbone dynamics of an entire RNA. In addition, it is possible to monitor the temperature-dependent reactivity of the RNA with nucleotide resolution using this method to yield local information about folding events (Wilkinson et al. 2005). The general usefulness of this form of probing is certain to increase in coming years, as the Weeks laboratory is developing improved reagents that will broaden the scope of questions that can be addressed (Mortimer and Weeks 2007).

To gain insights into the nature of local ligand-dependent folding events in the three-way junction, we have applied SHAPE chemistry to the aptamer domain of the *B. subtilis xpt-pbuX* guanine riboswitch (Stoddard et al. 2008). In this experiment, the RNA's structure was probed with NMIA in the presence and absence of 10 µM hypoxanthine between 20 and 80°C. Reactivity for each nucleotide in the RNA was obtained by quantifying primer extension stops in the sequencing gel so that a temperature-dependent reactivity profile could be generated. Fitting these data to a two-state transition model yields a melting temperature (T_m) for each residue in the RNA that corresponds to the midpoint between its unreactive (folded) and reactive (unfolded) states. Comparison of the melting transitions for each nucleotide in the free and bound RNA reveals that only a few residues show a strong ligand-dependent change in T_m. These residues are found exclusively in the J1/2 and J2/3 strands and are most apparent in the residues surrounding the turn in J2/3 (Fig. 10.2a, red).

These data reveal two features of the ligand binding pocket. First, the purine nucleobase accesses the interior of the junction via a gateway created by a flexible J2/3. Nucleotides in J2/3 involved in forming interactions with the ligand and base triples with the P1 helix (nucleotides 49–51) undergo the most substantial conformational change between the bound and unbound states (Fig. 10.2b, red). Second, in the absence of ligand, J3/1 and two base triples adjacent to P2 and P3 are significantly ordered relative to the rest of the junction. This conformationally restricts the junction in a fashion to provide the nucleobase with an initial docking site as well as minimizing the entropic penalty of productive binding (Gilbert et al. 2006). This behavior has also been observed by NMR spectroscopy, in which residual dipolar couplings were used to infer that two base triples in the junction are formed in the absence of ligand (Ottink et al. 2007).

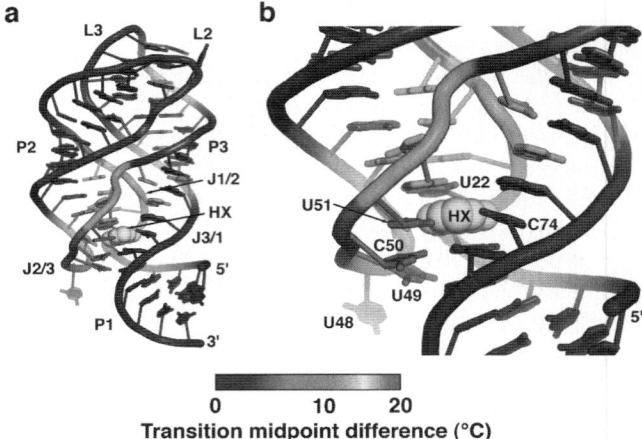

Transition midpoint difference (°C)

Fig. 10.2 NMIA probing of the purine riboswitch (adapted from (Stoddard et al. 2008)). (**a**) Global structure of the *xpt-pbuX* aptamer domain RNA colored to reflect the difference in the melting temperature (ΔT_m) for each nucleotide between the hypoxanthine bound and unbound states. Blue colors reflect nucleotide positions whose T_m is unaffected by ligand binding whereas red reflects significant changes in the T_m between the two states (see color bar). (**b**) Close-up view of the three-way junction emphasizing that most of the nucleotides affected by the presence of the ligand are centered about the turn in J2/3 (U48–U51, red) whereas J3/1 appears to be static (blue) (See figure insert for color reproduction)

10.3.2   2-Aminopurine as a Probe of the Local Chemical Environment

2-aminopurine (2AP) is an analog of adenine that experiences strong fluorescence quenching upon base stacking (Walter et al. 2001). This property as well as its ability to base pair with uridine makes it an ideal probe for monitoring conformational changes in RNA. More recently, it has been demonstrated that an analog of guanine, 8-azaguanine, exhibits similarly favorable properties and will likely increase in importance as a probe since its presence is likely to be less perturbing to RNA structure (Da Costa et al. 2007). Site-specific incorporation of 2AP or 8-azaguanine into RNA via chemical synthesis allows for observing changes in local chemical environments upon ligand binding by measuring changes in the probe's intrinsic fluorescence.

2-aminopurine incorporated at specific sites throughout the *V. vulnificus add* adenine riboswitch has been used to monitor local conformational changes due to Mg^{2+} and/or adenine (Rieder et al. 2007). Incorporation of 2AP into L2 and L3 reveals magnesium-dependent changes in fluorescence that are further augmented by the addition of adenine, consistent with smFRET and NMR analyses (Lemay et al. 2006; Noeske et al. 2007). Within the ligand-binding pocket, Mg^{2+} and adenine have variable effects on fluorescence; reporters in J1/2 are primarily affected by the addition of magnesium whereas in J2/3 fluorescence is mostly altered by adenine (Rieder et al. 2007). This observation further supports the view that portions of the three-way junction are organized to some degree prior to ligand binding.

The most interesting aspect of these studies investigated RNAs that encompassed not only the aptamer domain, but also downstream sequences corresponding to the expression platform. Placing 2AP at a residue in J2/3 that becomes flipped out into solvent in the bound structure provides an ideal means to assess ligand binding (Gilbert et al. 2006) and its effect on the structure of the expression platform (Rieder et al. 2007). This experiment, along with a similar one performed by the Lafontaine group on the *B. subtilis pbuE* adenine riboswitch (Lemay et al. 2006), allows for a comparison between full-length adenine riboswitches proposed to operate at the level of translation initiation (*add*) or transcription termination (*pbuE*). The translationally controlled RNA appears to be in a dynamic equilibrium between the free and bound states indicating that this mRNA element is capable of responding to adenine concentrations as a reversible switch (Rieder et al. 2007). However, the transcriptional regulator is locked into an adenine-unresponsive state, suggesting that this RNA's function may be coupled to transcription (Lemay et al. 2006). Thus, the relationship between the aptamer domain and expression platform is dependent upon the process being regulated. To understand how transcription and riboregulation are coupled, the kinetics of these processes must be examined.

10.4 Kinetic Analyses of Purine Riboswitch Function

Much of the discussed work strongly supports an induced fit binding mechanism for ligand recognition by the purine riboswitch. This mechanism often has a significant affect upon the rate of ligand binding such that typical observed association rates (k_{on}) are 3–4 orders of magnitude slower than diffusion controlled (10^8–$10^9 \, M^{-1} \, s^{-1}$) (Wickiser et al. 2005; Gilbert et al. 2006). A slow bimolecular association rate has crucial implications for genetic control, particularly for transcriptional regulation where it is linked to the rate of transcription by RNA polymerase (Wickiser et al. 2005). Specifically, the ligand must bind and induce the formation of the correct secondary structure in the expression platform before the polymerase escapes beyond the sequence that forms the rho-independent transcriptional terminator. In this fashion, riboswitches are considered to be at least partially kinetically controlled, rather than purely thermodynamically controlled as is most often the case for transcriptional regulation by protein repressors (Winkler and Breaker 2005). Thus, the rate of ligand association with the aptamer domain influences genetic regulation.

To determine the rates at which ligand binds the purine riboswitch aptamer domain, several independent studies employed stopped-flow kinetics with either 2AP as the target ligand or riboswitch RNA substituted with 2AP in the three-way junction (Wickiser et al. 2005; Gilbert et al. 2006). Kinetic experiments on 2AP binding the *B. subtilis pbuE* adenine riboswitch and the *B. subtilis xpt-pbuX* guanine riboswitch, bearing a C74U mutation that confers adenine-responsiveness, measured similar values for an observed bimolecular association rate constant, $\sim 10^5 \, M^{-1} \, s^{-1}$, consistent with an induced-fit mechanism. A further study using a photocaged hypoxanthine derivative in combination with time-resolved NMR spectroscopy yielded data supporting the fluorescence-based measurements (Buck et al. 2007).

Additionally, these studies indicate that binding is a multi-step process. Kinetic measurements of binding at ligand concentrations above 75 µM reveal a new rate-limiting unimolecular process occurring at a rate of ~30 s^{-1} (Gilbert et al. 2006). This unimolecular step is interpreted as the result of a heterogeneous population of free RNAs containing interconverting binding-competent and -incompetent states. This mode of binding was previously described in a study of theophylline, a purine derivative, binding its vitro selected aptamer (Jucker et al. 2003). The presence of a heterogeneous free RNA population is reinforced by the observation of a temperature-dependent heat capacity of ligand binding by isothermal titration calorimetry (Gilbert et al. 2006). Another step is revealed by NMR spectroscopy, in which an initial "docked" state is observed followed by conversion to the folded state seen by X-ray crystallography (Buck et al. 2007). The docked state, which must occur as an intermediate prior to folding of the RNA around the ligand to encapsulate it, likely reflects the initial recognition of the ligand by the specificity pyrimidine C74 (or the equivalent uridine in adenine riboswitches).

Kinetic studies on the purine riboswitch aptamer domain alone present a substantial conundrum for understanding efficient regulation. The observed rates of binding are sufficiently slow that at physiological purine concentrations the rate of transcription is such that the polymerase would have long escaped the riboswitch before the binding to the aptamer is complete (Wickiser et al. 2005). Furthermore, studies of the *pbuE* adenine riboswitch containing both the aptamer domain and expression platform demonstrate that stable formation of the antiterminator stem-loop completely abrogates ligand binding (Lemay et al. 2006; Rieder et al. 2007). Therefore, the ligand binding must act prior to transcription and formation of the antiterminator element.

How this can be achieved was illuminated in a kinetic analysis of both RNA transcription and ligand binding for the FMN riboswitch (Wickiser et al. 2005). These studies point to a series of programmed pauses in the riboswitch consisting of short tracks of uridines that temporarily stall the polymerase at key decision-making points. Pausing first occurs right after transcription of the aptamer domain (Fig. 10.3, pause 1), allowing for the slow ligand-binding process to occur if there is a sufficient concentration of ligand present. A second pause (Fig. 10.3, pause 2) occurs immediately after the antiterminator sequence, allowing for a second decision process. This pause favors the slow secondary structural rearrangement of the switching sequence from being incorporated into the aptamer domain to being incorporated into the antiterminator if no ligand is bound. Thus, sequences within the riboswitch that modulate transcription rates are a critical feature of efficient genetic regulation by riboswitches (Wickiser et al. 2005). At least for riboswitches that regulate at the transcriptional level, it is clear that their regulatory mechanism is a complex convolution of kinetic and thermodynamic parameters dictated by ligand concentration, the rate of ligand binding and dissociation, and the influence of RNA sequence on transcription.

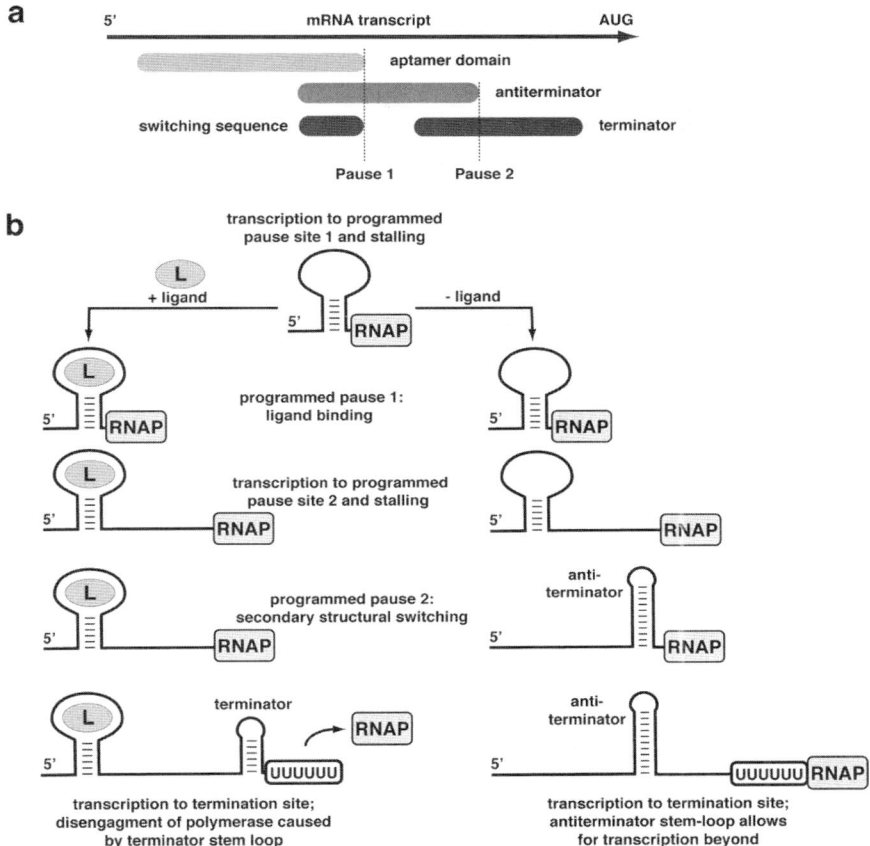

Fig. 10.3 (**a**) Cartoon representation of functional elements of a riboswitch contained within an mRNA (adapted from (Wickiser et al. 2005)). The switching sequence is the basis for the mutually exclusive formation of the antiterminator and terminator elements. Programmed pause sites are located at critical checkpoints to control gene expression. (**b**) Flow chart of the regulatory events that occur during a transcriptionally controlled riboswitch. Upon transcription of the aptamer domain (*top, center*), the presence or absence of ligand will favor distinct structural outcomes in the downstream expression platform (terminator or antiterminator formation). Pause site one favors formation of an RNA-ligand complex if ligand is present in sufficient quantities. A second pause further downstream allows secondary structural switching between the aptamer domain and expression platform. The structural state of the expression platform is then communicated to RNA polymerase resulting in the continuation or termination of transcription

10.5 Concluding Remarks

Biophysical studies of the purine riboswitch are continuing to yield an increasingly deep understanding of how conformational changes are harnessed by riboswitches during ligand binding to effect genetic regulation. As we have discussed, the influence of the L2–L3 interaction, the encapsulation of the purine nucleobase within the three-way junction, and incorporation of the switching sequence into one of two mutually exclusive structures all influence the secondary and tertiary structural changes in the mRNA associated with genetic regulation. Furthermore, studies of the FMN riboswitch have yielded new insights into how these conformational changes are coupled to the transcription of the mRNA. However, challenges remain in the development of a complete picture of genetic regulation by the purine riboswitch. In particular, emerging studies of the adenine riboswitch that observe events on the nanosecond timescale (Eskandari et al. 2007) are certain to yield new insights into the early events of ligand–RNA interactions. Also, further exploration of the direct linkage between ligand binding and switching during transcription is required to provide a detailed kinetic model of this form of riboregulation. As this review has illustrated, the purine riboswitch is rapidly becoming an important model system for understanding RNA-mediated processes, and yet we have just begun to reveal how RNA serves as an effective regulator of cellular metabolism.

Acknowlegments We would like to thank all of the members of the Batey laboratory who have provided valuable discussions and insights into our work on riboswitches. Due to space limitations, we could not comprehensively cover all of the research performed on the purine riboswitch in the last few years and apologize to those whose contributions were not discussed in this review. This work was funded by a grant from the National Institutes of Health (GM 073580) to R.T.B.

References

Barrick JE, Corbino KA, Winkler WC, Nahvi A, Mandal M, Collins J, Lee M, Roth A, Sudarsan N, Jona I, Wickiser JK, Breaker RR (2004) New RNA motifs suggest an expanded scope for riboswitches in bacterial genetic control. Proc Natl Acad Sci U S A 101:6421–6426

Batey RT, Gilbert SD, Montange RK (2004) Structure of a natural guanine-responsive riboswitch complexed with the metabolite hypoxanthine. Nature 432:411–415

Blouin S, Lafontaine DA (2007) A loop–loop interaction and a K-turn motif located in the lysine aptamer domain are important for the riboswitch gene regulation control. RNA 13:1256–1267

Buck J, Furtig B, Noeske J, Wohnert J, Schwalbe H (2007) Time-resolved NMR methods resolving ligand-induced RNA folding at atomic resolution. Proc Natl Acad Sci U S A 104:15699–15704

Cheah MT, Wachter A, Sudarsan N, Breaker RR (2007) Control of alternative RNA splicing and gene expression by eukaryotic riboswitches. Nature 447:497–500

Cromie MJ, Shi Y, Latifi T, Groisman EA (2006) An RNA sensor for intracellular Mg(2+). Cell 125:71–84

Da Costa CP, Fedor MJ, Scott LG (2007) 8-Azaguanine reporter of purine ionization states in structured RNAs. J Am Chem Soc 129:3426–3432

Dann CE III, Wakeman CA, Sieling CL, Baker SC, Irnov I, Winkler WC (2007) Structure and mechanism of a metal-sensing regulatory RNA. Cell 130:878–892

Ehresmann C, Baudin F, Mougel M, Romby P, Ebel JP, Ehresmann B (1987) Probing the structure of RNAs in solution. Nucleic Acids Res 15:9109–9128

Eskandari S, Prychyna O, Leung J, Avdic D, O'Neill MA (2007) Ligand-directed dynamics of adenine riboswitch conformers. J Am Chem Soc 129:11308–11309

Gilbert SD, Batey RT (2005) Riboswitches: natural selexion. Cell Mol Life Sci 62:2401–2404

Gilbert SD, Stoddard CD, Wise SJ, Batey RT (2006) Thermodynamic and kinetic characterization of ligand binding to the purine riboswitch aptamer domain. J Mol Biol 359:754–768

Griffiths-Jones S, Moxon S, Marshall M, Khanna A, Eddy SR, Bateman A (2005) Rfam: annotating non-coding RNAs in complete genomes. Nucleic Acids Res 33:D121–D124

Grundy FJ, Henkin TM (1998) The S box regulon: a new global transcription termination control system for methionine and cysteine biosynthesis genes in gram-positive bacteria. Mol Microbiol 30:737–749

Irnov A, Winkler WC (2006) Genetic control by cis-acting regulatory RNAs in *Bacillus subtilis*: general principles and propects for discovery. Cold Spring Harb Symp Quant Biol 71: 239–249

Jucker FM, Phillips RM, McCallum SA, Pardi A (2003) Role of a heterogeneous free state in the formation of a specific RNA-theophylline complex. Biochemistry 42:2560–2567

Lemay JF, Penedo JC, Tremblay R, Lilley DMJ, Lafontaine DA (2006) Folding of the adenine riboswitch. Chem Biol 13:857–868

Leulliot N, Varani G (2001) Current topics in RNA-protein recognition: control of specificity and biological function through induced fit and conformational capture. Biochemistry 40: 7947–7956

Mandal M, Breaker RR (2004) Adenine riboswitches and gene activation by disruption of a transcription terminator. Nat Struct Mol Biol 11:29–35

Mandal M, Boese B, Barrick JE, Winkler WC, Breaker RR (2003) Riboswitches control fundamental biochemical pathways in *Bacillus subtilis* and other bacteria. Cell 113:577–586

Mandal M, Lee M, Barrick JE, Weinberg Z, Emilsson GM, Ruzzo WL, Breaker RR (2004) A glycine-dependent riboswitch that uses cooperative binding to control gene expression. Science 306:275–279

McDaniel BA, Grundy FJ, Henkin TM (2005) A tertiary structural element in S box leader RNAs is required for *S*-adenosylmethionine-directed transcription termination. Mol Microbiol 57:1008–1021

Miranda-Rios J, Navarro M, Soberon M (2001) A conserved RNA structure (thi box) is involved in regulation of thiamin biosynthetic gene expression in bacteria. Proc Natl Acad Sci U S A 98:9736–9741

Mironov AS, Gusarov I, Rafikov R, Lopez LE, Shatalin K, Kreneva RA, Perumov DA, Nudler E (2002) Sensing small molecules by nascent RNA: a mechanism to control transcription in bacteria. Cell 111:747–756

Mortimer SA, Weeks KM (2007) A fast-acting reagent for accurate analysis of RNA secondary and tertiary structure by SHAPE chemistry. J Am Chem Soc 129:4144–4145

Nahvi A, Sudarsan N, Ebert MS, Zou X, Brown KL, Breaker RR (2002) Genetic control by a metabolite binding mRNA. Chem Biol 9:1043

Noeske J, Richter C, Grundl MA, Nasiri HR, Schwalbe H, Wohnert J (2005) An intermolecular base triple as the basis of ligand specificity and affinity in the guanine- and adenine-sensing riboswitch RNAs. Proc Natl Acad Sci U S A 102:1372–1377

Noeske J, Buck J, Furtig B, Nasiri HR, Schwalbe H, Wohnert J (2007a) Interplay of 'induced fit' and preorganization in the ligand induced folding of the aptamer domain of the guanine binding riboswitch. Nucleic Acids Res 35:572–583

Noeske J, Schwalbe H, Wohnert J (2007b) Metal-ion binding and metal-ion induced folding of the adenine-sensing riboswitch aptamer domain. Nucleic Acids Res 35:5262–5273

Ottink OM, Rampersad SM, Tessari M, Zaman GJ, Heus HA, Wijmenga SS (2007) Ligand-induced folding of the guanine-sensing riboswitch is controlled by a combined predetermined induced fit mechanism. RNA 13:2202–2212

Rieder R, Lang K, Graber D, Micura R (2007) Ligand-induced folding of the adenosine deaminase A-riboswitch and implications on riboswitch translational control. Chembiochem 8:896–902

Rodionov DA, Vitreschak AG, Mironov AA, Gelfand MS (2003) Regulation of lysine biosynthesis and transport genes in bacteria: yet another RNA riboswitch? Nucleic Acids Res 31: 6748–6757

Serganov A, Yuan YR, Pikovskaya O, Polonskaia A, Malinina L, Phan AT, Hobartner C, Micura R, Breaker RR, Patel DJ (2004) Structural basis for discriminative regulation of gene expression by adenine- and guanine-sensing mRNAs. Chem Biol 11:1729–1741

Stoddard CD, Batey RT (2006) Mix-and-match riboswitches. ACS Chem Biol 1:751–754

Stoddard CD, Gilbert SD, Batey RT (2008) Ligand-dependent folding of the three-way junction in the purine riboswitch. RNA 14:675–668

Sudarsan N, Barrick JE, Breaker RR (2003a) Metabolite-binding RNA domains are present in the genes of eukaryotes. RNA 9:644–647

Sudarsan N, Wickiser JK, Nakamura S, Ebert MS, Breaker RR (2003b) An mRNA structure in bacteria that controls gene expression by binding lysine. Genes Dev 17:2688–2697

Wachter A, Tunc-Ozdemir M, Grove BC, Green PJ, Shintani DK, Breaker RR (2007) Riboswitch control of gene expression in plants by splicing and alternative 3′ end processing of mRNAs. Plant Cell 19(11):3437–3450

Walter NG, Harris DA, Pereira MJ, Rueda D (2001) In the fluorescent spotlight: global and local conformational changes of small catalytic RNAs. Biopolymers 61:224–242

Weinberg Z, Barrick JE, Yao Z, Roth A, Kim JN, Gore J, Wang JX, Lee ER, Block KF, Sudarsan N, Neph S, Tompa M, Ruzzo WL, Breaker RR (2007) Identification of 22 candidate structured RNAs in bacteria using the CMfinder comparative genomics pipeline. Nucleic Acids Res 35:4809–4819

Wickiser JK, Cheah MT, Breaker RR, Crothers DM (2005a) The kinetics of ligand binding by an adenine-sensing riboswitch. Biochemistry 44:13404–13414

Wickiser JK, Winkler WC, Breaker RR, Crothers DM (2005b) The speed of RNA transcription and metabolite binding kinetics operate an FMN riboswitch. Mol Cell 18:49–60

Wilkinson KA, Merino EJ, Weeks KM (2005) RNA SHAPE chemistry reveals nonhierarchical interactions dominate equilibrium structural transitions in tRNA(Asp) transcripts. J Am Chem Soc 127:4659–4667

Wilkinson KA, Merino EJ, Weeks KM (2006) Selective 2′-hydroxyl acylation analyzed by primer extension (SHAPE): quantitative RNA structure analysis at single nucleotide resolution. Nat Protoc 1:1610–1616

Williamson JR (2000) Induced fit in RNA-protein recognition. Nat Struct Biol 7:834–837

Winkler WC, Breaker RR (2005) Regulation of bacterial gene expression by riboswitches. Annu Rev Microbiol 59:487–517

Winkler WC, Grundy FJ, Murphy BA, Henkin TM (2001) The GA motif: an RNA element common to bacterial antitermination systems, rRNA, and eukaryotic RNAs. RNA 7:1165–1172

Winkler WC, Cohen-Chalamish S, Breaker RR (2002) An mRNA structure that controls gene expression by binding FMN. Proc Natl Acad Sci U S A 99:15908–15913

Winkler WC, Nahvi A, Roth A, Collins JA, Breaker RR (2004) Control of gene expression by a natural metabolite-responsive ribozyme. Nature 428:281–286

Chapter 11
Ligand Binding and Conformational Changes in the Purine-Binding Riboswitch Aptamer Domains

Jonas Noeske, Janina Buck, Jens Wöhnert(✉), and Harald Schwalbe(✉)

Abstract Riboswitches are highly structured mRNA elements that regulate gene expression upon specific binding of small metabolite molecules. The purine-binding riboswitches bind different purine ligands by forming both canonical Watson–Crick and non-canonical intermolecular base pairs, involving a variety of hydrogen bonds between the riboswitch aptamer domain and the purine ligand. Here, we summarize work on the ligand binding modes of both purine-binding aptamer domains, their conformational characteristics in the free and ligand-bound forms, and their ligand-induced folding. The adenine- and guanine-binding riboswitch aptamer domains display different conformations in their free forms, despite nearly identical nucleotide loop sequences that form a loop–loop interaction in the ligand-bound forms. Interestingly, the stability of helix II is crucial for the formation of the loop–loop interaction in the free form. A more stable helix II in the guanine riboswitch leads to a preformed loop–loop interaction in its free form. In contrast, a less stable helix II in the adenine riboswitch results in a lack of this loop–loop interaction in the absence of ligand and divalent cations.

11.1 Riboswitches: Gene Regulation by mRNA Structure Rearrangement

Gene regulation enables bacteria to adjust their metabolic pathways quickly and efficiently to changing environmental conditions and metabolite availability. Different mechanisms of gene regulation are known but they exclusively depend on protein

H. Schwalbe
Institut für Organische Chemie und Chemische Biologie, Zentrum für Biomolekulare Magnetische Resonanz, Johann Wolfgang Goethe-Universität, Max-von-Laue-Strasse 7, N160-314, 60438 Frankfurt am Main, Germany
e-mail: schwalbe@nmr.uni-frankfurt.de

J. Wöhnert
Institut für Molekulare Biowissenschaften, Zentrum für Biomolekulare Magnetische Resonanz, Johann Wolfgang Goethe-Universität, Max-von-Laue-Strasse 9, N200-2.04, 60438 Frankfurt am Main, Germany
e-mail: woehnert@bio.uni-frankfurt.de

N.G. Walter et al. (eds.) *Non-Protein Coding RNAs*
doi: 10.1007/978-3-540-70840-7_11, © Springer-Verlag Berlin Heidelberg 2009

factors. Within the last 5 years, a novel mechanism of gene regulation has been discovered that is based on RNA structural rearrangements in mRNAs. This mechanism exploits the binding of an effector molecule of low molecular weight to a structured RNA domain, called riboswitch. Ligand binding modulates gene expression by inducing a conformational change (Winkler and Breaker 2005). Riboswitches are mainly found in the 5′-untranslated region (5′-UTR) of many bacterial mRNAs where they modulate expression of downstream genes by transcription termination or inhibition of translation initiation in response to the cellular level of a certain effector molecule. In this process, the RNA directly senses the free concentration of the effector molecule. This is in contrast to all other known gene regulation systems where a protein factor functions as the sensor. Until today, more than 15 different riboswitches have been discovered that sense a wide variety of small molecules such as adenosylcobalamine (Nahvi et al. 2002; Nou and Kadner 2000), flavine mononucleotide (FMN) (Mironov et al. 2002; Winkler et al. 2002b), thiamine pyrophosphate (TPP) (Winkler et al. 2002a), the purine nucleobases guanine (Mandal et al. 2003) and adenine (Mandal and Breaker 2004), the nucleoside 2′-deoxyguanosine (Kim et al. 2007), S-adenosylmethionine (SAM) (Epshtein et al. 2003; McDaniel et al. 2003; Winkler et al. 2003), the amino acids lysine (Sudarsan et al. 2003b) and glycine (Mandal et al. 2004), glucosamine-6-phosphate (GlcN6P) (Winkler et al. 2004), the queosine precursor preQ1 (Meyer et al. 2008; Roth et al. 2007), and Mg^{2+} (Cromie et al. 2006; Dann et al. 2007). Common to all riboswitches is the modulation of expression of those genes that are involved in the biosynthesis and/or the transport of the metabolite that triggers riboswitch function. Riboswitches therefore use the elegant mechanism of feedback regulation in fundamental metabolic pathways, a mechanism already known for proteins.

Riboswitches are generally composed of two modular domains, the aptamer domain and the expression platform. The aptamer domain is the metabolite binding region in a riboswitch. It shows a very high degree of conservation between different organisms in both its nucleotide sequence and secondary structure. The expression platform is located downstream to the metabolite binding aptamer domain. The sequence of the expression platform does not show a high degree of conservation; it determines, however, the mechanism by which the riboswitch regulates gene expression. In a transcriptional gene regulatory mechanism, the expression platform contains a nucleotide sequence that can alternatively fold into either a transcriptional terminator hairpin structure or a transcriptional antiterminator structure. The transcriptional terminator is a GC-rich nucleotide stretch that folds into a hairpin structure followed by a stretch of U residues which, when coupled, leads to premature abortion of transcription. The transcriptional antiterminator is a secondary structure that forms in order to prevent the formation of the terminator hairpin, thereby allowing transcription to proceed. The two mRNA secondary structures, the terminator and antiterminator conformation, are mutually exclusive and therefore exist in a structural equilibrium that is determined by the metabolite binding status of the aptamer domain. Metabolite binding to the aptamer domain favors one or the other secondary structure (Fig. 11.1).

On the translational level, the riboswitch-mediated gene regulatory mechanism involves a secondary structure that exposes or occludes the Shine–Dalgarno

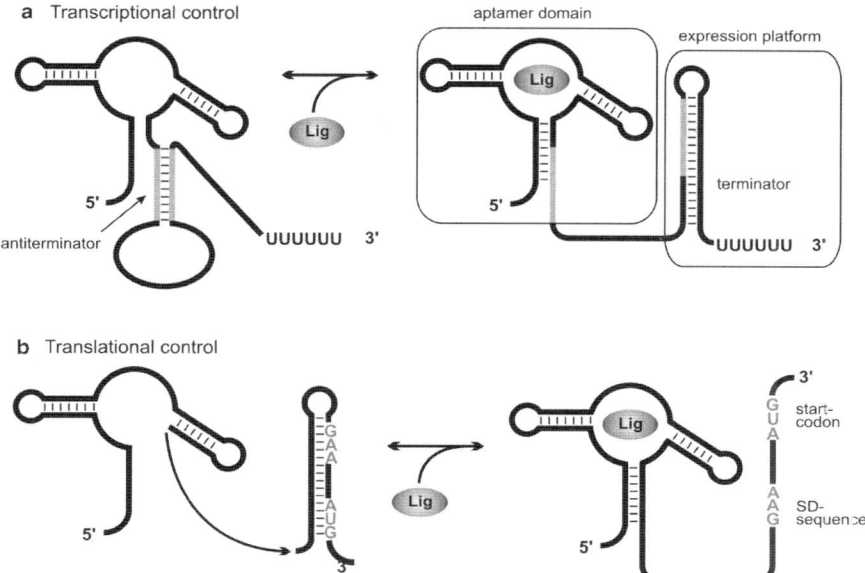

Fig. 11.1 Schematic representation of transcriptional and translational control exerted by riboswitches. (**a**) In the absence of ligand an antiterminator structure forms that prevents formation of a transcriptional terminator hairpin. At elevated cellular concentration of ligand, the ligand binds to the aptamer domain and impairs the formation of the antiterminator, thereby favoring the terminator hairpin. The presence of the terminator hairpin leads to premature abortion of transcription. (**b**) In the absence of ligand a structure is formed that occludes the SD-sequence and the start codon. Ribosomal access to the SD-sequence is prevented resulting in inhibition of translation initiation. At high cellular concentrations of ligand, ligand-binding to the aptamer domain induces a structural rearrangement so that the SD-sequence and the start codon are accessible to the ribosome for translation initiation

(SD)-sequence, depending on the cellular metabolite concentration. The inability of the ribosome to access the SD sequence, which is masked in a helical structure, leads to repression of gene expression by inhibiting translation initiation (Fig. 11.1).

Besides transcriptional termination and inhibition of translation initiation it was found that certain riboswitches can also exert their gene regulatory activity by mediation of mRNA-processing. For example, the GlcN6P-dependent riboswitch was shown to undergo metabolite-induced self-cleavage (Winkler et al. 2004), as opposed to mRNA secondary structure rearrangements, in the course of gene regulation, making this riboswitch a natural ribozyme. The presence of GlcN6P induces a cleavage reaction in the *glmS*-mRNA upstream of the metabolite binding site with kinetics that are accelerated 1,000-fold compared to the absence of GlcN6P (Winkler et al. 2004). The 3'-cleavage product containing a 5'-hydroxy terminus is further subject to RNase J1 degradation resulting in downregulation of *glmS*-gene expression in *B. subtilis* (Collins et al. 2007).

Initially, riboswitches were found in bacteria and it was believed that this gene regulation mechanism is a distinct feature of bacteria. However, mRNA sequence analysis revealed the occurrence of the TPP-riboswitch motif in eukaryotes (Sudarsan et al. 2003a). Recently, it was shown that the filamentous fungus *Neurospora crassa* controls splicing of the *NMR1*-gene by TPP-dependent intron selection. In the presence of TPP, splicing generates a transcript that contains upstream and out-of-frame translation initiation codons, resulting in a decrease of gene expression (Cheah et al. 2007). In several plant species, the TPP-riboswitch motif was found in the 3'-UTR of the thiamine biosynthetic gene *THIC* such that intron selection becomes TPP-dependent. In the presence of TPP, splicing leads to a transcript that is processed differently compared to the transcript produced in the absence of TPP. This distinct mRNA-processing pattern leads to a higher transcript turn-over rate and thereby downregulates gene expression (Wachter et al. 2007).

Metabolite binding to their cognate riboswitch aptamer domains occurs with affinities from 5 nM to 30 μM for SAM and glycin, respectively (Mandal et al. 2004; Winkler et al. 2003). The strength of these RNA-small molecule interactions is comparable to previously observed protein–small molecule interactions. However, the dissociation constant of certain riboswitch–ligand complexes measured at thermodynamic equilibrium are found to differ significantly from the metabolite concentration needed to result in a gene regulatory effect in vitro (McDaniel et al. 2003). Such observations indicate that, at least in some cases, the metabolite-binding mechanism that triggers gene regulation might be kinetically driven.

The interaction between the metabolite and the riboswitch RNA is very intricate. In general, most of the ligand's functional groups are recognized by the RNA leading to a high degree of ligand specificity. The SAM-binding riboswitch distinguishes precisely between its natural ligand SAM and *S*-adenosyl homocysteine (SAH), which lacks a single methyl group and is therefore bound 100-fold less tightly than SAM (Winkler et al. 2003). Interestingly, negatively charged metabolites such as TPP, FMN, and GlcN6P bind to their cognate riboswitch RNA more tightly than their unphosphorylated and hence uncharged counter parts (Winkler et al. 2002a, b, 2004). This observation seems contradictory given the polyanionic character of RNA. The X-ray structures of the aptamer domain of the TPP-binding riboswitch in complex with TPP as well as the *glmS* ribozyme in complex with GlcN6P explain these findings. Here, the negatively charged phosphate group of the metabolite is bridged by a Mg^{2+}-ion to functional groups of the RNA, thereby creating attractive forces (reviewed in Schwalbe et al. 2007).

11.2   Evolution of Riboswitches

The high sequence conservation of riboswitch aptamer domains between different organisms hints at an early origin in evolution. The occurrence of the TPP-binding riboswitch motif in all three domains of life suggests that this motif had already existed when eukaryotes evolved from bacteria about 1.5 billion years ago. In addition, the coenzymes adenosylcobalamine, FMN, SAM, and TPP that are ligands of certain riboswitch RNAs could have played an important role in catalytic processes in a hypothetical RNA world (White 1976). In such an RNA world that might have existed billions of years ago before the evolution of proteins, all events of life, such as the catalysis of pivotal processes and storage of genetic information, would have been mediated by RNA molecules (Joyce 1989). However, it is also possible that the known riboswitch classes are not primordial relics but rather represent the best suited system to carry out this specific biological function. According to such reasoning, riboswitch motifs could have evolved independently and have converged to the currently known riboswitch classes. This hypothesis is favored by the fact that three different classes of SAM-binding riboswitches are known that share no structural similarities (Corbino et al. 2005; Fuchs et al. 2006; Winkler et al. 2003). It also appears plausible that a combination of both evolutionary processes could have taken place to give rise to the currently known riboswitches.

The high degree of homology in primary and secondary structure between the adenine- and guanine-binding riboswitches, together with the specific mutation of a single nucleotide at position 74 that is involved in the formation of an intermolecular Watson–Crick base pair with the bound ligand, suggests that these two riboswitch classes could have evolved from a common evolutionary ancestor. Since the guanine riboswitch apparently occurs more frequently than the adenine riboswitch, the adenine riboswitch could have evolved from the guanine riboswitch by a spontaneous C74U mutation. The sole mutation of a specific nucleotide is sufficient to switch the specificity of the guanine riboswitch to make it responsive to adenine (Mandal and Breaker 2004).

11.2.1   The Class of Purine-Binding Riboswitches

The class of the purine-binding riboswitches binds specifically the nucleobases guanine (Mandal et al. 2003) and adenine (Mandal and Breaker 2004) and, along with the preQ1-binding riboswitch (Roth et al. 2007), belongs to the smallest riboswitches found to date. Purine-binding riboswitches are found in different proteobacteria; they modulate gene expression either by transcription termination or inhibition of translation initiation (Fig. 11.1). The metabolite-binding aptamer domain of the purine-binding riboswitches has a length of up to 80 nucleotides and forms three helices that are connected by a three-way junction. Helices II and III favor a length of six to seven base pairs and allow for one to two base mismatches. Each helix has a seven nucleotide loop that can form a pseudoknot structure. The helical loops and the three-way junction show a phylogenetic conservation of greater than 90% (Mandal et al. 2003; Mandal and Breaker 2004). Despite a high degree of conservation in their primary structure and an identical secondary structure, the guanine- and the adenine-binding riboswitches strictly differ in nucleotide 74, where the guanine-binding riboswitch bears a cytidine and the adenine-binding

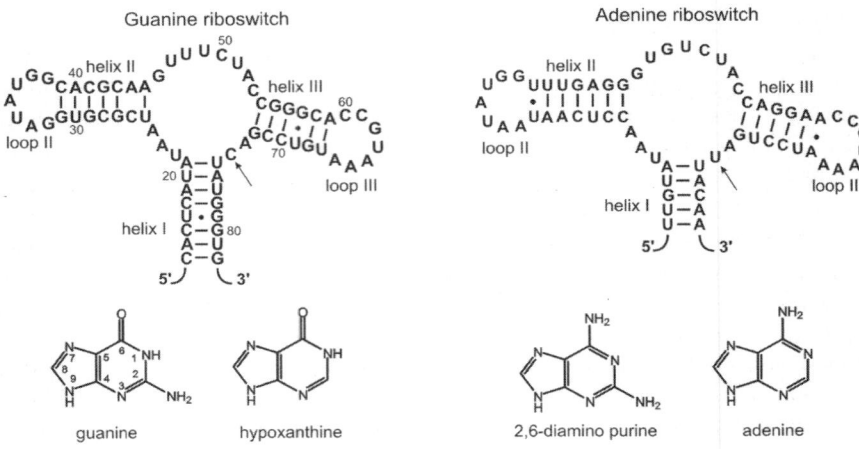

Fig. 11.2 Secondary structures of the *xpt-pbuX* guanine- and the *pbuE* adenine-binding riboswitch aptamer domains (both from *B. subtilis*) and the chemical structure of their respective ligands. Ligand specificity is mediated by a pyrimidine residue at position 74 (position indicated by an arrow). The guanine-binding riboswitch bears a cytidine at this position whereas the adenine-binding riboswitch bears a uridine. Guanine binds tighter to the guanine-binding riboswitch than hypoxanthine due to its additional 2-amino group which can form additional intermolecular hydrogen bonds with the RNA. The same rationale applies for 2,6-diamino purine binding as opposed to adenine binding to the adenine-binding riboswitch. Long-range base pairing interactions can be formed between loops II and III leading to the formation of loop–loop interactions in both purine-binding riboswitches

riboswitch bears a uridine (Fig. 11.2). Biochemical analysis showed that the *xpt-pbuX* guanine-binding riboswitch from *B. subtilis* specifically binds guanine as well as hypoxanthine, an intermediate in purine-metabolism. In contrast to the guanine-binding riboswitch, the *pbuE* adenine-binding riboswitch from *B. subtilis* binds adenine and the closely related 2,6-diaminopurine with high affinity. It was shown that the nucleotide 74 in the aptamer domain of the purine-binding riboswitch determines the specificity of the respective riboswitch by forming an intermolecular Watson–Crick base pair with the bound metabolite.

11.3 NMR-Based Investigation of the Purine-Binding Riboswitches

NMR spectroscopy is a powerful tool to study RNA structure, dynamics, conformational changes, and the interaction of RNA with ligands such as proteins, other RNA-molecules, small molecules, and ions (Fürtig et al. 2003). The most prominent disadvantage in NMR spectroscopic investigation of RNA is its size limitation. Increasing size of the RNA leads to poorly resolved resonances of the RNA due to spectral overlap. So far, RNAs with a size of up to 100 nucleotides have been studied by NMR spectroscopy with the help of selective and segmental isotopic labeling strategies or by subdividing the RNA into smaller domains that could be studied separately (Chen et al. 2006; D'Souza et al. 2004; Lukavsky et al. 2003). For the structure determination of an RNA molecule, short-range structural restraints are derived from ^{1}H,^{1}H-NOESY-spectra and from three bond scalar couplings that give rise to short-range through-space distance restraints and torsion angle restraints, respectively. However, for larger RNA molecules that contain long helical segments or multiple domains separated by bulge regions the incorporation of long-range structural restraints in structure calculation is necessary to accurately define the degree of helical bending and the relative orientation of different helical regions in an RNA molecule. These long-range structural restraints are derived from residual dipolar couplings (RDCs) that yield information about the relative orientation of specific bond vectors spread over the entire RNA molecule. Next to the overall molecular geometry, RDCs can also give valuable information about inter-domain conformational motions and dynamics (Zhang et al. 2006). In general, a range of different NMR-spectroscopic parameters allows probing RNA dynamics over a wide time range, including fast librational motions and base pair opening and closing processes, which occur on a time scale of picoseconds and seconds, respectively, to name a few (Ferner et al. 2008). Besides RNA-structure elucidation and RNA-dynamic studies, NMR spectroscopy can also study the course of chemical reactions and ligand binding processes to a target RNA in real time, with kinetics in the range from seconds to minutes and hours (Furtig et al. 2007).

Furthermore, NMR spectroscopy is applicable to study weak binding processes and transient interactions (Fig. 11.3). RNA-ligand binding processes can be traced by observing NMR resonances either of the target RNA or of the ligand. In case of binding of a divalent cation to a target RNA molecule with low affinity the NMR chemical shift of a nucleus involved in the binding gradually changes from its values in the free form to the value of the cation-bound form in the course of a titration experiment. Tracing the cation-induced chemical shift perturbations as a function of the cation's concentration gives rise to the equilibrium binding constant (K_D) of the binding process after non-linear regression. In this way, a Mg^{2+}-binding site was localized in the loop–loop interaction of the adenine-binding riboswitch aptamer domain and the K_D for this interaction could be determined to be in the lower millimolar range (Noeske et al. 2007b) (Fig. 11.3a). Weak and transient cation binding induces chemical shifts that can be observed in NMR spectra. Typically, saturation curves are observed that show that cation binding is in fast exchange on the NMR time scale. Structure induction at remote sites often leads to additional signals that acquire exchange protection due to tertiary rearrangements (Wu and Tinoco 1998). Such signals are typically in slow exchange on the NMR time scale and do not lead to a continuous shift at increasing Mg^{2+}-concentration but to a population weighted increase in signal intensity. Thereby, local (transient) binding of Mg^{2+} and tertiary structure induction can often be differentiated in ^{1}H,^{15}N-HSQC spectra.

Observing the NMR signal of the ligand also allows studying low-affinity and non-specific binding processes by NMR spectroscopy since the line width of the

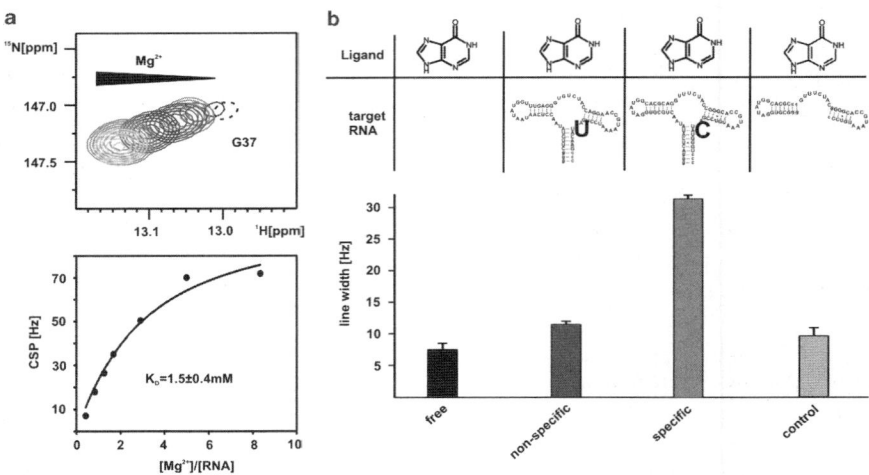

Fig. 11.3 Revealing low affinity and non-specific binding processes by NMR spectroscopy. (**a**) Imino group resonances of the RNA can be used to reveal and localize low-affinity cation-binding sites in a target RNA. Tracing chemical shift perturbations (CSP) of the imino group resonance as a function of the cation's concentration gives rise to the equilibrium binding constant (K_D) of the cation to its binding site in the target RNA after non-linear regression. (**b**) Revealing a non-specific binding process by following the line widths of the non-exchangeable NMR signals of the natural guanine ligand analogue hypoxanthine in the presence of no RNA, the adenine-binding riboswitch RNA as a non-specific binder, the guanine-binding riboswitch RNA as a specific binder, and the guanine-binding riboswitch lacking the ligand binding pocket as control. The line width of the ligand hypoxanthine increases with its increasing apparent time-averaged molecular weight. Therefore, the line width of the free ligand is narrow in contrast to a large line width in the presence of the high molecular guanine-riboswitch aptamer domain, reflecting the formation of a high affinity RNA–ligand complex. The line width of hypoxanthine in the presence of the non-specific binding adenine-binding riboswitch RNA is significantly larger compared to the control and the free ligand representing the non-specific association of the ligand with the high molecular weight target RNA to a low-affinity encounter complex

ligand's resonance is a very sensitive reporter that indicates whether the ligand shows no, weak or tight interaction with a target RNA. A ligand freely tumbling in solution shows an NMR signal with narrow line widths according to the ligand's low molecular weight. Weak interactions with a high molecular weight target RNA increase the time-averaged molecular weight of the ligand resulting in slightly increased line widths compared to the non-interacting free ligand. The formation of a tight RNA–ligand complex gives rise to largely broadened line widths for the ligand that now represent the high molecular weight of the whole complex (Fig. 11.3b). For example, observing the non-exchangeable NMR signals of the ligand hypoxanthine revealed the presence of a low-affinity encounter complex preceding specific hypoxanthine binding to the guanine-binding riboswitch aptamer domain (Buck et al. 2007).

Conformational changes in RNA usually result in new base pairing interactions which can be traced by the appearance of their imino group NMR signals. The imino group resonances present a probe that supplies valuable information about RNA conformational changes since they are spread all over the RNA molecule. A prerequisite for such studies is the assignment of all imino group resonances based on $^1H,^1H$-NOESY experiments that allow short-range through-space correlation of protons. In RNA, the imino protons of uridines and guanosines resonate at a proton frequency range of 10–15 ppm. In this spectral region, exclusively these RNA resonances are observable so that this is a well resolved region. Imino group resonances are only detectable if the imino group is involved in a hydrogen bond and is therefore protected from exchange with the solvent water. In canonical Watson–Crick base pairing interactions the guanosine N1H1 and the uridine N3H3 imino group form a hydrogen bond with the cytidine N3 and the adenosine N1, respectively. These base pairing interactions render the imino groups detectable by NMR and, conversely, imino groups that are not involved in base pairing interactions are undetectable.

There are only very few experimental parameters that allow for the direct observation of hydrogen bonds. Usually, they are deduced from spatial proximity in NMR spectroscopic or X-ray crystallographic structures. However, NMR spectroscopy is a technique to provide direct evidence for hydrogen bonds involved in canonical base pairings (Dingley and Grzesiek 1998), non-canonical base pairings such as reverse Hoogsteen A:U base pairs, and G:A mismatches (Wöhnert et al. 1999), and all other hydrogen bonds of the type N-H$\cdots$N, where an imino group is the hydrogen bond donor and a nitrogen is the hydrogen bond acceptor. These hydrogen bonds are detectable due to their cross hydrogen bond scalar coupling ($^{2h}J_{NN}$) from the donor to the acceptor nitrogen atom. The detection of hydrogen bonds in canonical and non-canonical base pairs gives valuable information about RNA secondary and tertiary structures and in the case of the purine-binding riboswitch aptamer domains it was used to determine that the purine ligands bind to the riboswitch-RNAs by an intermolecular Watson–Crick base pair (Noeske et al. 2005) (Fig. 11.4a).

11.4   Characteristics of the Ligand Binding Mechanism to the Purine-Binding Riboswitch RNAs

Both the aptamer domains of the adenine- and the guanine-binding riboswitches bind their respective purine ligand in a very similar way. In both cases, two base pairing interactions are observed between the purine ligand and the riboswitch RNA. One of these is an intermolecular Watson–Crick base pair: a G:C base pair for guanine bound to the guanine-binding riboswitch aptamer domain and an A:U base pair for adenine bound to the adenine-binding riboswitch aptamer domain. The second intermolecular base pairing is an unprecedented interaction between the N3/N9 edge of the purine ligand and a uridine residue of the RNA, leading to the formation of at least two additional intermolecular hydrogen bonds. Those two intermolecular hydrogen bonds are formed between the N9H9 imino group and the N3 of the purine ligand and the C4 carbonyl group and the N3H3 imino group of a uridine residue of

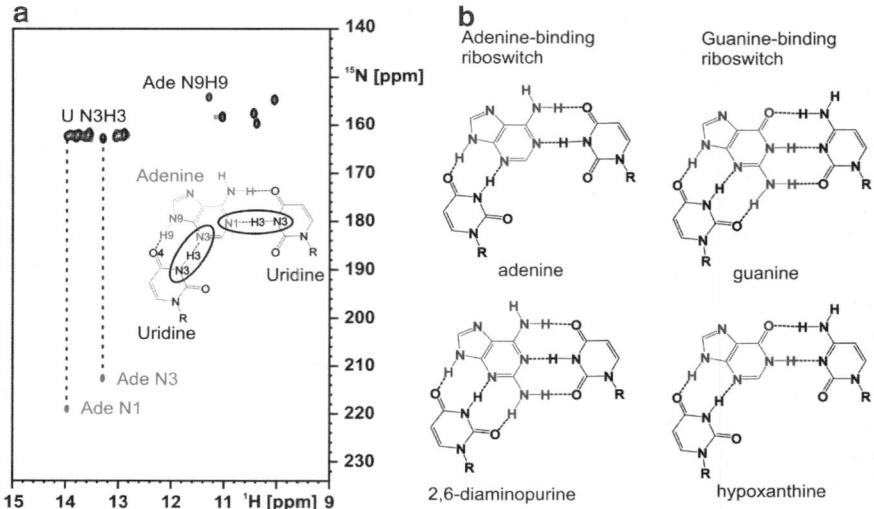

Fig. 11.4 Revealing the ligand binding mode of the purine-binding riboswitch RNAs by detecting intermolecular hydrogen bonds between the ligand and the riboswitch RNA by NMR spectroscopy. (**a**) NMR spectroscopy can directly detect N–H⋯N type hydrogen bonds by an HNN-COSY experiment. The spectrum shown reveals that the ligand adenine forms an intermolecular Watson–Crick A:U base pair with a uridine of the RNA and that the N3 of the ligand adenine is recognized by the N3H3 of an additional uridine of the RNA. Inset: Therefore, one could directly conclude that the ligand adenine binds the adenine-binding riboswitch RNA by an intermolecular base triple where the N3H3s of two different uridines (black) function as hydrogen bond donors and the N1 and the N3 of the adenine (grey) function as hydrogen bond acceptors. The detected hydrogen bonds are encircled. (**b**) Summarizing the binding modes of adenine and 2,6-diamino purine to the adenine-binding riboswitch and guanine and hypoxanthine to the guanine-binding riboswitch. The existence of each intermolecular hydrogen bond indicated was unambiguously detected by NMR spectroscopy

the RNA, respectively (Fig. 11.4b). The basis for the discrimination of the purine ligand by the respective riboswitch RNA is the formation of an intermolecular Watson–Crick base pairing interaction. In contrast, the recognition motif of the N3/N9 edge of the purine ligand by a uridine of the riboswitch RNA is a structural motif that is common to both riboswitch RNAs where the RNA recognizes either adenine or guanine (Batey et al. 2004; Noeske et al. 2005; Serganov et al. 2004). The combination of these two recognition motifs rationalizes the previous observation that the mutation of a single cytidine to a uridine residue switches the specificity of the guanine-binding riboswitch from guanine to adenine and vice versa without altering the binding affinity of the interaction (Mandal and Breaker 2004).

The base pairing interaction between the purine ligand and the riboswitch RNA (Fig. 11.4b) also accounts for the higher affinity to the adenine-binding riboswitch RNA of the closely related synthetic adenine analogue 2,6-diamino purine compared to adenine (Mandal and Breaker 2004). Here, the additional 2-amino group (Fig. 11.2) can form two additional hydrogen bonds with two C2 carbonyl groups of the RNA. Similar arguments could account for the comparable affinity of 2-amino purine and adenine to the adenine-binding riboswitch RNA. Here, the loss of one hydrogen bond typical for A:U Watson–Crick base pairing interactions due to the loss of the amino group at position 6 of adenine is compensated by two hydrogen bonds formed by the additional 2-amino group of the ligand. The affinity loss of hypoxanthine to the guanine-binding riboswitch RNA compared to guanine can be explained in a similar way.

Interestingly, no interactions could be observed by NMR spectroscopy between the Hoogsteen edge of the purine ligand and the riboswitch RNA. The Hoogsteen edge of guanine and adenine differ in that guanine bears two hydrogen bond acceptors (N7 and O6) whereas adenine bears one hydrogen bond acceptor (N7) and one hydrogen bond donor (6-amino group). However, there is only one nucleotide in the ligand-binding core of the purine riboswitch RNA that systematically differs between the guanine- and adenine-binding riboswitches (Mandal et al. 2003; Mandal and Breaker 2004). This nucleotide is the one involved in the formation of the intermolecular Watson–Crick base pairing interaction. If the purine ligands were to form an interaction with the RNA including both functional groups of the Hoogsteen edge, it would require a systematic difference of the adenine- and guanine-binding riboswitches in the sequence of their ligand-binding core, which is not observed (Mandal et al. 2003; Mandal and Breaker 2004). Alternatively, an RNA–ligand interaction involving the Hoogsteen edge of the purine ligand could be restricted to the N7-position of the purine ligand. Using the guanine analogue 7-deaza guanine it was ruled out by NMR that either an imino group or an amino group of the RNA specifically recognizes the N7 of the purine ligand. Due to the higher affinity to the guanine-binding riboswitch RNA of guanine compared to 7-deaza guanine it was hypothesized that a 2′-OH group of the RNA, the only remaining functional group that could act as a hydrogen bond donor, forms a hydrogen bond to the N7 of the purine ligand in a similar way in both purine-binding riboswitch RNAs. This hypothesis was confirmed by the X-ray structures of the purine-binding riboswitch RNAs in complex with their respective purine ligand (Batey et al. 2004; Serganov et al. 2004).

The described base pairing interaction between the purine ligand and the purine-binding riboswitch RNA has been observed never before in an RNA structure or in an RNA–ligand complex. Neither the SAM-binding riboswitch (Montange and Batey 2006), which displays a similar affinity to its cognate ligand (Winkler et al. 2003), nor in vitro evolved aptamers that specifically bind GTP or ATP with similar and lower affinities, respectively, recognize the nucleobase moiety of their bound ligands by a classic Watson–Crick base pairing interaction (Carothers et al. 2006; Dieckmann et al. 1996; Jiang et al. 1996). Therefore, the high affinities of the purine-binding riboswitches towards their cognate ligands are not necessarily dependent on the formation of a Watson–Crick base pairing interaction between the RNA and the bound ligand. Rather, it appears that the unique ligand binding mechanism of the adenine- and guanine-binding riboswitch RNAs reflects the specific requirements for selectivity in order to maintain riboswitch regulatory function in the cellular environment. Considering this notion, the specific recognition of the N9H9 imino group of the bound purine-ligand seems to be favorable since this functional group is sterically blocked in nucleotides and nucleotide derivatives that are abundant within the cell and are therefore excluded from hydrogen bonding.

11.5   Conformational Changes in the Aptamer Domain of the Purine-Binding Riboswitches

Many target RNAs bind their ligand by an *induced fit* mechanism. In this case, the ligand binding pocket is mostly unstructured and highly dynamic in its free form compared to the ligand-bound form. Ligand binding to an unstructured and dynamic binding pocket brings about high entropic costs that negatively influence the free energy of the ligand binding process (Leulliot and Varani 2001). However, if ligand binding follows an *induced fit* mechanism, the ligand can be completely engulfed by the RNA which enables the RNA to specifically recognize a large number of the ligand's functional groups and therefore maximizing the formation of intermolecular interactions leading to a higher degree of specificity. Some RNA molecules lower the entropic contribution to the free energy of the binding process by displaying a preformed binding pocket to the ligand (Klein and Ferre-D'Amare 2006; Ohlenschläger et al. 2004; Stoldt et al. 1999). A general feature of riboswitch aptamer domains is the occurrence of conserved sequence elements that are not directly involved in ligand binding. Structural investigations of riboswitch RNA–ligand complexes have revealed that these elements participate in RNA tertiary structures that stabilize the global fold of the RNA (Batey et al. 2004; Edwards and Ferre-D'Amare 2006; Montange and Batey 2006; Noeske et al. 2007a, b; Serganov et al. 2004, 2006; Thore et al. 2006). In some cases it was shown that the elimination of these structural elements is detrimental for ligand binding, indicating that they are essential for riboswitch function (Batey et al. 2004; Lemay et al. 2006; McDaniel et al. 2005). Studies on natural ribozymes like the hammerhead (De la Pena et al. 2003; Khvorova et al. 2003), hairpin (Murchie et al. 1998), and *glmS* (Klein and Ferre-D'Amare 2006) ribozymes reveal that a specific global fold of these ribozymes is favored by tertiary structure interactions distant to the catalytic centre. The formation of tertiary structure increases the catalytic activity of the ribozyme and decreases its dependency on divalent cations.

Ligand-induced conformational changes of the aptamer domains of the purine-binding riboswitches were elucidated by liquid-state NMR spectroscopy. NMR spectroscopy is extremely powerful since it allows studying conformational changes in solution at atomic resolution. The NMR data show no evidence for a structured ligand binding core of the purine-binding aptamer domains in their free form (Noeske et al. 2007a, b) (Fig. 11.5). In this respect, no differences could be observed between the guanine- and adenine-binding riboswitches, which only differ in a pyrimidine residue at position 74 that forms the intermolecular Watson–Crick base pairing upon binding of the purine ligand. Therefore, both purine-binding riboswitch RNAs bind their cognate purine ligand by an *induced fit* mechanism where

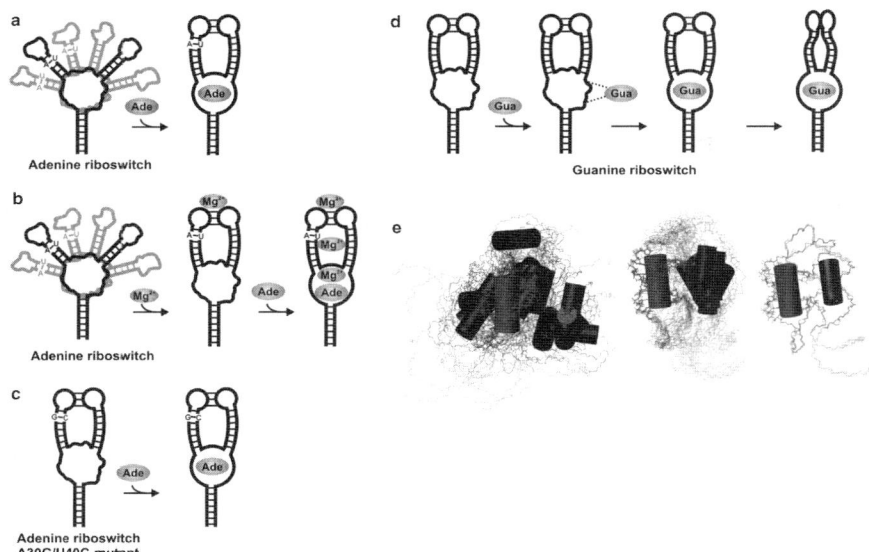

Fig. 11.5 Folding pathways of the adenine- and the guanine-binding aptamer domains. (**a**) The free form of the *pbuE* adenine riboswitch is an ensemble of heterogeneous and interconverting conformations. Adenine binds by an *induced fit* mechanism in both the ligand-binding core region and the loop–loop interactions. (**b**) In the presence of Mg^{2+}-ions the ensemble of heterogeneous and interconverting conformations of the free adenine-binding riboswitch converges into a single conformation in which the loop–loop interactions are formed. Subsequent binding of adenine only folds the ligand-binding core region. (**c**) A stabilizing mutation in the apical region of helix II leads to the formation of the loop–loop interactions in the free form of the adenine-binding aptamer domain in the absence of divalent cations. Adenine binds by an *induced fit* mechanism only in the ligand-binding core. (**d**) In contrast to the adenine riboswitch the loop–loop interactions are already formed in the free form of the *xpt-pbuX* guanine-binding riboswitch. Time-resolved NMR-spectroscopic kinetic investigations showed that the guanine analogue hypoxanthine binds in a two-step mechanism to the guanine-binding riboswitch, preceded by a low-affinity encounter complex. In a first step, hypoxanthine-induced folding of the ligand binding core is observed. The second step results in tightening of helices II and III. (**e**) By using MD simulations, the NMR derived kinetic results of the hypoxanthine-induced folding of the guanine-binding riboswitch are translated into structural information. The relative orientation of the helical elements fluctuates largely in the structural ensemble of the free form of the RNA despite helices II and III being anchored together at their tips by long-range base pairs between their capping loops. Hypoxanthine binding to the ligand binding core further stabilizes the RNA, resulting in less fluctuation between the helical elements. In the last step of the folding process helices II and III tighten, giving rise to the ligand bounded RNA with well defined interhelical angles similar to those in the X-ray structure. Ligand binding to both purine-binding aptamer domains is independent on divalent cations

the ligand is completely engulfed by the RNA, giving rise to intermolecular interactions involving all functional groups of the ligand. The formation of this high affinity RNA–ligand complex is independent of divalent cations in both purine-binding riboswitch RNAs.

In the ligand-bound form the loops capping helices II and III form loop-loop interactions in both purine-binding riboswitches as revealed by X-ray crystallography (Batey et al. 2004; Serganov et al. 2004). These loop–loop interactions mainly consist of two base quadruples that comprise five long-range base pairing interactions between nucleotides of loop II and III. Two of these long-range base pairings are canonical G:C Watson–Crick base pairs built by G38:C60 and G37:C61. Additionally, three non-canonical base pairs are formed by U34:A65, A33:A66, and A35:A64. The base pairings G37:C61 and U34:A65 as well as G38:C60 and A33:A66 each form a base quadruple, thereby stabilizing the tertiary structure motif of the loop-loop interaction. In the free form of the guanine-binding riboswitch, helices II and III are positioned parallel to each other with their tips anchored together by the loop–loop interactions. However, a certain degree of flexibility is observed for the helical elements in the free form of the guanine-binding riboswitch.

Time-resolved NMR studies of the binding kinetics of the natural guanine analogue hypoxanthine to the guanine-binding riboswitch aptamer domain identified a two-step binding mechanism (Buck et al. 2007). In the first step, an *induced fit* mechanism of the ligand binding core is revealed in the time range of 18.9–23.6 s. In a slower second step with a time range of 27.1–30.7 s the packing pattern of the helical elements is changed, leading to helical tightening. This rather slow binding mechanism is preceded by a low-affinity encounter complex in which hypoxanthine forms only weak and transient interactions with the RNA with millisecond off-rates.

The existence of a hypoxanthine-guanine riboswitch aptamer domain encounter complex preceding the actual specific binding process was revealed by comparing the line widths of non-exchangeable resonances of the hypoxanthine ligand in its free form, in the presence of the adenine-binding riboswitch as a non-specific binder, in the presence of the guanine-binding riboswitch as a specific binder, and in the presence of the guanine-binding riboswitch lacking the ligand binding pocket as a control (Buck et al. 2007). The ligand's line width is directly correlated to the apparent time-averaged molecular weight of the ligand or the complex to which the ligand is bound. Therefore, the line widths of the free hypoxanthine are narrow corresponding to its low molecular weight, whereas the line widths of the specific hypoxanthine-guanine riboswitch aptamer domain complex are large corresponding to the high molecular weight of this high affinity complex. The line widths of hypoxanthine in the presence of the adenine-binding riboswitch aptamer domain are significantly larger than its line widths either in the presence of the control RNA or in its free form indicating that hypoxanthine associates with the purine riboswitch RNA in a non-specific way preceding productive binding (Fig. 11.3b).

In order to provide insight into the structural transitions involved in the ligand-induced folding process of the guanine-binding riboswitch RNA, the kinetic results of the binding process of hypoxanthine to the RNA were translated into structural information using molecular dynamics (MD) simulation. By using NMR data as restraints in MD simulations, structural ensembles of the guanine riboswitch RNA could be described for each step in its hypoxanthine-induced folding process (Buck et al. 2007). The free RNA and the RNA in a low-affinity encounter complex are represented by structural ensembles in which the relative orientations of the helical elements largely fluctuate relatively to each other, despite helices II and III being anchored together at their tips by long-range base pairs between loops II and III. Hypoxanthine binding to the ligand binding core stabilizes the RNA, resulting in largely decreased fluctuations of the helical elements. This stabilization of the core region results in stabilization of the distant loop–loop interaction, which in turn leads to a well defined structure of the RNA–ligand complex similar to its X-ray structure (Fig. 11.5d, e).

In summary, the loop–loop interactions are preformed in the free form of the guanine-binding riboswitch, and ligand binding occurs by an *induced fit* mechanism of the ligand binding core and leads to tightening of helices II and III, thereby reducing their intrinsic dynamics in the ligand-bound form of the RNA (Fig. 11.5d, e). The presence of Mg^{2+}-ions further stabilizes the loop–loop interactions in the free form of the guanine-binding riboswitch, but is not essential for their formation (Noeske et al. 2007a).

In contrast to the *xpt-pbuX* guanine riboswitch aptamer domain, the closely related *pbuE* adenine riboswitch aptamer domain shows no tertiary structure interactions between the loops II and III in the free form. NMR data indicate that the free form of the adenine-binding aptamer domain is an ensemble of heterogeneous and interconverting conformations in which alternate base pairing interactions compete with the formation of the loop–loop interaction. Despite the loop–loop interaction being largely absent in the free form, addition of adenine to the adenine-binding aptamer domain results in complete RNA–ligand complex formation in the absence of divalent cations (Fig. 11.5a). This complex formation indicates that the presence of the loop–loop interaction is not a prerequisite for ligand binding to the purine-binding riboswitches, as shown by mutational studies (Gilbert et al. 2007; Lemay et al. 2006; Noeske et al. 2007a). Therefore, ligand binding to the aptamer domain of the adenine-binding riboswitch follows an *induced fit* mechanism in both the ligand binding core and the loop–loop interactions. However, in the absence of the purine ligand Mg^{2+}-ions can induce the loop–loop interactions as shown by fluorescence (Lemay et al. 2006; Rieder et al. 2007) and NMR (Noeske et al. 2007b) experiments. More specifically, Mg^{2+}-ions force the ensemble of interconverting conformations of the free form of the adenine-binding riboswitch aptamer domain to converge into a single conformation in which the loop–loop interactions are stably formed (Fig. 11.5b).

Since the sequences of loops II and III of the investigated adenine- and guanine-binding riboswitch RNAs are identical except for residue 32 in loop II (Fig. 11.2), which is not involved in any stabilizing hydrogen bonds (Batey et al. 2004; Serganov et al. 2004), the different behavior in the formation of the loop–loop interaction in the free form of these RNAs seems surprising. It suggests that the stability of the apical base pairs in helix II may be crucial for the formation of the loop–loop interaction in the free form of the purine riboswitch aptamer domain. Sequence analysis indicates that all known adenine-binding riboswitches only bear canonical A:U Watson–Crick or non-canonical base pairs in the three apical base pairs of helix II. In contrast, in more than 97% of the known guanine-binding riboswitches mostly canonical Watson–Crick base pairs with at least one canonical G:C base pair are found at these positions in helix II (Kim et al. 2007; Mandal et al. 2003). Therefore, the reduced stability of the apical base pairs of helix II might determine the absence of the loop–loop interactions in the free form of the adenine-binding riboswitch. A A30G/U40C mutant of the *pbuE* adenine-binding riboswitch in which a weaker A:U base pair in the apical region of helix II was replaced by a more stable G:C base pair indeed displayed the loop-loop interactions in the free form in the absence of any divalent cations (Noeske et al. 2007b), confirming that the stability of helix II is crucial for the formation of loop–loop interactions in the free form of the purine riboswitch RNAs (Fig. 11.5c). Interestingly, comparing the ligand affinities of different guanine-binding riboswitches reveals that the ligand affinity increases proportionally to the stability of helix II (Mulhbacher and Lafontaine 2007).

The different conformations of the free form of the adenine- and the guanine-binding riboswitches lead to different entropic penalties upon ligand binding due to changed folding pathways. The higher entropic cost for the ligand-induced folding of the adenine riboswitch might be compensated partially by the free energy of the formation of the loop–loop interactions.

These changed folding pathways could be used as a starting point for the design of selective inhibitors against purine-binding riboswitches (Blount and Breaker 2006). The high similarity of the aptamer-ligand complexes of the purine riboswitches will render it difficult to design drugs that specifically target only a subset of purine riboswitches such as those of a given bacterium. Yet, small sequence variations that give rise to different conformations of the free form of the riboswitch could provide a starting point for the design of selective drugs. For some riboswitches it may be interesting to design drugs that specifically inhibit the formation of the loop–loop interactions as an essential feature of the ligand-bound purine aptamer domain (Batey et al. 2004). Riboswitches with a preformed loop–loop interaction may then not be affected in their function.

References

Batey RT, Gilbert SD, Montange RK (2004) Structure of a natural guanine-responsive riboswitch complexed with the metabolite hypoxanthine. Nature 432:411–415

Blount KF, Breaker RR (2006) Riboswitches as antibacterial drug targets. Nat Biotechnol 24: 1558–1564

Buck J, Fürtig B, Noeske J, Wöhnert J, Schwalbe H (2007) Time-resolved NMR methods resolving ligand-induced RNA folding at atomic resolution. Proc Natl Acad Sci U S A 104:15699–15704

Carothers JM, Davis JH, Chou JJ, Szostak JW (2006) Solution structure of an informationally complex high-affinity RNA aptamer to GTP. RNA 12:567–579

Cheah MT, Wachter A, Sudarsan N, Breaker RR (2007) Control of alternative RNA splicing and gene expression by eukaryotic riboswitches. Nature 447:497–500

Chen Y, Fender J, Legassie JD, Jarstfer MB, Bryan TM, Varani G (2006) Structure of stem-loop IV of Tetrahymena telomerase RNA. EMBO J 25:3156–3166

Collins JA, Irnov I, Baker S, Winkler WC (2007) Mechanism of mRNA destabilization by the glmS ribozyme. Genes Dev 21:3356–3368

Corbino KA, Barrick JE, Lim J, Welz R, Tucker BJ, Puskarz I, Mandal M, Rudnick ND, Breaker RR (2005) Evidence for a second class of S-adenosylmethionine riboswitches and other regulatory RNA motifs in alpha-proteobacteria. Genome Biol 6:R70

Cromie MJ, Shi Y, Latifi T, Groisman EA (2006) An RNA sensor for intracellular Mg2+. Cell 125:71–84

D'Souza V, Dey A, Habib D, Summers MF (2004) NMR structure of the 101-nucleotide core encapsidation signal of the Moloney murine leukemia virus. J Mol Biol 337:427–442

Dann CE III, Wakeman CA, Sieling CL, Baker SC, Irnov I, Winkler WC (2007) Structure and mechanism of a metal-sensing regulatory RNA. Cell 130:878–892

De la Pena M, Gago S, Flores R (2003) Peripheral regions of natural hammerhead ribozymes greatly increase their self-cleavage activity. EMBO J 22:5561–5570

Dieckmann T, Suzuki E, Nakamura GK, Feigon J (1996) Solution structure of an ATP-binding RNA aptamer reveals a novel fold. RNA 2:628–640

Dingley AJ, Grzesiek S (1998) Direct observation of hydrogen bonds in nucleic acid base pairs by internucleotide 2JNN couplings. J Am Chem Soc 120:8293–8297

Edwards TE, Ferre-D'Amare AR (2006) Crystal structures of the thi-box riboswitch bound to thiamine pyrophosphate analogs reveal adaptive RNA-small molecule recognition. Structure 14:1459–1468

Epshtein V, Mironov AS, Nudler E (2003) The riboswitch-mediated control of sulfur metabolism in bacteria. Proc Natl Acad Sci U S A 100:5052–5056

Ferner J, Villa A, Duchardt E, Widjajakusuma E, Wöhnert J, Stock G, Schwalbe H (2008) NMR and MD studies of the temperature-dependent dynamics of RNA YNMG-tetraloops. Nucleic Acids Res 36:1928–1940

Fuchs RT, Grundy FJ, Henkin TM (2006) The S(MK) box is a new SAM-binding RNA for translational regulation of SAM synthetase. Nat Struct Mol Biol 13:226–233

Furtig B, Buck J, Manoharan V, Bermel W, Jaschke A, Wenter P, Pitsch S, Schwalbe H (2007) Time-resolved NMR studies of RNA folding. Biopolymers 86:360–383

Fürtig B, Richter C, Wöhnert J, Schwalbe H (2003) NMR spectroscopy of RNA. Chembiochem 4:936–962

Gilbert SD, Love CE, Edwards AL, Batey RT (2007) Mutational analysis of the purine riboswitch aptamer domain. Biochemistry 46:13297–13309

Jiang F, Kumar RA, Jones RA, Patel DJ (1996) Structural basis of RNA folding and recognition in an AMP-RNA aptamer complex. Nature 382:183–186

Joyce GF (1989) RNA evolution and the origins of life. Nature 338:217–224

Khvorova A, Lescoute A, Westhof E, Jayasena SD (2003) Sequence elements outside the hammerhead ribozyme catalytic core enable intracellular activity. Nat Struct Biol 10:708–712

Kim JN, Roth A, Breaker RR (2007) Guanine riboswitch variants from *Mesoplasma florum* selectively recognize 2'-deoxyguanosine. Proc Natl Acad Sci U S A 104:16092–16097

Klein DJ, Ferre-D'Amare AR (2006) Structural basis of glmS ribozyme activation by glucosamine-6-phosphate. Science 313:1752–1756

Lemay JF, Penedo JC, Tremblay R, Lilley DM, Lafontaine DA (2006) Folding of the adenine riboswitch. Chem Biol 13:857–868

Leulliot N, Varani G (2001) Current topics in RNA-protein recognition: control of specificity and biological function through induced fit and conformational capture. Biochemistry 40: 7947–7956

Lukavsky PJ, Kim I, Otto GA, Puglisi JD (2003) Structure of HCV IRES domain II determined by NMR. Nat Struct Biol 10:1033–1038

Mandal M, Breaker RR (2004) Adenine riboswitches and gene activation by disruption of a transcription terminator. Nat Struct Mol Biol 11:29–35

Mandal M, Boese B, Barrick JE, Winkler WC, Breaker RR (2003) Riboswitches control fundamental biochemical pathways in *Bacillus subtilis* and other bacteria. Cell 113:577–586

Mandal M, Lee M, Barrick JE, Weinberg Z, Emilsson GM, Ruzzo WL, Breaker RR (2004) A glycine-dependent riboswitch that uses cooperative binding to control gene expression. Science 306:275–279

McDaniel BA, Grundy FJ, Artsimovitch I, Henkin TM (2003) Transcription termination control of the S box system: direct measurement of *S*-adenosylmethionine by the leader RNA. Proc Natl Acad Sci U S A 100:3083–3088

McDaniel BA, Grundy FJ, Henkin TM (2005) A tertiary structural element in S box leader RNAs is required for *S*-adenosylmethionine-directed transcription termination. Mol Microbiol 57: 1008–1021

Meyer MM, Roth A, Chervin SM, Garcia GA, Breaker RR (2008) Confirmation of a second natural preQ1 aptamer class in Streptococcaceae bacteria. RNA 14:685–695

Mironov AS, Gusarov I, Rafikov R, Lopez LE, Shatalin K, Kreneva RA, Perumov DA, Nudler E (2002) Sensing small molecules by nascent RNA: a mechanism to control transcription in bacteria. Cell 111:747–756

Montange RK, Batey RT (2006) Structure of the S-adenosylmethionine riboswitch regulatory mRNA element. Nature 441:1172–1175

Mulhbacher J, Lafontaine DA (2007) Ligand recognition determinants of guanine riboswitches. Nucleic Acids Res 35:5568–5580

Murchie AI, Thomson JB, Walter F, Lilley DM (1998) Folding of the hairpin ribozyme in its natural conformation achieves close physical proximity of the loops. Mol Cell 1:873–881

Nahvi A, Sudarsan N, Ebert MS, Zou X, Brown KL, Breaker RR (2002) Genetic control by a metabolite binding mRNA. Chem Biol 9:1043

Noeske J, Richter C, Grundl MA, Nasiri HR, Schwalbe H, Wöhnert J (2005) An intermolecular base triple as the basis of ligand specificity and affinity in the guanine- and adenine-sensing riboswitch RNAs. Proc Natl Acad Sci U S A 102:1372–1377

Noeske J, Buck J, Fürtig B, Nasiri HR, Schwalbe H, Wöhnert J (2007a) Interplay of 'induced fit' and preorganization in the ligand induced folding of the aptamer domain of the guanine binding riboswitch. Nucleic Acids Res 35:572–583

Noeske J, Schwalbe H, Wöhnert J (2007b) Metal-ion binding and metal-ion induced folding of the adenine-sensing riboswitch aptamer domain. Nucleic Acids Res 35:5262–5273

Nou X, Kadner RJ (2000) Adenosylcobalamin inhibits ribosome binding to btuB RNA. Proc Natl Acad Sci U S A 97:7190–7195

Ohlenschläger O, Wöhnert J, Bucci E, Seitz S, Hafner S, Ramachandran R, Zell R, Görlach M (2004) The structure of the stemloop D subdomain of coxsackievirus B3 cloverleaf RNA and its interaction with the proteinase 3C. Structure 12:237–248

Rieder R, Lang K, Graber D, Micura R (2007) Ligand-induced folding of the adenosine deaminase A-riboswitch and implications on riboswitch translational control. Chembiochem 8: 896–902

Roth A, Winkler WC, Regulski EE, Lee BW, Lim J, Jona I, Barrick JE, Ritwik A, Kim JN, Welz R, et al. (2007) A riboswitch selective for the queuosine precursor preQ(1) contains an unusually small aptamer domain. Nat Struct Mol Biol 14:308–317

Schwalbe H, Buck J, Furtig B, Noeske J, Wohnert J (2007) Structures of RNA switches: insight into molecular recognition and tertiary structure. Angewandte Chemie (International ed) 46: 1212–1219

Serganov A, Yuan YR, Pikovskaya O, Polonskaia A, Malinina L, Phan AT, Hobartner C, Micura R, Breaker RR, Patel DJ (2004) Structural basis for discriminative regulation of gene expression by adenine- and guanine-sensing mRNAs. Chem Biol 11:1729–1741

Serganov A, Polonskaia A, Phan AT, Breaker RR, Patel DJ (2006) Structural basis for gene regulation by a thiamine pyrophosphate-sensing riboswitch. Nature 441:1167–1171

Stoldt M, Wöhnert J, Ohlenschläger O, Görlach M, Brown LR (1999) The NMR structure of the 5S rRNA E-domain-protein L25 complex shows preformed and induced recognition. EMBO J 18:6508–6521

Sudarsan N, Barrick JE, Breaker RR (2003a) Metabolite-binding RNA domains are present in the genes of eukaryotes. RNA 9:644–647

Sudarsan N, Wickiser JK, Nakamura S, Ebert MS, Breaker RR (2003b) An mRNA structure in bacteria that controls gene expression by binding lysine. Genes Dev 17:2688–2697

Thore S, Leibundgut M, Ban N (2006) Structure of the eukaryotic thiamine pyrophosphate riboswitch with its regulatory ligand. Science 312:1208–1211

Wachter A, Tunc-Ozdemir M, Grove BC, Green PJ, Shintani DK, Breaker RR (2007) Riboswitch control of gene expression in plants by splicing and alternative 3′ end processing of mRNAs. Plant Cell 19(11):3437–3450

White HBI (1976) Coenzymes as fossils of an earlier metabolic state. J Mol Evol 7:101–104

Winkler WC, Breaker RR (2005) Regulation of bacterial gene expression by riboswitches. Annu Rev Microbiol 59:487–517

Winkler W, Nahvi A, Breaker RR (2002a) Thiamine derivatives bind messenger RNAs directly to regulate bacterial gene expression. Nature 419:952–956

Winkler WC, Cohen-Chalamish S, Breaker RR (2002b) An mRNA structure that controls gene expression by binding FMN. Proc Natl Acad Sci U S A 99:15908–15913

Winkler WC, Nahvi A, Sudarsan N, Barrick JE, Breaker RR (2003) An mRNA structure that controls gene expression by binding S-adenosylmethionine. Nat Struct Biol 10:701–707

Winkler WC, Nahvi A, Roth A, Collins JA, Breaker RR (2004) Control of gene expression by a natural metabolite-responsive ribozyme. Nature 428:281–286

Wöhnert J, Dingley AJ, Stoldt M, Görlach M, Grzesiek S, Brown LR (1999) Direct identification of NH...N hydrogen bonds in non-canonical base pairs of RNA by NMR spectroscopy. Nucleic Acids Res 27:3104–3110

Wu M, Tinoco I Jr (1998) RNA folding causes secondary structure rearrangement. Proc Natl Acad Sci U S A 95:11555–11560

Zhang Q, Sun X, Watt ED, Al-Hashimi HM (2006) Resolving the motional modes that code for RNA adaptation. Science 311:653–656

Chapter 12
The RNA–Protein Complexes of *E. coli* Hfq: Form and Function

Taewoo Lee and Andrew L. Feig(✉)

Abstract *E. coli* Hfq is an RNA binding protein that has received significant attention due to its role in post-transcriptional gene regulation. Hfq facilitates the base-pairing between mRNAs and ncRNAs leading to translational activation, translational repression and/or degradation of mRNAs – the bacterial analog of the RNA interference pathway. Hfq is the bacterial homolog of the Sm and Lsm proteins and has a similar doughnut-shaped structure. This review summarizes what is known about the diverse physiological roles of Hfq and how its structure facilitates a diverse array of RNA–protein and protein–protein interactions. These interactions are put into context to explain the models of how Hfq is thought to help facilitate post-transcriptional gene regulation by non-coding RNAs in bacteria.

12.1 Introduction

Escherichia coli Hfq is a small protein composed of 102 amino acids, originally identified as a host factor required for the replication of bacteriophage Qβ (Franze de Fernandez et al. 1968, 1972; Shapiro et al. 1968). At that time, since phage replication was its only known physiological function, the intrinsic importance of Hfq was underestimated. Two decades later, Hfq received renewed attention when it was discovered that *hfq* mutants had broadly pleiotropic phenotypes including decreased growth rate, stress sensitivity, and reduced virulence (Tsui et al. 1994; Robertson and Roop 1999). Subsequent studies revealed that these strains displayed altered expression patterns of approximately 50 proteins, suggesting that Hfq is integrally involved in the regulation of many genes (Muffler et al. 1996, 1997). Part of the misregulation in the Δ*hfq* strain resulted from inefficient translation of *rpoS*, a gene which encodes an alternative sigma factor expressed in response to stress or at the entry of stationary phase. However, a mechanistic explanation of how Hfq affected *rpoS* translation remained elusive.

A.L. Feig

Department of Chemistry, Wayne State University, 5101 Cass Avenue Detroit, MI 48202, USA
e-mail: afeig@chem.wayne.edu

N.G. Walter et al. (eds.) *Non-Protein Coding RNAs*   249
doi: 10.1007/978-3-540-70840-7_12, © Springer–Verlag Berlin Heidelberg 2009

The discovery of the non-coding RNA (ncRNA) OxyS was a major turning point in the history of Hfq and its biological function. OxyS was shown to be expressed after oxidative stress and to downregulate *rpoS* translation in an Hfq-dependent manner (Altuvia et al. 1997; Zhang et al. 1998). As a result, the search began for other ncRNAs with stress-induced expression patterns. These RNomics analyses revealed a wide range of small regulatory RNAs in bacteria responsible for post-transcriptional gene regulation (Wassarman 2002; Masse et al. 2003b; Gottesman 2005; Storz et al. 2005). While the expression of these ncRNAs is Hfq independent, many of them fail to regulate their mRNA targets in the absence of this important RNA-binding protein.

Hfq is broadly (but not universally) conserved across the bacterial kingdom. Hfq homologs are usually 70–110 amino acids, with a highly variable C-terminal extension that typically shows little homology across species (Fig. 12.1). Bioinformatics and structural studies further showed that Hfq is evolutionarily related to a large family of

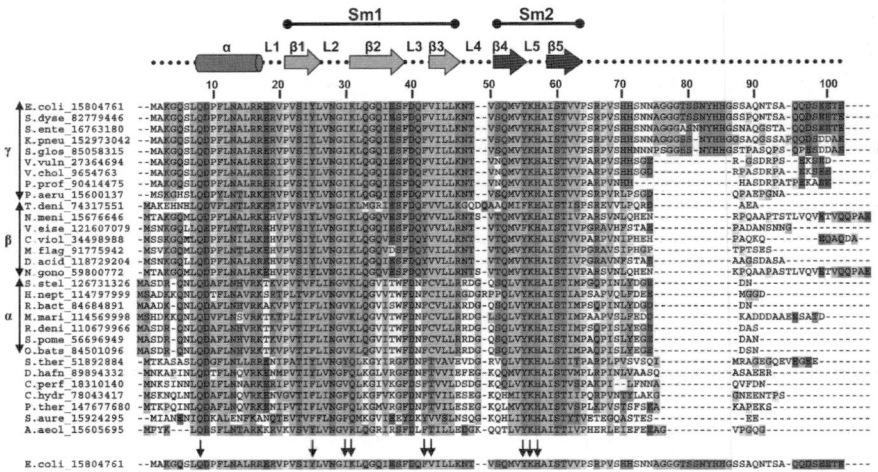

Fig. 12.1 Multiple sequence alignment of Hfq proteins from 30 bacteria. Organisms corresponding to sequences are listed with their accession numbers. Letters α, β, and γ on the left of the organism list indicate α-, β-, and γ-proteobacteria, respectively. Other organisms belong to firmicutes and aquificae phyla. The numbering at the top corresponds to *E. coli* sequence. Secondary structure of Hfq is schematized based on the *E. coli* Hfq crystal structure. Sm1 and Sm2 motifs are colored in green and blue, respectively. *E. coli* Hfq sequence is presented once more at the bottom to indicate those residues that are specifically mentioned in the text (*arrows*). Species (from top to bottom): γ-proteobacteria: *Escherichia coli, Shigella dysenteriae, Salmonella enterica, Klebsiella pneumoniae, Sodalis glossinidius, Vibrio vulnificus, Vibrio cholerae, Photobacterium profundum, Pseudomonas aeruginosa.* β-proteobacteria: *Thiobacillus denitrificans, Neisseria meningitides, Verminephrobacter eiseniae, Chromobacterium violaceum, Methylobacillus flagellatus, Delftia acidovorans, Neisseria gonorrhoeae.* α-proteobacteria: *Sagittula stellata, Hyphomonas neptunium, Rhodobacterales bacterium, Maricaulis maris, Roseobacter denitrificans, Silicibacter pomeroyi, Oceanicola batsensis.* Firmicutes: *Symbiobacterium thermophilum, Desulfitobacterium hafniense, Clostridium perfringens, Carboxydothermus hydrogenoformans, Pelotomaculum thermopropionicum, Staphylococcus aureus.* Aquificae: *Aquifex aeolicus* (See figure insert for color reproduction)

Sm and Lsm proteins (Arluison et al. 2002; Moller et al. 2002; Schumacher et al. 2002; Sun et al. 2002; Zhang et al. 2002; Sauter et al. 2003; Wilusz and Wilusz 2005). Sm and Lsm proteins are ubiquitous RNA-binding proteins found throughout eukaryotes and archaea. The Sm proteins form a heteroheptameric ring shaped structure, bind single-stranded uridine-rich sequences of the small nuclear RNA (snRNA), and are important for snRNP biogenesis and function (Battle et al. 2006). Lsm proteins perform more diverse roles in biology and have greater heterogeneity due to the many Lsm genes. Like their Sm cousins, eukaryotic Lsm proteins also form heteroheptameric ring-shaped structures. Here however, the identity of the polypeptide subunits varies depending on the function and cellular localization of the Lsm complex. This diversity among Lsm complexes is further enriched through protein–protein interactions such that they recruit appropriate binding partners to build intricate assemblies involved in various aspects of RNA metabolism including the processing of nuclear RNAs, RNA localization and RNA turnover (Beggs 2005; Khusial et al. 2005). The structural similarity between Lsm proteins and Hfq makes it possible to hypothesize that these proteins share many common physiological functions, but direct data linking Hfq to some of the more diverse activities of the Lsm proteins is still lacking.

In this chapter, the structure and function of Hfq are reviewed, with a particular emphasis on those roles associated with the biology of regulatory RNAs. We pay particular attention to the shape of the Hfq hexamer and how this topology assists its function by providing distinct interaction surfaces. This topology allows it to simultaneously bind two or more RNAs and various accessory proteins. This distribution of binding surfaces is critical to its function in post-transcriptional gene regulation and the stimulation of stress responses in vivo.

12.2 Hfq Facilitates Post-Transcriptional Gene Regulation

12.2.1 Overview of the Pathway

Hfq participates in post-transcriptional gene regulation (Lease et al. 1998; Majdalani et al. 1998; Masse and Gottesman 2002) filling a physiological niche similar to the components of the eukaryotic RNA interference (RNAi) pathway. Figure 12.2 illustrates this analogy. In eukaryotic systems, precursors of the small regulatory RNAs derive from nuclear Pol II transcription, exogenous sources, or in the case of transitive gene regulation, cytoplasmic RNA-dependent RNA polymerases (Zamore 2002; Tang 2005; Chu and Rana 2007). The primary microRNA (miRNA) transcripts are typically processed first by Drosha in the nucleus and then by Dicer in the cytoplasm to form the mature 19–25 nt double-stranded RNAs that are substrates for incorporation into the RNA induced silencing complex (RISC). This extensive trail of processing and transport is required to differentiate functional ncRNAs from various splicing-derived degradation intermediates constantly being produced in the nucleus.

In contrast, the primary transcripts of bacterial ncRNAs are fully functional in bacteria. These RNAs typically vary in length from 80 to about 400 nucleotides and

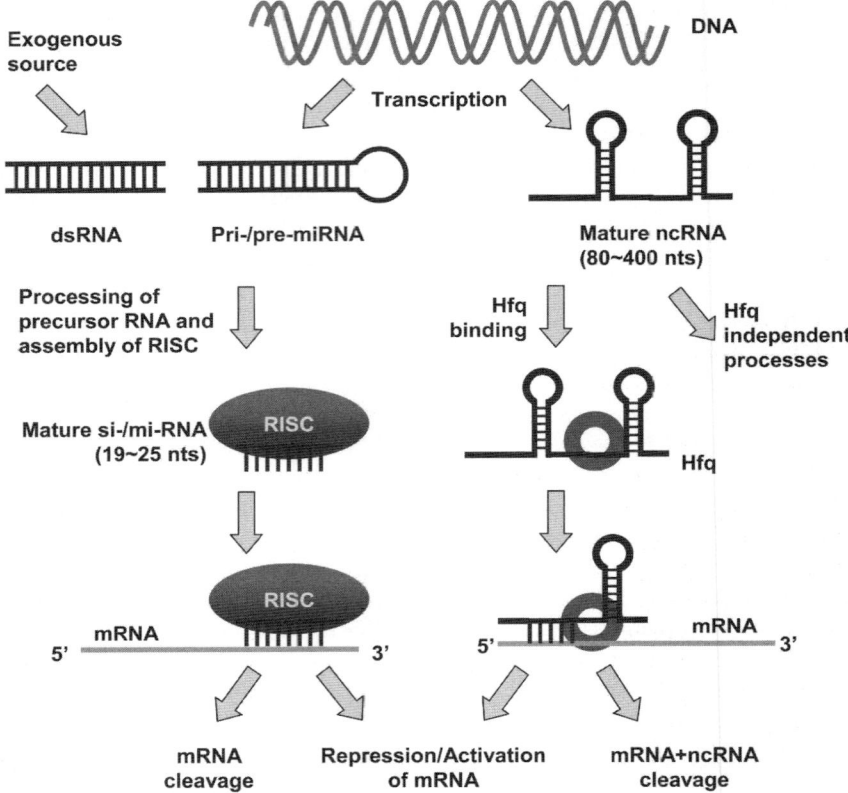

Fig. 12.2 Comparison of RNA interference to bacterial post-transcriptional gene regulation. Biogenesis of mi-/siRNA in eukaryotes is programmed genetically (miRNA) or can be triggered by double-stranded RNA from an exogenous source (siRNA). The initial precursors have to be cleaved to 19–25 nt double strand RNAs by Dicer, a multidomain enzyme of the RNase III family, and they are assembled into an RNA induced silencing complex (RISC) mediated by RISC loading complex (RLC). When RISC is loaded with RNA, only one strand of RNA is chosen. In contrast, ncRNAs are transcribed into mature single stranded form from DNA in response to stresses. The length of ncRNAs varies from 80 to 400 nts. Some of ncRNAs bind to Hfq to regulate the gene expression through base-pairing with their target mRNAs. ncRNAs usually anneal to the 5′ UTR in the proximity of ribosomal binding site (RBS) to repress/activate the expression of mRNA or to lead to the degradation of mRNA. mi-/siRISC, however, bind to the 3′ UTR to induce mRNA cleavage or to repress translation (See figure insert for color reproduction)

commonly fold into two or more hairpin structures with an intervening single-stranded region that is A/U-rich (Fig. 12.3) (Moller et al. 2002; Zhang et al. 2002; Brescia et al. 2003; Geissmann and Touati 2004). About 80 ncRNAs in *E. coli* have been discovered by biochemical and computational search, and more than 20 are known to bind Hfq while the others participate in Hfq independent processes (Gottesman 2004,

Fig. 12.3 Representative examples of the secondary structures of ncRNAs. A/U rich single-stranded sequence is located adjacent to a helix, and this region was shown to be a binding site of Hfq. The binding motif is indicated on DsrA, RyhB, OxyS and Spot42

2005). As with other RNA-guided processes, these ncRNAs have significant sequence complementarity with their mRNA targets. Hfq binds both the ncRNAs as well as the mRNAs and facilitates intermolecular base-pairing. Upon pairing, both eukaryotic and bacterial post-transcriptional gene regulation systems produce similar outcomes, most commonly translational repression or mRNA degradation with a few examples of translational activation. In the case of RISC, the programmed complex can act catalytically to degrade many copies of the targeted mRNAs whereas complexes involved in translational repression act stoichiometrically.

Similar questions have been raised in the case of Hfq. Hfq was proposed to induce the base-pairing of two RNAs and be released from the ternary complex for recycling in a catalytic model (Lease and Woodson 2004; Arluison et al. 2007). It is not yet clear whether Hfq forms such transient complexes or remains bound to these RNA to further facilitate the interaction with additional proteins that act on the mRNA after recruitment of the appropriate ncRNA (Mohanty et al. 2004; Morita et al. 2006). In neither case is the system catalytic with respect to the ncRNA, as RNA degradation is in the RISC complex (Masse et al. 2003a). More work is required to define the specific mechanisms by which Hfq induces these downstream effects on gene regulation.

12.2.2 Translational Repression

One example of translational repression by an ncRNA involves the action of DsrA on *hns* (Lease et al. 1998; Lease and Belfort 2000). DsrA, an 87 nt ncRNA from the *E. coli* genome, is expressed at low temperature and H–NS is an abundant nucleoid protein. DsrA is complementary to a region surrounding the ribosomal binding site (RBS) of *hns*, such that annealing of DsrA to *hns* blocks the RBS, effectively

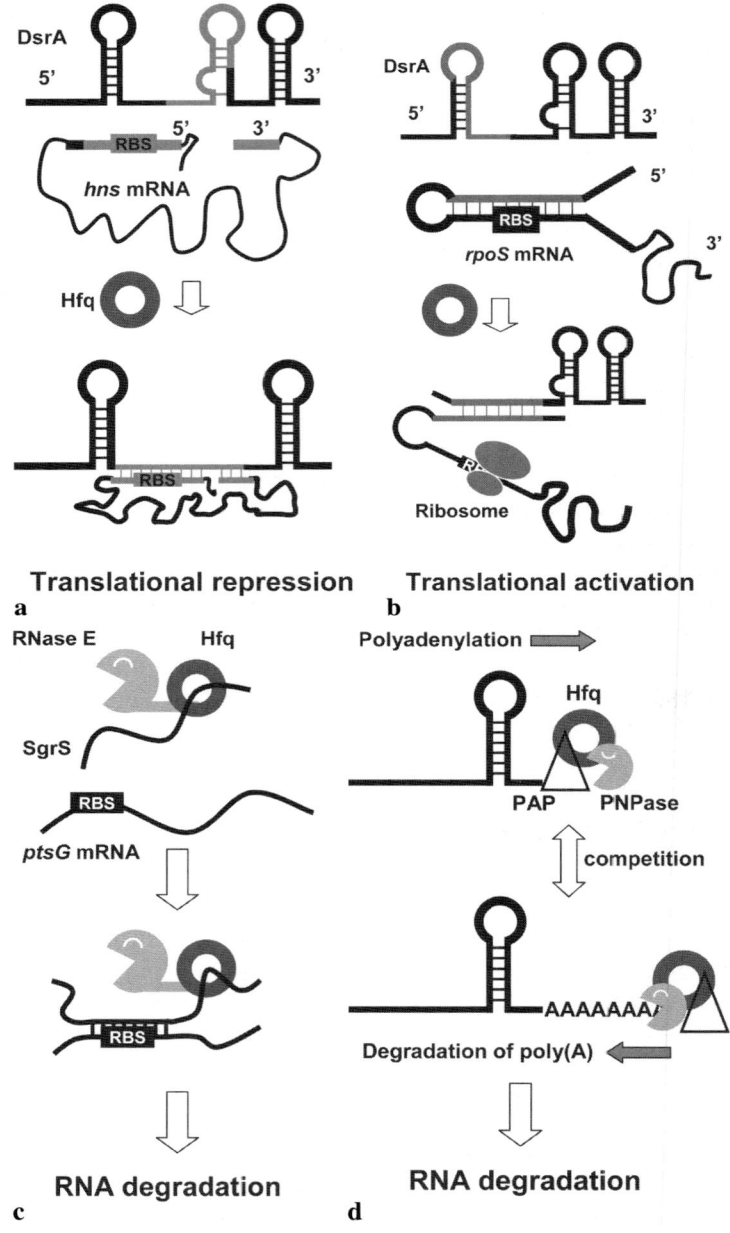

Fig. 12.4 Models for the physiological roles of Hfq. **a** Translational repression of *hns* by DsrA by facilitating base-pairing between them. Annealing of DsrA stem II region on RBS of *hns* prevents ribosome binding. **b** The RBS of *rpoS* mRNA is blocked by folding its 5′ UTR. Pairing between stem I of DsrA and an upstream region of the *rpoS* mRNA refolds this region leading to derepression of translation. **c** Target specific RNA decay is performed by an RNase E-Hfq-ncRNA ribonucleoprotein complex. The ncRNAs acts as a guide sequence to target specific mRNA

preventing translation initiation (Fig. 12.4a). In this case, steric occlusion of the regulatory sequences is sufficient to explain the behavior of the regulatory RNA.

12.2.3   Translational Activation

During cold shock, several regulatory events are coordinated through the action of DsrA on multiple mRNAs. Downregulation of *hns* translation has been described earlier. Concomitantly, DsrA interacts with *rpoS,* leading to its upregulation. The 5′-UTR region of *rpoS* folds into a translationally repressed structure that precludes ribosome binding (Fig. 12.4b). With the help of Hfq, DsrA anneals to the upstream region of *rpoS,* altering this structure and releasing the RBS for ribosome binding and translation initiation (Majdalani et al. 1998). Whether the complex is further delivered to the ribosome for translation is unclear. Together, the *rpoS* and *hns* interactions illustrate the bifunctional nature of DsrA and other bacterial ncRNAs. They have the ability to target multiple messages and to induce individualized outcomes. This mechanism differs somewhat from the RNAi paradigm where seed matches are critical for setting specificity and incomplete complementarity is sufficient beyond that region (Tang 2005; Chu and Rana 2007), but it reflects the nature of the longer ncRNAs typical in the bacterial systems. Here, the targeting sequences can be distinct or can overlap, but do not rely heavily on an identifiable core within the ncRNA sequence. Thus, the manner in which the complementary sequence is presented to the target likely varies between Hfq dependent RNPs and the RISC system.

12.2.4   mRNA Degradation

Hfq is also known to be involved in RNA decay. Two types of decay processes have been reported. One is the targeted mRNA decay coupled to post transcriptional gene silencing, which is similar to eukaryotic siRNAs. The other is a poly(A) polymerase I (PAP I)-dependent mRNA turnover mechanism, which is more akin to nuclear RNA surveillance systems employing 3′ poly(A) tailing of RNAs (Kadaba et al. 2004; LaCava et al. 2005). Both mechanisms likely occur in parallel, but the signals that differentiate the two pathways are not yet clear.

The SgrS:*ptsG* system exemplifies the targeted mRNA decay in bacteria (Fig. 12.4c) (Morita et al. 2004; Kawamoto et al. 2005; Aiba 2007). SgrS induction occurs upon accumulation of glucose-6-phosphate and represses the translation of *ptsG* which encodes a glucose transporter (Vanderpool and Gottesman 2004). Initial repression of *ptsG* by SgrS is analogous to *hns* repression by DsrA, requiring

Fig. 12.4 (Continued) degradation. **d** Hfq is involved in PAP I/PNPase dependent mRNA decay pathway. In this pathway, repetitive cycles of poly(A) tailing of mRNA by PAP I and decay of the poly(A) tail by PNPase leads to degradation of some mRNAs having Rho-independent terminators

Hfq-dependent base-pairing between the two RNAs. At this point, however, the pathways diverge and RNase E is recruited to the complex, simultaneously degrading both the ncRNA and the mRNA (Masse et al. 2003a). This mechanism contrasts with RISC-dependent gene silencing in which Argonaut cleaves only the targeted mRNA, leaving the programmed RISC intact, allowing it to act catalytically. Eventually the RISC complex can be disassembled and the components reprogrammed, but the mechanism of that process is currently unknown.

The role of Hfq in generalized mRNA decay in bacteria has been looked at primarily in the case of RNAs containing Rho-independent terminators, stem loop structures followed by a string of uridine residues (Mohanty et al. 2004). These structures can be challenging for single-stranded $3'\rightarrow5'$ exonucleases to degrade, presumably because they have difficulty invading the terminal helix. In this pathway PAP I triggers decay by adding poly(A) tails to the 3' terminus of the RNA followed by degradation by polynucleotide phosphorylase (PNPase) (Fig. 12.4d) (O'Hara et al. 1995; Mohanty and Kushner 1999; Kushner 2004). Hfq recognizes the 3' termini of these mRNAs and facilitates the interaction with PAP I. In the absence of Hfq, PAP I fails to efficiently tail these messages which consequently have short tails and longer half-lives (Mohanty et al. 2004). This result suggests that Hfq participates in the generalized mRNA decay process by regulating mRNA polyadenylation. Hfq may also serve to weaken the duplex at the base of the stem to improve nuclease accessibility or allow switching back and forth between elongation and decay cycles that give the nucleases multiple attempts to attack the terminal helix.

Recent work has uncovered a nuclear RNA surveillance pathway in yeast that functions in a manner remarkably similar to that described above. Polyadenylation-assisted decay efficiently captures hypomodified tRNAs and other structured RNAs that fail to mature properly (Kadaba et al. 2004, 2006; LaCava et al. 2005; Vanacova et al. 2005; Alexandrov et al. 2006). As with bacterial RNA degradation, polyadenylation is followed by recruitment of the degradation machinery, in this case, the nuclear exosome (Kadaba et al. 2004; LaCava et al. 2005). This pathway has also been linked to the clearance of RNAs required for heterochromatin assembly through RNA-induced transcriptional silencing (Wang et al. 2008). It is not yet clear whether nuclear Lsm complexes participate in this process in any way.

With these functions in mind, we can now consider how the structure of Hfq directly facilitates the diverse biochemical pathways in which it participates.

12.3   Hfq Structure

12.3.1   Hfq Forms Doughnut-Shaped Hexamers

The structures of Hfq homologs from three organisms (*Staphylococcus aureus*, *Escherichia coli*, and *Pseudomonas aeruginosa*) have been solved by X-ray crystallography (Schumacher et al. 2002; Sauter et al. 2003; Nikulin et al. 2005). All show Hfq in its doughnut-shaped hexameric form (Fig. 12.5a), consistent with the earlier

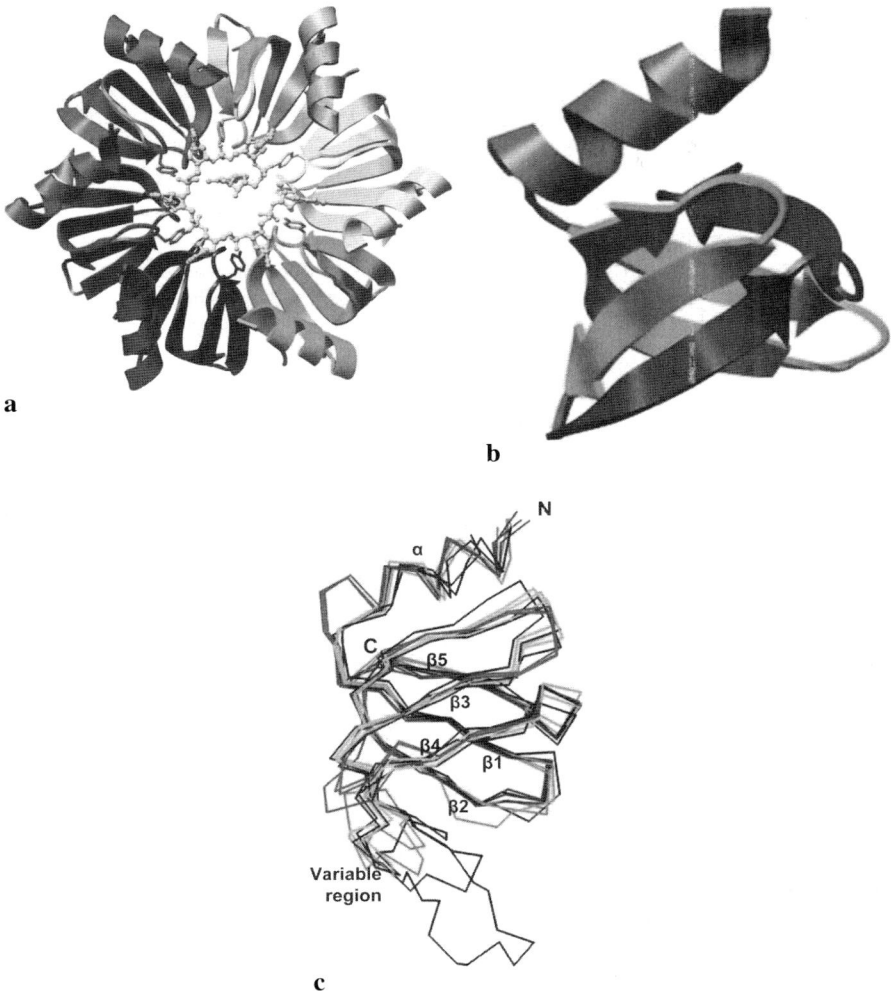

Fig. 12.5 Hfq structures. **a** Hexameric form of Hfq from *S. aureus* is displayed with AU₅G bound in the central cavity. For clarity each subunit is colored differently. Y55 residues are shown from each subunit to illustrate the stacking interactions with RNA bases (PDB ID: 1KQ2). **b** Isolated monomeric unit from the crystal structure of *E. coli* Hfq shows the Sm fold containing an N-terminal α-helix followed by five β-strands forming a barrel-shaped structure. Sm1 and Sm2 motifs are colored in green and blue, respectively. Loop 4, which lies between the Sm1 and Sm2 motifs has variable length in Sm/Lsm proteins, is colored in magenta (PDB ID: 1HK9). **c** Superimposition of the monomeric units from bacterial Hfq, archaeal Lsm and eukaryotic Sm proteins clearly shows the structural similarity especially within Sm1 and Sm2 motifs of these proteins. Hfq monomers from *E.coli, Staphylococcus aureus,* and *Pseudomonas aeruginosa* are colored red. Lsm proteins from *Archaeoglobus fulgidus* (AF-Sm2), *Methanobacterium thermo-autotrophicum*, and *Pyrobaculum aerophilum* are cyan, and Sm proteins from human (SmB and SmD3) and yeast (SmF) are navy blue (See figure insert for color reproduction)

mass spectrometry and electron microscopy results (Moller et al. 2002; Zhang et al. 2002). The torus is 65 Å across and 23 Å thick with a central cavity 12 Å in diameter in the absence of RNA. Each monomer displays a classical Sm fold, characterized by an N-terminal α-helix followed by four β strands (Fig. 12.5b) (Kambach et al. 1999; Mura et al. 2001; Toro et al. 2001). Although Hfq shares little sequence homology with the Sm2 motif (Schumacher et al. 2002), the superposition of the Sm, Lsm and Hfq structures shows superb overlap (Fig. 12.5c), validating the earlier predictions that Hfq was the bacterial Lsm protein.

The *S. aureus* Hfq was co-crystallized in the presence of the short RNA oligonucleotide 5'-AU$_5$G-3'. This RNA is bound along one face of the central pore as shown in Fig. 12.5a (Schumacher et al. 2002). The cavity enlarges slightly to 15 Å upon binding, but is still too small to allow a helical stem to pass through the hole. Since the Hfq binding motif typically involves single-stranded regions flanked by helices (Fig. 12.3) and the hexamer is extremely stable and unlikely to open up and reassemble around an RNA substrate, it is unlikely that RNAs can thread through the cavity.

12.3.2   Oligomerization

The Sm fold allows the formation of exceptionally stable toroidal structures because each monomer contributes to a continuous anti-paralleled β-sheet around the ring. The subunit interface involves contact of strand β4 with β5' of the adjacent monomer (Fig. 12.6) (Kambach et al. 1999; Mura et al. 2001; Toro et al. 2001; Schumacher et al. 2002; Sauter et al. 2003; Nikulin et al. 2005). The identical contact surface is observed in both the hexameric structure of Hfq and the heptameric ring forms of the Sm and the Lsm proteins. The mean plane of the beta sheet adjusts to accommodate the different oligomerization states. Since Sm proteins often yield inactive hexameric forms when over-expressed in bacteria (Zaric et al. 2005) and the heptameric form has never been observed in Hfq, it is likely that the hexamer is the thermodynamically more stable form of this fold. In vivo, the eukaryotic Sm proteins are actively assembled into their proper heptameric form through the action of the SMN complex. This ordered, ATP-dependent assembly helps to prevent incorrect isoforms from assembling on an snRNA (Meister et al. 2001). After assembly, the Sm ring is permanently attached to the snRNP. Active assembly has not been observed for either Hfq or Lsm proteins, nor do these proteins act as permanent partners of the RNAs to which they bind, setting them apart from their Sm relatives.

To explain these differences in the oligomerization state, several groups have focused on the structural features of the monomer. One model proposed that differences in the variable loop (L4) located between the Sm1 and Sm2 motifs (Fig. 12.5c) drove the assembly process. It was speculated that Hfq's short variable loop may allow for the compact hexameric oligomerization, while longer loops predispose the Sm fold toward heptameric structures (Schumacher et al. 2002). However, as Sauter and colleagues point out, this rationale does not account for the oligomeric structures of AF-Sm1 and AF-Sm2, which form heptameric and hexameric structures in vitro respectively, despite the same length of variable loop (Toro

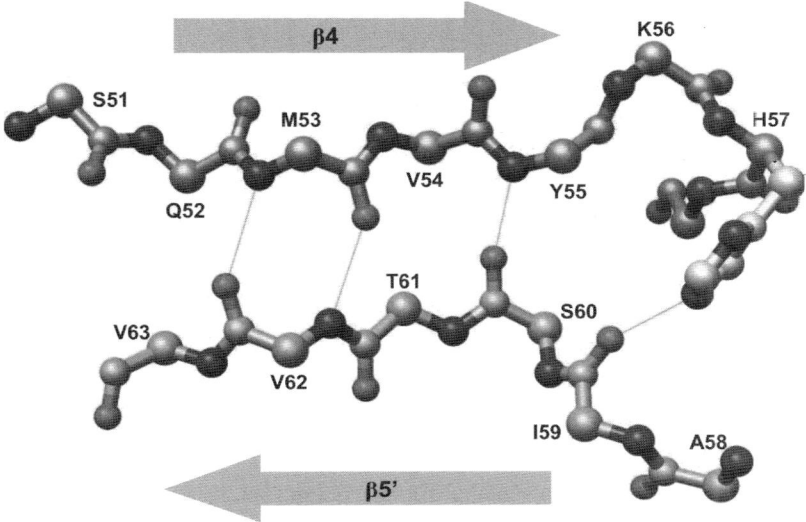

Fig. 12.6 Hydrogen bonding between two anti-parallel β strands stabilizes the subunit interface. Three hydrogen bonds are indicated between the backbone of β4 and β5′. H57 residue of β4 also participates in the hydrogen bonding across the interface

et al. 2002; Sauter et al. 2003). Instead, Sauter proposed that the difference might result from the effect of hydrophilic and hydrophobic interactions between amino acid side chains on the alpha helix and beta stands. Many electrostatic interactions found in AF-Sm1 are missing in AF-Sm2, giving rise to significantly different electrostatic surfaces on the structures.

Nikulin et al. introduced a third potential model to explain the hexamer versus heptamer conundrum. They proposed that hydrogen bonding between H57 and the peptide backbone at I59 of the adjacent subunit of *E. coli* Hfq gives rise to the hexameric state (Fig. 12.6) (Nikulin et al. 2005). These residues are part of the highly conserved YKHAI motif of loop 5 and are located in the central cavity. This additional hydrogen bonding was proposed to tweak the relative orientation of one monomer to another to have a hexameric (60° per monomer) instead of heptameric (51.4° per monomer) ring structure. The H57A mutant of Hfq, however, is functionally active in vivo (Mikulecky et al. 2004), implying the formation of stable hexamers. Thus, the origin of the oligomerization state remains an unanswered question requiring further biophysical analysis.

12.3.3   C-Terminal Region Stabilizes Oligomer

Homology between Hfq genes covers only the first ~70 amino acids. When C-terminal extensions are present, they appear to be unique, unrelated sequences. Long C-terminal regions appear to be common to the β- and γ-proteobacteria (Fig. 12.1) (Arluison et al. 2004). Little structural information is available about these extensions as they were either absent or unresolved in the proteins used for structural analysis. Several studies have looked at the effect of truncated Hfq in vivo and in vitro, and the functional significance has been greatly debated over the past few years. One report indicated a significant effect on the structural stability ($\Delta\Delta G$ of $+1.8$ kcal mol^{-1}) upon truncation of the 30 amino acid extension of *E. coli* Hfq (Arluison et al. 2004). It is not clear how this region stabilizes the hexamer, but it was proposed to contribute to the thermodynamic stability by protecting the inter-subunit β strands.

12.4   Hfq–RNA Interactions

12.4.1   RNA Binding Motif

Footprinting analysis has shown that ncRNAs that bind Hfq typically contain an A/U-rich single-stranded sequence that serves as the primary contact surface (Zhang et al. 2002; Geissmann and Touati 2004). The presence of an adjacent helical region located either upstream or downstream of the A/U-sequence was shown to further enhance binding affinity (Moller et al. 2002; Brescia et al. 2003). While this "conventional Hfq binding RNA motif" (Fig. 12.3) has been established, other secondary structural regions have been shown to bind Hfq as well. Hairpin motifs of *rpoS* were shown to bind Hfq based on nuclease footprinting (Lease and Woodson 2004), as was a loop region connecting two stems in the pseudoknot structure of RydC (Antal et al. 2005). RyhB contains a conventional Hfq binding motif (Geissmann and Touati 2004), but protection was also observed in a loop sequence (Moll et al. 2003) indicative of binding at more than one site. Taken together, these findings imply the existence of different mechanisms by which Hfq can interact with RNA substrates. Surprisingly, Hfq was also recently shown to bind quite specifically to tRNAs (Lee and Feig 2008). The interaction was mapped to two sites located on the T- and D-stems of tRNA. These tRNA binding sites have no sequence or structural similarities with the conventional binding motif so the specifics of the interaction are likely to be dramatically different from the one observed in the crystal structures. The structural details of this interaction remain to be elucidated.

12.4.2   Hfq has Independent RNA Binding Sites

The presence of two independent RNA binding sites on Hfq was first demonstrated during site-directed mutagenesis studies on the two flat surfaces of the hexamer (Mikulecky et al. 2004; Sun and Wartell 2006). These sites have specificity for different types of RNAs which is of functional importance for Hfq. Mutations on one surface were found to reduce affinity for DsrA and other ncRNAs (Mikulecky et al. 2004; Sun and Wartell 2006). This result was consistent with binding to the crystallographically observed site along the central cavity. Two separate nomenclatures have been used for this surface. The L4 face (Mura et al. 2003) and proximal face (Mikulecky et al. 2004) both refer to this side of the torus. In the former case, the nomenclature is referenced to the protein secondary structure whereas in the latter case it is referenced to the crystallographically observed RNA binding site. Mutations on the opposite face (L1 or distal face) had no effect on the binding affinity of any ncRNAs tested but instead altered the binding of poly(A) RNAs (Mikulecky et al. 2004; Sun and Wartell 2006).

The *S. aureus* Hfq crystal structure was solved in the presence of a small RNA with the sequence AU_5G (Schumacher et al. 2002). In the Hfq–RNA co-crystal structure, the RNA is bound in a circular manner around the basic central cavity. Figure 12.5a shows some of the details of this binding mode. The aromatic ring of Y42 from each monomer stacks on the adjacent base as the nucleotide backbone winds its way around the central core. This conserved tyrosine residue resides in loop 3 and is a component of the Sm1 motif (Fig. 12.1). This binding mode is similar to that observed in the AF-Sm1:RNA complex although in the latter case a histidine replaces the tyrosine (Toro et al. 2001; Thore et al. 2003). The stacking of Y42 with RNA bases is the most striking structural feature of the Hfq–RNA interaction, but the F42A mutation of *E. coli* Hfq (which corresponds to Y42A in the *S. aureus* structure) does not significantly reduce the affinity of AU_5G or DsrA (Mikulecky et al. 2004; Sun and Wartell 2006). This result suggests the existence of an alternative mode of binding on the proximal surface and that the observed base stacking does not contribute significantly to the thermodynamic stabilization of the RNA-protein complex (Brennan and Link 2007).

12.4.3 Importance of Residues Y55, K56 and H57 Within the Sm2 Motif

Residues K57 and H58 in *S. aureus*, located in the highly conserved YKH sequence within the Sm2 motif, also participate in the RNA binding interactions through hydrogen bonding to both the uracil base and the backbone (Schumacher et al. 2002). In good agreement with the structural analysis, affinity for AU_5G diminished dramatically in the K56A and H57A mutants of *E. coli* Hfq in vitro (Mikulecky et al. 2004). In addition, the role of K56 in RNA binding was further supported by in vivo assays showing that K56A was severely deficient in post-transcriptional gene regulation and lost the ability to stabilize ncRNAs against degradation (Mikulecky et al. 2004). Mutagenesis experiments done by another group confirmed the specified role of K56 on the proximal face in binding ncRNAs (Sonnleitner et al. 2004). In their assay, K56 mutations had little effect on phage Qβ replication with few exceptions. This result was expected because Hfq binds to an adenine rich portion of Qβ RNA (Senear and Steitz 1976), and it was discovered that Hfq uses its distal surface to bind poly(A) RNAs (Mikulecky et al. 2004; Sun and Wartell 2006).

The effect on RNA binding of Y55 from the YKH motif is interesting because this residue appears to contribute to RNA binding by stabilizing the hexameric structure thermodynamically and/or by making direct contact to RNAs. This residue (Y56 in *S. aureus*) was originally proposed to reinforce the inter-subunit interaction through hydrogen bonding to Y63 from the adjacent subunit in *S. aureus* (Schumacher et al. 2002). In the case of *E. coli* and *P. aeruginosa* Y55 interacts with Q8 from the adjacent subunit rather than the tyrosine, but the interaction is similar in nature (Nikulin et al. 2005). The formation of this hydrogen bond in solution is supported by fluorescence measurement of Q8A mutants of *E. coli* Hfq. This substitution leads to a 140% increase in intrinsic fluorescence from the tyrosine residues relative to wild-type Hfq (there are no tryptophans in wild-type *E. coli* Hfq), implying that the release of Y55's hydroxyl group from this hydrogen bond restores its quenched fluorescence (Sun and Wartell 2006). However, the Q8A mutant is not significantly impaired with respect to RNA binding or in vivo activation of gene expression (Mikulecky et al. 2004), so the thermodynamic importance of this hydrogen bond is likely to be modest.

In contrast to this model where Y55 plays a structural role in the hexamer, another group proposed that Y55 may make direct contacts with the RNA upon binding (Sauter et al. 2003). Mutation of Y55 decreased the binding ability more than eightfold toward AU_5G (Mikulecky et al. 2004). A complementary mutation of Q8, however, did not show any negative effect on RNA binding, implying the direct involvement of Y55 in RNA binding (Mikulecky et al. 2004). Nonetheless, the loss of binding activity may not be purely attributed to its direct involvement in RNA contact. In contrast to other single point mutations on the proximal face, Y55A also decreases poly(A) binding on the opposite face by threefold (Mikulecky et al. 2004), implying that deletion of Y55 may induce a conformational change of the hexamer.

Additional evidence for the importance of Y55 and K56 for RNA binding came from the analysis of mutations at V43. Whereas V43C showed wild-type activity, V43R revealed a number of defects typical of inactive Hfq, including decreased RNA binding affinity and inability to activate poly(A) polymerase activity (Ziolkowska et al. 2006). Since V43 had neither a known role in RNA binding nor was it thought to contribute to the structural stability of Hfq hexamers, molecular dynamics simulations were performed to examine the possible effects of this modification. The study indicated that the substitution of a small hydrophobic residue to a large charged side chain induced an internal structural change that shifted the position of Y55 and K56. Thus, the defects resulted from the misalignment of the critical residues in the Sm2 motif despite the actual modification being rather distant in the primary structure.

12.4.4 Poly(A) Binding Occurs on the Distal Face

Poly(A) RNA was shown to bind Hfq by de Haseth and Uhlenbeck (1980). They showed that addition of poly(A) RNAs of various length could saturate the Hfq binding site, now known to be on the distal face. The amount of poly(A) required for the saturation was the same in terms of moles of adenine bases regardless of the length of poly(A) tested, indicating that Hfq had a fixed number of binding pockets for adenine residues. In addition, the length and the geometry of poly(A) RNA was reported to alter its affinity. Circularized poly(A) RNA had lower affinity compared to its linear counterpart if the length was less than 15 nucleotides. For longer sequences such as A_{18}, the circular form binds Hfq more tightly than the linear form. This result implies that the putative adenine binding pockets on the distal face are arranged in a circular geometry. Mutagenesis studies identified several residues on the distal face associated with poly(A) RNA binding, including Y25, I30 and K31 (Mikulecky et al. 2004; Sun and Wartell 2006). One potential pocket was recently modeled by Brennan and Link (2007). This model involves interaction with a hydrophobic pocket formed by the side chains of Y25, I30 and L32. The side chain of K31 points into the solvent and interacts with Y25. However, several aspects of this model remain to be tested experimentally.

12.4.5   A Role for the C-Terminal Extension in mRNA Binding

In Sect. 12.3.3 a possible role for the C-terminal extension in stabilizing the hexameric structure was described. Another recent study proposed that this region might constitute a third independent RNA binding surface (Vecerek et al. 2008). Two truncated variants of *E. coli* Hfq were prepared lacking either the last 27 (Hfq_{75}) or 37 amino acids (Hfq_{65}). No significant differences were detected between these mutants and the wild-type protein in terms of their ability to replicate bacteriophage Qβ RNA, implying that the highly conserved N-terminal region, especially the distal face of the hexamer, is responsible for this function (Sonnleitner et al. 2002, 2004). Hfq_{65}, however, had reduced affinity for certain regulated mRNAs such as *hfq, sodB* and *rpoS* in vitro, and it was unable to regulate the expression of these mRNAs in vivo (Vecerek et al. 2008). Based on this observation, the C-terminal extension, and in particular the peptide between amino acids 65 and 75 of Hfq, was proposed to facilitate mRNA binding (Vecerek et al. 2008). However, not all mRNAs appear to bind this region as Hfq_{65} was able to regulate expression of *ompA* mRNA (Sonnleitner et al. 2002, 2004). In addition, this result is difficult to reconcile with the previous report of cross-species complementation assays wherein Hfq from *P. aeruginosa* (82 aa) effectively stimulates the expression of *rpoS* as well as replaces the endogenous stress sensitivity of Δ*hfq* despite having limited sequence homology after residue 68 (Fig. 12.1) (Sonnleitner et al. 2002). Thus, while the ability of the C-terminal extension to participate in RNA binding is intriguing, the mechanism of this interaction and its widespread impact on Hfq function require additional study.

12.4.6   Dynamic Modulation of RNA–RNA Interactions

The structural and mutagenesis studies described above illustrate how Hfq interacts with RNAs and what residues from Hfq play critical roles for the interactions.

These studies, however, did not reveal the mechanism by which Hfq enhances the base-pairing of ncRNAs with their target mRNAs. Recently, two studies explored DsrA and *rpoS* to investigate the influence of Hfq on the strand displacement process. By using gel shift experiments, Lease et al. showed that Hfq forms a stable ternary complex with DsrA and *rpoS* (Lease and Woodson 2004). The formation of a ternary complex is a prerequisite for Hfq to mediate annealing, and similar complexes had been previously identified in other ncRNA-mRNA systems (Moller et al. 2002; Zhang et al. 2002). However, Hfq had little effect on the rate of base-pairing, inducing only a twofold increase on annealing rate (Lease and Woodson 2004).

More recently, Arluison et al. used fluorescence resonance energy transfer (FRET) to study the facilitation of base-pairing by Hfq using a three-piece RNA construct. From their data, they proposed a cycle consisting of two main steps: (1) fast binding of Hfq to both DsrA and *rpoS* and (2) slow annealing of the two RNAs (Arluison et al. 2007). In the absence of Hfq, spontaneous base-pairing between DsrA and *rpoS* at low temperature (15°C) was negligible, but the addition of Hfq into the pool of DsrA and *rpoS* dramatically accelerated the formation of the ternary complex. After the initial rapid binding, the RNAs annealed to each other in a slow step that required several minutes to complete. These observations support the idea that Hfq increases the local concentration of RNAs to build the favorable environment for the slow annealing step.

One interesting feature of this process is that the annealing step between DsrA and *rpoS* was reversible. Melting of pre-annealed duplex of DsrA and *rpoS* was also observed in the presence of Hfq. In addition, Hfq was shown to be involved in melting of the initial *rpoS* stem (Fig. 12.4b) but was unable to mediate the reverse annealing process of the same stem. In other words, the ability to anneal or melt RNAs is target specific. Based on these observations, Arluison et al. proposed a model for post-transcriptional gene regulation by Hfq (Arluison et al. 2007). This proposed mechanism begins with rapid association of DsrA and *rpoS* onto Hfq, followed by an internal structural rearrangement of *rpoS*. Slow annealing between DsrA and *rpoS* can then occur. Once formed, Hfq can dissociate from the duplex due to reduced affinity for the DsrA:*rpoS* product. The released Hfq can mediate new rounds of pairing or induce melting of the existing DsrA-*rpoS* duplex depending on the physiological environment.

Some of these observations however, are inconsistent with previous work. The stability of DsrA-*rpoS* duplex measured by gel shift experiments showed that Hfq made the duplex more stable rather than induced the melting of it (Lease and Woodson 2004). In addition, footprinting experiments revealed that Hfq binding to *rpoS* did not change the secondary structure, which contradicts the irreversible melting of the *rpoS* stem (Lease and Woodson 2004). The origin of these discrepancies is not clear, but they may result from the differences between the RNAs constructs used. The FRET study used sequence fragments representing the inter-molecular regulatory region of DsrA and *rpoS* rather than the entire sequence, which might release the steric strain otherwise imposed by structural elements that have been removed.

12.5 Hfq–Protein Interactions

Lsm proteins make use of their heterogeneity to form higher order complexes on their RNA substrates. In addition, these complexes cooperate with other proteins to take part in various riboregulatory processes (Khusial et al. 2005; Wilusz and Wilusz 2005). Considering the diverse roles of Hfq, it is not surprising that Hfq can interact with other proteins to overcome its structural homogeneity. In good agreement with this speculation, recent studies revealed that Hfq binds many protein partners (Butland et al. 2005). Approximately 50 proteins were proposed to interact with Hfq in this study. These proteins can be categorized into several groups based on their function (Fig. 12.7) (Wilusz and Wilusz 2005). In addition to the pull-down experiments to map the Hfq–protein interactions, several specific proteins have been studied extensively, based on the knowledge of the diverse functions of Hfq. The drawback of these studies involving the protein–protein interactions of Hfq, as well as several others discussed below, is the relative ambiguity with respect to the RNAs associated with Hfq. Recent work has shown that the proteins bound to Hfq in the context of RNA-dependent particles vary significantly as a function of the ncRNA to which Hfq is bound (Lee et al., submitted for publication).

12.5.1 Interactions with Ribosome and S1 Protein

At least 24 ribosomal proteins have been found to interact with Hfq (Kajitani et al. 1994; Azam et al. 2000; Butland et al. 2005). These studies showed no preference

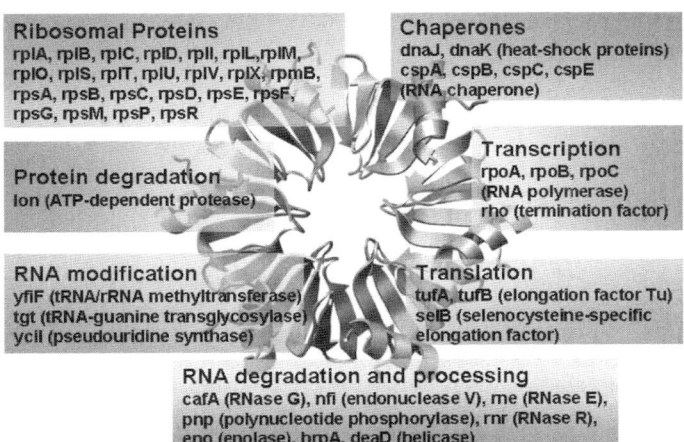

Fig. 12.7 Proteins identified to interact with Hfq in the pull-down experiments (Butland et al. 2005) can be categorized into several groups based on their biochemical function (Wilusz and Wilusz 2005)

with respect to binding large or small subunit proteins, and the biological significance of these interactions is not yet clear. The best studied of these interactions is that between Hfq and small ribosomal subunit protein S1. S1 protein co-purifies with Hfq and was proposed to mediate an interaction between Hfq and RNA polymerase (RNAP) (Sukhodolets and Garges 2003). In the same study, it was proposed that Hfq modulated RNAP activities. This is an attractive model for increasing the efficiency of post-transcriptional gene regulation by providing the ability to further couple transcription and translation. It is unclear, however, how such a process would work when regulation requires two independent transcriptional events, one for the mRNA and a second for the ncRNA. More likely, specific regulated messages may get Hfq preloaded on them to facilitate their regulation by ncRNAs if and when the regulatory RNAs are transcribed in response to an environmental cue.

12.5.2   Interactions with PAP I and PNPase

PAP I and PNPase play important roles in RNA degradation and RNA quality control (Dreyfus and Regnier 2002). PNPase, which is one of components of RNA degradation machinery called degradosome (Carpousis et al. 1994; Miczak et al. 1996; Py et al. 1996), is involved in RNA degradation and maturation. Given the role of Hfq in regulating post-transcriptional gene expression via ncRNAs, a connection between Hfq and these enzymes would be understandable. The Kushner lab therefore probed directly for this binding through reciprocal immunoprecipitation and co-purification experiments (Mohanty et al. 2004). Hfq was shown to bind independently to both proteins. As this association occurs without the involvement of intact degradosome, they are thought to constitute a distinct and independent complex (Mohanty et al. 2004). Although PAP I was not detected in more generalized pull-down experiments (Butland et al. 2005), this might be due to its low cellular concentration.

12.5.3   Interaction with RNase E

Morita et al. demonstrated that RNase E forms a complex with Hfq through protein-protein contacts (Morita et al. 2005). RNase E is an essential endonuclease in *E. coli*, which is not only involved in RNA turnover but also in processing of RNA precursors (Kushner 2002). RNase E is largely divided into N-terminal catalytic domain and C-terminal scaffold region. When RNase E is found as part of the degradosome, the other degradosomal constituents are known to interact with the C-terminal region (Kaberdin et al. 1998; Vanzo et al. 1998). However, when Hfq binds the C-terminal region of RNase E, no additional components of the degradosome can bind, allowing a specific Hfq complex to form. The Hfq:RNase E complex has the additional capability of binding a ncRNA such as RyhB or SgrS. Such a particle now looks more like a RISC complex capable of performing targeted mRNA degradation (Fig. 12.4c, see above). RNase E dependent decay using ncRNAs such as RyhB is known to induce ncRNA-mRNA coupled degradation (Masse et al. 2003a), wherein both the ncRNA and mRNA are destroyed concomitantly. RNase E is however, a single strand endonuclease. How it manages to cleave both strands remains intriguing. One could argue that the entire complex dimerizes to form a higher order structure and thus has the ability to use multiple RNase E active sites during the targeted degradation reaction. Alternatively, significant rearrangements might be required to allow sequential cleavage of the two strands in the lone active site. The final stage of this degradation process needs greater analysis to unravel these discrepancies.

12.5.4   C-Terminal Region is a Plausible Protein–Protein Interacting Site

The next question is where protein–protein interactions occur on Hfq and how many proteins can bind to it simultaneously. Since Hfq is able to form discrete complexes with PAP I and PNPase, but PNPase was not detected when Hfq was bound to RNase E (Morita et al. 2005), Hfq appears to use a common protein-binding interface for several binding partners. Some Lsm proteins use their C-terminal region to bind other proteins (Khusial et al. 2005). Therefore, long C-terminal extensions, like that of *E. coli* Hfq, are thought be the primary binding site. However, the data supporting this claim remain somewhat circumstantial. Sonnleitner et al. proposed that the reason Hfq_{65} showed enhanced DsrA lifetimes relative to wild-type Hfq was that the truncated species failed to recruit PAP I to the ncRNA to facilitate its degradation (Sonnleitner et al. 2004). Since the C-terminal region has also been proposed to bind mRNAs and promote overall stabilization of the hexamer (see above), it seems unlikely that all these roles can occur simultaneously. Establishing appropriate systems to test the function of this region of the protein should be a high priority.

12.6   Conclusions

A flurry of activity has surrounded Hfq over the past decade or so. To a great extent, this has been a result of the rising interest in non-coding RNAs, the proteins that bind to them and their role in post-transcriptional gene regulation. The structural work provided a firm foundation upon which to build interesting and detailed mechanistic analysis of this enigmatic protein. While some aspects of the biochemistry have become clear, there remain large gaps in our understanding of this system. Details of the binary complexes between Hfq and a variety of RNAs are pretty well characterized. Unfortunately, these species do not tell the whole story. Higher order complexes of Hfq are now known to form but how they function remains largely unknown. The diversity of contacts between Hfq and RNAs and other proteins is staggering and the mechanism by which these contacts are regulated will keep the field occupied for several years to come. The isolation of distinct Hfq RNP complexes and the identification of their components will be a good starting point for the studies of these particles.

Acknowledgments This work was supported by a grant from the NIH (GM-075068) to A.L.F.

References

Aiba H (2007) Mechanism of RNA silencing by Hfq-binding small RNAs. Curr Opin Microbiol 10:134–139

Alexandrov A, Chernyakov I, Gu W, Hiley SL, Hughes TR, Grayhack EJ, Phizicky EM (2006) Rapid tRNA decay can result from lack of nonessential modifications. Mol Cell 21:87–96

Altuvia S, Weinstein-Fischer D, Zhang A, Postow L, Storz G (1997) A small, stable RNA induced by oxidative stress: role as a pleiotropic regulator and antimutator. Cell 90:43–53

Antal M, Bordeau V, Douchin V, Felden B (2005) A small bacterial RNA regulates a putative ABC transporter. J Biol Chem 280:7901–7908

Arluison V, Derreumaux P, Allemand F, Folichon M, Hajnsdorf E, Regnier P (2002) Structural modelling of the Sm-like protein Hfq from *Escherichia coli*. J Mol Biol 320:705–712

Arluison V, Folichon M, Marco S, Derreumaux P, Pellegrini O, Seguin J, Hajnsdorf E, Regnier P (2004) The C-terminal domain of *Escherichia coli* Hfq increases the stability of the hexamer. Eur J Biochem 271:1258–1265

Arluison V, Hohng S, Roy R, Pellegrini O, Regnier P, Ha T (2007) Spectroscopic observation of RNA chaperone activities of Hfq in post-transcriptional regulation by a small non-coding RNA. Nucleic Acids Res 35:999–1006

Azam TA, Hiraga S, Ishihama A (2000) Two types of localization of the DNA-binding proteins within the *Escherichia coli* nucleoid. Genes Cells 5:613–626

Battle DJ, Kasim M, Yong J, Lotti F, Lau CK, Mouaikel J, Zhang Z, Han K, Wan L, Dreyfuss G (2006) The SMN complex: an assembly machine for RNPs. Cold Spring Harb Symp Quant Biol 71:313–320

Beggs JD (2005) Lsm proteins and RNA processing. Biochem Soc Trans 33:433–438

Brennan RG, Link TM (2007) Hfq structure, function and ligand binding. Curr Opin Microbiol 10:125–133

Brescia CC, Mikulecky PJ, Feig AL, Sledjeski DD (2003) Identification of the Hfq-binding site on DsrA RNA: Hfq binds without altering DsrA secondary structure. RNA 9:33–43

Butland G, Peregrin-Alvarez JM, Li J, Yang W, Yang X, Canadien V, Starostine A, Richards D, Beattie B, Krogan N, Davey M, Parkinson J, Greenblatt J, Emili A (2005) Interaction network containing conserved and essential protein complexes in *Escherichia coli*. Nature 433:531–537

Carpousis AJ, Van Houwe G, Ehretsmann C, Krisch HM (1994) Copurification of *E. coli* RNAase E and PNPase: evidence for a specific association between two enzymes important in RNA processing and degradation. Cell 76:889–900

Chu CY, Rana TM (2007) Small RNAs: regulators and guardians of the genome. J Cell Physiol 213:412–419

de Haseth PL, Uhlenbeck OC (1980) Interaction of *Escherichia coli* host factor protein with oligoriboadenylates. Biochemistry 19:6138–6146

Dreyfus M, Regnier P (2002) The poly(A) tail of mRNAs: bodyguard in eukaryotes, scavenger in bacteria. Cell 111:611–613

Franze de Fernandez MT, Eoyang L, August JT (1968) Factor fraction required for the synthesis of bacteriophage Q$_\beta$-RNA. Nature 219:588–590

Franze de Fernandez MT, Hayward WS, August JT (1972) Bacterial proteins required for replication of phage Q ribonucleic acid. Purification and properties of host factor I, a ribonucleic acid-binding protein. J Biol Chem 247:824–831

Geissmann TA, Touati D (2004) Hfq, a new chaperoning role: binding to messenger RNA determines access for small RNA regulator. EMBO J 23:396–405

Gottesman S (2004) The small RNA regulators of *Escherichia coli*: roles and mechanisms. Annu Rev Microbiol 58:303–328

Gottesman S (2005) Micros for microbes: non-coding regulatory RNAs in bacteria. Trends Genet 21:399–404

Kaberdin VR, Miczak A, Jakobsen JS, Lin-Chao S, McDowall KJ, von Gabain A (1998) The endoribonucleolytic N-terminal half of *Escherichia coli* RNase E is evolutionarily conserved in *Synechocystis* sp. and other bacteria but not the C-terminal half, which is sufficient for degradosome assembly. Proc Natl Acad Sci U S A 95:11637–11642

Kadaba S, Krueger A, Trice T, Krecic AM, Hinnebusch AG, Anderson J (2004) Nuclear surveillance and degradation of hypomodified initiator tRNA^{Met} in *S. cerevisiae*. Genes Dev 18:1227–1240

Kadaba S, Wang X, Anderson JT (2006) Nuclear RNA surveillance in *Saccharomyces cerevisiae*: Trf4p-dependent polyadenylation of nascent hypomethylated tRNA and an aberrant form of 5S rRNA. RNA 12:508–521

Kajitani M, Kato A, Wada A, Inokuchi Y, Ishihama A (1994) Regulation of the *Escherichia coli hfq* gene encoding the host factor for phage Qβ. J Bacteriol 176:531–534

Kambach C, Walke S, Young R, Avis JM, de la Fortelle E, Raker VA, Luhrmann R, Li J, Nagai K (1999) Crystal structures of two Sm protein complexes and their implications for the assembly of the spliceosomal snRNPs. Cell 96:375–387

Kawamoto H, Morita T, Shimizu A, Inada T, Aiba H (2005) Implication of membrane localization of target mRNA in the action of a small RNA: mechanism of post-transcriptional regulation of glucose transporter in *Escherichia coli*. Genes Dev 19:328

Khusial P, Plaag R, Zieve GW (2005) LSm proteins form heptameric rings that bind to RNA via repeating motifs. Trend Biochem Sci 30:522–528

Kushner SR (2002) mRNA decay in *Escherichia coli* comes of age. J Bacteriol 184:4658–4665

Kushner SR (2004) mRNA decay in prokaryotes and eukaryotes: different approaches to a similar problem. IUBMB Life 56:585–594

LaCava J, Houseley J, Saveanu C, Petfalski E, Thompson E, Jacquier A, Tollervey D (2005) RNA degradation by the exosome is promoted by a nuclear polyadenylation complex. Cell 121:713–724

Lease RA, Belfort M (2000) A trans-acting RNA as a control switch in *Escherichia coli*: DsrA modulates function by forming alternative structures. Proc Natl Acad Sci U S A 97:9919–9924

Lease RA, Woodson SA (2004) Cycling of the Sm-like protein Hfq on the DsrA small regulatory RNA. J Mol Biol 344:1211–1223

Lease RA, Cusick ME, Belfort M (1998) Riboregulation in *Escherichia coli*: DsrA RNA acts by RNA:RNA interactions at multiple loci. Proc Natl Acad Sci U S A 95:12456–12461

Lee T, Feig AL (2008) The RNA binding protein Hfq interacts specifically with tRNAs. RNA 14:514–523

Majdalani N, Cunning C, Sledjeski D, Elliott T, Gottesman S (1998) DsrA RNA regulates translation of RpoS message by an anti-antisense mechanism, independent of its action as an antisilencer of transcription. Proc Natl Acad Sci U S A 95:12462–12467

Masse E, Gottesman S (2002) A small RNA regulates the expression of genes involved in iron metabolism in *Escherichia coli*. Proc Natl Acad Sci U S A 99:4620–4625

Masse E, Escorcia FE, Gottesman S (2003a) Coupled degradation of a small regulatory RNA and its mRNA targets in *Escherichia coli*. Genes Dev 17:2374–2383

Masse E, Majdalani N, Gottesman S (2003b) Regulatory roles for small RNAs in bacteria. Curr Opin Microbiol 6:120–124

Meister G, Buhler D, Pillai R, Lottspeich F, Fischer U (2001) A multiprotein complex mediates the ATP-dependent assembly of spliceosomal U snRNPs. Nat Cell Biol 3:945–949

Miczak A, Kaberdin VR, Wei C-L, Lin-Chao S (1996) Proteins associated with RNase E in a multicomponent ribonucleolytic complex. Proc Natl Acad Sci U S A 93:3865–3869

Mikulecky PJ, Kaw MK, Brescia CC, Takach JC, Sledjeski DD, Feig AL (2004) *Escherichia coli* Hfq has distinct interaction surfaces for DsrA, *rpoS* and poly(A) RNAs. Nat Struct Mol Biol 11:1206–1214

Mohanty BK, Kushner SR (1999) Analysis of the function of *Escherichia coli* poly(A) polymerase I in RNA metabolism. Mol Microbiol 34:1094–1108

Mohanty BK, Maples VF, Kushner SR (2004) The Sm-like protein Hfq regulates polyadenylation dependent mRNA decay in *Escherichia coli*. Mol Microbiol 54:905–920

Moll I, Afonyushkin T, Vytvytska O, Kaberdin VR, Blasi U (2003) Coincident Hfq binding and RNase E cleavage sites on mRNA and small regulatory RNAs. RNA 9:1308–1314

Moller T, Franch T, Hojrup P, Keene DR, Bachinger HP, Brennan RG, Valentin-Hansen P (2002) Hfq: a bacterial Sm-like protein that mediates RNA–RNA interaction. Mol Cell 9:23–30

Morita T, Kawamoto H, Mizota T, Inada T, Aiba H (2004) Enolase in the RNA degradosome plays a crucial role in the rapid decay of glucose transporter mRNA in the response to phosphosugar stress in *Escherichia coli*. Mol Microbiol 54:1063–1075

Morita T, Maki K, Aiba H (2005) RNase E-based ribonucleoprotein complexes: mechanical basis of mRNA destabilization mediated by bacterial noncoding RNAs. Genes Dev 19:2176–2186

Morita T, Mochizuki Y, Aiba H (2006) Translational repression is sufficient for gene silencing by bacterial small noncoding RNAs in the absence of mRNA destruction. Proc Natl Acad Sci U S A 103:4858–4863

Muffler A, Fischer D, Hengge-Aronis R (1996) The RNA-binding protein HF-I, known as a host factor for phage Qβ RNA replication, is essential for *rpoS* translation in *Escherichia coli*. Genes Dev 10:1143–1151

Muffler A, Traulsen DD, Fischer D, Lange R, Hengge-Aronis R (1997) The RNA-binding protein HF-I plays a global regulatory role which is largely, but not exclusively, due to its role in expression of the σ^S subunit of RNA polymerase in *Escherichia coli*. J Bacteriol 179:297–300

Mura C, Cascio D, Sawaya MR, Eisenberg DS (2001) The crystal structure of a heptameric archaeal Sm protein: implications for the eukaryotic snRNP core. Proc Natl Acad Sci U S A 98:5532–5537

Mura C, Phillips M, Kozhukhovsky A, Eisenberg D (2003) Structure and assembly of an augmented Sm-like archaeal protein 14-mer. Proc Natl Acad Sci U S A 100:4539–4544

Nikulin A, Stolboushkina E, Perederina A, Vassilieva I, Blaesi U, Moll I, Kachalova G, Yokoyama S, Vassylyev D, Garber M, Nikonov S (2005) Structure of *Pseudomonas aeruginosa* Hfq protein. Acta Crystallogr D Biol Crystallogr 61:141–146

O'Hara EB, Chekanova JA, Ingle CA, Kushner ZR, Peters E, Kushner SR (1995) Polyadenylylation helps regulate mRNA decay in *Escherichia coli*. Proc Natl Acad Sci U S A 92:1807–1811

Py B, Higgins CF, Krisch HM, Carpousis AJ (1996) A DEAD-box RNA helicase in the *Escherichia coli* RNA degradosome. Nature 381:169–172

Robertson GT, Roop RM, Jr (1999) The *Brucella abortus* host factor I (HF-I) protein contributes to stress resistance during stationary phase and is a major determinant of virulence in mice. Mol Microbiol 34:690–700

Sauter C, Basquin J, Suck D (2003) Sm-like proteins in eubacteria: the crystal structure of the Hfq protein from *Escherichia coli*. Nucleic Acids Res 31:4091–4098

Schumacher MA, Pearson RF, Moller T, Valentin-Hansen P, Brennan RG (2002) Structures of the pleiotropic translational regulator Hfq and an Hfq–RNA complex: a bacterial Sm-like protein. EMBO J 21:3546–3556

Senear AW, Steitz JA (1976) Site-specific interaction of Qβ host factor and ribosomal protein S1 with Qβ and R17 bacteriophage RNAs. J Biol Chem 251:1902–1912

Shapiro L, Franze de Fernandez MT, August JT (1968) Resolution of two factors required in the Q_β-RNA polymerase reaction. Nature 220:478–480

Sonnleitner E, Moll I, Blasi U (2002) Functional replacement of the *Escherichia coli hfq* gene by the homologue of *Pseudomonas aeruginosa*. Microbiology 148:883–891

Sonnleitner E, Napetschnig J, Afonyushkin T, Ecker K, Vecerek B, Moll I, Kaberdin VR, Blasi U (2004) Functional effects of variants of the RNA chaperone Hfq. Biochem Biophy Res Comm 323:1017–1023

Storz G, Altuvia S, Wassarman KM (2005) An abundance of RNA regulators. Annu Rev Biochem 74:199–217

Sukhodolets MV, Garges S (2003) Interaction of *Escherichia coli* RNA polymerase with the ribosomal protein S1 and the Sm-like ATPase Hfq. Biochemistry 42:8022–8034

Sun X, Wartell RM (2006) *Escherichia coli* Hfq binds A_{18} and DsrA domain II with similar 2:1 Hfq_6/RNA stoichiometry using different surface sites. Biochemistry 45:4875–4887

Sun X, Zhulin I, Wartell RM (2002) Predicted structure and phyletic distribution of the RNA-binding protein Hfq. Nucleic Acids Res 30:3662–3671

Tang G (2005) siRNA and miRNA: an insight into RISCs. Trends Biochem Sci 30:106–114

Thore S, Mayer C, Sauter C, Weeks S, Suck D (2003) Crystal structures of the *Pyrococcus abyssi* Sm core and its complex with RNA. Common features of RNA binding in archaea and eukarya. J Biol Chem 278:1239–1247

Toro I, Thore S, Mayer C, Basquin J, Seraphin B, Suck D (2001) RNA binding in an Sm core domain: X-ray structure and functional analysis of an archaeal Sm protein complex. EMBO J 20:2293–2303

Toro I, Basquin J, Teo-Dreher H, Suck D (2002) Archaeal Sm proteins form heptameric and hexameric complexes: crystal structures of the Sm1 and Sm2 proteins from the hyperthermophile *Archaeoglobus fulgidus*. J Mol Biol 320:129–142

Tsui HC, Leung HC, Winkler ME (1994) Characterization of broadly pleiotropic phenotypes caused by an *hfq* insertion mutation in *Escherichia coli* K-12. Mol Microbiol 13:35–49

Vanacova S, Wolf J, Martin G, Blank D, Dettwiler S, Friedlein A, Langen H, Keith G, Keller W (2005) A new yeast poly(A) polymerase complex involved in RNA quality control. PLoS Biol 3:e189

Vanderpool CK, Gottesman S (2004) Involvement of a novel transcriptional activator and small RNA in post-transcriptional regulation of the glucose phosphoenolpyruvate phosphotransferase system. Mol Microbiol 54:1076–1089

Vanzo NF, Li YS, Py B, Blum E, Higgins CF, Raynal LC, Krisch HM, Carpousis AJ (1998) Ribonuclease E organizes the protein interactions in the *Escherichia coli* RNA degradosome. Genes Dev 12:2770–2781

Vecerek B, Rajkowitsch L, Sonnleitner E, Schroeder R, Blasi U (2008) The C-terminal domain of *Escherichia coli* Hfq is required for regulation. Nucleic Acids Res 36:133–143

Wang SW, Stevenson AL, Kearsey SE, Watt S, Bahler J (2008) Global role for polyadenylation-assisted nuclear RNA degradation in posttranscriptional gene silencing. Mol Cell Biol 28:656–665

Wassarman KM (2002) Small RNAs in bacteria: diverse regulators of gene expression in response to environmental changes. Cell 109:141–144

Wilusz CJ, Wilusz J (2005) Eukaryotic Lsm proteins: lessons from bacteria. Nat Struct Mol Biol 12:1031–1036

Zamore PD (2002) Ancient pathways programmed by small RNAs. Science 296:1265–1269

Zaric B, Chami M, Remigy H, Engel A, Ballmer-Hofer K, Winkler FK, Kambach C (2005) Reconstitution of two recombinant LSm protein complexes reveals aspects of their architecture, assembly, and function. J Biol Chem 280:16066–16075

Zhang A, Altuvia S, Tiwari A, Argaman L, Hengge-Aronis R, Storz G (1998) The OxyS regulatory RNA represses *rpoS* translation and binds the Hfq (HF-I) protein. EMBO J 17:6061–6068

Zhang A, Wassarman KM, Ortega J, Steven AC, Storz G (2002) The Sm-like Hfq protein increases OxyS RNA interaction with target mRNAs. Mol Cell 9:11–22

Ziolkowska K, Derreumaux P, Folichon M, Pellegrini O, Regnier P, Boni IV, Hajnsdorf E (2006) Hfq variant with altered RNA binding functions. Nucleic Acids Res 34:709–720

Chapter 13
Assembly of the Human Signal Recognition Particle

Elena Menichelli and Kiyoshi Nagai(✉)

Abstract Large RNA-protein complexes (ribonucleoprotein particles or RNPs) control fundamental biological processes. Their correct assembly is essential for function and occurs by the ordered addition of proteins to the RNA. A good model system for studying RNP assembly is provided by the Signal Recognition Particle (SRP), an RNP conserved from bacteria to humans, with different degrees of complexity. Human SRP, composed of a single RNA molecule and six proteins, is responsible for the co-translational targeting of secretory and membrane proteins to the endoplasmic reticulum membrane. In vitro studies reveal that the SRP proteins need to be added to the RNA sequentially. If the order of addition is altered, non-native particles are formed. The sequential association of proteins causes conformational changes in the RNA, allowing binding of other proteins. The in vivo assembly is regulated by the translocation of precursors between different cellular compartments. In this chapter we review the current understanding of the human SRP assembly mechanism.

Abbreviations EM electron microscopy; ER endoplasmic reticulum; Fh fifty four homolog; GFP green fluorescent protein; GTP guanosine triphosphate; GTPase gua-nosine triphosphatase; RNC ribosome-nascent-chain; RNP ribonucleoprotein; rRNA ribosomal RNA; SR SRP receptor; SS signal sequence; SRP signal recognition particle

13.1 Introduction

Large RNA-protein complexes (ribonucleoprotein particles or RNPs) are essential in fundamental biological processes such as translation (ribosome) and splicing (spliceosome). The assembly of these ribonucleoprotein particles is a complex

K. Nagai
Structural Studies Division, MRC Laboratory of Molecular Biology, Hills Road, Cambridge CB2 2QH, UK
e-mail: kn@mrc-lmb.cam.ac.uk

N.G. Walter et al. (eds.) *Non-Protein Coding RNAs*
doi: 10.1007/978-3-540-70840-7_13, © Springer-Verlag Berlin Heidelberg 2009

process both in vivo and in vitro, requiring proper folding of RNA and ordered association of proteins. This is usually achieved by translocating precursors between different organelles or cellular compartments. In vitro reconstitution of 30S ribosomal subunits by Nomura and colleagues in the 1960s revealed a hierarchical pathway of protein assembly on ribosomal RNA (rRNA). Only a subset of 30S ribosomal proteins (primary proteins) was found to assemble on the 16S rRNA autonomously. The binding sites for other protein subunits (secondary proteins) are created through conformational changes in the rRNA only upon binding of these primary proteins. It is thought that parallel and sequential binding events relieve the RNA of incorrectly folded structures and stabilize the native 30S conformation (Talkington et al. 2005).

The cytoplasmic signal recognition particle (SRP) provides an ideal system to study conformational changes of RNA during the assembly of a ribonucleoprotein particle. SRP is one of the best characterized RNPs since the functional particle can readily be assembled in vitro. Over time a large amount of biochemical and biophysical data have become available (Maity and Weeks 2007; Menichelli et al. 2007; Rose and Weeks 2001). This data, as well as structures determined by X-ray crystallography and cryo-electron microscopy (cryo-EM) (Hainzl et al. 2002; Halic et al. 2004; Kuglstatter et al. 2002; Oubridge et al. 2002), have provided new insights into how SRP assembles and performs its function. In this chapter we review the current understanding of the structural rearrangements which occur on the RNA upon protein binding and how they enable the binding of other proteins. The biological relevance of the proposed in vitro assembly mechanism is also discussed.

13.1.1   The SRP Targeting Cycle

SRP mediates the co-translational targeting of secretory and membrane proteins to the endoplasmic reticulum (ER) membrane. SRP binds to the signal peptide of secretory and membrane proteins as it emerges from translating ribosomes, halting the elongation of the polypeptide chain (elongation arrest). The ribosome-nascent chain (RNC)-SRP complex is then targeted to the ER membrane through the interaction of SRP with its receptor (SR). Upon arrival at the membrane, SRP releases the RNC, which binds to the translocation channel (translocon), and dissociates from the receptor. SRP is then free to start a new targeting cycle and the ribosome can resume protein synthesis. Membrane proteins are inserted into the membrane while secretory proteins are translated across it. This strict coupling between protein synthesis and translocation ensures that the nascent polypeptides do not misfold in the cytoplasm. The SRP targeting cycle, schematically represented in Fig. 13.1, is finely regulated by the guanosine triphosphatase (GTPase) activity of SRP54 and the SR (Doudna and Batey 2004; Egea et al. 2005; Keenan et al. 2001; Luirink and Sinning 2004; Nagai et al. 2003).

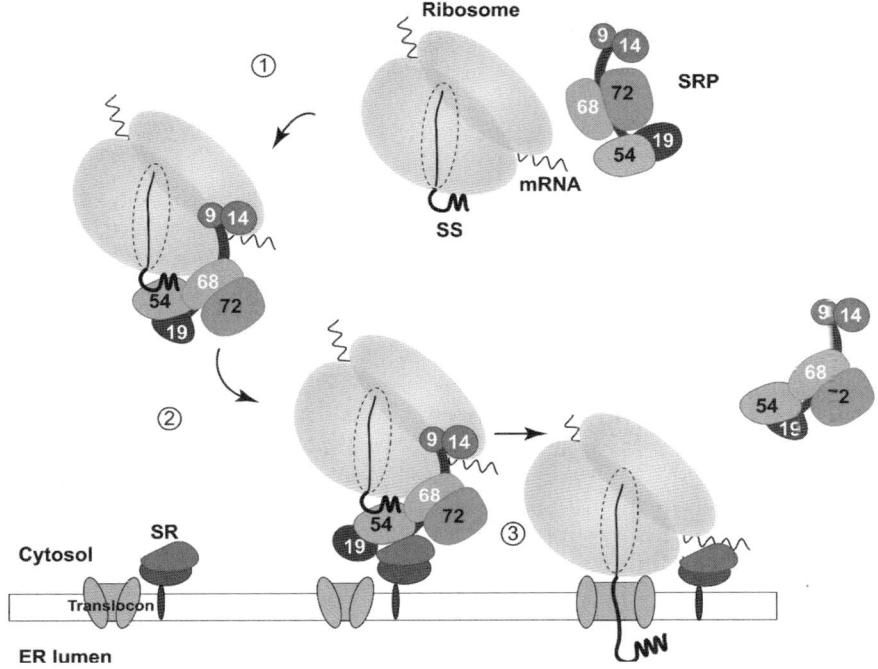

Fig. 13.1 Functional cycle of mammalian SRP. (**1**) Mammalian SRP is composed of a single RNA molecule, the 7SL RNA and six proteins (SRP72, SRP68, SRP54, SRP19, SRP14 and SRP9, shown as ovals). SRP binds to the signal sequence (SS) of nascent secretory and membrane proteins as it emerges from the ribosome and halts the elongation. (**2**) The ribosome-nascent-chain (RNC)-SRP complex is targeted to the ER membrane where SRP interacts with the SRP receptor (SR) in a GTP dependent manner. (**3**) The ribosome is transferred to the translocon SRP is released and protein synthesis resumes (figure adapted from Schwartz and Blobel 2003) See figure insert for color reproduction)

13.1.2 *Mammalian SRP*

SRP is found in all organisms, but its RNA has become larger and its subunit composition more complex during evolution. Mammalian SRP is the most complex particle. It consists of a single RNA molecule of approximately 300 nucleotides, the 7SL RNA, and six protein subunits (SRP72, SRP68, SRP54, SRP19, SRP14 and SRP9), named according to their apparent molecular weight in kilodaltons. Bacterial SRP is much simpler. *Escherichia coli* (*E. coli*) SRP is composed of a single protein called Ffh (*F*ifty *F*our *H*omolog) and the 4.5S RNA (Poritz et al. 1990). The SRP54 homolog is the only protein conserved from bacteria to humans. The 4.5S RNA contains a universally conserved region of 50 nucleotides (corresponding to mammalian helix 8) that includes the primary binding site for Ffh (Lentzen et al. 1996). Archaeal SRP is larger and more complex than its bacterial counterpart, containing one additional protein component, SRP19. Archaeal SRP

RNA is much longer than *E. coli* 4.5S RNA and its secondary structure highly resembles its human counterpart (Zwieb et al. 1996).

The mammalian SRP particle can be separated into two functional domains- the Alu and S domains- by mild digestion with micrococcal nuclease. The Alu domain is responsible for the elongation arrest activity of SRP, while signal sequence recognition and protein translocation activities reside in the S domain (Siegel and Walter 1986).

The Alu domain comprises the smallest SRP proteins, SRP9 and SRP14, in addition to roughly half the 7SL RNA, including its 5′ and 3′ termini (Weichenrieder et al. 1997). The Alu domain RNA is related to the human and rodent families of Alu interspersed repetitive sequences and includes helices 2, 3, 4 and part of helix 5 (Fig. 13.2). Cross-linking experiments (Terzi et al. 2004) and the molecular model generated from the cryo-EM structure of SRP bound to the RNC complex (Halic et al. 2004) explains how SRP induces elongation arrest. Upon signal sequence recognition, the Alu domain is located at the ribosomal subunit interface and competes with the binding of the elongation factors. In addition to its role in translational control, this domain is crucial for transcription, maturation, localization and transport of 7SL RNA (paragraph (1).3).

The S domain of SRP comprises the remaining four proteins and the central region of the 7SL RNA (Siegel and Walter 1986). The roughly Y-shaped S domain RNA contains helices 5, 6, 7 and 8 connected by a three-way junction. The S domain recognizes the signal sequence through SRP54 (Krieg et al. 1986; Kurzchalia et al. 1986), which also contains the GTPase domain responsible for the GTP dependent interaction with the SRP receptor. The cryo-EM structure of the

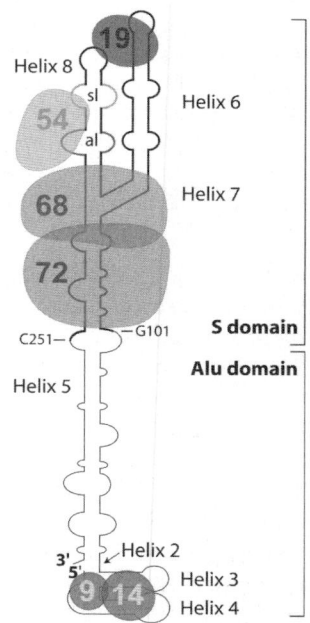

Fig. 13.2 Schematic representation of mammalian SRP. The Alu domain comprises the first Alu 100 nucleotides of the 7SL RNA and approximately 50 nucleotides of the 3′-end, in addition to the SRP9/14 heterodimer. The S domain is composed of the remaining four proteins and the central region of the 7SL RNA (nucleotides G101–C251, thick line), which form helices 5 (nucleotides 101–128 and 222–251), 6, 7 and 8. Symmetric loop (sl) and asymmetric loop (al) in helix 8 are indicated (See figure insert for color reproduction)

RNC–SRP complex has confirmed that the location of the S domain on the large ribosomal subunit is near the peptide exit channel (Halic et al. 2004), as predicted by cross-linking experiments (Pool et al. 2002).

SRP54 is a multi-domain protein that contains a C-terminal methionine-rich domain (M domain), a Ras-like GTPase domain (G domain) and an N-terminal four-helix bundle (N domain). The N and G domains are closely associated, forming the NG unit. The M domain anchors SRP54 to helix 8 of 7SL RNA (Bernstein et al. 1993; Lütcke et al. 1992; Romisch et al. 1990; Zopf et al. 1990; Kuglstatter et al. 2002) and is connected to the NG domain through a flexible linker. Signal sequence recognition is thought to occur within a hydrophobic groove closed by a finger loop in the M domain, although the NG domain may also influence the efficiency of signal sequence recognition (Clerico et al. 2007; Zopf et al. 1993). The details of signal sequence recognition are still not understood and will only be revealed by the atomic structure of the signal-peptide bound SRP54.

SRP19 is a single domain protein that plays an essential role in the assembly of eukaryotic SRP since it is required for the binding of SRP54 (Walter and Blobel 1983). SPR19 homologues are present in all organisms in which the RNA contains helix 6, the primary binding site of SRP19 (Zwieb and Larsen 1997).

Although essential for SRP function, SRP68 and SRP72 are the least characterized SRP proteins. They form a stable heterodimer that cannot be dissociated under native conditions, but is released from the 7SL RNA by high ionic strength (Scoulica et al. 1987; Walter and Blobel 1983). SRP68/72 is required for both SRP functions and possibly coordinates the actions of the S and Alu domains (Andreazzoli and Gerbi 1991; Halic et al. 2004). The domain composition of SRP68/72 has been investigated by proteolysis with elastase (Scoulica et al. 1987) and by RNA binding assays of recombinant SRP68 and SRP72 fragments (Iakhiaeva et al. 2005; Lütcke et al. 1993). Limited proteolysis experiments generated an approximately 30 kDa fragment of SRP68. This N-terminal fragment remained associated with the 7SL RNA, indicating that the RNA binding domain is located in the N-terminus. Analysis of amino acid sequence revealed a cluster of positive charged residues near the N-terminus, which also contains a glycine-rich region present in other RNP proteins (Herz et al. 1990; Lütcke et al. 1993). Central and C-terminal regions of SRP68 contain sequences similar to conserved motifs found in guanine nucleotide dissociation stimulators of Ras-related GTPases (Althoff et al. 1994). The C-terminus of the protein is required for the interaction with SRP72 (Lütcke et al. 1993). Limited proteolysis of SRP72 released a 54 kDa fragment, while a 14 kDa fragment remained bound to the particle. Iakhiaeva et al. (2005) have shown that a 63 amino acid C-terminal region interacts with 7SL RNA.

13.2   In Vitro Assembly of SRP

Although known for 25 years that the in vitro assembly of mammalian SRP is ordered and cooperative (Walter and Blobel 1983), the molecular details of this mechanism have been elucidated only recently. The assembly mechanism of

SRP can be inferred by comparing the crystal structures of assembly intermediates and from RNA probing data. Since not all the assembly intermediates have yet yielded crystals, we have to rely on biochemical data for these complexes. In vitro, the assembly of the Alu domain is independent of S domain assembly (Janiak et al. 1992) and, therefore, the two domains can be investigated separately.

13.2.1   Assembly of the Alu Domain

In vitro studies have demonstrated that SRP9 and SRP14 form a stable heterodimer and that neither protein can bind to the Alu domain RNA autonomously (Strub and Walter 1990). The structure of the two proteins is similar. Both adopt a three-stranded amphipathic β-sheet structure with two α-helices packed on one side. Within the heterodimer, a curved six-stranded antiparallel β-sheet surface is formed, which functions as the primary RNA binding site. The four α-helices are packed on the convex side of the β-sheet surface (Weichenrieder et al. 2000). The SRP9/14 heterodimer binds to helices 3 and 4 at the 5'-end of the Alu domain RNA, inducing and/or stabilizing the stacking of RNA helices 3 and 4. This early binding event is thought to occur co-transcriptionally and to help correctly fold the RNA. In the current model, after termination of transcription, the flexibly linked 3'-end of the 7SL RNA folds back by 180° to contact the 5' domain complex (Huck et al. 2004; Weichenrieder et al. 2001).

13.2.2   Assembly of the S Domain

Walter and Blobel (1983) showed that only SRP19 and the SRP68/72 heterodimer bind to the 7SL RNA independently. SRP54 does not bind tightly to the free RNA and requires the presence of SRP19. The two proteins do not form a complex in the absence of RNA and the order of addition is essential for correct complex formation (Maity and Weeks 2007). In the crystal structures of *Methanococcus jannaschii* (*M. jannaschii*) SRP19 bound to human or *M. jannaschii* S domain RNA, the RNA is in an elongated conformation (Hainzl et al. 2002; Oubridge et al. 2002). SRP19 binds to the tips of helices 6 and 8, and enables extensive interactions between the tetraloops of two helices. Binding of SRP19 to the 7SL RNA is believed to occur, at least, in two steps. Initial rapid contact between SRP19 and helix 6 is followed by a slower step corresponding to conformational rearrangements of the RNA. If bases mediating the tetraloop-tetraloop interaction are mutated, the second step does not occur (Maity and Weeks 2007).

Binding of SRP54 to the pre-formed SRP19-RNA complex induces a major conformational change of helix 8 (Kuglstatter et al. 2002). Four nucleotides in the asymmetric loop (A184, C185, C186 and A183) protrude from the helix to form a

Fig. 13.3 Structural changes of 7SL RNA in two assembly intermediates revealed by crystallography (**a**) Binary complex with SRP19; (**b**) Ternary complex with SRP19 and the M domain of SRP54 (Kuglstatter et al 2002; Oubridge et al. 2002) (See figure insert for color reproduction)

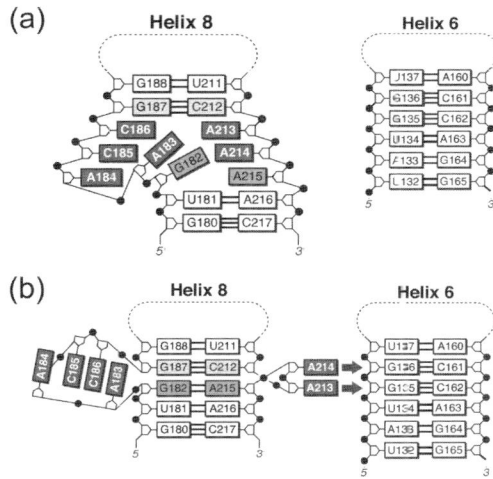

platform that interacts with the M domain of SRP54 (Fig. 13.3), causing helix 8 to collapse. The collapsed structure of helix 8 is stabilized by the formation of two A-minor motifs. The adenine bases of A213 and A214 in the short strand of the asymmetric loop interact with the adjacent helix 6. The structural rearrangements that occur in the RNA upon binding of SRP54 are corroborated by RNA probing experiments (Menichelli et al. 2007; Oubridge et al. 2003).

At present, there is no high-resolution structure of any assembly intermediate containing SRP68/72. Notwithstanding the lack of the atomic structure, their role in the assembly process can be inferred from the results of RNA probing experiments (Menichelli et al. 2007). Since the probing experiments were performed with the SRP68/72 heterodimer, it is not possible to distinguish between individual proteins.

The two largest protein subunits of SRP bind to the central region of the RNA around the three-way junction formed by helices 5, 6, 7 and 8 (Siegel and Walter 1988, Yin et al. 2007). Binding of SRP68/72 to the three-way junction brings the lower part of helices 6 and 8 in close proximity, while the tips of the helices retain their flexibility (Menichelli et al. 2007). Interestingly, SRP68/72 also protects regions that bind SRP54 from chemical and enzymatic modification, indicating that the binding site of SRP54 might be occluded when only SRP68/72 is bound. The combined addition of SRP68/72 and SRP19 to RNA causes a major rearrangement of helix 8 asymmetric loop. The SRP54 binding platform protrudes from the helix in this complex and becomes accessible to chemical and enzymatic probes. Data indicates the simultaneous binding of SRP19 and SRP68/72 reorganizes the RNA in a SRP54 binding competent state. These results suggest the following order of events: SRP68/72 clamps the lower part of helix 6 and 8, while SRP19 facilitates the interactions between the tips of the helices. The SRP54 binding platform is also formed. SPR54 is presumably the last protein to assemble onto the particle. This order of binding is likely to occur in vivo, where compartmentalization directs the sequential addition of the proteins, as discussed in the following paragraph.

13.3   SRP Biogenesis

Although SRP performs its function in the cytoplasm, the assembly of mammalian SRP is initiated in the nucleolus. This membrane-free subdomain of the nucleus is known to be the site for rRNA transcription, rRNA processing and ribosomal assembly (Pederson 1998).

7SL RNA is transcribed by RNA polymerase III (Zieve et al. 1977), which also transcribes a number of small untranslated RNAs, -5S rRNA, tRNAs, U6 and H1 RNAs. 7SL RNA is post-transcriptionally modified at the 3′-end, where three terminal uridylate residues encoded by the CTT(T) pol III terminator are removed and a single adenine is added (Chen et al. 1998; Sinha et al. 1999).

Localization of 7SL RNA has been investigated by microinjection of fluorescent RNA in the nucleus of mammalian cells. Upon injection, the labelled RNA rapidly localized in the nucleolus (Jacobson and Pederson 1998) before appearing in the cytoplasm. The use of mutant 7SL RNAs has shown that the elements responsible for its localization to the nucleolus are associated with protein binding sites in the Alu domain and helix 8 in the S domain (Jacobson and Pederson 1998).

SRP proteins SRP19, SRP68 and SRP72 also localize in the nucleolus (Politz et al. 2000) and in the cytoplasm, as shown by studies on rat kidney fibroblasts (NRK) transfected with DNA encoding green fluorescent protein (GFP)-tagged proteins. GFP-tagged SRP54 was only detected in the cytoplasm. Moreover, SRP14 has also been found in the nucleolus of HeLa cells (Andersen et al. 2002). Similar results were obtained with yeast, where all the SRP protein subunits localized in the nucleolus, except for the SRP54 homologue (Grosshans et al. 2001).

The finding that both 7SL RNA and the SRP proteins co-localize in the nucleolus indicates that the assembly of mammalian SRP might initiate at this site. Partially assembled SRP is exported to the cytoplasm where SRP54 finally associates and SRP can start its targeting cycle. The export of partially assembled SRP from the nucleus to the cytoplasm requires functional CRM1 (Xpo1p in yeast), the export receptor for proteins containing the leucine-rich nuclear export signal (NES), which is also involved in the export of the ribosomal subunits. It is as yet unclear how CRM1 is recruited to export SRP. Putative NESs have been identified in SRP72 (Grosshans et al. 2001), although all the SRP proteins contain leucine-rich stretches. The Alu domain is necessary for nuclear export of the human SRP and deletions that abolish SRP9/14 binding also abolish export (Chen et al. 1998; Lütcke et al. 1995). In yeast, the nuclear export of the pre-SRP requires the presence of SRP68, SRP72, SRP14 homologues, Srp21p and an intact RNA 3′-end (He et al. 1994). Disruption of any of these proteins and 3′-end variations cause nuclear accumulation of the pre-SRP.

13.4 Summary

The in vivo and in vitro studies have provided a model for the biogenesis of SRP, which is schematically shown in Fig. 13.4. SRP9/14 binds to the 7SL RNA co-transcriptionally and functions as a checkpoint for correctly folded RNA. SRP19 and SRP68/72 are the next proteins that bind in the nucleolus. SRP68/72 clamps the lower part of helices 6 and 8 and occludes the SRP54 binding site, possibly providing a 'quality control' mechanism for correct assembly. Based upon studies in yeast, we can speculate the complexes lacking SRP19 could be exported to the cytoplasm. SRP68/72 would then prevent the binding of SRP54 to these incorrectly assembled particles. SRP19 binds the tips of the helices and together with SRP68/72 causes the formation of the SRP54 binding platform. The pre-SRP is exported to the cytoplasm where SRP54 associates to form the active complex.

Although much has been learnt about the assembly of the human SRP, there are still gaps in our knowledge and further structural and in vivo studies are required.

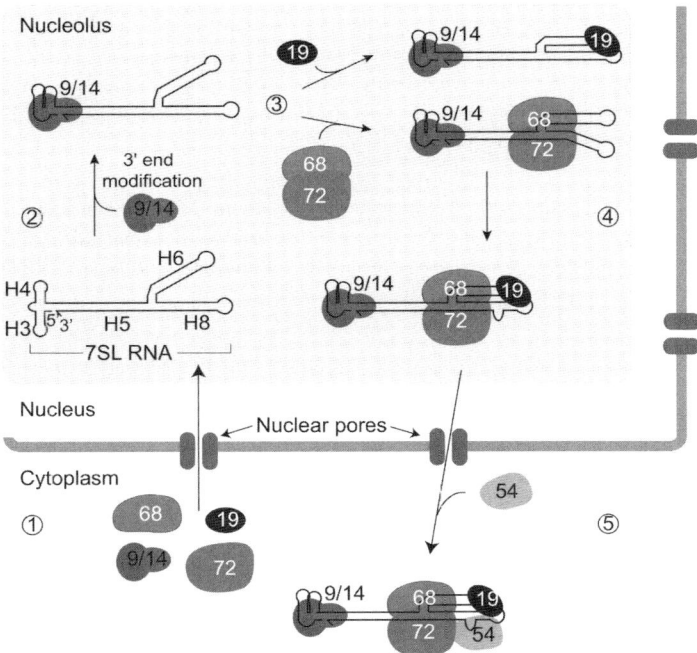

Fig. 13.4 Proposed mechanism for SRP biogenesis and assembly. (**1**) All SRP proteins, except SRP54, are imported into the nucleolus, where assembly of SRP initiates. (**2**) The binding of SRP9/14 to the 5'-end of 7SL RNA is thought to occur co-transcriptionally. (**3**) SRP68/72 binds near the three-way junction, SRP19 binds to the tips of helices 6 and 8. (**4**) Simultaneous binding of SRP68/72 and SRP19 pre-organizes the binding site of SRP54. (**5**) SRP becomes active after the pre-SRP is exported to the cytoplasm and bound by SRP54 (See figure insert for color reproduction)

Acknowledgments We thank Blair Szymczyna and Goran Pljevaljcic for comments on the manuscript.

References

Althoff S, Selinger D, Wise JA (1994) Molecular evolution of SRP cycle components: functional implications. Nucleic Acids Res 22:1933–1947

Andersen JS, Lyon CE, Fox AH, Leung AK, Lam YW, Steen H, Mann M, Lamond AI (2002) Directed proteomic analysis of the human nucleolus. Curr Biol 12:1–11

Andreazzoli M, Gerbi SA (1991) Changes in 7SL RNA conformation during the signal recognition particle cycle. EMBO J 10:767–777

Bernstein HD, Zopf D, Freymann DM, Walter P (1993) Functional substitution of the signal recognition particle 54-kDa subunit by its *Escherichia coli* homolog. Proc Natl Acad Sci U S A 90:5229–5233

Chen Y, Sinha K, Perumal K, Gu J, Reddy R (1998) Accurate 3′ end processing and adenylation of human signal recognition particle RNA and Alu RNA in vitro. J Biol Chem 273:35023–35031

Clerico EM, Maki JL, Gierasch LM (2008) Use of synthetic signal sequences to explore the protein export machinery. Biopolymers 90:307–319

Doudna JA, Batey RT (2004) Structural insights into the signal recognition particle. Annu Rev Biochem 73:539–557

Egea PF, Stroud RM, Walter P (2005) Targeting proteins to membranes: structure of the signal recognition particle. Curr Opin Struct Biol 15:213–220

Grosshans H, Deinert K, Hurt E, Simos G (2001) Biogenesis of the signal recognition particle (SRP) involves import of SRP proteins into the nucleolus, assembly with the SRP–RNA, and Xpo1p-mediated export. J Cell Biol 153:745–762

Hainzl T, Huang S, Sauer-Eriksson AE (2002) Structure of the SRP19 RNA complex and implications for signal recognition particle assembly. Nature 417:767–771

Halic M, Becker T, Pool MR, Spahn CM, Grassucci RA, Frank J, Beckmann R (2004) Structure of the signal recognition particle interacting with the elongation-arrested ribosome. Nature 427:808–814

He XP, Bataille N, Fried HM (1994) Nuclear export of signal recognition particle RNA is a facilitated process that involves the Alu sequence domain. J Cell Sci 107(Pt 4):903–912

Herz J, Flint N, Stanley K, Frank R, Dobberstein B (1990) The 68 kDa protein of signal recognition particle contains a glycine-rich region also found in certain RNA-binding proteins. FEBS Lett 276:103–107

Huck L, Scherrer A, Terzi L, Johnson AE, Bernstein HD, Cusack S, Weichenrieder O, Strub K (2004) Conserved tertiary base pairing ensures proper RNA folding and efficient assembly of the signal recognition particle Alu domain. Nucleic Acids Res 32:4915–4924

Iakhiaeva E, Yin J, Zwieb C (2005) Identification of an RNA-binding domain in human SRP72. J Mol Biol 345:659–666

Jacobson MR, Pederson T (1998) Localization of signal recognition particle RNA in the nucleolus of mammalian cells. Proc Natl Acad Sci U S A 95:7981–7986

Janiak F, Walter P, Johnson AE (1992) Fluorescence-detected assembly of the signal recognition particle: binding of the two SRP protein heterodimers to SRP RNA is noncooperative. Biochemistry 31:5830–5840

Keenan RJ, Freymann DM, Stroud RM, Walter P (2001) The signal recognition particle. Annu Rev Biochem 70:755–775

Krieg UC, Walter P, Johnson AE (1986) Photocrosslinking of the signal sequence of nascent preprolactin to the 54-kilodalton polypeptide of the signal recognition particle. Proc Natl Acad Sci U S A 83:8604–8608

Kuglstatter A, Oubridge C, Nagai K (2002) Induced structural changes of 7SL RNA during the assembly of human signal recognition particle. Nat Struct Biol 9:740–744

Kurzchalia TV, Wiedmann M, Girshovich AS, Bochkareva ES, Bielka H, Rapoport TA (1986) The signal sequence of nascent preprolactin interacts with the 54K polypeptide of the signal recognition particle. Nature 320:634–636

Lentzen G, Moine H, Ehresmann C, Ehresmann B, Wintermeyer W (1996) Structure of 4.5S RNA in the signal recognition particle of Escherichia coli as studied by enzymatic and chemical probing. RNA 2:244–253

Luirink J, Sinning I (2004) SRP-mediated protein targeting: structure and function revisited. Biochim Biophys Acta 1694:17–35

Lütcke H (1995) Signal recognition particle (SRP), a ubiquitous initiator of protein translocation. Eur J Biochem 228:531–550

Lütcke H, High S, Romisch K, Ashford AJ, Dobberstein B (1992) The methionine-rich domain of the 54 kDa subunit of signal recognition particle is sufficient for the interaction with signal sequences. EMBO J 11:1543–1551

Lütcke H, Prehn S, Ashford AJ, Remus M, Frank R, Dobberstein B (1993) Assembly of the 68- and 72-kD proteins of signal recognition particle with 7S RNA. J Cell Biol 121:977–985

Maity TS, Weeks KM (2007) A threefold RNA-protein interface in the signal recognition particle gates native complex assembly. J Mol Biol 369:512–524

Menichelli E, Isel C, Oubridge C, Nagai K (2007) Protein-induced conformational changes of RNA during the assembly of human signal recognition particle. J Mol Biol 367:187–203

Nagai K, Oubridge C, Kuglstatter A, Menichelli E, Isel C, Jovine L (2003) Structure, function and evolution of the signal recognition particle. EMBO J 22:3479–3485

Oubridge C, Kuglstatter A, Jovine L, Nagai K (2002) Crystal structure of SRP19 in complex with the S domain of SRP RNA and its implication for the assembly of the signal recognition particle. Mol Cell 9:1251–1261

Oubridge C, Isel C, Kuglstatter A, Nagai H (2003) Reply to "complex formations and crystal contacts". Nat Struct Biol 10:494–495

Pederson T (1998) The plurifunctional nucleolus. Nucleic Acids Res 26:3871–3876

Politz JC, Yarovoi S, Kilroy SM, Gowda K, Zwieb C, Pederson T (2000) Signal recognition particle components in the nucleolus. Proc Natl Acad Sci U S A 97:55–60

Poritz MA, Bernstein HD, Strub K, Zopf D, Wilhelm H, Walter P (1990) An E. coli ribonucleoprotein containing 4.5S RNA resembles mammalian signal recognition particle. Science 250:1111–1117

Pool MR, Stumm J, Fulga TA, Sinning I, Dobberstein B (2002) Distinct modes of signal recognition particle interaction with the ribosome. Science 297:1345–1348

Romisch K, Webb J, Lingelbach K, Gausepohl H, Dobberstein B (1990) The 54-kD protein of signal recognition particle contains a methionine-rich RNA binding domain. J Cell Biol 111:1793–1802

Rose MA, Weeks KM (2001) Visualizing induced fit in early assembly of the human signal recognition particle. Nat Struct Biol 8:515–520

Scoulica E, Krause E, Meese K, Dobberstein B (1987) Disassembly and domain structure of the proteins in the signal-recognition particle. Eur J Biochem 163:519–528

Schwartz T, Blobel G (2003) Structural basis for the function of the beta subunit of the eukaryotic signal recognition particle receptor. Cell 112:793–803

Siegel V, Walter P (1986) Removal of the Alu structural domain from signal recognition particle leaves its protein translocation activity intact. Nature 320:81–84

Siegel V, Walter P (1988) Binding sites of the 19-kDa and 68/72-kDa signal recognition particle (SRP) proteins on SRP RNA as determined in protein-RNA "footprinting". Proc Natl Acad Sci U S A 85:1801–1805

Sinha K, Perumal K, Chen Y, Reddy R (1999) Post-transcriptional adenylation of signal recognition particle RNA is carried out by an enzyme different from mRNA Poly(A) polymerase. J Biol Chem 274:30826–30831

Strub K, Walter P (1990) Assembly of the Alu domain of the signal recognition particle (SRP): dimerization of the two protein components is required for efficient binding to SRP RNA. Mol Cell Biol 10:777–784

Talkington MW, Siuzdak G, Williamson JR (2005) An assembly landscape for the 30S ribosomal subunit. Nature 438:628–632

Terzi L, Pool MR. Dobberstein B, Strub K (2004) Signal recognition particle Alu domain occupies a defined site at the ribosomal subunit interface upon signal sequence recognition. Biochemistry 43:107–117

Walter P, Blobel G (1983) Disassembly and reconstitution of signal recognition particle. Cell 34:525–533

Weichenrieder O, Kapp U, Cusack S, Strub K (1997) Identification of a minimal Alu RNA folding domain that specifically binds SRP9/14. RNA 3:1262–1274

Weichenrieder O, Wild K, Strub K, Cusack S (2000) Structure and assembly of the Alu domain of the mammalian signal recognition particle. Nature 408:167–173

Weichenrieder O, Stehlin C, Kapp U, Birse DE, Timmins PA, Strub K, Cusack S (2001) Hierarchical assembly of the Alu domain of the mammalian signal recognition particle. RNA 7:731–740

Yin J, Iakhiaeva E, Menichelli E, Zwieb C (2007) Identification of the RNA binding regions of SRP68/72 and SRP72 by systematic mutagenesis of human SRP RNA. RNA Biol 4(3):154–159

Zieve G, Benecke BJ, Penman S (1977) Synthesis of two classes of small RNA species in vivo and in vitro. Biochemistry 16:4520–4525

Zopf D, Bernstein HD, Johnson AE, Walter P (1990) The methionine-rich domain of the 54 kD protein subunit of the signal recognition particle contains an RNA binding site and can be crosslinked to a signal sequence. EMBO J 9:4511–4517

Zopf D, Bernstein HD, Walter P (1993) GTPase domain of the 54-kD subunit of the mammalian signal recognition particle is required for protein translocation but not for signal sequence binding. J Cell Biol 120:1113–1121

Zwieb C, Larsen N (1997) The Signal Recognition Particle Database (SRPDB). Nucleic Acids Res 25:107–108

Zwieb C, Muller F, Larsen N (1996) Comparative analysis of tertiary structure elements in signal recognition particle RNA. Fold Des 1:315–324

Chapter 14
Forms and Functions of Telomerase RNA

Kathleen Collins

Abstract Telomerase adds single-stranded telomeric DNA repeats to chromosome ends. Unlike other polymerases involved in genome replication, telomerase synthesizes DNA without use of a DNA template. Instead, the enzyme active site copies a template carried within the integral RNA subunit of the telomerase ribonucleoprotein (RNP) complex. In addition to providing a template, telomerase RNA has non-template motifs with critical functions in the catalytic cycle of repeat synthesis. In its complexity of structure and function, telomerase RNA resembles the non-coding RNAs of RNP machines like the ribosome and spliceosome that evolved from catalytic RNAs of the RNA World. However, unlike these RNPs, telomerase evolved its RNP identity after advent of the Protein World. Insights about telomerase have broad significance for understanding non-coding RNA biology as well as chromosome end maintenance and human disease.

14.1 Telomerase Biological Function and Regulation

In all but few eukaryotes, nuclear chromosome ends terminate in a tandem array of simple-sequence repeats. Groups of eukaryotes have different sequences and lengths of telomeric repeat array, such as ~1,000 repeats of T_2AG_3 at an average human cell telomere or ~50 repeats of T_2G_4 at an average telomere in the single-celled model ciliate. The array of telomeric repeats serves at least two conserved functions (Gilson and Geli 2007; Verdun and Karlseder 2007). (1) telomeric repeats and their associated proteins form chromatin structures that protect authentic chromosome termini from fusion or degradation. (2) telomeric repeats provide a mechanism for telomere maintenance that can compensate for incomplete terminus replication by DNA-templated DNA polymerases.

K. Collins
Department of Molecular and Cell Biology, University of California at Berkeley, Berkeley, CA 94720-3200, USA
e-mail: kcollins@berkeley.edu

N.G. Walter et al. (eds.) *Non-Protein Coding RNAs*
doi: 10.1007/978-3-540-70840-7_14, © Springer-Verlag Berlin Heidelberg 2009

Eukaryotic chromosome end maintenance relies on new telomeric repeat synthesis by the specialized polymerase telomerase. The telomerase ribonucleoprotein (RNP) complex carries an internal RNA template, complementary to one or more copies of the telomeric repeat. In the presence of a single-stranded DNA oligonucleotide primer in vitro or an appropriate chromosome 3′ end in vivo, telomerase uses its internal template to accomplish de novo synthesis of telomeric DNA repeats (Greider and Blackburn 1989; Yu et al. 1990). Reconstitution of core telomerase enzyme activity in vitro requires template-containing telomerase RNA (TER) and telomerase reverse transcriptase protein (TERT). TERT shares the active site motifs of viral reverse transcriptases (RTs), including the aspartic acids that coordinate catalytic magnesium ions (Autexier and Lue 2006). However, unlike other RT or DNA polymerase polypeptides, the catalytic activity of TERT depends both on the TER template and TER motifs physically separable from the template (Miller and Collins 2002). Non-template RNA motifs play key roles in establishing telomerase's specialized repeat synthesis activity, as described in detail below.

In single-celled eukaryotes, telomerase extends at least a subset of telomeres in each cell cycle. The shortest telomeres preferentially recruit telomerase, giving rise to telomere length homeostasis (Hug and Lingner 2006). In contrast, in cells of multicellular eukaryotes, telomerase activation and telomere maintenance are highly restricted. All human somatic cells express and accumulate TER, but few of these cells express TERT and even fewer show evidence of telomerase-dependent telomere maintenance (Collins and Mitchell 2002; Cong et al. 2002). With ongoing cell proliferation in the absence of telomerase, telomeric repeat tracts become too short to protect chromosome ends from recognition as DNA double-stranded breaks. The presence of critically short telomeres induces cells to cease dividing and instead undergo proliferative senescence or apoptosis (Verdun and Karlseder 2007). The general repression of TERT in somatic cells acts as tumor suppression mechanism (Serrano and Blasco 2007). Cancer cells dramatically up-regulate telomerase to gain extended proliferative capacity (Shay and Wright 2006). Inherited or spontaneous gene mutations that cause as little as 50% deficiency in human TER or TERT cause premature bone marrow failure and a spectrum of other proliferative defects (Wong and Collins 2003; Garcia et al. 2007; Vulliamy and Dokal 2007).

14.2   Telomerase Subunit Composition

Success in the heterologous reconstitution of active telomerase was first achieved using rabbit reticulocyte lysate to express and assemble human TER and TERT (Weinrich et al. 1997). The same approach was successful in reconstitution of the catalytic core of *T. thermophila* telomerase (Collins and Gandhi 1998) and reconstitution of *Saccharomyces cerevisiae* telomerase from a minimized version of *S. cerevisiae* TER and TERT/Est2 (Zappulla et al. 2005). Against these successes are many failures to reconstitute active RNP from the combination of TERT and TER of other species. All telomerase in vitro reconstitution systems require crude eukaryotic cell extract for assembly of TER and TERT, due to the need for folding and assembly chaperones (Holt et al. 1999; Licht and Collins 1999). How in vitro reconstitution conditions either fulfill or bypass the physiological requirements for active RNP assembly is not yet understood.

The physiologically assembled telomerase RNP complexes with telomeric repeat synthesis activity are designated telomerase holoenzymes. Telomerase holoenzymes harbor numerous proteins other than TERT. Telomerase catalytic activity in a ciliate, vertebrate, other cell extract fractionates by gel filtration with an apparent mass of ~500 kDa or more, with differences depending on the cells and methods used (Collins 1999; Lingner et al. 1997; Schnapp et al. 1998; Aigner et al. 2003; Muñoz and Collins 2004). At the molecular level, telomerase holoenzyme proteins differ between ciliates, vertebrates, and yeasts. Despite this divergence in identity among telomerase-associated proteins, the biological roles of holoenzyme proteins have some similarity. Known roles of telomerase-associated proteins include folding and stabilization of TER, localization of TER and/or TERT, and telomerase activation on telomere substrates (Collins 2006; Zappulla and Cech 2006; Bertuch and Lundblad 2006; Collins 2008; Gallardo et al. 2008). Much remains to be learned about how TER and TERT are folded, assembled, and regulated in function in vivo.

14.3 A Multistep Catalytic Cycle of Repeat Synthesis

Telomerase has an atypically complex catalytic cycle for a nucleic acid polymerase. The overall features of the catalytic cycle can be illustrated using the *T. thermophila* enzyme as an example, because it is intensively studied with regard to enzyme mechanism. The template region of *T. thermophila* TER harbors a 1.5-copy complement of the organisms' telomeric repeat (Fig. 14.1). The 3′ end of a primer can be positioned relative to the template by two mechanisms. If the primer 3′ end is complementary to the template, it can base-pair to create a primer-template hybrid (Fig. 14.1a). Primer alignment by hybridization ensures that any permutation of telomeric repeat at a primer 3′ end is recognized and extended accordingly, to maintain a precise register of telomeric repeats in the product DNA. Alternately, some telomerases including the *T. thermophila* enzyme can position non-telomeric sequence primers at the template "default" position without hybridization. For *T. thermophila* telomerase, the template 3′ cytidine is copied first, resulting in addition of a -G4T2 permutation of the telomeric repeat. This specificity of default template alignment is observed in vitro and also in vivo, when telomerase heals chromosome ends generated by developmentally programmed chromosome fragmentation (Yu and Blackburn 1991; Wang and Blackburn 1997).

Nucleotide addition processivity (NAP) describes the efficiency of copying each successive template position required for synthesis of a telomeric repeat (Fig. 14.1a–c). For the *T. thermophila* enzyme, copying to template 5′ boundary gives product with the 3′ permutation G3T2G-3′ (Fig. 14.1c). This differs from the 3′ telomeric repeat permutation of chromosome ends in vivo, suggesting that cellular factors either modulate telomerase NAP or trim the released product DNA (Jacob

Fig. 14.1 The telomerase catalytic cycle. *T. thermophila* TER template and flanking sequences are depicted over the course of repeat synthesis. Template region nucleotides including the redundant 3′ primer alignment site are shown in bold. Primer and product nucleotides forming the template hybrid are also indicated in bold, along with the putative primer 5′ region bound to TERT

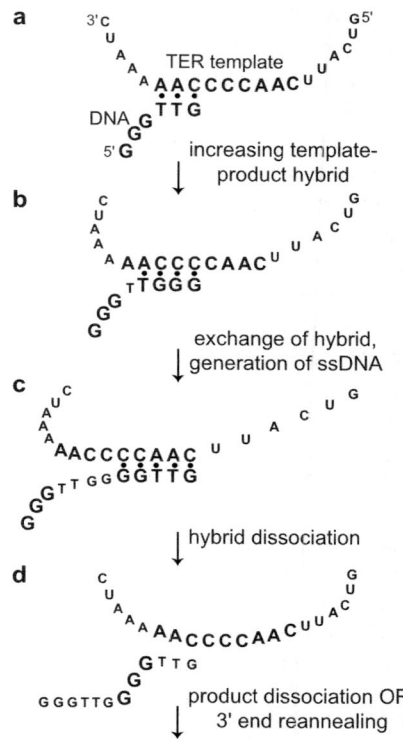

et al. 2003). Notably, primer or product positioning to copy 5′ template positions requires a greater length of template hybrid than to copy 3′ template positions: although 2–3 base-pairs are sufficient to uniquely specify addition of the second G (Fig. 14.1a), at least four base-pairs are required to unambiguously specify addition of the fourth G (Fig. 14.1b). Experiments using sequence-altered templates or chemical modification suggest that the template 3′ end must unpair from product prior to copying the template 5′ end (Fig. 14.1c), limiting the number of base-pairs of template-product hybrid (Wang et al. 1998; Förstemann and Lingner 2005). This feature of the catalytic cycle may serve to equalize the binding affinity of primers with different 3′ permutations or promote product-template dissociation after repeat synthesis (Collins and Greider 1993; Hammond and Cech 1998).

In vitro telomerase activity assays are typically conducted with a large excess of primer. Under these conditions, many telomerases will add multiple repeats to a bound substrate before dissociation. Repeat addition processivity (RAP) describes the efficiency of successive complete repeat additions to each bound primer. RAP requires product to release from pairing with the template and yet remain bound to the enzyme, such that the product 3′ end can be repositioned at the 3′ end of the template (Fig. 14.1d). The *T. thermophila* telomerase holoenzyme assayed in cell extract has over 95% probability of next repeat addition (Greider 1991). In comparison, recombinant

T. thermophila TER + TERT has a maximum ~25% probability of repeat addition and unlike the case for holoenzyme, this relatively limited RAP requires allosteric stimulation by dGTP (Hardy et al. 2001). The extent of telomerase RAP in vivo is unclear (see below). The cellular conditions that determine telomerase RAP in vivo may not be recapitulated in assays of single-stranded primer DNA elongation in vitro.

Telomerase enzymes have additional biochemical activities that are of uncertain biological significance. *T. thermophila* and other telomerases can catalyze endonucleolytic cleavage of bound substrates or products, stimulated by conditions that stall forward DNA synthesis near the template 5′ end (Autexier and Lue 2006). *T. thermophila* and other ciliate telomerase holoenzymes will also readily polymerize a poly-guanosine ladder instead of a ladder of telomeric repeats (Collins 1999). Homopolymer synthesis is not detected using dTTP, perhaps due to the higher Km for the pyrimidine nucleotide.

14.4 Biochemical Differences In Telomerase Holoenzymes

Telomerase activity assays using cell extracts from different organisms have revealed heterogeneity of telomerase catalytic properties, varying around the general theme of template-directed telomeric repeat synthesis described above. A major difference between telomerase holoenzymes of different species is in their relative NAP and RAP. *S. cerevisiae* telomerase predominantly acts with low NAP and low RAP, with product synthesis in vitro limited by slow turnover of the product-template hybrid (Prescott and Blackburn 1997). The degenerate TG₁–3 telomeric repeats cloned from telomeres in vivo are produced by frequent copying of a variable, incomplete subset of template positions (Singer and Gottschling 1994; McEachern and Blackburn 1995). In contrast, other yeast telomerases efficiently copy the entire template; some even display substantial RAP (Cohn and Blackburn 1995). Mammalian telomerase holoenzymes differ dramatically in RAP, ranging from high RAP for human telomerase to little or no RAP for telomerase from mice (Morin 1989; Prowse et al. 1993). A synthetic drug that reduces human telomerase RAP in vitro induces telomere shortening in vivo (Pascolo et al. 2002). Likewise, some *T. thermophila* TER variants that reduce holoenzyme RAP in vitro maintain shorter telomeres than wild-type when expressed in vivo (Cunningham and Collins 2005). However, other *T. thermophila* TER variants that reduce RAP maintain telomeres longer than wild-type (Cunningham and Collins 2005). These studies suggest that the balance of telomere length homeostasis depends on specific features of the telomerase catalytic cycle, but the details of this relationship remain to be understood.

More surprising than telomerase holoenzyme differences in NAP and RAP are deviations in the fidelity of primer alignment and template copying. Apparently, for some telomerases, infidelity has a crucial biological role in telomerase function. The variable telomeric repeats of the fission yeast *Schizosaccharomyces pombe* require telomerase to elongate telomere substrates that are 3′-mismatched with the TER template (Leonardi et al. 2008; Webb and Zakian 2008). Plasmodium telomerase generates mixed telomeric repeats of T2(T/C)AG3 by inserting either dTTP or dCTP across from a particular template guanosine (Chakrabarti et al. 2007). Telomerases from different Paramecium species copy the same A2C4 TER template with different fidelity: *P. caudatum* telomeres are expected repeats of T2G4, but *P. tetraurelia* telomeres are ~75% T2G4 and ~25% T3G3 (McCormick-Graham and Romero 1996). Expression of *P. caudatum* TER in *P. tetraurelia* does not increase fidelity of telomeric repeat synthesis in vivo (McCormick-Graham et al. 1997), suggesting that *P. tetraurelia* TERT or another telomere factor is responsible for reduced fidelity of repeat synthesis.

14.5   TER

The primary sequence of TER is divergent across eukaryotes. Genome sequencing projects have enabled TERT identification in diverse organisms, but identification of TER has been more challenging. Three major phylogenetic groups have been the dominant model systems for molecular studies of TER: ciliates, vertebrates, and budding yeasts. TERs from these organisms are described in greater detail below. Secondary structure models have been derived by comparative sequence analysis. High-resolution structures have been reported for individual motifs within *T. thermophila* or human TER (Theimer and Feigon 2006; Legassie and Jarstfer 2006). Knowledge of global tertiary structure of any TER is still lacking.

14.5.1   Ciliate TERs

Relative abundance of ciliate telomerases has facilitated TER biochemical and molecular characterization. TERs have been cloned from 28 ciliates, ranging in length from 147 to 209 nt (Romero and Blackburn 1991; Lingner et al. 1994; McCormick-Graham and Romero 1996; Ye and Romero 2002). Ciliate TER is an unprocessed primary transcript of RNA Polymerase III with conserved secondary structure (Fig. 14.2a), including a template/pseudoknot region closed by stem I (P1) and a 3′ stem (P4). P2 is absent in some ciliate TERs, and others have an extra stem-loop immediately following the pseudoknot.

  Studies of *T. thermophila* telomerase reconstituted in vitro and in vivo have characterized numerous roles for TER motifs in the catalytic cycle (Fig. 14.2a). Substitutions 5′ of the template in the template boundary element (TBE) allow copying of TER positions beyond the normal template 5′ boundary (Autexier and Greider 1995; Lai et al. 2002). The TBE also provides a high-affinity binding site for TERT. TERT contact with the 5′ half of the *T. thermophila* TBE is most critical for high-affinity interaction, while contact with the 3′ half is most critical for template boundary definition; however, all TER substitutions that compromise TBE function reduce TERT binding and vice-versa (Lai et al. 2002). Substitutions 3′ of the template in the template recognition element (TRE) impose a modest defect in NAP and a more

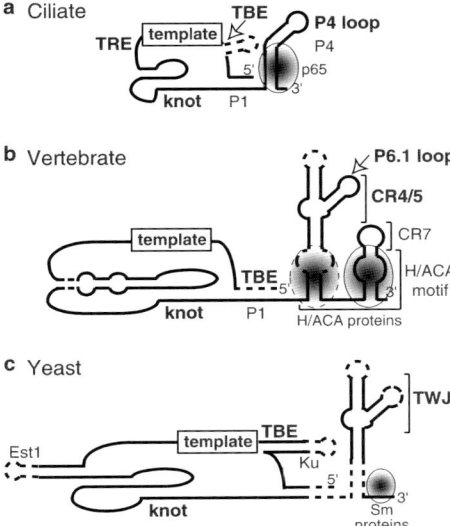

Fig. 14.2 TER secondary structure and functional motifs. TER secondary structures are illustrated with dashed lines representing regions that are variable within the phylogenetic group. Pseudoknots are labeled as knot; other functional motifs are labeled with abbreviations explained in the text: *TWJ* three way junction; *TBE* template boundary element; *TRE* template recognition element. In yeast TER, the three arms extending from the template/pseudoknot core are not drawn to scale; they are truncated in this illustration. TER motifs crucial for biogenesis of a biologically stable RNP are shown with a shaded oval depicting the location of the protein(s) that bind the motif, which are p65, the H/ACA proteins, and the Sm proteins in ciliates, vertebrates, and yeasts, respectively. Whether H/ACA proteins bind to one or both stems of the TER H/ACA motif is not yet established

severe one in RAP (Licht and Collins 1999; Lai et al. 2001; Cunningham and Collins 2005). Substitutions in the P4 loop drastically reduce overall activity (Licht and Collins 1999; Sperger and Cech 2001; O'Connor et al. 2005). Unpairing or deletion of either pseudoknot stem induces a surprisingly modest defect in RAP and RNP stability (Lai et al. 2003; Cunningham and Collins 2005).

The P4 stem and 3′ polyuridine tail are the binding site for a ciliate telomerase holoenzyme protein p65 (Prathapam et al. 2005; O'Connor and Collins 2006). In vivo, p65 is essential for accumulation of TER and TERT (Witkin and Collins 2004). Reconstitution of p65-dependent telomerase RNP assembly in vitro demonstrated that p65 binding to TER induces a stable kink in central P4, which allows subsequent assembly of TERT to generate a unique, conformationally stable p65-TER-TERT ternary complex (Stone et al. 2007). The TER TBE and P4 loop are brought closer upon p65 binding and even closer when TERT joins the p65-TER complex (Stone et al. 2007). These observations are consistent with a model that the TBE/template/TRE region and the P4 loop each bind TERT and perhaps also make a tertiary structure contact, as proposed for the vertebrate TER template region and loop of P6.1 (see below).

An RNA oligonucleotide containing the *T. thermophila* template and 3′-flanking TRE can be positioned in the TERT active site and copied with NAP, if the template-less remainder of the telomerase RNA is supplied as a separate fragment (Miller and Collins 2002). The low NAP and RAP of TER with a P4 loop substitution or deletion can also be fully rescued by addition of the P4 stem-loop as a separate oligonucleotide (Lai et al. 2003; Mason et al. 2003). These and related findings from circular permutation of the TER 5′ and 3′ ends (Miller and Collins 2002) suggest that non-template TER motifs have modular function. Some non-template motifs act in direct physical linkage with the template (for example, the TBE and TRE) while others can function without covalent connection (for example, the P4 loop). Division of labor among TER non-template motifs suggests that these RNAs could have gained their functional complexity in an incremental manner over evolution (see Conclusions below).

14.5.2   Vertebrate TERs

Vertebrate TERs are transcribed by RNA Polymerase II as 3′-extended precursors, which are processed to create mature, functional TER (Feng et al. 1995; Mitchell et al. 1999a). Mature TER also bears a post-transcriptionally added 5′ trimethylguanosine (TMG) cap (Fu and Collins 2006). Phylogenetic sequence comparison of 35 vertebrate TERs ranging from 382 to 559 nt in length revealed elements of shared secondary structure (Chen et al. 2000), including a template/pseudoknot region closed by stem P1 (Fig. 14.2b). The vertebrate TER pseudoknot has much greater sequence conservation than its ciliate counterpart and vastly increased stability of folding, due in part to base-triple interactions formed by one stem and the crossing uridine-rich loop (Theimer et al. 2005). Beyond the template/pseudoknot region, vertebrate TERs possess a 3′ H/ACA (hairpin-Hinge-hairpin-ACA) motif and two Conserved Regions (CRs) of primary sequence, CR4/5 and CR7 (Fig. 14.2b). The 5′ half of the CR7 loop harbors a non-essential CAB box for RNA localization to Cajal bodies (Jády et al. 2004; Fu and Collins 2007; Cristofari et al. 2007).

The vertebrate TER H/ACA motif and a motif on the 3′ side of the CR7 loop termed BIO box are both essential for TER accumulation in vivo (Mitchell et al. 1999a; Martin-Rivera and Blasco 2001; Fu and Collins 2003). The H/ACA motif is shared by a large family of eukaryotic H/ACA-motif RNAs and binds a heterotrimer of dyskerin, NOP10, and NHP2 proteins, each of which is essential for accumulation of mature RNA in vivo and remains associated with mature RNA in the biologically stable RNP (Fu and Collins 2007; Cohen et al. 2007; Collins 2008). Single amino acid substitutions of dyskerin identified to cause the X-linked form of the bone marrow failure syndrome dyskeratosis congenita (DC) reduce the accumulation of TER but not other H/ACA-motif RNAs and impose a telomere maintenance deficiency (Mitchell et al. 1999b; Wong and Collins 2006). Heterozygous TER mutations that either delete a large portion of the H/ACA motif or disrupt BIO box function were identified in patients with autosomal dominant DC (Vulliamy et al. 2001). As expected, these mutations prevent mutant TER accumulation in vivo (Fu and Collins 2003; Westin et al. 2007).

Vertebrate TER motifs required for telomerase catalytic activity have been investigated by both forward mutagenesis and reverse analysis of disease-linked TER variants identified in patients with DC, aplastic anemia, and other disorders (Garcia et al. 2007; Vulliamy and Dokal 2007). The template/pseudoknot region and CR4/5 are both important for activity reconstituted in vivo or in vitro (Mitchell and Collins 2000; Martin-Rivera and Blasco 2001). Each region alone can bind TERT (Mitchell and Collins 2000; Chen et al. 2002). As expected from the high level of vertebrate TER pseudoknot sequence conservation, pseudoknot stem mutations or mutation of the uridines involved in triple-helix formation inhibit activity assayed in vitro (Fu and Collins 2003; Ly et al. 2003; Chen and Greider 2005). Changes in P6.1 stem pairing, stem length, or loop sequence within the CR4/5 region also inhibit activity (Chen et al. 2002). Short-range cross-links between P6.1 and the template region have been detected using recombinant human TER (Ueda and Roberts 2004). This finding suggests the possibility that in the vertebrate as well as ciliate telomerase RNP, a stem-loop structure distant from the template in primary sequence (vertebrate P6.1, ciliate P4) may be brought in close proximity to the template in the active site of the RNP.

Surprisingly, TERs from mouse and some other rodents lack P1 and instead have only 2 nt 5′ of the template (Chen et al. 2000). These rodent TERs also lack the template 3′ region that is the site of primer hybridization for RAP. This change in TER structure may have evolved in collaboration with a gain in function by a regulatory factor such as a telomere-associated helicase, which could have assumed biological control of template/product dissociation. In vitro reconstitution assays swapping segments between human and mouse TERs have examined the basis for some activity differences between the two species' telomerase enzymes (Chen and Greider 2003) Human TER with a mouse-like 5′ end only 2 nt before the start of the template supports RAP, but lack of P1 abolishes correct specification of the template 5′ boundary. Mouse TERT combined with the human template/pseudoknot has RAP, while human TERT combined with the mouse template/pseudoknot is more inactive than merely lacking NAP or RAP. These and other findings suggest that there has been rapid evolutionary fine-tuning of telomerase activity and regulation among mammals.

14.5.3   Yeast TERs

Yeast TERs are transcribed by RNA Polymerase II and, like vertebrate TERs, gain a 5′ TMG cap (Seto et al. 1999; Leonardi et al. 2008; Webb and Zakian 2008). Yeast TERs are also 3′-processed like vertebrate TERs to form the functional mature RNA, but unlike the case of vertebrates, the yeast TER precursor is implicated to be a polyadenylated transcript (Chapon et al. 1997; Leonardi et al. 2008). Mature yeast TERs range in length from 930 to 1,540 nt, larger than all other known TERs except that of Plasmodium (Chakrabarti et al. 2007). Partial secondary structure has been predicted for budding yeast TERs by phylogenetic sequence comparison and modeling (Chen and Greider 2004; Tzfati et al. 2003). In contrast with ciliate and vertebrate TERs, the yeast TER template/pseudoknot region forms from non-contiguous primary sequence elements (Fig. 14.2c). However, there is strong evidence for a pseudoknot with the same type of triple-helix discovered in human TER (Shefer et al. 2007). As in ciliate and human TERs, the yeast TER template 5′ end is adjacent to a base-paired stem that acts to prevent copying past the template 5′ boundary (Tzfati et al. 2000; Seto et al. 2003).

Three base-paired arms extend from the template/pseudoknot core (Fig. 14.2c). These harbor a binding site for the telomerase holoenzyme protein Est1, a binding site for the chromosome end-binding complex Ku (in Saccharomyces only), or a three-way junction (TWJ) potentially analogous to vertebrate CR4/5 (Peterson et al. 2001; Seto et al. 2002; Chen and Greider 2004; Brown et al. 2007). In vivo reconstitution assays in *S. cerevisiae* indicate that TERT interaction occurs with the central template/pseudoknot region (Livengood et al. 2002; Chappell and Lundblad 2004; Lin et al. 2004). The TWJ arm includes a binding site for Sm proteins near the mature RNA 3′ end. In both *S. cerevisiae* and in *S. pombe*, mutations that disrupt the TER Sm site result in loss of mature RNA (Seto et al. 1999; Leonardi et al. 2008).

Yeast TER motifs important for catalytic activity have been assayed predominantly by in vivo reconstitution, with the exception of a recent rabbit reticulocyte lysate-based reconstitution assay using TERT and a minimized *S. cerevisiae* TER (Zappulla et al. 2005). Due to the large size of yeast TERs, these assays have focused on defining general regions required for function and protein interactions described above. Extensive mutagenesis has been done within the pseudoknot region and TWJ motif to establish more detail about the structures and sequences important for enzyme activity (Chappell and Lundblad 2004; Lin et al. 2004; Shefer et al. 2007; Brown et al. 2007).

14.6   TERT

TERTs have four structural or functional domains (Fig. 14.3): (1) a far-N terminal (TEN) domain, (2) a high-affinity telomerase RNA binding domain (TRBD), (3) reverse transcriptase active site motifs (RT), and (4) a C-terminal extension (TEC). High-resolution structures have been solved for a large portion of the *T. thermophila* TEN domain (Jacobs et al. 2006) and a truncated version of the *T. thermophila* TRBD lacking TER binding activity (Rouda and Skordalakes 2007). How these and the other TERT domains fold relative to each other in the overall active RNP architecture has not been established. *T. thermophila* TERT and TER assemble in vitro and in vivo to form an active RNP with a monomer of each subunit (Witkin and Collins 2004; Cunningham and Collins 2005; Bryan et al. 2003). Human TERT + TER RNPs reconstituted in vitro are said to contain a dimer or higher-order multimer of TER and/or TERT, while reconstitution in vivo produces an active RNP architecture with subunit monomers (Autexier and Lue 2006; Errington et al. 2008).

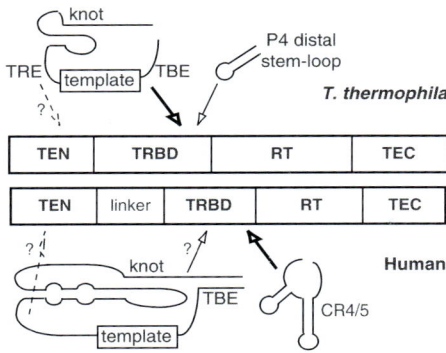

Fig. 14.3 TERT domains and TERT–TER interactions. See text for explanation

Yeast TERs purified from cell extract do not appear to be dimerized (Livengood et al. 2002). However, cooperation between telomerase active sites may occur during telomere elongation (Yu and Blackburn 1991; Prescott and Blackburn 1997).

The TRBD domain of *T. thermophila* or human TERT is sufficient to mediate a specific, high-affinity interaction with TER (Lai et al. 2001). Amino acid substitutions or deletions within the TRBD greatly inhibit TERT assembly with TER in ciliate, vertebrate, and yeast systems (Friedman and Cech 1999; Bryan et al. 2000; Armbruster et al. 2001; Lai et al. 2002; Bosoy et al. 2003). Purified recombinant *T. thermophila* TRBD binds to TER dependent on the TBE motif 5′ of the template, but the TBE alone is not sufficient for high-affinity interaction (O'Connor et al. 2005). In the physiological context of a p65-TER complex, *T. thermophila* TRBD interaction with TER involves both the TBE and the distal stem-loop of P4 (Fig. 14.3). Likewise, the human TRBD can bind both the human TER template/pseudoknot region and CR4/5 (Fig. 14.3). Vertebrate TERT binding to CR4/5 occurs with higher affinity than binding to the template/pseudoknot region (Mitchell and Collins 2000; Chen et al. 2002), and the TRBD is sufficient for this interaction (Mitchell and Collins 2000; Moriarty et al. 2004).

Cross-linking and mutagenesis-based structure/function assays establish the *T. thermophila* TEN domain as a site of DNA interaction in the active RNP (Romi et al. 2007; Jacobs et al. 2006). Similar studies implicate human and *S. cerevisiae* TEN domains in primer binding as well (Lee et al. 2003; Lue 2005). The TEN domain also has RNA binding activity (O'Connor et al. 2005; Jacobs et al. 2006; Moriarty et al. 2004), though the relatively weak TER interaction affinity compared to that mediated by the TRBD has hampered study of its functional significance. High-resolution structure of the *T. thermophila* TEN domain suggests a possible surface for RNA binding apart from the prominent groove for interaction with single-stranded DNA (Jacobs et al. 2006). One tantalizing model for TEN domain function would be to coordinate template positioning in the active site with the positioning of single-stranded product DNA. It will be necessary to develop methods

for monitoring dynamic protein–RNA and protein–DNA interactions across the
catalytic cycle in order to understand the complex interplay of protein and nucleic
acid associations that drive telomeric repeat synthesis.

14.7 Conclusions: An Evolutionary Perspective on TER Functional Complexity

Studies of structures and functions of TER illuminate many roles for non-coding
RNA in the telomerase RNP enzyme, both in the enzyme's complex catalytic cycle
and the biological regulation of telomerase function at telomeres. When the prede-
cessor of modern-day telomerase began elongating genome 3′ ends, it may have
copied template RNAs only loosely associated with the active site. Specificity
would have been improved by anchoring a unique RNA into a stable RNP complex
(Fig. 14.4a). The original TER may have harbored only a template and a TERT
binding site, but evolution would have subsequently utilized this physical connection to specialize telomerase through changes in TER and TERT. TER-TERT interactions in template-flanking regions would have refined the precision of repeat
synthesis (Fig. 14.4b). Additional TER structures could have evolved to improve
template use and elongation processivity (Fig. 14.4c). The order of these gains in
TER function is purely speculative, but the expanding size and structural complexity

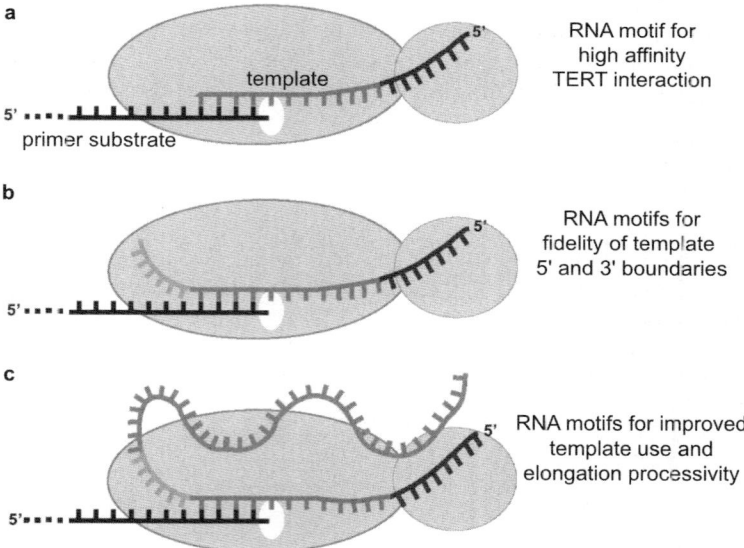

Fig. 14.4 A speculative evolutionary route for gains of TER non-template motif functions. See
text for explanation (See figure insert for color reproduction)

of TER may have in turn promoted evolution of a hierarchical RNP assembly pathway with ordered steps of RNA folding and protein–RNA interaction (O'Connor and Collins 2006; Stone et al. 2007). Continuing studies of telomerase RNA form and function will illuminate the mechanism of a fascinating RNP enzyme and also provide insights into the diversity of biological roles of non-coding RNA beyond its ancestral roles as reaction catalyst.

Acknowledgments I thank Collins lab members particularly James Mitchell, Michael Miller, Jill Licht, Cary Lai, Doreen Cunningham, Dragony Fu, Catherine O'Connor, Emily Egan and Aaron Robart for their contributions towards understanding TER motif structure and function. I also thank Peter Baumann and Virginia Zakian for sharing unpublished data during the early preparation of this chapter.

References

Aigner S, Postberg J, Lipps HJ, Cech TR (2003) The *Euplotes* La motif protein p43 has properties of a telomerase-specific subunit. Biochemistry 42:5736–5747

Armbruster BN, Banik SSR, Guo C, Smith AC, Counter CM (2001) N-terminal domains of the human telomerase catalytic subunit required for enzyme activity in vivo. Mol Cell Biol 22:7775–7786

Autexier C, Greider CW (1995) Boundary elements of the *Tetrahymena* telomerase RNA template and alignment domains. Genes Dev 9:2227–2239

Autexier C, Lue NF (2006) The structure and function of telomerase reverse transcriptase. Annu Rev Biochem 75:493–517

Bertuch AA, Lundblad V (2006) The maintenance and masking of chromosome termini. Curr Opin Cell Biol 18:247–253

Bosoy D, Peng Y, Mian IS, Lue NF (2003) Conserved N-terminal motifs of telomerase reverse transcriptase required for ribonucleoprotein assembly in vivo. J Biol Chem 278:3882–3890

Brown Y, Abraham M, Pearl S, Kabaha MM, Elboher E, Tzfati Y (2007) A critical three-way junction is conserved in budding yeast and vertebrate telomerase RNAs. Nucleic Acids Res 35:6280–6289

Bryan TM, Goodrich KJ, Cech TR (2000) Telomerase RNA bound by protein motifs specific to telomerase reverse transcriptase. Mol Cell 6:493–499

Bryan TM, Goodrich KJ, Cech TR (2003) *Tetrahymena* telomerase is active as a monomer. Mol Biol Cell 14:4794–4804

Chakrabarti K, Pearson M, Grate L, Sterne-Weiler T, Deans J, Donohue JP, Ares M. (2007) Structural RNAs of known and unknown function identified in malaria parasites by comparative genomics and RNA analysis. RNA 13:1923–1939

Chapon C, Cech T, Zaug A (1997) Polyadenylation of telomerase RNA in budding yeast. RNA 3:1337–1351

Chappell AS, Lundblad V (2004) Structural elements required for association of the *Saccharomyces cerevisiae* telomerase RNA with the Est2 reverse transcriptase. Mol Cell Biol 24:7720–7736

Chen JL, Greider CW (2003) Determinants in mammalian telomerase RNA that mediate enzyme processivity and cross-species incompatibility. EMBO J 22:304–314

Chen JL, Greider CW (2004) An emerging consensus for telomerase RNA structure. Proc Natl Acad Sci USA 101:14683–14684

Chen JL, Greider CW (2005) Functional analysis of the pseudoknot structure in human telomerase RNA. Proc Natl Acad Sci U S A 102:8080–8085

Chen JL, Blasco MA, Greider CW (2000) Secondary structure of vertebrate telomerase RNA. Cell 100:503–514

Chen JL, Opperman KK, Greider CW (2002) A critical stem-loop structure in the CR4-CR5 domain of mammalian telomerase RNA. Nucleic Acids Res 30:592–597

Cohn M, Blackburn EH (1995) Telomerase in yeast. Science 269:396–400

Collins K (1999) Ciliate telomerase biochemistry. Annu Rev Biochem 68:187–218

Collins K (2006) The biogenesis and regulation of telomerase holoenzymes. Nat Rev Mol Cell Biol 7:484–494

Collins K (2008) Physiological assembly and activity of human telomerase complexes. Mech Ageing Dev 129:91–98

Collins K, Gandhi L (1998) The reverse transcriptase component of the *Tetrahymena* telomerase ribonucleoprotein complex. Proc Natl Acad Sci U S A 95:8485–8490

Collins K, Greider CW (1993) Nucleolytic cleavage and non-processive elongation catalyzed by *Tetrahymena* telomerase. Genes Dev 7:1364–1376

Collins K, Mitchell JR (2002) Telomerase in the human organism. Oncogene 21:564–579

Cong YS, Wright WE, Shay JW (2002) Human telomerase and its regulation. Microbiol Mol Biol Rev 66:407–425

Cristofari G, Adolf E, Reichenbach P, Sikora K, Terns RM, Terns MP, Lingner J (2007) Human telomerase RNA accumulation in Cajal bodies facilitates telomerase recruitment to telomeres and telomere elongation. Mol Cell 27:882–889

Cunningham DD, Collins K (2005) Biological and biochemical functions of RNA in the *Tetrahymena* telomerase holoenzyme. Mol Cell Biol 25:4442–4454

Errington TM, Fu D, Wong JM, Collins K (2008) Disease-associated human telomerase RNA variants show loss of function for telomere synthesis without dominant-negative interference. Mol Cell Biol PMID 18710936

Feng J, Funk WD, Wang S, Weinrich SL, Avilion AA, Chiu C, Adams RR, Chang E, Allsopp RC, Yu J, Le S, West M, Harley CB, Andrews WH, Greider CW, Villeponteau B (1995) The RNA component of human telomerase. Science 269:1236–1241

Förstemann K, Lingner J (2005) Telomerase limits the extent of base pairing between template RNA and telomeric DNA. EMBO Rep 6:361–366

Friedman KL, Cech TR (1999) Essential functions of amino-terminal domains in the yeast telomerase catalytic subunit revealed by selection for viable mutants. Genes Dev 13:2863–2874

Fu D, Collins K (2003) Distinct biogenesis pathways for human telomerase RNA and H/ACA small nucleolar RNAs. Mol Cell 11:1361–1372

Fu D, Collins K (2006) Human telomerase and Cajal body ribonucleoproteins share a unique specificity of Sm protein association. Genes Dev 20:531–536

Fu D, Collins K (2007) Purification of human telomerase complexes identifies factors involved in telomerase biogenesis and telomere length regulation. Mol Cell 28:773–785

Gallardo F, Olivier C, Dandjinou AT, Wellinger RJ, Chartrand P (2008) TLC1 RNA nucleo-cytoplasmic trafficking links telomerase biogenesis to its recruitment to telomeres. EMBO J 27:748–757

Garcia CK, Wright WE, Shay JW (2007) Human diseases of telomerase dysfunction: insights into tissue aging. Nucleic Acids Res 35:7406–7416

Gilson E, Geli V (2007) How telomeres are replicated. Nat Rev Mol Cell Biol 8:825–838

Greider CW (1991) Telomerase is processive. Mol Cell Biol 11:4572–4580

Greider CW, Blackburn EH (1989) A telomeric sequence in the RNA of *Tetrahymena* telomerase required for telomere repeat synthesis. Nature 337:331–337

Hammond PW, Cech TR (1998) *Euplotes* telomerase: evidence for limited base-pairing duing primer elongation and dGTP as an effector of translocation. Biochemistry 37:5162–5172

Hardy CD, Schultz CS, Collins K (2001) Requirements for the dGTP-dependent repeat addition processivity of recombinant *Tetrahymena* telomerase. J Biol Chem 276:4863–4871

Holt SE, Aisner DL, Baur J, Tesmer VM, Dy M, Ouellette M, Trager JB, Morin GB, Toft DO, Shay JW, Wright WE, White MA (1999) Functional requirement of p23 and Hsp90 in telomerase complexes. Genes Dev 13:817–826

Hug N, Lingner J (2006) Telomere length homeostasis. Chromosoma 115:413–425

Jacob NK, Kirk KE, Price CM (2003) Generation of telomeric G strand overhangs involves both G and C strand cleavage. Mol Cell 11:1021–1032

Jacobs SA, Podell ER, Cech TR (2006) Crystal structure of the essential N-terminal domain of telomerase reverse transcriptase. Nat Struct Mol Biol 13:218–225

Jády BE, Bertrand E, Kiss T (2004) Human telomerase RNA and box H/ACA scaRNAs share a common Cajal body-specific localization signal. J Cell Biol 164:647–652

Lai CK, Mitchell JR, Collins K (2001) RNA binding domain of telomerase reverse transcriptase. Mol Cell Biol 21:990–1000

Lai CK, Miller MC, Collins K (2002) Template boundary definition in *Tetrahymena* telomerase. Genes Dev 16:415–420

Lai CK, Miller MC, Collins K (2003) Roles for RNA in telomerase nucleotide and repeat addition processivity. Mol Cell 11:1673–1683

Lee SR, Wong JM, Collins K (2003) Human telomerase reverse transcriptase motifs required for elongation of a telomeric substrate. J Biol Chem 278:52531–52536

Legassie JD, Jarstfer MB (2006) The unmasking of telomerase. Structure 14:1603–1609

Leonardi J, Box JA, Bunch JT, Baumann P (2008) TER1, the RNA subunit of fission yeast telomerase. Nat Struct Mol Biol 15:26–33

Licht JD, Collins K (1999) Telomerase RNA function in recombinant *Tetrahymena* telomerase. Genes Dev 13:1116–1125

Lin J, Ly H, Hussain A, Abraham M, Pearl S, Tzfati Y, Parslow TG, Blackburn EH (2004) A universal telomerase RNA core structure includes structured motifs required for binding the telomerase reverse transcriptase protein. Proc Natl Acad Sci U S A 101:14713–14718

Lingner J, Hendrick LL, Cech TR (1994) Telomerase RNAs of different ciliates have a common secondary structure and a permuted template. Genes Dev 8:1984–1998

Lingner J, Cech TR, Hughes TR, Lundblad V (1997) Three ever shorter telomere (*EST*) genes are dispensable for in vitro yeast telomerase activity. Proc Natl Acad Sci U S A 94:11190–11195

Livengood AJ, Zaug AJ, Cech TR (2002) Essential regions of *Saccharomyces cerevisiae* telomerase RNA: separate elements for Est1p and Est2p interaction. Mol Cell Biol 22:2366–2374

Lue NF (2005) A physical and functional constituent of telomerase anchor site. J Biol Chem 280:26586–26591

Ly H, Blackburn EH, Parslow TG (2003) Comprehensive structure-function analysis of the core domain of human telomerase RNA. Mol Cell Biol 23:6849–6856

Martin-Rivera L, Blasco MA (2001) Identification of functional domains and dominant negative mutations in vertebrate telomerase RNA using an in vivo reconstitution system. J Biol Chem 276:5856–5865

Mason DX, Goneska E, Greider CW (2003) Stem-loop IV of *tetrahymena* telomerase RNA stimulates processivity in trans. Mol Cell Biol 23:5606–5613

McCormick-Graham M, Romero DP (1996) A single telomerase RNA is sufficient for the synthesis of variable telomeric DNA repeats in ciliates of the genus *Paramecium*. Mol Cell Biol 16:1871–1879

McCormick-Graham M, Haynes WJ, Romero DP (1997) Variable telomeric repeat synthesis in *Paramecium tetraurelia* is consistent with misincorporation by telomerase. EMBO J 16:3233–3242

McEachern MJ, Blackburn EH (1995) Runaway telomere elongation caused by telomerase RNA gene mutations. Nature 376:403–409

Miller MC, Collins K (2002) Telomerase recognizes its template by using an adjacent RNA motif. Proc Natl Acad Sci USA 99:6585–6590

Mitchell JR, Collins K (2000) Human telomerase activation requires two independent interactions between telomerase RNA and telomerase reverse transcriptase in vivo and in vitro. Mol Cell 6:361–371

Mitchell JR, Cheng J, Collins K (1999a) A box H/ACA small nucleolar RNA-like domain at the human telomerase RNA 3′ end. Mol Cell Biol 19:567–576

Mitchell JR, Wood E, Collins K (1999b) A telomerase component is defective in the human disease dyskeratosis congenita. Nature 402:551–555

Moriarty TJ, Marie-Egyptienne DT, Autexier C (2004) Functional organization of repeat addition processivity and DNA synthesis determinants in the human telomerase multimer. Mol Cell Biol 24:3720–3733

Morin GB (1989) The human telomere terminal transferase enzyme is a ribonucleoprotein that synthesizes TTAGGG repeats. Cell 59:521–529

Muñoz DP, Collins K (2004) Biochemical properties of *Trypanosoma cruzi* telomerase. Nucleic Acids Res 32:5214–5222

O'Connor CM, Collins K (2006) A novel RNA binding domain in *Tetrahymena* telomerase p65 initiates hierarchical assembly of telomerase holoenzyme. Mol Cell Biol 26:2029–2036

O'Connor CM, Lai CK, Collins K (2005) Two purified domains of telomerase reverse transcriptase reconstitute sequence-specific interactions with RNA. J Biol Chem 280:17533–17539

Pascolo E, Wenz C, Lingner J, Hauel N, Priepke H, Kauffmann I, Garin-Chesa P, Rettig WJ, Damm K, Schnapp A (2002) Mechanism of human telomerase inhibition by BIBR1532, a synthetic, non-nucleosidic drug candidate. J Biol Chem 277:15566–15572

Peterson SE, Stellwagen AE, Diede SJ, Singer MS, Haimberger ZW, Johnson CO, Tzoneva M, Gottschling DE (2001) The function of a stem-loop in telomerase RNA is linked to the DNA repair protein Ku. Nat Genet 27:64–67

Prathapam R, Witkin KL, O'Connor CM, Collins K (2005) A telomerase holoenzyme protein enhances telomerase RNA assembly with telomerase reverse transcriptase. Nat Struct Mol Biol 12:252–257

Prescott J, Blackburn EH (1997) Functionally interacting telomerase RNAs in the yeast telomerase complex. Genes Dev 11:2790–2800

Prowse KR, Avilion AA, Greider CW (1993) Identification of a nonprocessive telomerase activity from mouse cells. Proc Natl Acad Sci U S A 90:1493–1497

Romero DP, Blackburn EH (1991) A conserved secondary structure for telomerase RNA. Cell 67:343–353

Romi E, Baran N, Gantman M, Shmoish M, Min B, Collins K, Manor H (2007) High-resolution physical and functional mapping of the template adjacent DNA binding site in catalytically active telomerase. Proc Natl Acad Sci U S A 104:8791–8796

Rouda S, Skordalakes E (2007) Structure of the RNA-binding domain of telomerase: Implications for RNA recognition and binding. Structure 13:1403–1412

Schnapp G, Rodi H-P, Rettig WJ, Schnapp A, Damm K (1998) One-step affinity purification protocol for human telomerase. Nucleic Acids Res 26:3311–3313

Serrano M, Blasco MA (2007) Cancer and ageing: convergent and divergent mechanisms. Nat Rev Mol Cell Biol 8:715–722

Seto AG, Zaug AJ, Sobel SG, Wolin SL, Cech TR (1999) *Saccharomyces cerevisiae* telomerase is an Sm small nuclear ribonucleoprotein particle. Nature 401:177–180

Seto AG, Livengood AJ, Tzfati Y, Blackburn EH, Cech TR (2002) A bulged stem tethers Est1p to telomerase RNA in budding yeast. Genes Dev 16:2800–2812

Seto AG, Umansky K, Tzfati Y, Zaug AJ, Blackburn EH, Cech TR (2003) A template-proximal RNA paired element contributes to *Saccharomyces cerevisiae* telomerase activity. RNA 9:1323–1332

Shay JW, Wright WE (2006) Telomerase therapeutics for cancer: challenges and new directions. Nat Rev Drug Discov 5:577–584

Shefer K, Brown Y, Gorkovoy V, Nussbaum T, Ulyanov NB, Tzfati Y (2007) A triple helix within a pseudoknot is a conserved and essential element of telomerase RNA. Mol Cell Biol 27:2130–2143

Singer MS, Gottschling DE (1994) TLC1: template RNA component of *Saccharomyces cerevisiae* telomerase. Science 266:404–409

Sperger JM, Cech TR (2001) A stem-loop of *Tetrahymena* telomerase RNA distant from the template potentiates RNA folding and telomerase activity. Biochemistry 40:7005–7016

Stone MS, Mihalusova M, O'Connor CM, Prathapam R, Collins K, Zhuang X (2007) Stepwise protein-mediated RNA folding directs assembly of telomerase ribonucleoprotein. Nature 446:458–461

Theimer CA, Feigon J (2006) Structure and function of telomerase RNA. Curr Opin Struct Biol 16:307–318

Theimer CA, Blois CA, Feigon J (2005) Structure of the human telomerase RNA pseudoknot reveals conserved tertiary interactions essential for function. Mol Cell 17:671–682

Tzfati Y, Fulton TB, Roy J, Blackburn EH (2000) Template boundary in a yeast telomerase specified by RNA structure. Science 288:863–867

Tzfati Y, Knight Z, Roy J, Blackburn EH (2003) A novel pseudoknot element is essential for the action of a yeast telomerase. Genes Dev 17:1779–1788

Ueda CT, Roberts RW (2004) Analysis of a long-range interaction between conserved domains of human telomerase RNA. RNA 10:139–147

Verdun RE, Karlseder J (2007) Replication and protection of telomeres. Nature 447:924–931

Vulliamy TJ, Dokal I (2007) Dyskeratosis congenita: the diverse clinical presentation of mutations in the telomerase complex. Biochimie 90:122–130

Vulliamy T, Marrone A, Goldman F, Dearlove A, Bessler M, Mason PJ, Dokal I (2001) The RNA component of telomerase is mutated in autosomal dominant dyskeratosis congenita. Nature 413:432–435

Wang H, Blackburn EH (1997) De novo telomere addition by *Tetrahymena* telomerase in vitro. EMBO J 16:866–879

Wang H, Gilley D, Blackburn EH (1998) A novel specificity for the primer-template pairing requirement in *Tetrahymena* telomerase. EMBO J 17:1152–1160

Webb CJ, Zakian VA (2008) Identification and characterization of the *Schizosaccharomyces pombe* TER1 telomerase RNA. Nat Struct Mol Biol 15:34–42

Weinrich SL, Pruzan R, Ma L, Ouellette M, Tesmer VM, Holt SE, Bodnar AG, Lichsteiner S, Kim NW, Trager JB, Taylor RD, Carlos R, Andrews WH, Wright WE, Shay JW, Harley CB, Morin GB (1997) Reconstitution of human telomerase with the template RNA component hTR and the catalytic protein subunit hTRT. Nat Genet 17:498–502

Westin ER, Chavez E, Lee KM, Gourronc FA, Riley S, Lansdorp PM, Goldman FD, Klingelhutz AJ (2007) Telomere restoration and extension of proliferative lifespan in dyskeratosis congenita fibroblasts. Aging Cell 6:383–394

Witkin KL, Collins K (2004) Holoenzyme proteins required for the physiological assembly and activity of telomerase. Genes Dev 18:1107–1118

Wong JMY, Collins K (2003) Telomere maintenance and disease. Lancet 362:983–988

Wong JMY, Collins K (2006) Telomerase RNA level limits telomere maintenance in X-linked dyskeratosis congenita. Genes Dev 20:2848–2858

Ye AJ, Romero DP (2002) Phylogenetic relationships amongst tetrahymenine ciliates inferred by a comparison of telomerase RNAs. Int J Syst Evol Microbiol 52:2297–2302

Yu G, Blackburn EH (1991) Developmentally programmed healing of chromosomes by telomerase in *Tetrahymena*. Cell 67:823–832

Yu G, Bradley JD, Attardi LD, Blackburn EH (1990) In vivo alteration of telomere sequences and senescence caused by mutated *Tetrahymena* telomerase RNAs. Nature 344:126–132

Zappulla DC, Cech TR (2006) RNA as a flexible scaffold for proteins: yeast telomerase and beyond. Cold Spring Harb Symp Quant Biol 71:217–224

Zappulla DC, Goodrich K, Cech TR (2005) A miniature yeast telomerase RNA functions in vivo and reconstitutes activity in vitro. Nat Struct Mol Biol 12:1072–1077

Chapter 15
Ribosomal Dynamics: Intrinsic Instability of a Molecular Machine

Haixiao Gao, Jamie LeBarron, and Joachim Frank(✉)

Abstract Ribosomes are molecular machines that translate genetic message into nascent peptides, through a complex dynamics interplay with mRNAs, tRNAs, and various protein factors. A prominent example of ribosomal dynamics is the rotation of small ribosomal subunit with respect to a large subunit, characterized as the "ratchet motion," which is triggered by the binding of several translation factors. Here, we analyze two kinds of ribosomal ratchet motions, induced by the binding of EF-G and RF3, respectively, as previously observed by cryo-electron microscopy. Using the flexible fitting technique (real-space refinement) and an RNA secondary structure display tool (coloRNA), we obtained quasi-atomic models of the ribosome in these ratchet-motion-related functional states and mapped the observed differences onto the highly conserved RNA secondary structure. Comparisons between two sets of ratchet motions revealed that, while the overall patterns of the RNA displacement are very similar, several local regions stand out in their differential behavior, including the highly conserved GAC (GTPase-associated-center) region. We postulate that these regions are important in modulating general ratchet motion and bestowing it with the dynamic characteristics required for the specific function.

15.1 Introduction

The ribosome is a highly complex molecular machine dedicated to the process of translating the genetic message into protein. This is accomplished on the ribosome by the lining-up of aminoacyl-transfer RNAs along the messenger RNA through

J. Frank
Howard Hughes Medical Institute, Department of Biochemistry and Molecular Biophysics and Department of Biological Sciences, Columbia University, 650 West 168th Street, NY 10032, USA
e-mail: joachim@wordsworth.org

N.G. Walter et al. (eds.) *Non-Protein Coding RNAs*
doi: 10.1007/978-3-540-70840-7_15, © Springer-Verlag Berlin Heidelberg 2009

template-anti-template matching. Amino acids carried by the tRNAs are strung together into a polypeptide in the precise sequence dictated by the genetic message. The polypeptide then folds into the energetically most favored conformation, forming active protein. Advances in X-ray crystallography have resulted in the structure of the ribosome being now known to atomic resolution (Ban et al. 2000; Schluenzen et al. 2000; Wimberly et al. 2000; Harms et al. 2001; Schuwirth et al. 2005; Korostelev et al. 2006; Selmer et al. 2006). Despite this remarkable feat of structure research, the precise mechanism of different steps of protein synthesis has remained largely elusive. This is because the process involves a complex dynamic interplay between the ribosome and a number of ligands during which the system goes through a sequence of states, each of which must be known in atomic detail. Difficulties in obtaining highly-ordered crystals have made it impossible, except in a few cases to obtain X-ray structures for such states (Petry et al. 2005; Borovinskaya et al. 2007; Weixlbaumer et al. 2007). However, great strides have been made in the exploration of structural dynamics of translation through recent advances in electron microscopy and image reconstruction. These techniques yield "snapshot" density maps in the resolution range that allow the construction of quasi-atomic models by flexible fitting of the X-ray structural components into the EM density maps (Spahn et al. 2001; Gao et al. 2003; Tama et al. 2003; Klaholz et al. 2004; Halic et al. 2006; Trabuco et al. 2008). Recently, single-molecule FRET studies have complemented and enriched the picture of functional dynamics (Blanchard et al. 2004; Ermolenko et al. 2007; Munro et al. 2007).

An important concept emerging from EM studies is that many of the steps of translation involve a large change in ribosomal conformation, characterized as "ratchet motion." First described for the binding of EF-G in translocation (Agrawal et al. 1999; Frank and Agrawal 2000; Valle et al. 2003), it was subsequently observed for IF2 binding during initiation (Allen et al. 2005), RF3 binding during termination (Klaholz et al. 2004; Gao et al. 2007), and RRF binding during recycling (Gao et al. 2005; Barat et al. 2007). Upon binding of any of these factors, the ribosomal goes from a normal (termed macrostate I; Frank et al. 2007) to a ratcheted configuration (macrostate II), as the small subunit rotates counterclockwise with respect to the large subunit, around an axis which involves small ribosome proteins S8, S12, and helices 19, 20, 24, and 27 in the 16S rRNA (Tama et al. 2003), lying normal to the plane of the intersubunit space. Observations by cryo-EM on the 80S ribosome bound with or without eEF2 (Spahn et al. 2004; Taylor et al. 2007) confirm that the motion is a universal feature of translation.

The ratchet motion of the ribosome, ostensibly triggered by binding of the various translation factors, can also be understood as an instance of an "induced fit" mechanism since, as a rule, both ribosome and factor are seen to change their conformations. In the case of EF-G, domains III, IV, and V rotate jointly around a hinge relative to domains G,G' and II (Valle et al. 2003). Similarly in RF3, domains II and III rotates relative to domain I (Gao et al. 2007) (Fig. 15.1). In line with the emerging paradigm of molecular binding interactions, both molecules in the thermal

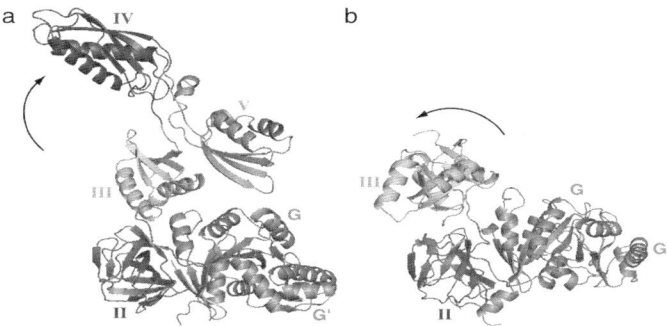

Fig. 15.1 Structural models of ribosome-bound EF-G and RF3 inferred by fitting X-ray structures into cryo-EM maps. (**a**) EF-G·GDPNP (1PN6, Valle et al. 2003). (**b**) RF3·GDPNP (2O0F, Gao et al. 2007). Arrows indicate domain motions in EF-G and RF3 upon binding of the factors to the ribosome

environment undergo structural changes probing their conformational spaces and the stable binding interaction will occur when the two molecules come into sterically matching conformations that are further stabilized by specific interactions such as hydrogen bonding, etc. The occurrence of ratchet motion even in the *absence* of a factor has been observed in recent FRET experiments by the Noller group (Ermolenko et al. 2007).

The notable absence of the aa-tRNA·EF-Tu·GTP ternary complex from the "ratchet motion" list confirms a rule spelt out by Valle et al. (2003): ratchet motion of the ribosome cannot occur unless the P-site tRNA is deacetylated. This rule makes perfect sense when one considers the mechanical properties of the system formed by the two subunits: torsional movement can take place in principle since all peripheral bridges are either elastically deformable or capable of disconnecting and gliding. However, this movement is blocked or greatly reduced by the existence of a mechanical link between the CCA arm of the P-site tRNA and the nascent polypeptide chain, which is the condition in the post-translocational state preceding the binding of the ternary complex.

Below, we will briefly summarize the steps from the collection of the EM data to the construction of what will be referred to as "quasi-atomic model." We will later explore, with the help of such models, how the ratchet motion manifests itself in terms of the structural reorganization of the ribosome. We will further investigate how conformational changes accompanying EF-G-induced ratchet motion differ from those observed with RF3, and how these differences might be linked to the distinct functional agendas of the two factors. This discovery process will thus be useful and informative for studying the roles of other factors binding to the ribosome, as well.

15.2 Construction of Dynamic Atomic Models of the Ribosome

The single-particle cryo-electron microscopy (cryo-EM) method, which is used to study ribosomal complexes in different translational states, can be described as a sequence of several steps. First, to prepare ribosomes in a particular functional state, an antibiotic or a non-hydrolyzable GTP analog is used to stop the reaction at a defined point in the translation process. These chemically "trapped" ribosomes are quickly frozen in a thin layer of vitreous ice, with the aid of a plunge-freezer, so that they are preserved in a close-to-native state (see (Grassucci et al. 2007)). Next, the cryo-sample is imaged in the electron microscope to collect many projections of the ribosomal complexes (see (Grassucci et al. 2008). On the assumption that all these experimental 2D projections are from complexes in the same functional state, a three-dimensional density map can be computationally generated using the single-particle reconstruction method (Frank 2006). When different states coexist in the same preparation, classification methods may be used to sort out the projection images into homogenous subsets, to be reconstructed separately (Valle et al. 2002; Gao et al. 2004; Klaholz et al. 2004; Fu et al. 2007; Scheres et al. 2007).

Nowadays, for cryo-EM study of ribosome complexes the practical resolution target is in the medium spatial resolution range (7–12 Å)- a limitation which is due mainly to the low electron dose applied and the low contrast of the molecule embedded in the matrix of ice. While detailed atomic structural information is not yet directly visible in this range, it is nevertheless feasible to interpret these density maps on the molecular scale via various flexible fitting techniques. With proper constraints, such as stereochemically restrained rigid blocks, atomic structural components can be accurately positioned into the EM density map. As a rule of thumb, the flexible fitting can generate quasi-atomic models having an accuracy of fourfold to fivefold better than the nominal experimental resolution (Baker and Johnson 1996; Rossmann 2000).

Of the various flexible fitting methods, a real-space refinement method – RSRef has been extensively used to build atomic models of ribosomal complexes in different functional states based on medium-resolution cryo-EM density maps. RSRef performs flexible fitting with multiple structural components, and optimizes a target function, assessing the deviations between the model and experimental observations in both density and stereo-chemical geometry (Chapman 1995; Fabiola and Chapman 2005). For the fitting of the ribosomal complexes, we have developed a progressive hierarchical multi-rigid-body refinement scheme, in which fidelity of the model is gradually increased by dividing the structure into smaller rigid blocks in the refinement, the number of blocks being guided by closely monitoring fitting quality in the local regions (Gao et al. 2003; Gao and Frank 2005). The final rigid-body assignment for the ribosomal RNA includes over 100 pieces, in which the integrity of the double-helical regions is kept intact (Fig. 15.2).

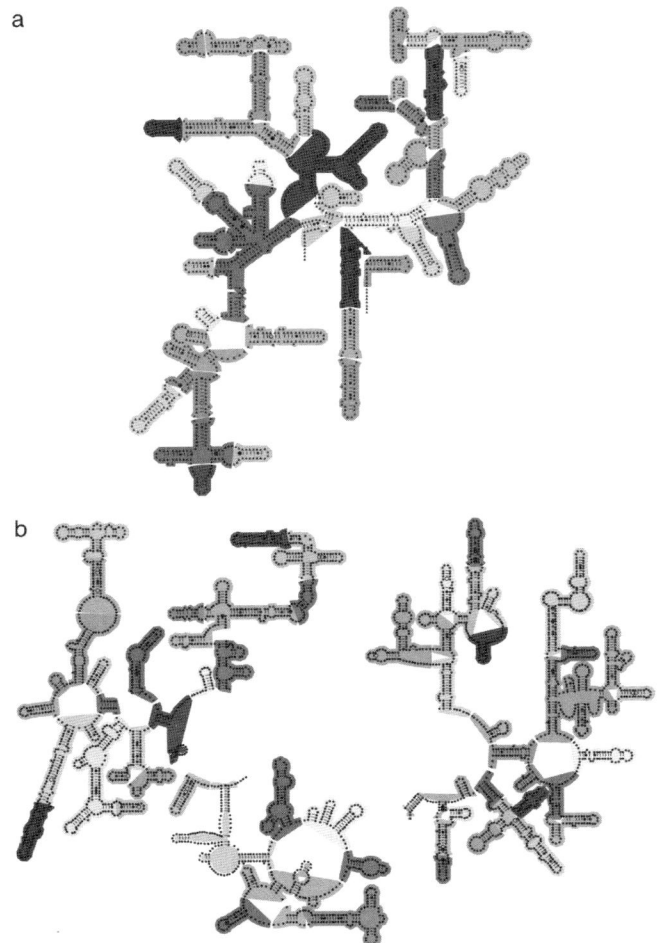

Fig. 15.2 Rigid-body assignment of *E. coli* 70S ribosomal RNAs used in the real-space refinement. (**a**) 16S rRNA (43 pieces). (**b**) 23S rRNA (62 pieces) (See figure insert for color reproduction)

15.3   Dynamic Interplay of the Ribosome with Translation Factors EF-G and RF3

In this work, we focus on the structural dynamics of the ribosome associated with the binding of two translational GTPases – EF-G and RF3. Functions of EF-G and RF3 in translation are quite different – EF-G promotes translocation of both mRNA and tRNA on the ribosome during the elongation cycle, while RF3, as a class-II release factor, plays a catalytic role in removing the class-I release factors, RF1 and RF2 from the ribosome during the termination process. Nevertheless, the GTPase activities of both EF-G and RF3 are regulated by the same parameter, namely absence vs. presence of a peptide on the tRNA in the ribosomal P site (Zavialov and Ehrenberg 2003). Binding of either EF-G·GDPNP (non-hydrolyzable GTP analog) or RF3·GDPNP to the ribosome requires the release of peptide from the P-site tRNA, which unlocks the ribosome, and allows it to undergo large conformational changes, i.e., from macrostate I to macrostate II (Valle et al. 2003; Gao et al. 2007).

Evidently, both in translocation and termination, in the interplay with EF-G and RF3, the ribosome acts through a unified mechanism. As observed from the ratchet motion upon binding of IF2 (Allen et al. 2005) and RRF (Gao et al. 2005), the same mechanism must also apply to initiation and recycling. Alexander Spirin, who proposed relative motion of loosely coupled subunits as an essential requirement for translocation even before much was known about ribosome structure (Spirin 1968), was also the first to speculate that the intersubunit motion observed by cryo-EM might be due to thermal agitation (Spirin 2002). One is forced to this conclusion when trying to understand the observed factor-free translation (see Frank et al. 2007). Subsequently, it was shown that the ratchet motion constitutes the most prominent mode in normal mode analysis of the ribosome structure, apparently intrinsically tied to its architecture (Tama et al. 2003; Wang et al. 2004). Thus, the ribosome has probably developed a particular architecture, over the course of evolution, which is able to harness ambient energy in the thermal environment toward dynamics changes required in several functional processes (Frank et al. 2007). Since each of these processes is quite different in nature, and involves binding of a different factor, the question arises whether there are in fact local differences in the conformational changes associated with the ratchet motion that lend themselves to different functions.

With this in mind, we sought to compare the ratchet motion observed upon the binding of both EF-G·GDPNP and RF3·GDPNP from two previous cryo-EM studies. Real-space refinement was used to build atomic models based on cryo-EM density maps; these models were then analyzed for similarities and differences in their detailed dynamic behavior. To this end, four cryo-EM maps obtained from previous studies were investigated, namely: (1) a pretranslocation complex (70S tRNA^{+met}-fMet-Phe-tRNA^{Phe}; Valle et al. 2003); (2) an EF-G-bound translocation complex (70S·MFTI-tRNA^{Ile}-EF-G·GDPNP·puromycin; Valle et al. 2003); (3) an RF1-bound termination complex (70S·MFTI-tRNA^{Ile}·UAA·RF1; Rawat et al. 2006); and (4) an RF3-bound termination complex (70S·MFTI-tRNA^{Ile}·UAA·RF3·GDPNP·puromycin; Gao et al. 2007). Maps #1 and #2 represent two states in the elongation cycle, in which map #1 shows a non-ratcheted state (macrostate I) and map #2 a ratcheted state (macrostate II) due to the bind of EF-G·GDPNP. Similarly, maps #3 and #4 represent two states in the termination process, in which map #3 is a non-ratcheted state (macrostate I) and map #4 shows a ratcheted conformation (macrostate II) seen in the presence of RF3·GDPNP. Thus, in future, we associate maps #1 and #2 with the EF-G-related ratchet motion and maps #3 and #4 with the RF3-related ratchet motion.

According to previous cryo-EM studies, the EF-G- and RF3-related ratchet motions are very similar based on an overall comparison of the density maps. To perform a detailed analysis at the molecular level, all four maps were fitted using the 3.5Å crystal structure of the 70S ribosome from *E. coli* (Schuwirth et al. 2005) A multi-rigid-body refinement scheme was applied as described above, in which ribosomal RNAs were divided into rigid pieces as shown in Fig. 15.2, and each ribosomal protein was treated as a rigid piece. For a start, a CNS (Brunger et al. 1998) implementation of the RSRef was employed to perform conjugate gradient energy minimization and Cartesian molecular dynamics to minimize the number of initial non-bonded steric violations in the initial model. The multi-rigid-body refinement was then performed as in Gao et al. (2003) to obtain the best fit of model into the EM density map. The final fitting statistics are listed in Table 15.1, showing all four maps are fitted with similar quality.

Comparing the atomic models obtained by real-space refinement, we can interpret both the EF-G-related and the RF3-related ratchet motions in terms of the RNA 3D structure. Overall, maximal displacement ranges of 16S rRNA are ~20 Å in the EF-G-related ratchet motion and ~16 Å in the RF3-related ratchet motion. For atomic interpretation of ratchet motion, ribosomal RNAs are colored according to displacement of each residue between non-ratcheted and ratcheted states (Fig. 15.3). Both for the EF-G related and RF3-related ratchet motions, high mobility of the 16S rRNA with respect to 23S rRNA emerges in very similar fashion. However, the displacements of the head (h39, h41, h42) and beak regions (h33) in the EF-G-related ratchet motion are significantly larger than those in the RF3-related ratchet motion. In contrast, in the 23S rRNA, maximal displacement ranges are quite similar between the EF-G related and RF3-related ratchet motions, about 8Å, and largest motions are mainly localized in the L1 stalk and the GAC (GTPase-associated-center) regions.

To facilitate analysis of ratchet motion-related changes at the sites of RNA components, the three-dimensional displacement of each residue onto the secondary structure of the ribosomal RNAs, is further mapped using a recently developed RNA visualization tool, coloRNA (LeBarron et al. 2007). With this, any conformational change in a local RNA region can be easily traced to the associated region in the secondary structure, allowing the stability of secondary structure elements and their connectivity to be taken into consideration.

Table 15.1 Fitting statistics of ribosomal complexes

Complex	Resolution (Å)	Correlation coefficient	Real-space R factor
Pretranslocation complex[a]	10	0.72	0.23
EF-G-bound translocation complex[a]	10.8	0.77	0.20
RF1-bound termination complex[b]	12.8	0.74	0.2
RF3-bound termination complex[c]	15.5	0.72	0.21

Complexes with "a, b, and c" were described in Valle et al. (2003), Rawat et al. (2006), and Gao et al. (2007), respectively.

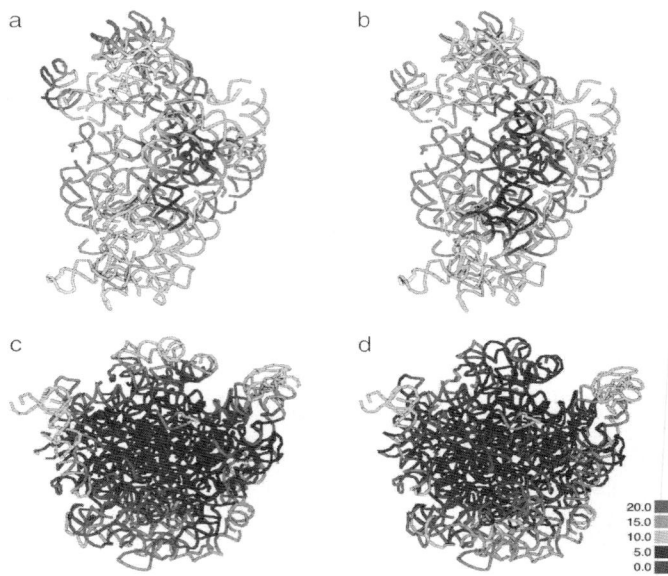

Fig. 15.3 Extent of factor-related ratchet motions depicted on the RNA 3D structures. (**a** and **b**) 16S rRNA for EF-G-related and RF3-related ratchet motions. (**c** and **d**) 23S rRNA for EF-G-related and RF3-related ratchet motions. Colors are assigned for corresponding residues based on distance between non-ratcheted and ratcheted states

This analysis shows that, while the overall displacement of the 16S rRNA in the EF-G-related ratchet motion is larger than for RF3, detailed pattern of displacement is quite consistent in the two motions (Fig. 15.4). In both cases, domain III of the 16S rRNA has highest mobility, and movements of several helices in domain I (h6, h9, h10) and II (h17) stand out, as well. For the 23S rRNA, total displacement range is very similar in the EF-G- and RF3-related conformational changes (Fig. 15.5). Interestingly, several RNA regions, including helices 43 and 44 in domain II, helices 47, 56, 58, 60 in domain III, and helices 97, 100, 101 in domain VI, exhibit larger displacements in the RF3-related than in the EF-G-related ratchet motion- a behavior which is different from the overall trends of the ratchet motion.

We can postulate that the unusual behaviors of these regions are important in modulating general ratchet motion and bestowing it with the characteristics required for the specific function. A case in point is in the region of helices 43–44 of 23S rRNA, which has a larger displacement in the RF3-related (5–6 Å) but a smaller one in the EF-G-related ratchet motion (~3 Å). Known as the 58-nucleotides region (1051–1108), helices 43 and 44 of 23S rRNA form the highly conserved GAC region on the 50S subunit along with protein L11. During termination, the GAC region undergoes a two-phase motion associated with interplays between the ribosome and two types of release factors. (1), triggered by the binding of class-I release factors (RF1

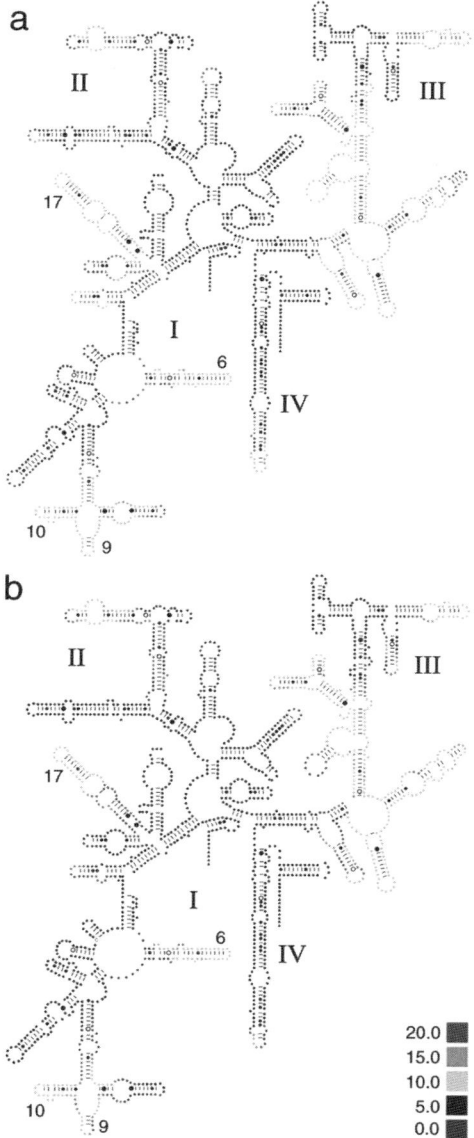

Fig. 15.4 Extent of factor-related ratchet motions depicted on the secondary structure of 16S rRNA. (**a**) EF-G-related ratchet motion. (**b**) RF3-related ratchet motion. Note that only original secondary structures are used here and the possible changes in the secondary structures are not taken into account

and RF2) to the ribosome, the GAC moves 6 Å inwards toward the central protuberance (CP) (Rawat et al. 2003; Rawat et al. 2006). (2), upon binding of class-II release factor RF3, the GAC undergoes a 6 Å outward movement, exactly reversing the previous motion. The functional importance of this outward movement of the

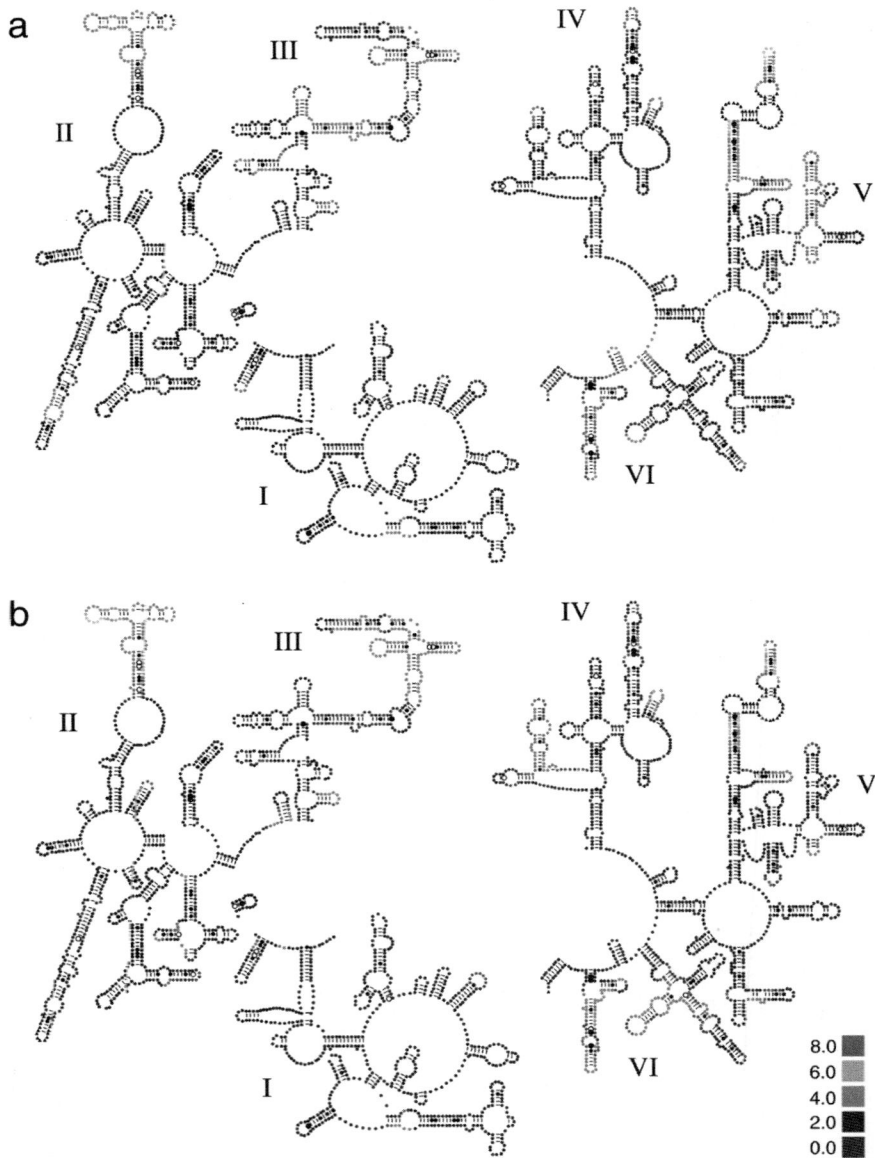

Fig. 15.5 Extent of factor-related ratchet motions depicted on the secondary structure of 23S rRNA. (**a** and **c**) EF-G-related ratchet motion with color bar on an expanded scale. (**b** and **d**) RF3-related ratchet motion with color bar on an expanded scale. (See figure insert for color reproduction)

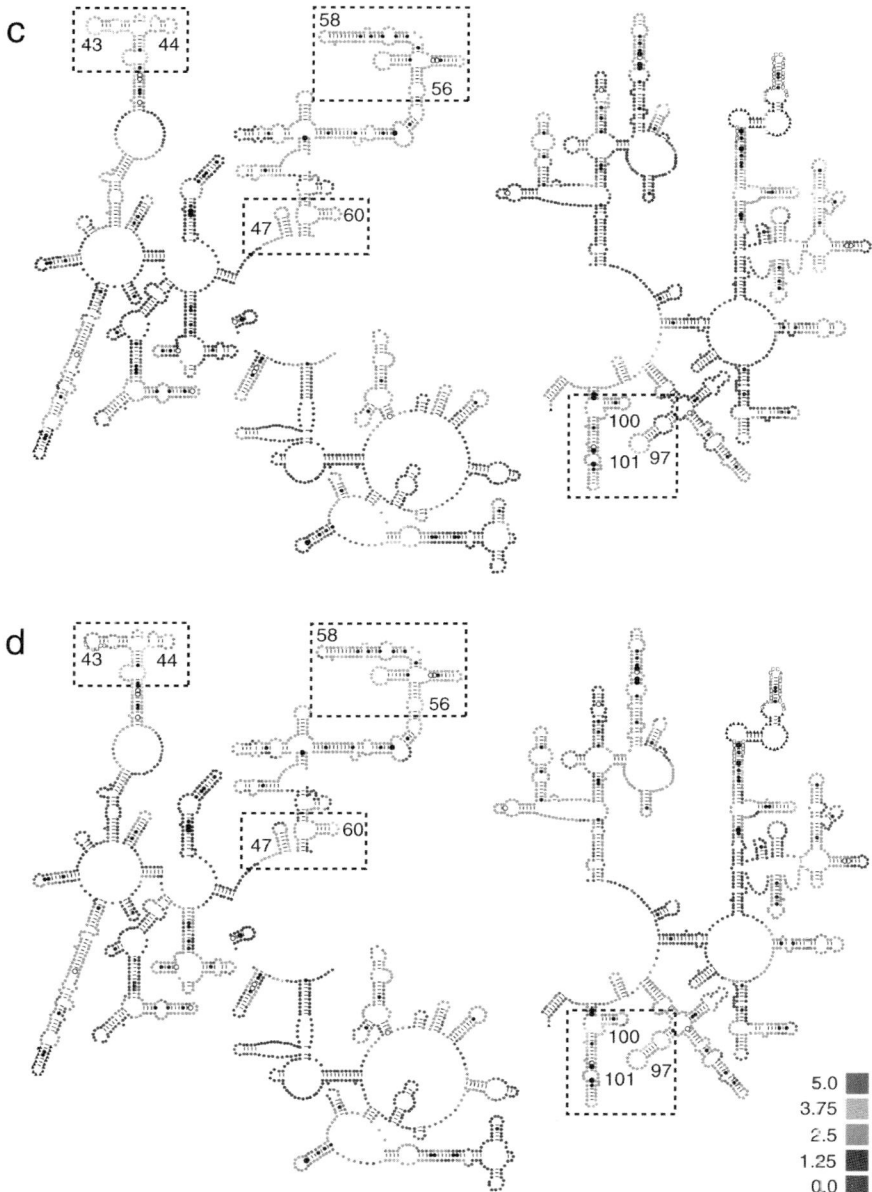

Fig. 15.5 (Continued)

GAC is to disrupt its interaction with the domain I of class-I release factors, and promote the release of the class-I release factors from the ribosome (Gao et al. 2007). In contrast, in the EF-G-related ratchet motion, the smaller movement of the GAC was associated with its interaction with domain V of EF-G (Gao et al. 2003). Together, it is evident that larger movement of the GAC observed in the RF3-related ratchet motion, compared with the one in the EF-G-related ratchet motion, reflects its specific functional role in the dissociation of class-I release factors during the process of translation termination.

The above example was presented as a paradigm for exploration and making maximum use of structural information acquired from various sources, namely X-ray crystallography, cryo-EM, and examination of secondary structure. In this case, quasi-atomic models of the ribosome were first obtained in different functional states by flexibly fitting crystal structures into EM density maps. Further analysis was performed to trace back important conformational changes to elements of the highly conserved secondary structure of ribosomal RNA. This approach provided mappings and rationalizations of conformational changes in the ribosomal RNA at the molecular level. We believe similar analysis of dynamical steps, by comparing cryo-EM reconstructions of successive functional complexes, will significantly advance the understanding of translation.

Accession Numbers

Atomic models of 16S and 23S ribosomal RNAs have been deposited in the Protein Data Bank under the following accession codes: 3DG0 (EF-G·GDPNP-bound state), 3DG2 (pretranslocation state); 3DG4 (RF1-bound state); and 3DG5 (RF3·GDPNP-bound state).

Acknowledgments We would like to thank Michael Watters for assistance with the preparation of the illustrations. This work is supported by HHMI and NIH grants P41 RR01219 and R37 GM29169 (to J.F.).

References

Agrawal RK, Heagle AB, Penczek P, Grassucci RA, Frank J (1999) EF-G-dependent GTP hydrolysis induces translocation accompanied by large conformational changes in the 70S ribosome. Nat Struct Biol 6:643–647

Allen GS, Zavialov A, Gursky R, Ehrenberg M, Frank J (2005) The cryo-EM structure of a translation initiation complex from Escherichia coli. Cell 121:703–712

Baker TS, Johnson JE (1996) Low resolution meets high: towards a resolution continuum from cells to atoms. Curr Opin Struct Biol 6:585–594

Ban N, Nissen P, Hansen J, Moore PB, Steitz TA (2000) The complete atomic structure of the large ribosomal subunit at 2.4 A resolution. Science 289:905–920

Barat C, Datta PP, Raj VS, Sharma MR, Kaji H, Kaji A, Agrawal RK (2007) Progression of the ribosome recycling factor through the ribosome dissociates the two ribosomal subunits. Mol Cell 27:250–261

Blanchard SC, Gonzalez RL, Kim HD, Chu S, Puglisi JD (2004) tRNA selection and kinetic proofreading in translation. Nat Struct Mol Biol 11:1008–1014

Borovinskaya MA, Pai RD, Zhang W, Schuwirth BS, Holton JM, Hirokawa G, Kaji H, Kaji A, Cate JH (2007) Structural basis for aminoglycoside inhibition of bacterial ribosome recycling. Nat Struct Mol Biol 14:727–732

Brunger AT, Adams PD, Clore GM, DeLano WL, Gros P, Grosse-Kunstleve RW, Jiang JS, Nilges N, Pannu NS, Read RJ, Rice LM, Simonson T, Warren GL (1998) Crystallography and NMR system (CNS): a new software system for macromolecular structure determination. Acta Cryst D 54:905–921

Chapman MS (1995) Restrained real-space macromolecular atomic refinement using a new resolution-dependent electron-density function. Acta Cryst A 51:69–80

Ermolenko DN, Majumdar ZK, Hickerson RP, Spiegel PC, Clegg RM, Noller HF (2007) Observation of intersubunit movement of the ribosome in solution using FRET. J Mol Biol 370:530–540

Fabiola F, Chapman MS (2005) Fitting of high-resolution structures into electron microscopy reconstruction images. Structure 13:389–400

Frank J (2006) Three-dimensional electron microscopy of macromolecular assemblies. Oxford University Press, New York

Frank J, Agrawal RK (2000) A ratchet-like inter-subunit reorganization of the ribosome during translocation. Nature 406:318–322

Frank J, Gao H, Sengupta J, Gao N, Taylor DJ (2007) The process of mRNA-tRNA translocation. Proc Natl Acad Sci USA 104:19671–19678

Fu J, Gao H, Frank J (2007) Unsupervised classification of single particles by cluster tracking in multi-dimensional space. J Struct Biol 157:226–239

Gao H, Frank J (2005) Molding atomic structures into intermediate-resolution cryo-EM density maps of ribosomal complexes using real-space refinement. Structure 13:401–406

Gao H, Sengupta J, Valle M, Korostelev A, Eswar N, Stagg SM, Van Roey P, Agrawal RK, Harvey SC, Sali A, Chapman MS, Frank J (2003) Study of the structural dynamics of the E. coli 70S ribosome using real space refinement. Cell 113:789–801

Gao H, Valle M, Ehrenberg M, Frank J (2004) Dynamics of EF-G interaction with the ribosome explored by classification of a heterogeneous cryo-EM dataset. J Struct Biol 147:283–290

Gao N, Zavialov AV, Li W, Sengupta J, Valle M, Gursky RP, Ehrenberg M, Frank J (2005) Mechanism for the disassembly of the posttermination complex inferred from cryo-EM studies. Mol Cell 18:663–674

Gao H, Zhou Z, Rawat U, Huang C, Bouakaz L, Wang C, Cheng Z, Liu Y, Zavialov A, Gursky R, Sanyal S, Ehrenberg M, Frank J, Song H (2007) RF3 induces ribosomal conformational changes responsible for dissociation of class I release factors. Cell 129:929–941

Grassucci RA, Taylor DJ, Frank J (2007) Preparation of macromolecular complexes for cryo-electron microscopy. Nat Protocols 2:3239–3246

Grassucci RA, Taylor D, Frank J (2008) Visualization of macromolecular complexes using cryo-electron microscopy with FEI Tecnai transmission electron microscopes. Nat Protocols 3:330–339

Halic M, Blau M, Becker T, Mielke T, Pool MR, Wild K, Sinning I, Beckmann R (2006) Following the signal sequence from ribosomal tunnel exit to signal recognition particle. Nature 444:507–511

Harms J, Schlunzen F, Zarivach R, Bashan A, Gat S, Agmon I, Bartels H, Franceschi F, Yonath A (2001) High resolution structure of the large ribosomal subunit from a mesophilic eubacterium. Cell 107:679–688

Klaholz BP, Myasnikov AG, van Heel M (2004) Visualization of release factor 3 on the ribosome during termination of protein synthesis. Nature 427:862–865

Korostelev A, Trakhanov S, Laurberg M, Noller HF (2006) Crystal structure of a 70S ribosome-tRNA complex reveals functional interactions and rearrangements. Cell 126:1065–1077

LeBarron J, Mitra K, Frank J (2007) Displaying 3D data on RNA secondary structures: coloRNA. J Struct Biol 157:262–270

Munro JB, Altman RB, O'Connor N, Blanchard SC (2007) Identification of two distinct hybrid state intermediates on the ribosome. Mol Cell 25:505–517

Petry S, Brodersen DE, Murphy FV, Dunham CM, Selmer M, Tarry MJ, Kelley AC, Ramakrishnan V (2005) Crystal structures of the ribosome in complex with release factors RF1 and RF2 bound to a cognate stop codon. Cell 123:1255–1266

Rawat UBS, Zavialov AV, Sengupta J, Valle M, Grassucci RA, Linde J, Vestergaard B, Ehrenberg M, Frank J (2003) A cryo-electron microscopic study of ribosome-bound termination factor RF2. Nature 421:87–90

Rawat U, Gao H, Zavialov AV, Gursky R, Ehrenberg M, Frank J (2006) Interactions of the release factor RF1 with the ribosome as revealed by cryo-EM. J Mol Biol 357:1144–1153

Rossmann MG (2000) Fitting atomic models into electron-microscopy maps. Acta Cryst D56:1341–1349

Scheres SH, Gao H, Valle M, Herman GT, Eggermont PP, Frank J, Carazo JM (2007) Disentangling conformational states of macromolecules in 3D-EM through likelihood optimization. Nat Methods 4:27–29

Schluenzen F, Tocilj A, Zarivach R, Harms J, Gluehmann M, Janell D, Bashan A, Bartels H, Agmon I, Franceschi F, Yonath A (2000) Structure of functionally activated small ribosomal subunit at 3.3 angstroms resolution. Cell 102:615–623

Schuwirth BS, Borovinskaya MA, Hau CW, Zhang W, Vila-Sanjurjo A, Holton JM, Cate JH (2005) Structures of the bacterial ribosome at 3.5 A resolution. Science 310:827–834

Selmer M, Dunham CM, Murphy FV, Weixlbaumer A, Petry S, Kelley AC, Weir JR, Ramakrishnan V (2006) Structure of the 70S ribosome complexed with mRNA and tRNA. Science 313:1935–1942

Spahn CM, Beckmann E, Eswar N, Penczek PA, Sali A, Blobel G, Frank J (2001) Structure of the 80S ribosome from *Saccharomyces cerevisiae* – tRNA-ribosome and subunit-subunit interactions. Cell 107:373–386

Spahn CMT, Gomez-Lorenzo MG, Grassucci GA, Jorgensen R, Andersen GR, Beckmann R, Penczek PA, Ballesta JPG, Frank J (2004) Domain movements of elongation factor eEF2 and the eukaryotic 80S ribosome facilitate tRNA translocation. EMBO J 23:1008–1019

Spirin AS (1968) How does the ribosome work? A hypothesis based on the two subunit construction of the ribosome. Curr Mod Biol 2:115–127

Spirin AS (2002) Ribosome as a molecular machine. FEBS Lett 514:2–10

Tama F, Valle M, Frank J, Brooks CL III (2003) Dynamic reorganization of the functionally active ribosome explored by normal mode analysis and cryo-electron microscopy. Proc Natl Acad Sci U S A 100:9319–9323

Taylor DJ, Nilsson J, Merrill AR, Andersen GR, Nissen P, Frank J (2007) Structures of modified eEF2 80S ribosome complexes reveal the role of GTP hydrolysis in translocation. EMBO J 26:2421–2431

Trabuco LG, Villa E, Mitra K, Frank J, Schulten K (2008) Flexible fitting of atomic structures into electron microscopy maps using molecular dynamics. Structure 16:673–683

Valle M, Sengupta J, Swami K, Grassucci RA, Burkhardt N, Nierhaus KH, Agrawal RK, Frank J (2002) Cryo-EM reveals an active role for the aminoacyl-tRNA in the accommodation process. EMBO J 21:3557–3567

Valle M, Zavialov AV, Sengupta J, Rawat U, Ehrenberg M, Frank J (2003) Locking and unlocking of ribosomal motions. Cell 114:123–134

Wang Y, Rader AJ, Bahar I, Jernigan RL (2004) Global ribosome motions revealed with elastic network model. J Struct Biol 147:302–314

Weixlbaumer A, Petry S, Dunham CM, Selmer M, Kelley AC, Ramakrishnan V (2007) Crystal structure of the ribosome recycling factor bound to the ribosome. Nat Struct Mol Biol 14:733–737

Wimberly BT, Brodersen DE, Clemons WM Jr., Morgan-Warren RJ, Carter AP, von Rhein C, Hartsch T, Ramakrishnan V (2000) Structure of the 30S ribosomal subunit. Nature 407:327–339

Zavialov AV, Ehrenberg M (2003) Peptidyl-tRNA regulates the GTPase activity of translation factors. Cell 114:113–122

Chapter 16
Biophysical Analyses of IRES RNAs from the *Dicistroviridae*: Linking Architecture to Function

Jeffrey S. Kief

Abstract Internal ribosome entry sites (IRES) are non-protein coding RNAs that can drive translation initiation using RNA in place of protein factors and the modified nucleotide cap. IRESs are critical for successful infection by many viruses and may be an important means of regulating gene expression in healthy cells, yet our understanding of their structure-based mechanism of action remains incomplete. A critical part of understanding their function is knowledge of the biophysical properties of the free IRES RNA itself, as the architecture of the unbound RNA is likely to be a key functional determinant. This chapter presents the application of several biophysical methods to the study of an IRES RNA. Comparing the results of these studies to those performed on a different IRES RNA suggests the biophysical properties of these two IRES RNAs both reflect and predict their mode of interacting with the ribosome.

Abbreviations IRES internal ribosome entry site; UTR untranslated region; ORF open reading frame; IGR intergenic region; HCV hepatitis C virus; eIF eukaryotic initiation factor; mRNA messenger RNA; ITAF IRES trans-activating factors; EMCV encephalomyocarditis virus; HAV hepatitis A virus; FMDV foot-and-mouth disease virus; HIV-1 human immunodeficiency virus-1; DMS dimethysulfate; Kethoxyl β-ethoxy-α-ketobutyraldehyde; CMCT 1-cyclohexyl-3-(morpholinoethyl)–carbodiimide metho-p-toluene sulfonate; RNase ribonuclease; SHAPE selective 2′-hydroxyl acylation analyzed by primer extension; ENU ethylnitrosourea; tRNA transfer RNA; cryo-EM cryo-electron microscopy; SV/AUC sedimentation velocity analytical ultracentrifugation.

J.S. Kief

Department of Biochemistry and Molecular Genetics, Denver School of Medicine, University of Colorado, 12801 East 17th Ave, Rm L18-9110, Aurora, CO 80045, USA

e-mail: Jeffrey.Kief@ucdenver.edu

N.G. Walter et al. (eds.) *Non-Protein Coding RNAs*
doi: 10.1007/978-3-540-70840-7_16, © Springer-Verlag Berlin Heidelberg 2009

16.1 Introduction

Protein synthesis (translation) is central to life as we know it. The phase of translation during which the protein-making machinery assembles is called "translation
initiation;" it is the primary step at which translation is regulated (Mathews et al.
2000). The canonical mechanism of translation initiation in eukaryotes is a multi-
step, orchestrated process dominated by proteins (Kapp and Lorsch 2004; Jackson
2005). In total, at least ten protein factors and multiple ATP and GTP hydrolysis
events are needed to assemble an 80S ribosome at the start codon in the canonical
cap- and scanning-dependent pathway (Preiss and Hentze 2003). Given the necessity for eIF proteins in canonical cap-dependent eukaryotic translation initiation, it
is somewhat remarkable that this set of protein factors can be replaced by an RNA
sequence that drives translation initiation in an essentially protein-free, non-canonical mechanism. The RNA capable of this is described later in this chapter; it is one
of a larger group called internal ribosome entry site (IRES) RNAs, which are
defined by their ability to initiate translation without recognizing the 5′ cap or end
of the mRNA in a process called *internal initiation of translation* (Fig. 16.1). IRESs
recruit, position, and activate the translation machinery using specific RNA
sequences that are usually found in mRNA's or viral RNA's untranslated region
(UTR), directly upstream of the open reading frame. Despite their shared ability
to drive cap-independent translation, IRES RNAs have diverse sequences and

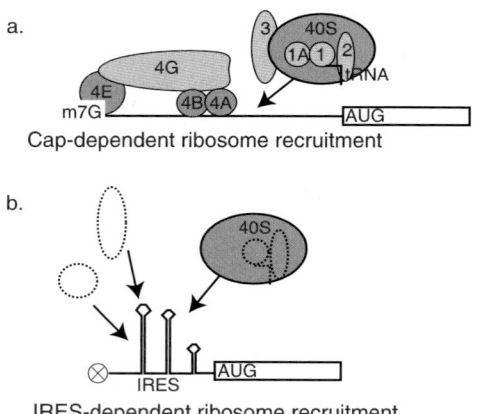

Fig. 16.1 Comparison of two modes of translation initiation in eukaryotes. (**a**) In canonical cap-
dependent translation, the 40S subunit is recruited to the message by recognition of the 7-methyl
guanosine cap and many protein factors, which are depicted here as shaded objects. The 40S
subunit then scans the mRNA to find the start codon. (**b**) In cap-independent, IRES-driven initiation, the cap is not needed and the 40S subunit is recruited to the message internally. In some
cases, a subset of the initiation factors or other trans-activating factors are needed (shown as
dashed objects), but in other cases, IRES RNA structure is sufficient to recruit the ribosome

requirements for factors (Jackson and Kaminski 1995; Jackson 2005). While some IRESs do not require any eIF proteins, others use a subset of these factors and still others must use additional proteins that are not part of the canonical translation machinery, called IRES trans-activating factors (ITAFs).

IRESs were originally discovered in the poliovirus and encephalomyocarditis virus (EMCV) (Jang et al. 1988; Pelletier and Sonenberg 1988), and they have since been reported in many others, such as hepatitis C virus (HCV) (Tsukiyama-Kohara et al. 1992; Wang et al. 1993), hepatitis A virus (HAV) (Glass and Summers 1992), rhinovirus (Borman and Jackson 1992), foot-and-mouth disease virus (FMDV) (Kuhn et al. 1990), and human immunodeficiency virus-1 (HIV-1) (Buck et al. 2001; Brasey et al. 2003). In addition, IRESs have been found in diverse cellular mRNAs involved in processes like differentiation, proliferation, growth, and apoptosis (Hellen and Sarnow 2001; Stoneley and Willis 2004). In short, IRESs are important non-protein coding RNAs in both healthy and diseased cells, motivating studies to understand their structure and function in detail.

16.1.1 General Characteristics of IRES RNAs

Soon after their discovery, it was apparent that RNA sequences responsible for cap-independent translation initiation retained function when removed from their natural viral RNA context and placed upstream of reporter genes. This suggested that the majority of IRESs are self-contained functional RNAs. Application of a variety of biochemical approaches has defined the minimal sequence requirements and produced secondary structure models for many viral IRESs and some cellular IRESs (Fig. 16.2). In recent years, it has been recognized that the three-dimensional conformation of some IRESs may be critical for function and thus biophysical methods have played an increasingly prominent role in the study of these RNAs. The first step of the IRES-driven mechanism is an RNA folding event and even if the final active conformation of the IRES differs from the initial folded structure, the free RNA must be initially recognized by the translation machinery or other factors. Whether an IRES assumes a stable compact fold, is a largely unstructured scaffold, or has no discernable higher-order fold at all, the biophysical characteristics of the free IRES RNA are determinants of its overall mechanism of action. Even if the IRES RNA is ultimately part of a large multi-component complex, these complexes are only fully understood if we understand the biophysical characteristics (i.e., size, shape, fold, and folding requirements) of the IRES RNA itself.

16.1.2 The Dicistroviridae Intergenic Region IRESs

The most streamlined IRES mechanism belongs to the intergenic region (IGR) IRESs of the *Dicistroviridae* family of viruses (hereafter referred to as the "IGR IRESs")

a. *Plautia stali* intestine virus IGR IRES (*Dicistroviridae*)

b. Hepatitis C virus (*Flaviviridae*)

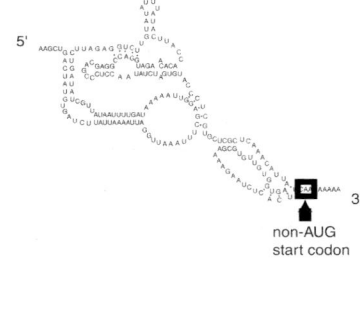

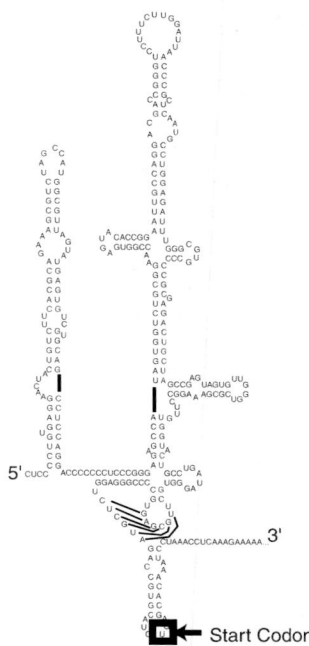

Fig. 16.2 Comparison of secondary structures of the two viral IRS RNA discussed in this chapter (**a**) *Plautia stali* intestine virus intergenic region (IGR) IRES. (**b**) Hepatitis C virus IRES

(review: (Jan 2006)). These single-stranded, positive-sense RNA viruses contain two open reading frames (ORFs), the downstream ORF is under the control of an IRES sequence found in the non-protein coding intergenic sequence between the two cistrons (Fig. 16.3a). The IGR IRESs do not require proteins or GTP hydrolysis to recruit the ribosomal subunits to form IRES-80S ribosome complexes. In addition, these IRESs do not need initiator tRNA as they initiate from the A-site of the ribosome and start protein synthesis from a non-AUG codon ((Jan 2006), references therein). The IGR IRES mechanism is shown in Fig. 16.3b. Briefly, the IRES binds directly to the two subunits to form an 80S ribosome, within which a pseudoknot of the IRES RNA is placed into the small subunit's P-site, occupying the space normally used by initiator tRNA^{met}. Delivery of an aminoacylated elongator tRNA to the A-site by eukaryotic elongation factor 1A (eEF1A) leads to pseudotranslocation, so called because it is not preceded by formation of a peptide bond. At this point, the ribosome receives another tRNA in the A-site and peptide bond formation and elongation proceed normally.

The ability of the IGR IRESs to functionally replace the initiation factor proteins raises a number of questions regarding biophysical characteristics of this non-protein coding RNA sequence, including: Does this RNA have a functionally important fold and what is the nature of the fold? Within the putative fold, which

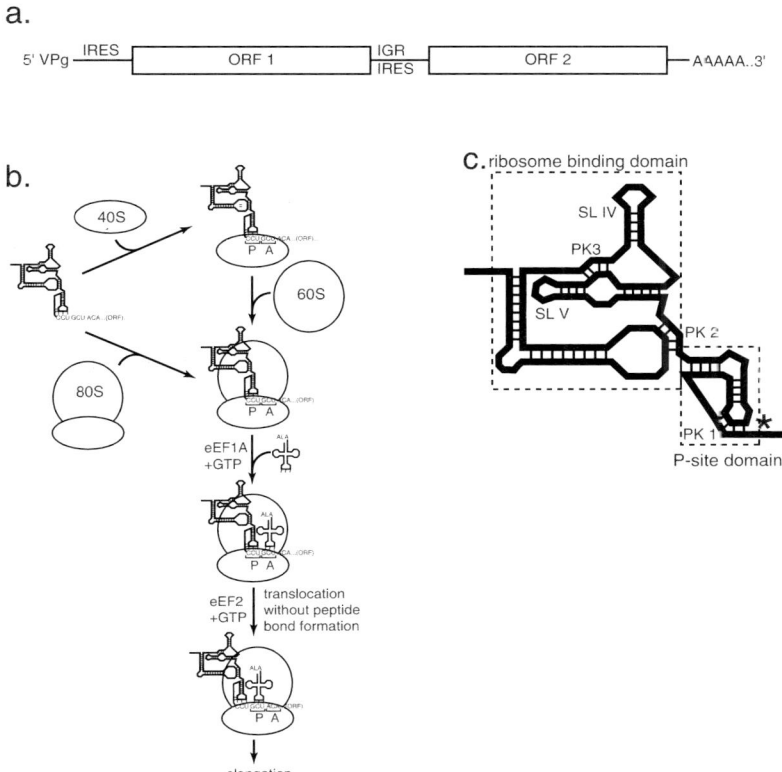

Fig. 16.3 The IGR IRESs of the *Dicistroviridae*. (**a**) Organization of the single-stranded positive-sense RNA of the *Dicistroviridae*. The RNA contains two open reading frames, with the intergenic region (IGR) IRES between the two. The viral RNA has a VPg peptide on its 5′ end and an IRES lies in the 5′ untranslated region that drives translation of the first ORF. The IGR IRES drives translation of the second ORF. (**b**) Mechanism of IGR IRES recruitment of two ribosome subunits and translocation without peptide bond formation (pseudotranslocation). Briefly, the IRES first recruits the 40S subunit directly, then the 60S subunit to make an 80S ribosome without initiator tRNA or GTP hydrolysis. Alternately, the IRES can bind a preassembled 80S ribosome. Delivery of tRNA to the A-site leads to translocation without peptide bond formation, and subsequent tRNA addition to the A-site begins elongation. Figure is based on an illustration from Jan (2006). (**c**) Secondary structure cartoon of an IGR IRES RNA showing important structural features and the location of two independently folded domains and their roles in initiation

elements stabilize the structure and which extend into solution to be recognized by the ribosome? Are there areas of stable IRES RNA structure and other areas that are more flexible and perhaps unstructured? Answers to these and other questions were pursued by a variety of biophysical methods, treating the IGR IRES as a three-dimensional object with characteristics crucial to its mechanism. This chapter presents an overview of these biophysical studies, briefly comparing IGR IRES RNAs to the HCV IRES RNA. Special emphasis is placed on analyses of the free (unbound) state of the IGR IRES and what they reveal information about the mechanism of these non-protein coding RNAs.

16.2 Architecture of the Unbound IGR IRESs

Two classes of IGR IRESs exist, and in both the secondary structure contains helices, loops, junctions, "single-stranded" segments, and three pseudoknots (Kanamori and Nakashima 2001; Hatakeyama et al. 2004; Cevallos and Sarnow 2005; Jan 2006) (Fig. 16.3c). Biochemical experiments showed that IRES contains several regions, likely possesses a higher-order fold, and that different regions of the IRES interact with each of the two ribosomal subunits (Jan et al. 2001, 2003; Jan and Sarnow 2002; Nishiyama et al. 2003). However, these experiments still left the questions presented in the previous section unanswered, mandating the application of other biophysical methods.

16.2.1 Hydroxyl Radical Probing

Hydroxyl radicals generated in solution (by Fenton chemistry using an Fe(II)-EDTA complex, using peroxynitrous acid, or by X-ray radiation) cleave the RNA backbone independent of sequence ((Tullius and Greenbaum 2005), references therein). The degree of cleavage by the solvent-borne radicals at a given backbone location depends on the solvent accessibility; portions of the RNA involved in tightly packed structures are protected, whereas portions extended into solution are cleaved. As this method can be used under a wide variety of buffer, salt, and temperature conditions, it is often used to establish both the nature of the RNA fold and conditions necessary to form that fold. Using this method, time-resolved RNA folding experiments can be done even at the millisecond timescale (Brenowitz et al. 2002; Shcherbakova et al. 2006; Shcherbakova and Brenowitz 2008).

Information obtained from hydroxyl radical probing is different from that obtained from other chemical or enzymatic probing experiments (Ehresmann et al. 1987). The reactivity of RNA to reagents such as dimethysulfate (DMS), β-epoxy-α-ketobutyraldehyde (kethoxyl), and 1-cyclohexyl-3-(morpholinoethyl)-carbodiimide metho-p-toluene sulfonate (CMCT) display sensitivity to the hydrogen bonding formed between base pairs, and larger ribonuclease (RNase) probes (~11 kDa) may also depend on base-pairing and can be sterically occluded from some regions of the surface. Selective 2′-hydroxyl acylation analyzed by primer extension (SHAPE) probes RNA secondary structure but not backbone solvent accessibility (Merino et al. 2005; Wilkinson et al. 2006). Ethylnitrosourea (ENU) targets the backbone of the RNA, but unlike hydroxyl radicals, it modifies phosphates not involved in hydrogen bonding or cation coordination, and because it is larger than a hydroxyl radical (MW ~ 117) ENU may not penetrate all solvent accessible surfaces (Ehresmann et al. 1987). Hydroxyl radicals therefore remain the best probe to detect and explore tight packing of RNA backbones.

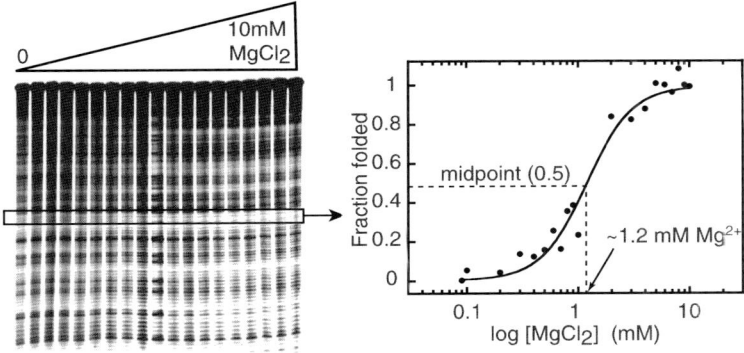

Fig. 16.4 IGR IRES folding monitored by hydroxyl radical probing. At left is a gel in which the IGR IRES RNA was probed with hydroxyl radicals in the presence of increasing amounts of MgCl$_2$ (0–10 mM). Each lane on the gel represents a different concentration of MgCl$_2$. Subtle changes in the cleavage pattern indicated folding of the RNA. The degree of cleavage in the boxed region was quantified for each lane, used to calculate fraction of folded RNA in each lane and plotted as a function of [MgCl$_2$]. The plot shows magnesium concentration required to induce tight folding of RNA, with the midpoint of the transition (0.5 fraction folded) shown with a dashed line. Figure is an adaptation of a figure in (Kieft et al. 2007)

Hydroxyl radical probing of the *Plautia stali* intestine virus (PSIV, a member of the *Dicistroviridae*) IGR IRES in the presence and absence of magnesium showed differences in the cleavage pattern between the two conditions (Costantino and Kieft 2005). This result established that the IRES undergoes a magnesium-dependent conformational change involving folding of the IRES into a structure with tightly packed RNA backbone. This result alone was revealing as it established the IRES is a compact, globular structure and not an extended scaffold-like molecule. Next by monitoring the degree of radical protection as a function of [Mg^{2+}], the IRES RNA folding was shown to occur at 1–2 mM Mg^{2+}, which is in the range of physiological cation concentrations and magnesium concentrations at which the IRES is active (Fig. 16.4). This suggested that magnesium-dependent compactly-folded architecture was biologically relevant and functionally necessary.

Additional evidence of the importance of the folded structure of the unbound IGR IRES was found in continued application of hydroxyl radical probing to other members of the IGR IRES family (Pfingsten et al. 2007). These experiments showed that the pattern of cleavage and protection from radicals for each of four different IGR IRESs tested was similar despite only ~25% sequence identity between members of the IGR IRES family. This suggested that all fold in the presence of magnesium into similar compact structures. This was true even for the class 2 IGR IRESs, whose secondary structure differs from class 1 (Kanamori and Nakashima 2001; Hatakeyama et al. 2004; Cevallos and Sarnow 2005; Jan 2006). These biophysical results thus support the idea that the IGR IRES' three-dimensional architecture and specifically shaped pre-folded structure are critical determinants of function. Since the fold of IGR IRESs is evolutionarily maintained despite sequence changes, the fold is likely conserved for a functional reason. The presence of the conserved compact fold hints at the IGR IRES mechanism of ribosome binding by suggesting that the IRES probably first folds and then occupies a relatively small space on the ribosome, an idea corroborated by cryo-EM structures of IRES•ribosome complexes (Spahn et al. 2004; Schuler et al. 2006). This description of folded the architecture of the IRES agreed well with results from biochemical assays exploring the roles of various IGR IRES structural elements in ribosome binding (Jan and Sarnow 2002; Nishiyama et al. 2003).

Analysis of cleavage patterns of the folded IGR IRES RNA not only showed the nature of the fold and conditions in which it folds, but also revealed details of overall architecture of the RNA as a three-dimensional object and insight into how parts of the structure are arrayed relative to one another (Costantino and Kieft 2005). Specifically, two highly conserved stem-loop structures (SL IV, SL V, Fig. 16.3) were cleaved more in the folded structure than in the unfolded state, demonstrating that they extend away from the compact part of the IRES. As the apical loops of these stem-loops were known to make contact with the 40S subunit (Jan and Sarnow 2002; Nishiyama et al. 2003), this result revealed how the folded core structure presents these recognition surfaces to be bound by the translation machinery, directly linking function to pre-folded structure.

The probing pattern also revealed the elements likely involved in holding the structure together through extensive intermolecular interactions and tight backbone packing; these were protected from solvent and found in several different locations of the sequence. Some of the protected portions are pseudoknots, suggesting that these would be critical determinants of structure. These hypotheses regarding determinants of folding were then directly addressed by coupling the hydroxyl radical probing with mutagenesis. If mutation of a certain part of the IRES causes a loss in solvent protection, it can be concluded that the element is involved in folding. Indeed, when pseudoknot 2 and pseudoknot 3 (PK 2 and PK 3) in the IGR IRES were mutated, protections in the folded core disappeared, establishing that these interactions stabilize the fold (Costantino and Kieft 2005). Conversely, mutation of two stem-loops that extend into solution did not affect the cleavage pattern, confirming they are not involved in stabilization of the fold, but are positioned by the fold.

Finally, by probing truncated IRES RNAs, it was discovered that the IGR IRESs contain two independently folded domains. Here, a domain is defined as a group of secondary structure elements that fold into a higher-order stable three-dimensional structure with specific tight RNA backbone packing, while regions are defined as a collection of logically connected secondary structure elements that may or may not fold into a higher-order structure. One of these two IGR IRES domains is sufficient to recruit an 80S ribosome by itself (Jan and Sarnow 2002; Nishiyama et al. 2003; Costantino and Kieft 2005); this ribosome binding domain contains regions 1 and region 2 and is responsible for initial positioning of the IRES on the ribosome. The other domain comprises region 3 and was shown to dock into the P-site where it supports subsequent initiation from the A-site (Wilson et al. 2000). Hence, the two independently folded structural domains of the IRES correspond with two partially independent roles in initiation, providing another link between biophysical characteristics and function. Taken together, these biophysically-directed hydroxyl radical probing experiments established the overall three-dimensional architecture of the RNA as it prepares to interact with the ribosome, revealed the presence of separate folded domains, and showed the sequence and conditional determinants of the fold.

16.2.2 Analytical Ultracentrifugation

To expand upon the hydroxyl radical probing results and learn about how the IGR IRES RNAs' size and shape change as the RNA goes from an unfolded to a folded state, sedimentation velocity by analytical ultracentrifugation (SV/AUC) was used (Lebowitz et al. 2002). SV/AUC measures the rate at which soluble macromolecules move through a chamber (the sample cell) during ultracentrifugation (Fig. 16.5a). At the start of the experiment (at rest), molecules are uniformly distributed within the chamber of the sealed sample cell. During ultracentrifugation, the RNA macromolecules sediment towards the outside of the cell and this is monitored by UV absorbance at 260 nm, measured as a function of position in the cell (Fig. 16.5b). The sedimenting RNA (in this case, the IGR IRES) forms a boundary that is observed in

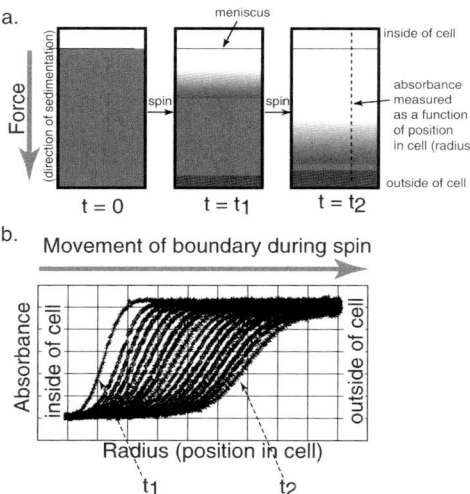

Fig. 16.5 Illustration of SV/AUC. (**a**) Diagram of SV/AUC method. Briefly, the solution containing the macromolecule of interest (in this case, an IRES RNA) is sealed in a cell. At this point (time = 0, or t_0), the macromolecule is evenly distributed in the cell. As the sample cell is spun in the ultracentrifuge, the macromolecule sediments to the outside of the cell, so at t_1 and t_2 there is more macromolecule at the outside of the cell than at the inside. The effect is analogous to sand settling in a jar of water that has been shaken and then put to rest. The distribution of the macromolecule in the cell is monitored though a clear window by measuring the UV absorbance. This gives rise to a set of boundary traces. (**b**) A set of boundary traces obtained from SV/AUC on an IGR IRES RNA. Analysis of this set of boundaries leads to information about the hydrodynamic properties, and hence the size and shape, of the RNA

the UV trace and which moves over time. The rate of movement and shape of the boundary reflect the hydrodynamic properties of the particle and depend on molecular weight, size, and shape. Analysis of a set of boundary traces collected over time yields the sedimentation and diffusion coefficients (S and D, respectively). In a general sense, the sedimentation coefficient describes the rate at which the molecule moves to the outside of the cell, and for a given molecular weight, this number increases with molecules that are more compact. Likewise, the diffusion coefficient describes the degree to which the molecules diffuse (broadening the boundary), and this also increases with more compact molecules (assuming molecular weight as a constant). S and D are used with the partial specific volume and molecular mass of the RNA to derive other parameters, including the Stokes radius (R_h), which is defined as radius of a sphere with the same molecular weight and same S as the molecule of interest (including associated waters). The larger the R_h, the more elongated the molecule (all else being equal). Related to R_h is the parameter f/f_o, which is the ratio of observed frictional coefficient of the molecule to frictional coefficient of a spherical molecule; the closer f/f_o is to a value of 1, the more spherical the molecule. A more detailed review of this method and the meaning of these parameters can be found elsewhere (Lebowitz et al. 2002); this chapter concentrates on what this biophysical method revealed about the IGR IRES RNAs.

Probing methods established that the IGR IRES RNAs fold in the presence of 1–2 mM Mg^{2+} and this folded state is compact (Costantino and Kieft 2005), suggesting the folding transition would be accompanied by a large change in the size and shape of the RNA. To test this idea, a [Mg^{2+}] titration experiment was performed and changes in the hydrodynamic properties of IRES were monitored by SV/AUC (Costantino and Kieft 2005). If the IGR IRES RNA sample was structurally and chemically homogeneous and if the molecules behaved close to ideality (no interaction between individual RNA molecules), the values for S and D could be obtained and interpreted to reflect the shape of the molecule under different conditions. In fact, the boundaries obtained from SV/AUC on the IGR IRES fit well to a single-species model, suggesting that in the presence of Mg^{2+}, all molecules fold into a homogeneous population. This uniformity of folding is a characteristic not readily obtained from probing or biochemical analyses, but apparent using this biophysical method. These experiments also showed that addition of Mg^{2+} to IGR IRES RNAs caused an increase in the S and D coefficients consistent with RNA changing from an extended unfolded state to more compact, folded state. Likewise, the derived parameters R_h decreased and f/f_o became closer to one, showing a change in the structure of the IRES consistent with a more compact structure (Costantino and Kieft 2005). When other compactly folded RNAs have been analyzed by SV/AUC, similar changes in hydrodynamic properties were associated with the folding of the RNA into its native state (e.g., (Deras et al. 2000; Takamoto et al. 2002)).

Additional insight into the folding characteristics of the IGR IRES came from the observation that the midpoint of the folding transition as monitored by SV/AUC was ~0.2 mM Mg^{2+}, which is ~10-fold lower than the amount needed to fold the RNA into its tightly folded structure as monitored by hydroxyl radical probing (1–2 mM) (Costantino and Kieft 2005). These seemingly disparate biophysical results were reconciled by a model in which at lower cation concentrations, the RNA forms its secondary structure and perhaps certain parts of the tertiary structure also form transiently. However, higher cation concentrations are needed, to form the tightly folded core that excludes solvent. This behavior has precedent in other folded RNAs such as the self-splicing group I intron, suggesting the IGR IRES's structure and fold is akin to this RNA (Deras et al. 2000). However, the self-splicing intron's fold forms a specifically configured catalytic active site, whereas the IGR IRES's fold must arrange recognition surfaces in space. Therefore, in some ways the IGR IRES RNA might be compared to transfer RNAs (tRNAs), which are also stably folded RNAs having exterior surfaces arrayed to interact with various proteins such as aminoacyl synthetases and elongation factors, and interior specific intramolecular interactions that stabilize the three-dimensional shape. One must be cautious in extending these comparisons too far, but nonetheless these biophysical characteristics help to place the architecture of the IGR IRES in context with other folded RNAs.

16.2.3 Biophysical Comparison of the IGR IRESs and HCV IRES RNAs

The biophysical analyses described above revealed characteristics of the folded IGR IRES RNAs that were compared with the biophysical properties of the HCV IRES RNA (Kieft et al. 1999; Costantino and Kieft 2005). Hydroxyl radical probing of the HCV IRES RNA showed that although parts of the RNA backbone are protected from solvent in the presence of Mg^{2+}, these protections were found in discrete and isolated locations (Kieft et al. 1999). Specifically, there was protection from radicals in and around two four-way junction regions, but no evidence of extended backbone packing between helices. This suggested that in contrast to the IGR IRESs, the HCV IRES has an extended conformation in the presence of Mg^{2+} (Costantino and Kieft 2005). These probing results predicted IGR IRESs and the HCV IRES would show different behavior in the analytical ultracentrifuge. Indeed, the change in the hydrodynamic properties associated with HCV IRES folding was very different than those of the IGR IRESs (Costantino and Kieft 2005). This confirmed that the HCV IRES does not undergo similar structural transition as the IGR IRESs, but remains extended.

Biophysical methods revealed fundamental differences in the ion-induced folds of the functionally related IGR and HCV IRES RNAs, perhaps reflecting differences in their modes of action and particularly in how they recruit the ribosome. Although both of these RNAs directly bind the 40S subunit, biophysics show they use different three-dimensional architectures to achieve this end (Kieft et al. 1999; Costantino and Kieft 2005). As shown by cryo-EM, the compact IGR IRES binds directly over and within the decoding groove, "tucking into" the space between the two subunits and making contact to a relatively small portion of the ribosome (Spahn et al. 2004; Schuler et al. 2006). In contrast, the extended HCV IRES contacts a larger area on the ribosome, including the E-site between the two subunits and the solvent accessible side of the 40S subunit (Spahn et al. 2001; Boehringer et al. 2005), interacting with ribosomal proteins arrayed across a large surface of the 40S subunit (Otto et al. 2002). In addition, the HCV IRES must bind eIF3 and requires initiator tRNA and eIF2 to function (Pestova et al. 1998; Siridechadilok et al. 2005). The need for the HCV IRES to interact with, and perhaps coordinate action of, several components of the translation machinery may require the extended scaffold-like structure of the HCV IRES RNA. Thus, the architecture of these two IRES RNAs as revealed by biophysical analysis makes sense in light of their differing mechanisms of interacting with the translation machinery. This also suggests that by understanding the biophysical nature of other IRESs, we might be able to make predictions about their interacting with the translation machinery and their mechanisms. However the HCV and IGR IRESs are only two of many diverse IRES RNAs and so before this idea can be generalized, additional biophysical analysis of a greater variety of IRESs is needed.

16.3   Crystal Structure of the IGR IRESs

X-ray crystallography can, in some cases, reveal near-atomic resolution structures of the molecule of interest. If this method could be applied to IGR IRESs, a more detailed model of IRES function could potentially be constructed. However, crystallography of this and other RNAs is not trivial and the major hurdle is obtaining diffracting crystals. The biophysical characterization of the IGR IRESs provided crucial information to rationally engineer the RNA and obtain crystals suitable for structure determination. The fact that the IGR IRES folds into two independent domains mandated a strategy in which each domain was crystallized independently. Knowledge gained from biophysical studies regarding which IRES elements are critical for forming the fold and which parts can be modified without affecting the fold was likewise instrumental in designing constructs for crystallization (Kieft et al. 2007). Knowledge of the overall architecture of RNA, the relationship of different structural elements to one another, and the function of each element made the design of RNA constructs for crystallization much more rational, and enhanced the chance of success.

Ultimately, the crystal structures of the ribosome binding domain from the *Plautia stali* intestinal virus (PSIV) and the P-site domain from the cricket paralysis virus (CrPV) were solved (Pfingsten et al. 2006; Costantino et al. 2008) (Fig. 16.6a). Because probing experiments show the members of the IGR IRES family adopt the same fold, these structures can be regarded as representative of the group as a whole (Pfingsten et al. 2007). One of the key conclusions gleaned from the crystal structures of the IGR IRES RNA was that architectural features predicted by biophysical analysis were found in the structure (Costantino and Kieft 2005; Pfingsten et al. 2006) (Fig. 16.6b). Specifically, the pattern of solvent accessibility observed by hydroxyl radical probing matched the crystal structure, and the compact structure predicted by SV/AUC was observed. These biophysical and structural results validated and supported each other, providing a coherent description of pre-folded IGR IRES RNAs.

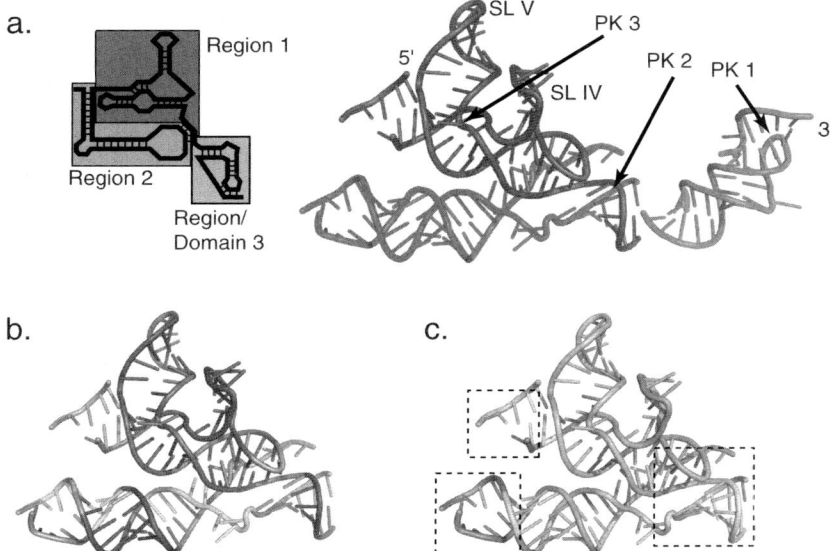

Fig. 16.6 Structural features of the IGR IRES from crystal structures. (**a**) A diagram of the secondary structure of an IGR IRES RNA is shown at left, with the three regions colored. Region 1 and 2 form the ribosome binding domain. At right are ribbon representations of the two crystal structures of the two domains, colored to match the diagram. Important structural features mentioned in the text are noted. (**b**) Hydroxyl radical probing data overlaid on a ribbons representation of the ribosome binding domain of the PSIV IGR IRES. Red indicates enhanced cleavage upon folding and solvent accessibility. Green indicates reduced cleavage upon folding and protection from solvent. Lighter green areas were protected, but to a lesser extent. The pattern agrees well with the crystal structure, linking the folded free form of the RNA to the crystal structure. (**c**) Ribbon representation of the ribosome binding domain of the PSIV IGR IRES, colored to reflect crystallographic B-factors. "Cool" colors (blue/green) denote low B-factors and "warm" colors (red/orange/yellow) have higher B-factors. These B-factors reflect disorder in the crystal structure; higher B-factors are more disordered and these portions are boxed (See figure insert for color reproduction)

The crystal structures of the two IGR IRES domains provided a wealth of new knowledge, but these structures are static snapshots of molecules whose dynamic conformational changes may be important for function. However, when crystal structures are linked to the results obtained from other biophysical and biochemical methods, it is possible to interpret these static structures and develop testable functional hypotheses that include dynamic conformational changes. Thus, while dynamic changes in structure that may accompany a functioning RNA cannot be directly observed in the crystal structure, they can sometimes be predicted. The structure of the ribosome binding domain of PSIV IGR IRES provides an example of how the static picture obtained by crystallography can be interpreted as part of a larger dynamic process. Within the PSIV IGR IRES ribosome binding domain structure, parts of the molecule are involved in stable structure, and many of these are parts that are tucked into the folded core (Fig. 16.6b). These stable domains appear to pre-position the two stem-loops that interact with the small ribosomal subunit during the first steps of IGR IRES-driven translation initiation. Hence this stable part of the IRES is likely forms to drive these first intermolecular recognition events. In contrast to the stable core, other parts of the IRES had electron density that was missing, or had high crystallographic temperature factors (B-factors) that suggest they are more mobile (Fig. 16.6c). Care must be taken in interpreting these observations, as the process of crystallization can give rise to higher B-factors. Nonetheless, the presence of disordered or more mobile RNA elements in this crystal structure allows speculation that these regions are also more mobile in solution and might be important in dynamic processes during IRES-driven translation. In fact, the regions of IGR IRES that are poorly ordered in the crystal structure also appear to be so when examined by other methods (Spahn et al. 2004; Schuler et al. 2006). Thus, in this case the static structure of the IGR IRES provides new hypotheses regarding conformational changes and their coupling to specific events in the IRES mechanism. For a more in-depth analysis of crystal structures, the reader is directed to the primary literature (Pfingsten et al. 2006; Costantino et al. 2008).

16.4   Cryo-EM Reconstructions of IGR IRES-Ribosome Complexes

This chapter has focused primarily on the biophysical characteristics of the unbound IGR IRES RNA and the architectural and functional implications of those characteristics. The hypothesis that the prefolded structure of the IRES reflects its ribosome-bound structure can be tested directly using cryo-electron microscopy (cryo-EM). Because cryo-EM can give structures of IRES RNAs bound to the ribosome (and to individual subunits) that cannot be obtained by other means, it is a biophysical method of special importance to the IRES structure field. To date, cryo-EM reconstructions of the CrPV IRES bound to various ribosome complexes have been reported and in all cases the location and overall conformation of the IRES RNA is readily observed (Spahn et al. 2004; Schuler et al. 2006). In these structures, the position of the IRES correlates well with biochemical data such as directed hydroxyl radical probing and toeprinting (Wilson et al. 2000; Pestova and Hellen 2003; Pestova et al. 2004). Many conclusions are gleaned from these reconstructions; one of the most pertinent for the topic of this chapter is that the CrPV IGR IRES appears on the ribosome as a compact structure comprised of two domains, one of which is tucked into the P-site of the 40S subunit. Hence, the architecture of the unbound IRES observed by biophysical methods is preserved when bound to the ribosome. The similarity between the free and bound states of the IGR IRES allows the crystal structure to be docked into the cryo-EM reconstructions and also for cryo-EM based models of the IRES to be built (Pfingsten et al. 2006; Schuler et al. 2006). These docked structures reveal what IRES elements interact with what features of the ribosome, but also show where potentially dynamic elements (that were identified by other biophysical means) might lie and what they might contribute to the function.

Are the IGR IRESs unique in their ability to pre-fold into a conformation with mechanistic significance? As previously discussed, similar behavior is apparent in the cryo-EM structures of the HCV IRES bound to 40S and 80S ribosomes as well as to eIF3 (Spahn et al. 2001; Boehringer et al. 2005; Siridechadilok et al. 2005). In this case, the IRES is bound to the ribosome in an extended conformation mirroring the extended conformation present in the free IRES and detected by biophysical methods (Kieft et al. 1999). Thus, in both the HCV IRES and IGR IRESs, there is strong evidence for the importance of the pre-folded structure of the IRES that determines mechanism of action. The degree to which this theme can be extended to other IRES RNAs awaits biophysical and structural analysis of more IRESs.

16.5 Concluding Remarks

IRESs represent an interesting class of non-protein coding RNAs that functionally replace proteins and other structures to drive translation initiation using specific RNA structures and sequences. Given that IRES function is an RNA-driven process, understanding the biophysical characteristics of the IRES RNA is fundamental to understanding the mechanisms by which these IRESs operate. Combining biophysical studies with other approaches is proving particularly powerful to understanding these RNAs, and in developing new models for their function.

Acknowledgments Thank you to the members of the Kieft Lab for many valuable discussions regarding IRES RNAs, and to David Costantino, John Hammond, Amanda Keel, and Jennifer Pfingsten for critical reading of this manuscript. IRES RNA studies in the Kieft Lab are supported by grants R01 GM072560 and R03 AI072187 from the National Institutes of Health and Research Scholar Grant 0805801GMC from the American Cancer Society.

References

Boehringer D, Thermann R, et al. (2005) Structure of the hepatitis C Virus IRES bound to the human 80S ribosome: remodeling of the HCV IRES. Structure 13:1695–1706

Borman A, Jackson RJ (1992) Initiation of translation of human rhinovirus RNA: mapping the internal ribosome entry site. Virology 188:685–696

Brasey A, Lopez-Lastra M, et al. (2003) The leader of human immunodeficiency virus type 1 genomic RNA harbors an internal ribosome entry segment that is active during the G2/M phase of the cell cycle. J Virol 77:3939–3949

Brenowitz M, Chance MR, et al. (2002) Probing the structural dynamics of nucleic acids by quantitative time-resolved and equilibrium hydroxyl radical "footprinting". Curr Opin Struct Biol 12:648–653

Buck CB, Shen X, et al. (2001) The human immunodeficiency virus type 1 gag gene encodes an internal ribosome entry site. J Virol 75:181–191

Cevallos RC, Sarnow P (2005) Factor-independent assembly of elongation-competent ribosomes by an internal ribosome entry site located in an RNA virus that infects penaeid shrimp. J Virol 79:677–683

Costantino D, Kieft JS (2005) A preformed compact ribosome-binding domain in the cricket paralysis-like virus IRES RNAs. RNA 11:332–343

Costantino DA, Pfingsten JS, et al. (2008) tRNA-mRNA mimicry drives translation initiation from a viral IRES. Nat Struct Mol Biol 15:57–64

Deras ML, Brenowitz M, et al. (2000) Folding mechanism of the Tetrahymena ribozyme P4-P6 domain. Biochemistry 39:10975–10985

Ehresmann C, Baudin F, et al. (1987) Probing the structure of RNAs in solution. Nucleic Acids Res 15:9109–9128

Glass MJ, Summers DF (1992) A cis-acting element within the hepatitis A virus 5′-non-coding region required for in vitro translation. Virus Res 26:15–31

Hatakeyama Y, Shibuya N, et al. (2004) Structural variant of the intergenic internal ribosome entry site elements in dicistroviruses and computational search for their counterparts. RNA 10:779–786

Hellen CU, Sarnow P (2001) Internal ribosome entry sites in eukaryotic mRNA molecules. Genes Dev 15:1593–1612

Jackson RJ (2005) Alternative mechanisms of initiating translation of mammalian mRNAs. Biochem Soc Trans 33:1231–1241

Jackson RJ, Kaminski A (1995) Internal initiation of translation in eukaryotes: the picornavirus paradigm and beyond. RNA 1:985–1000

Jan E (2006) Divergent IRES elements in invertebrates. Virus Res 119:16–28

Jan E, Sarnow P (2002) Factorless ribosome assembly on the internal ribosome entry site of cricket paralysis virus. J Mol Biol 324:889–902

Jan E, Thompson SR, et al. (2001) Initiator Met-tRNA-independent translation mediated by an internal ribosome entry site element in cricket paralysis virus-like insect viruses. Cold Spring Harb Symp Quant Biol 66:285–292

Jan E, Kinzy TG, et al. (2003) Divergent tRNA-like element supports initiation, elongation, and termination of protein biosynthesis. Proc Natl Acad Sci U S A 100:15410–15415

Jang SK, Krausslich HG, et al. (1988) A segment of the 5′ nontranslated region of encephalomyocarditis virus RNA directs internal entry of ribosomes during in vitro translation. J Virol 62:2636–2643

Kanamori Y, Nakashima N (2001) A tertiary structure model of the internal ribosome entry site (IRES) for methionine-independent initiation of translation. RNA 7:266–274

Kapp LD, Lorsch JR (2004) The molecular mechanics of eukaryotic translation. Annu Rev Biochem 73:657–704

Kieft JS, Costantino DA, et al. (2007) Structural methods for studying IRES function. Methods Enzymol 430:333–371

Kieft JS, Zhou K, et al. (1999) The hepatitis C virus internal ribosome entry site adopts an ion-dependent tertiary fold. J Mol Biol 292:513–529

Kuhn R, Luz N, et al. (1990) Functional analysis of the internal translation initiation site of foot-and-mouth disease virus. J Virol 64:4625–4631

Lebowitz J, Lewis MS, et al. (2002) Modern analytical ultracentrifugation in protein science: a tutorial review. Protein Sci 11:2067–2079

Mathews MB, Sonenberg N, et al. (2000) Origins and principles of translational control. In: Sonenberg N, Hershey JWB and Mathews MB (eds.) Translational control of gene expression. Cold Spring Harbor Laboratory Press, Cold Spring Harbor, New York, pp. 1–31

Merino EJ, Wilkinson KA, et al. (2005) RNA structure analysis at single nucleotide resolution by selective 2′-hydroxyl acylation and primer extension (SHAPE). J Am Chem Soc 127:4223–4231

Nishiyama T, Yamamoto H, et al. (2003) Structural elements in the internal ribosome entry site of Plautia stali intestine virus responsible for binding with ribosomes. Nucleic Acids Res 31:2434–2442

Otto GA, Lukavsky PJ, et al. (2002) Ribosomal proteins mediate the hepatitis C virus IRES-HeLa 40S interaction. RNA 8:913–923

Pelletier J, Sonenberg N (1988) Internal initiation of translation of eukaryotic mRNA directed by a sequence derived from poliovirus RNA. Nature 334:320–325

Pestova TV, Hellen CU (2003) Translation elongation after assembly of ribosomes on the Cricket paralysis virus internal ribosomal entry site without initiation factors or initiator tRNA. Genes Dev 17:181–186

Pestova TV, Shatsky IN, et al. (1998) A prokaryotic-like mode of cytoplasmic eukaryotic ribosome binding to the initiation codon during internal translation initiation of hepatitis C and classical swine fever virus RNAs. Genes Dev 12:67–83

Pestova TV, Lomakin IB, et al. (2004) Position of the CrPV IRES on the 40S subunit and factor dependence of IRES/80S ribosome assembly. EMBO Rep 5:906–913

Pfingsten JS, Costantino DA, et al. (2006) Structural basis for ribosome recruitment and manipulation by a viral IRES RNA. Science 314:1450–1454

Pfingsten JS, Costantino DA, et al. (2007) Conservation and diversity among the three-dimensional folds of the Dicistroviridae intergenic region IRESes. J Mol Biol 370:856–869

Preiss T, Hentze MW (2003) Starting the protein synthesis machine: eukaryotic translation initiation. BioEssays 25:1201–1211

Schuler M, Connell SR, et al. (2006) Structure of the ribosome-bound cricket paralysis virus IRES RNA. Nat Struct Mol Biol 13:1092–1096

Shcherbakova I, Brenowitz M (2008) Monitoring structural changes in nucleic acids with single residue spatial and millisecond time resolution by quantitative hydroxyl radical footprinting. Nat Protoc 3:288–302

Shcherbakova I, Mitra S, et al (2006) Fast Fenton footprinting: a laboratory-based method for the time-resolved analysis of DNA, RNA and proteins. Nucleic Acids Res 34:e48

Siridechadilok B, Fraser CS, et al. (2005) Structural roles for human translation factor eIF3 in initiation of protein synthesis. Science 310:1513–1515

Spahn CM, Kieft JS, et al. (2001) Hepatitis C virus IRES RNA-induced changes in the conformation of the 40s ribosomal subunit. Science 291:1959–1962

Spahn CM, Jan E, et al. (2004) Cryo-EM visualization of a viral internal ribosome entry site bound to human ribosomes. The IRES functions as an RNA-based translation factor. Cell 118:465–475

Stoneley M, Willis AE (2004) Cellular internal ribosome entry segments: structures, trans-acting factors and regulation of gene expression. Oncogene 23:3200–3207

Takamoto K, He Q, et al. (2002) Monovalent cations mediate formation of native tertiary structure of the *Tetrahymena thermophila* ribozyme. Nat Struct Biol 9:928–933

Tsukiyama-Kohara K, Iizuka N, et al. (1992) Internal ribosome entry site within Hepatitis C virus RNA. J Virol 66:1476–1483

Tullius TD, Greenbaum JA (2005) Mapping nucleic acid structure by hydroxyl radical cleavage. Curr Opin Chem Biol 9:127–134

Wang C, Sarnow P, et al. (1993) Translation of human hepatitis C virus RNA in cultured cells is mediated by an internal ribosome-binding mechanism. J Virol 67:3338–3344

Wilkinson KA, Merino EJ, et al. (2006) Selective 2′-hydroxyl acylation analyzed by primer extension (SHAPE): quantitative RNA structure analysis at single nucleotide resolution. Nat Protoc 1:1610–1616

Wilson JE, Pestova TV, et al. (2000) Initiation of protein synthesis from the A site of the ribosome. Cell 102:511–520

Chapter 17
Structure and Gene-Silencing Mechanisms of Small Noncoding RNAs

Chia-Ying Chu and Tariq M. Rana(✉)

Abstract Small (19–31-nucleotides) noncoding RNAs were identified in the past 10 years for their distinct function in gene silencing. The best known gene-silencing phenomenon, RNA interference (RNAi), is triggered in a sequence-specific manner by endogenously produced or exogenously introduced small doubled-stranded RNAs. As knowledge of the structure and function of the RNAi machinery has expanded, this phenomenon has become a powerful tool for biochemical research; it has enormous potential for therapeutics. This chapter summarizes significant aspects of three major classes of small noncoding, regulatory RNAs: small interfering RNAs (siRNAs), microRNAs (miRNAs), and Piwi-interacting RNAs (piRNAs). Here, we focus on the biogenesis of these small RNAs, their structural features and coupled effectors as well as the mechanisms of each small regulatory RNA pathway which reveal fascinating ways by which gene silencing is controlled and fine-tuned at an epigenetic level.

17.1 Introduction

RNA interference (RNAi) is the mechanism by which double-stranded RNA (dsRNA) complexed with RNA-binding proteins triggers sequence-specific gene silencing. First details of RNA-mediated gene silencing were observed in the early 1990s in several studies of plants and flies. The potent and specific gene-silencing phenomenon induced by dsRNA was first described in *Caenorhabditis elegans* by the Fire and Mello groups (Fire et al. 1998). These findings released a flood of studies on how RNAs mediate gene silencing in different eukaryotic cell types, from yeast to humans. Triggers for gene silencing were identified not only as exogenously introduced dsRNA, but also different types of small (19–31-nucleotides [nt]) endogenous regulatory RNAs that divided into several classes according to their interacting proteins or their origins (Table 17.1). Studies on these small RNAs

T.M. Rana
Department of Biochemistry and Molecular Pharmacology, University of Massachusetts Medical School, Worcester, MA 01605, USA
e-mail: tariq.rana@umassmed.edu

N.G. Walter et al. (eds.) *Non-Protein Coding RNAs*
doi: 10.1007/978-3-540-70840-7_16, © Springer-Verlag Berlin Heidelberg 2009

Table 17.1 Classes of small RNAs involved in gene silencing

Class	Length (nt)	Mechanism	Organism	References
Small interfering RNA (siRNA)	19–21	Target mRNA cleavage	Vertebrates, fly, worm, plants, and yeast	Rana (2007)
MicroRNA (miRNA)	19–25	Translational repression, mRNA decay	Vertebrates, fly, worm, plants, and yeast	Bartel (2004)
Repeat-associated siRNA (rasiRNA) or piwi-interacting RNA (piRNA)	24–31	Transposon control and/or transcriptional silencing	Vertebrates, fly, worm, and plants	O'Donnell and Boeke (2007)
Trans-acting siRNA (tasiRNA)	21–22	mRNA cleavage	Fly, plants, and yeast	Peragine et al. (2004)
Small-scan RNA (scnRNA)	~28	DNA elimination	Tetrahymena	Mochizuki et al. (2002); Taverna et al. (2002)

and their function in sequence-specific silencing provide insights into the epigenetic regulation of gene expression in development and disease. As the RNAi field has quickly expanded, researchers have worked to identify key effectors and cofactors, to elucidate the mechanisms of sequence-specific silencing for cellular functions of RNAi, and to exploit it as a potential therapeutic agent.

This chapter summarizes significant aspects of three major classes of small non-coding, regulatory RNAs: small interfering RNAs (siRNAs), microRNAs (miRNAs), and Piwi-interacting RNAs (piRNAs). We will focus on their biogenesis, structural features, coupled effectors, and mechanisms of pathways involving small regulatory RNAs to bring out fascinating methods by which gene silencing is controlled and fine-tuned at the epigenetic level.

17.2 Small Interfering RNAs

17.2.1 RNA as the Trigger for Gene Silencing

RNAi, considered a conserved antiviral mechanism in many organisms, is a response to foreign dsRNAs (reviewed in Mello and Conte 2004). When long dsRNA enters eukaryotic cells, it is processed by a cytoplasmic RNase III, Dicer, which cleaves dsRNA to form a ~21 nt RNA duplex. Dicer cleaves not only viral RNA or viral replication intermediates, but some endogenous RNA transcripts with long stem-loop structures, such as microRNA precursors (see below). Cleavage of these dsRNA structures by Dicer results in a ~21 nt small interfering RNA (siRNA) with 19 nt base pairs, a 2 nt overhang at both 3′ ends, and a phosphate group at both

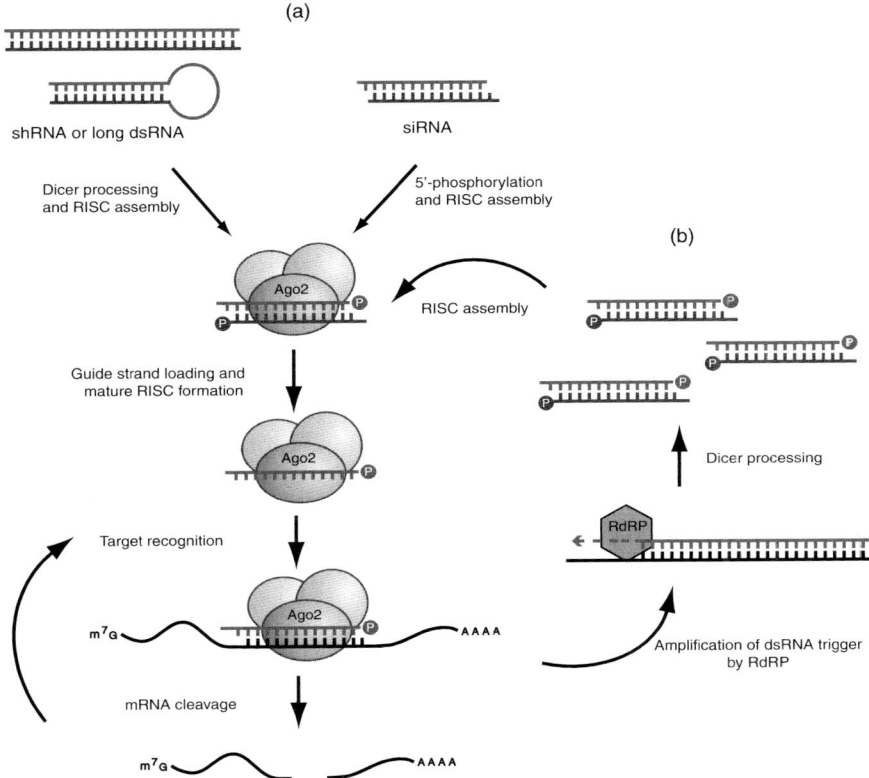

Fig. 17.1 Short-interfering RNA (siRNA)-mediated gene silencing. (**a**) In vertebrates, 21–22 rt siRNA duplexes are processed from long double-stranded RNAs (dsRNAs) or short hairpin RNA (shRNA) by the Dicer/TRBP complex and incorporated into RISC. Similar mechanism is used to introduce synthetic 21 nt siRNAs into cells and load them onto RISC. Active RISC containing guide strand of siRNA recognizes and cleaves complementary target mRNA. After target cleavage, RISC is recycled for multiple rounds of catalysis. (**b**) In plants and worms, the RNAi response is amplified by RNA-dependent RNA polymerases (RdRPs). RdRPs coupled with RISC generate dsRNA using target mRNA as the template. This long dsRNA is subsequently cleaved by Dicer to form secondary siRNAs that can target more mRNA and amplify the RNAi effect (See figure insert for color reproduction)

5′ ends (Fig. 17.1a). Since Dicer processes both linear dsRNA duplexes and hairpin RNAs, cells can be transfected with DNA constructs which encode short hairpin RNAs (shRNA) as an alternative method to induce RNAi. RNAi pathway in cells can also be stimulated by introducing synthetic 21 nt siRNAs with structures similar to post-Dicer cleavage products (reviewed in Carmell and Hannon 2004). However, these 21 nt siRNAs are less potent RNAi triggers than Dicer-substrate dsRNAs (Kim et al. 2005; Siolas et al. 2005), possibly because Dicer processing is coupled with other co-factors and facilitates siRNA loading into an RNA-protein complex called RNA-induced silencing complex (RISC).

Following Dicer cleavage, only one strand of siRNA duplex is assembled into the RNAi engine, i.e., the RISC. Since the functions of RISC are governed by its loaded small RNA, the RISCs involved in siRNA- and miRNA-mediated gene-silencing will be referred to as siRNA-induced silencing complex (siRISC) and miRNA-induced complex (miRISC), respectively. The strand incorporated into siRISC is known as the guide strand because it is complementary to the RNA targeted for silencing, thus serving as a guide for sequence-specific silencing of RNA. The other strand, called passenger strand, is eliminated during siRISC loading (Matranga et al. 2005; Rand et al. 2005). Strand selection, an important feature of siRISC loading, may be determined by relative thermodynamic stability of the two siRNA 5′ termini (Khvorova et al. 2003; Schwarz et al. 2003). With the assistance of other co-factors, the strand with less stability at the 5′ end is loaded onto siRISC, while the passenger stand is cleaved and eliminated. Guide-strand siRNA is bound to the core endonuclease of the active siRISC, the Argonaute protein, and directs cleavage of target mRNA. With perfect base pairing and formation of an A-form helix structure (see below) between the siRNA guide strand and target mRNA, siRISC can then cleave a single phosphodiester bond of the target mRNA between the tenth and eleventh nucleotides from the 5′ end of the paired guide-strand siRNA (Elbashir et al. 2001). The siRISC complex is then released for the next cycle of target cleavage, while the cleaved target mRNA is subsequently degraded by other exonucleases in the cytoplasm (Fig. 17.1).

Though RNAi in flies and humans involves a simple cycle of target recognition and cleavage, RNAi in plants and *C. elegans* can be amplified by a mechanism that requires a specific enzyme called RNA-dependent RNA polymerase (RdRP). RdRP has been shown to produce additional dsRNA by using target mRNAs as templates in primer-independent synthesis of complementary RNA. This RdRP-derived dsRNA can be subsequently processed by Dicer to generate more secondary siRNAs for enhanced RNAi- similar to the phenomenon triggered by primary siRNAs (reviewed in Mello and Conte 2004) (Fig. 17.1b).

17.2.2 Structural Features of siRNAs in RNAi Pathways

A fundamental question regarding the RNAi mechanism is why RNA, and not DNA, duplexes, can induce gene silencing. This phenomenon is explained by differences between dsRNA and dsDNA structures (reviewed in Rana 2007). DsRNA forms a right-handed A-form helix with 11 base pairs per helical turn, whereas dsDNA is a right-handed B-form helix with 10 base pairs per turn. The importance of the A-form helix in RNAi was examined by introducing bulge structures in synthetic siRNAs to abolish their A-form structure (Chiu and Rana 2002). Synthetic siRNA with a 2 nt bulge structure in the passenger strand, which disrupts its A-form helical geometry, can still be incorporated into RISC and induce RNAi. However, if the bulge is on the guide strand when it pairs with target mRNA, the RNAi effect is abolished, indicating helical geometry is critical for target recognition but not for the RISC loading stage (Chiu and Rana 2002). Further understanding of the siRNA structural features required for RNAi, came from studies of RNAi efficiency, using chemically modified siRNAs at various functional groups (Chiu and Rana 2002). Results confirmed a central A-form helix is required for target cleavage by siRISC in RNAi, and for providing guidelines for developing modified and stable siRNAs with high potency.

Besides A-form helical geometry between the guide strand and target mRNA, other structural features important for RNAi activity are terminal structures of siR-NAs. Since the termini of dsRNA or short hairpin RNA (shRNA) are processed by Dicer to generate siRNAs with a phosphate group at the 5′ end and a hydroxyl group at the 3′ end, synthetic siRNAs with the same features can function as exogenous triggers of RNAi. To understand the required siRNA terminal structures for RNAi, several modifications were engineered at the 3′ or 5′ end of both the guide and passenger strands (Chiu and Rana 2002). Results of these experiments showed that only the 5′ phosphate group of the guide strand is required. However, efficacy of siRNAs was not affected by adding other functional groups to the 5′ end of the passenger strand and the 3′ end of both strands. Interestingly, siRNAs with a 5′ hydroxyl group on the guide stand can still induce RNAi because the hydroxyl group can be phosphorylated by endogenous RNA kinases, resulting in the same structure as that of Dicer products (Chiu and Rana 2002). The requirement for a 5′ phosphate on the guide strand was confirmed by solving the crystal structure of siRNA-associated PIWI proteins (Ma et al. 2005). The structure showed the guide-strand siRNA is anchored to RISC via its 5′ phosphate binding to the phosphate-binding domain of PIWI proteins (Ma et al. 2005). The 3′ end of *Drosophila* siRNA has recently been shown to be 2′-*O*-methylated by the Hen1 protein, thereby increasing stability of siRNA after its loading onto RISC (Horwich et al. 2007; Saito et al. 2007). This finding is consistent with a previous report that 3′ modification does not affect RNAi activity in human cells (Chiu and Rana 2002).

These results together characterize structural requirements for siRNA and provide guidelines for designing modified siRNAs that are both stable and potent (Table 17.2 . Results of studies on chemically modified siRNAs have provided insight into application of RNAi not only as a powerful tool in biology research but also as a potentially potent therapeutic agent.

17.2.3 Argonaute Proteins: The Key Components in RISC

The effector protein that directly binds siRNAs in the siRISC complex is Argonaute. Argonaute (Ago) proteins perform the cleavage activity of RNAi and mediate translation repression in miRNA pathways (see below). Human Argonaute family contains eight protein members. Four of them are grouped as a germline-specific PIWI subset and associate with 24–30 nt RNAs called piRNAs (see below). The other four members, Ago1, Ago2, Ago3, and Ago4, are expressed ubiquitously; all are capable of binding to miRNAs or siRNAs. All four proteins have highly conserved amino

Table 17.2 Guidelines for chemically modified siRNAs

Modification at guide strand	Modification at passenger strand	RNAi efficiency	References
5′-End capping	Unmodified	−[a]	Chiu and Rana (2002), Schwarz et al. (2002), Czauderna et al. (2003)
Unmodified	5′-End capping	++[b]	Chiu and Rana (2002), Schwarz et al. (2002), Czauderna et al. (2003)
3′-end capping	Unmodified	++	Chiu and Rana (2002), Schwarz et al. (2002), Czauderna et al. (2003)
Unmodified	3′-End capping	++	Chiu and Rana (2002), Schwarz et al. (2002), Czauderna et al. (2003)
2′-deoxy	Unmodified	−	Chiu and Rana (2003)
Unmodified	2′-deoxy	+[c]	Chiu and Rana (2003)
2′-fluoro-Pyrimidines	2′-fluoro-Pyrimidines	++	Chiu and Rana (2003), Braasch et al. (2003), Prakash et al. (2005), Morrissey et al. (2005)
2′-O-methyl	Unmodified	−	Chiu and Rana (2003), Braasch et al. (2003), Prakash et al. (2005), Morrissey et al. (2005), Amarzguioui et al. (2003)
2′-O-methyl at position 2 from the 5′ end	Unmodified	++[a,d]	Jackson et al. (2006)
Unmodified	2′-O-methyl	+	Chiu and Rana (2003), Braasch et al. (2003), Prakash et al. (2005), Morrissey et al. (2005), Amarzguioui et al. (2003)
2′-O-(2-methoxyethyl)	Unmodified	−/+[e]	Prakash et al. (2005)
Unmodified	2′-O-(2-methoxyethyl)	++	Prakash et al. (2005)
3-methyl-U	Unmodified	−	Chiu and Rana (2003)
C5-Br-U	Unmodified	++	Chiu and Rana (2003)
C5-I-U	Unmodified	++	Chiu and Rana (2003)
Phosphorothioate backbone	Unmodified	+	Chiu and Rana (2003), Braasch et al. (2003), Prakash et al. (2005)
Unmodified	Phosphorothioate backbone	++	Chiu and Rana (2003), Braasch et al. (2003), Prakash et al. (2005)

[a] −: <20% silencing

[b] ++: >50% silencing

[c] +: 20–50% silencing

[d] Reduced off-target effect

[e] Efficiency depends on the position of the modified base

acid sequences and functional domains, but only Ago2 exhibits endonuclease activity to cleave target mRNA during RNAi (Liu et al. 2004; Meister et al. 2004).

Ago proteins, as shown by the crystal structure of full-length *Pyrococcus furiosus* Ago, have four domains: the N-terminal, PAZ, middle, and PIWI domains (reviewed in Tolia and Joshua-Tor 2007). PAZ domain recognizes the 3′ end of single-stranded RNA, whereas the other three domains create a unique structure, providing a positively charged groove for base-pairing between guide-strand RNA and target mRNA. PIWI domain is an RNase H-like one and functions as the catalytic core for substrate cleavage. Cleavage reaction requires the conserved Asp-Asp-His motif, with the assistance of Mg^{2+} ions and water molecules, through an SN2 reaction mechanism. Cleavage by the PIWI active motif leaves a 5′phosphate group and a 3′ hydroxyl group (Tolia and Joshua-Tor 2007).

17.2.4 Kinetic Properties of RISC

Kinetic properties of human siRISC have been analyzed in siRISC complexes assembled in three ways: by transfecting siRNAs into HeLa cells to produce in vivo-assembled holo-siRISC (Brown et al. 2005), by incubating siRNAs with immunopurified Ago2 complex to produce in vitro-assembled siRISC (Martinez and Tuschl 2004), or by assembling siRNAs with recombinant Ago2 (Rivas et al. 2005). While all these systems show siRISC is a multi-turnover enzyme complex, kinetic parameters are different in each case, indicating that the three siRISCs are catalytically different enzyme complexes (reviewed in Rana 2007). This difference are explained by the composition of each siRISC. Since the holo-siRISC is assembled in vivo, other proteins in the RISC complex may interact with Ago2, limiting its structural flexibility and resulting in lower enzymatic activity. Concentration of assembled RISC were unaffected by the presence of target mRNA in the programming stage (Brown et al. 2005). Data backs the concept that the two rate-limiting steps of siRNA function- RISC loading and target recognition- take place independently in different complexes.

The challenge in using RNAi as a research and medical tool is how to induce efficient sequence-specific silencing without increasing off-target effects. Research on molecular features of both siRNA sequences and target structures has provided insights into the design of siRNAs. Systematic studies on the functional characteristics of siRNAs targeting a single gene have led to several algorithms for designing more specific and effective siRNAs (Amarzguioui and Prydz 2004; Reynolds et al. 2004; Ui-Tei et al. 2004). Analysis of sequence-position specificity on the 21 nt duplex permitted determination of several criteria, for a better design for siRNA sequences (Birmingham et al. 2007). While the sequence requirement of siRNA are summarized, some exceptions exist for sequences that follow design rules, suggesting that other mechanisms influence RNAi efficacy.

The sequence features of siRNAs, not only target accessibility but has also been shown to dictate RNAi efficiency. The role of target accessibility in RNAi was probed using the transactivation responsive (TAR) element of HIV-1 RNA as a target for siRISC cleavage (Brown et al. 2005). As expected, the target's rigid stem-loop structure was only minimally cleaved, indicating the target RNA structure contributes to siRISC activity. Conversely, when the TAR structure was disrupted by clamping 2′-O-methyl antisense oligos complementary to the 5′ or 3′ regions of the target site, thus increasing its accessibility, the target RNA was cleaved. Since the clamps blocked the possibility of RISC scanning from either the 5′ or 3′

direction, these results also suggest that siRISC recognizes its target by diffusion instead of a scanning model as for ribosomes (Brown et al. 2005). More recent evidence for the dependence of RNAi on target accessibility is that the 5′ end of siRNA in RISC creates a thermodynamic threshold that determines a stable association between RISC and the target RNA (Ameres et al. 2007).

17.3   microRNAs

17.3.1   The miRNA Genome and Biogenesis

The first miRNA gene was discovered in 1993. The Ambros lab cloned the *C. elegans lin-4* gene which encodes two tiny (~22- and 61 nt) RNA products, complementary to the 3′UTR (untranslated region) of lin-14 mRNA (Lee et al. 1993). Subsequently, thousands of miRNA genes from different species have been identified using direct cloning and computational approaches. miRNA loci, found in both noncoding genes and introns of coding genes, are distributed among all chromosomes. The biogenesis of miRNAs starts in the nucleus, where the miRNA gene is transcribed by RNA polymerase II (pol II) into primary miRNAs (pri-miRNAs) with 5′ m^{7}G capping structures and 3′ poly-A tails (reviewed in Kim 2005; Fig. 17.2). Half of all identified miRNAs are grouped in clusters, generated as polycistronic primary transcripts. The long primary transcripts of miRNA genes (pri-miRNA), with an extended hairpin structure, are subsequently processed by the Drosha–DGCR8 (DiGeorge syndrome critical region gene-8) complex, also known as the Drosha–Pasha complex in *Drosophila* and *C. elegans*. Like other RNase III family endonucleases, Drosha contains two RNase III domains and a dsRNA-binding domain for catalytic function. However, DGCR8, a dsRNA-binding protein (dsRBP), functions as a ruler for this complex, measuring the cleavage site for Drosha (Han et al. 2006). The Drosha–DGCR8 complex cleaves pri-miRNAs at the eleventh nucleotide from the base of the stem-loop structure, resulting in a ~70 nt long, stem-loop structured miRNA precursor (pre-miRNA) with a 5′ phosphate end and a 2 nt 3′ overhang.

Mostly pri-miRNAs are processed by Drosha; some however are derived from a different pathway. In *Drosophila* and mammals, a specific group of miRNA loci called "mirtrons" has recently been localized in the introns of coding genes and shown to generate intronic hairpins during pre-mRNA splicing (Okamura et al. 2007; Ruby et al. 2007). Due to this RNA splicing, the intronic hairpin forms a pre-miRNA-like structure that bypasses Drosha cleavage. Mirtron-encoded pre-miRNAs are then exported and processed in a shared pathway with Drosha-cleaved pre-miRNAs.

After this primary processing step in the nucleus, pre-miRNAs are delivered to the cytoplasm for the next step of miRNA maturation. Transport of pre-miRNAs to the cytoplasm is mediated by the nuclear export factor, Exportin-5 (Exp5), which partners the GTP-bound cofactor, Ran protein, to drive pre-miRNA through the nuclear pore (Bohnsack et al. 2004; Lund et al. 2004; Yi et al. 2003). Exp5 mediates not only the nuclear export of pre-miRNAs, but also plays a minor role in the

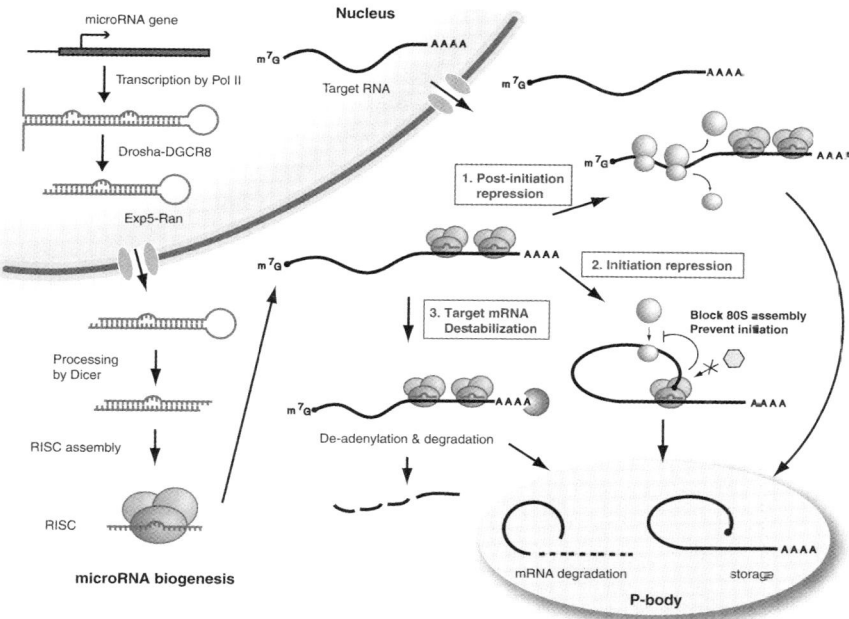

Fig. 17.2 microRNA-mediated gene silencing. miRNA genes are transcribed by RNA Pol II into primary miRNAs (pri-miRNAs), which are processed in the nucleus by Drosha/DGCR8 into pre-miRNAs and exported into the cytoplasm by Exp5-Ran. The hairpin miRNA precursors (pre-miR-NAs) are subsequently cleaved by Dicer/TRBP (transactivation-response element RNA-binding protein) to ~22 nt miRNA/miRNA* duplexes; these are then loaded into RISC. Active RISC then recognizes the 3′ UTR of target mRNAs with imperfect base-pairing and induces translation repression through three possible mechanisms: post-initiation repression, initiation repression, or target mRNA destabilization. In post-initiation repression, RISC would inhibit protein translation after the initiation step and prevent protein synthesis. To repress initiation, RISC would bind to the cap structure of target mRNA and block the initiation of protein translation by preventing the translation initiation factor, eIF4E, from association with the mRNA 5′ cap. In this case, RISC would also block assembly of the 80S ribosome. To destabilize the target mRNA, its recognition by RISC would induce deadenylation and decay of the target mRNA. Destabilization and increased turnover of target mRNA would result in reduced protein synthesis. In all three mechanisms, RISC-targeted mRNAs would be sequestered in P-bodies for storage or decay, although mRNA might also be degraded outside P-bodies (See figure insert for color reproduction)

nuclear export of transfer RNA (tRNAs) (Bohnsack et al. 2002). Exp5 also exports some viral RNAs with similar stem-loop structures and 3′ overhangs, like adenovirus RNA VA1 (Zeng and Cullen 2004). These clues led to identification of the structural requirement for Exp5-mediated delivery of pre-miRNA delivery to the cytoplasm as an RNA structure with a 16-base pair stem-loop and a short 3′ overhang (Zeng and Cullen 2004). Following export from the nucleus, pre-miRNAs are then processed by Dicer for miRNA maturation in the cytoplasm (Fig. 17.2).

As mentioned earlier for siRNA processing, Dicer is a cytoplasmic RNase III that cleaves long dsRNAs into ~22 nt duplexes. In miRNA, duplexes comprise a mature miRNA strand (the guide strand) and an miRNA passenger strand (miRNA*), each produced with a 5' phosphate and 2 nt overhangs at its 3' end. Dicer is highly conserved in most eukaryotic species, but in some organisms it is expressed as multiple homologs with distinct functions involved in different RNAi mechanisms. For instance, in *Drosophila* Dicer-1 plays a role in miRNA maturation, whereas Dicer-2 is required for the siRNA pathway (Lee et al. 2004). Like Drosha and other RNase III family proteins, Dicer also interacts with dsRNA-binding proteins that may facilitate miRNA processing. These proteins include the known human RISC components, transactivation-response element RNA-binding protein (TRBP; Chendrimada et al. 2005) and protein activator of the interferon-induced protein kinase (PACT; Lee et al. 2006). In *Drosophila,* Dicer-1 collaborates with the TRBP homologue, Loquacious (Loqs), to facilitate miRNA maturation (Forstemann et al. 2005; Saito et al. 2005). Since purified Dicer alone maintains cleavage activity in vitro, these Dicer-interacting proteins are believed to be nonessential for Dicer's RNase activity. Instead, the complex of Dicer and its interacting proteins play important roles in the next stage, where the miRNA/miRNA* duplex is unwound and loaded onto the functional complex, miRNA-induced silencing complex (miRISC).

Similar to the unwinding process of siRNA described earlier, only one strand of the miRNA/miRNA* duplex can be successfully incorporated into RISC, while the other strand is released and eliminated. The strand selected may be determined by the relative thermodynamic stability of the ends of miRNA duplexes the strand with less stability at the 5' end is loaded into RISC, while the other strand is released and degraded. Several interesting questions remain with regard to the miRISC assembly stage. First, the mechanism of duplex unwinding is still unknown. Although Dicer contains a DEAD-box helicase domain, no direct evidence shows that Dicer is required for RNA unwinding. In *Drosophila*, a putative RNA helicase, Armitage, is shown to enhance miRNA maturation. In human cells, RNA helicase A (RHA) is shown to interact with RISC and function as a co-factor that facilitates siRNA and miRNA loading into RISC but detailed mechanism for this loading step is unclear (Robb and Rana 2007). RHA may directly unwind dsRNA for RISC loading or it may also remodel RISC complexes to enhance miRNA loading.

The second interesting question about miRISC assembly is how cellular factors can distinguish the miRNA and siRNA pathways. Although miRISC contains several proteins such as Dicer, TRBP, PACT, and Gemin3, only Argonaute proteins have been directly associated with miRNAs. In *Drosophila,* Ago1 and Ago2 play distinct roles in two separate pathways: Ago1 is part of miRISC and Ago2 of siRISC (Okamura et al. 2004). The *Drosophila* miRNA/miRNA* or siRNA duplexes are sorted into different complexes with Ago1 or Ago2 according to the internal structures of each duplex (Förstemann et al. 2007; Tomari et al. 2007). The miRNA/miRNA* duplex usually forms bulge structures at the imperfectly matched sequence and binds Ago1 in miRISC, whereas perfectly matched siRNA duplex enters siRISC with Ago2. In human cells, the Ago family contains four proteins, Ago1–4, all of which can associate with miRNA, suggesting that the human miRNA pathway may be mediated by all four Ago proteins.

Table 17.3 Online sources of miRNA target-prediction programs

Program	Website	Organism	References
DIANA-microT	http://www.diana.pcbi.upenn.edu/	All	Kiriakidou et al. (2004)
Micro-Ispector	http://mirna.imbb.forth.gr/microinspector/	All	Rusinov et al. (2005)
MiRanda	http://www.mircorna.org/	Fly, zebrafish, and human	Enright et al. (2003); John et al. (2004)
MiRNA-Target gene prediction at EMBL	http://www.russell.embl.de/miRNAs/	Fly	Brennecke et al. (2005)
PicTar	http://pictar.bio.nyu.edu/	Worm, fly, and vertebrates	Grun et al. (2005); Krek et al. (2005); Lall et al. (2006)
RNAhybrid	http://bibiserv.techfak.uni-bielefeld.de/	All	Rehmsmeier et al. (2004)
TargetBoost	http://demo1.interagon.com/demo	Worm and fly	Saetrom et al. (2005)
TargetScanS	http://www.targetscan.org/	Worm, fly, and vertebrates	Lewis et al. (2005); Lewis et al. (2003)

17.3.2   Target Recognition by miRNAs

While siRNAs require a perfect match with their target mRNA for target cleavage, a key feature of the interaction between miRNAs and their target mRNA is imperfect base-pairing between the miRNA guide strand and target mRNA. Given that imperfectly matched nucleotides usually create bulge structures, sequence specificity for target recognition is determined by the base-paired nucleotides at positions 2–8 of the miRNA 5′ end, referred to as the "seed sequence" (Doench and Sharp 2004). The requirement for miRISC function is only 7 nt long; so one miRNA could possibly recognize multiple mRNAs because several mRNA sequences could harbor complementary sequences to a single miRNA seed sequence. Indeed, miRNAs have been predicted by computational analysis of genome databases and miRNA sequences to regulate the expression of more than one-third of all human proteins (Lewis et al. 2005).

Specificity of miRNA for targeting protein-encoding genes has been validated by using experimental and computational approaches to develop several algorithms for miRNA-target recognition (Table 17.3). In addition to the target site's perfect match with the miRNA seed region, miRNA targeting is enhanced by five other features of the target site (Grimson et al. 2007): (1) 3'UTRs with multiple miRNA target sites, (2) additional base-pairing to miRNA 5' nucleotides 13–16, (3) local AU-rich nucleotide context, (4) 3'UTR target sequence positioned at least 15 nt downstream from the stop codon, and (5) positions away from the center of long UTRs. Overall, a better knowledge of the mechanism of miRNA targeting helps biochemical and genetic studies of regulatory miRNA functions in different cellular pathways.

17.3.3 Mechanism of miRNA-Mediated Gene Silencing

In plants and a few animals, miRNAs can direct the cleavage of their target mRNAs if they are perfectly matched. However, most miRNAs in animals function as translational repressors. Once a miRNA recognizes the 3'UTR of its target mRNA, miRISC interferes with gene expression by several mechanisms. Evidence that miRISC regulates gene expression by accelerating mRNA degradation, repressing protein synthesis, and sequestering mRNA to storage compartments suggests several models for miRNA function (Fig. 17.2).

The first model suggests a post-initiation repression mechanism, where miRISC may attenuate or stop the elongation of translation. This hypothesis is supported by evidence that lin-4 miRNA downregulates *lin-14*-encoded protein in *C. elegans* larva without changing lin-14 mRNA levels (Olsen and Ambros 1999). Analysis of mRNA polysome profiles revealed that repressed lin-14 mRNA was in polysome fractions, suggesting miRNA mediates protein expression after initiation of translation. Similar results were observed in mammalian cells transfected to express reporter constructs containing miRNA target sites (Nottrott et al. 2006). Further support for a post-initiation repression mechanism came from studies of human cells, using a polycistronic reporter construct with both cap- and internal ribosome entry site (IRES)-dependent reporter genes showing both types of genes were associated with polysomes and can be repressed by miRNA and (Petersen et al. 2006). Finally, polysome-profile studies in HeLa cells have shown that miRNA, together with its target mRNAs, are found in polysome fractions of cell lysates, indicating miRNAs interact with translating mRNAs (Maroney et al. 2006). These findings suggest that miRNA blocks translation after the initiation step, but details of how miRNA induces translation repression is still unclear. One possible explanation could be after ribosomes are released from mRNAs, transcripts associated with miRISC are relocated to mRNA processing bodies (P-bodies) for storage or decay (see below).

The second hypothesis for miRNA-mediated translation repression is that miRISC functions at the initiation step. This hypothesis was first proposed in studies of human cells where translation initiation was inhibited by endogenous let-7 miRNA or by Ago2 tethered at the 3'UTR of reporter mRNA (Pillai et al. 2005). In contrast to data from other groups, this report showed cap-independent translation by IRES resists miRNA-mediated gene silencing, suggesting miRNA represses translation at initiation stage. Recently, in vitro recapitulations of miRNA-mediated repression in cell-free systems including rabbit reticulocyte lysates (Wang et al. 2006), *Drosophila* embryonic lysates (Thermann and Hentze 2007), human 293 cell lysates (Wakiyama et al. 2007), and mouse Krebs-2 ascites lysates (Mathonnet et al. 2007) revealed that miRNA function is mediated by the 5′ cap structure of target mRNA. Target mRNAs with an ApppG cap structure instead of a normal m7GpppG were resistant to miR-2-mediated repression in flies or *let-7*-mediated repression in mammals. In another approach, affinity purified TRBP was used to show eIF6 complexes with TRBP in human miRISC (Chendrimada et al. 2007). Further, depleting eIF6 abolished miRNA-mediated gene silencing in human cells and lin-4-mediated translation repression of lin-14 and lin-28 in *C. elegans* (Chendrimada et al. 2007). Given that eIF6 is a 60S ribosome-interacting protein which blocks the assembly of 80S ribosome during translational initiation, involvement of eIF6 in miRISC function provides further evidence for miRNA repressing translation by a mechanism that represses translation initiation.

Other support for this mechanism comes from recent studies indicating human Ago2 directly interacts with the mRNA cap structure and contains a cap-binding motif (MC) similar to that of eIF4E, a translation initiation factor that binds to the cap in normal translation initiation (Kiriakidou et al. 2007). Cap-binding efficacy was shown to be regulated by two phenylalanine residues, F470 and F505, in the MC domain of Ago2. Mutations at these two amino acid residues abolished the specific binding of Ago2 to m7GTP-sepharose and its function in translation repression, without affecting miRISC assembly and catalytic activity of Ago2 (Kiriakidou et al. 2007). These findings, are consistent with those from in vitro miRISC recapitulations. They support a model in which Ago2 competes with eIF4E for cap binding and miRISC functions at the initiation stage. Interestingly, sequence alignment and analysis of Argonaute proteins in flies reveal that Ago1 contains similar cap-binding domain, while Ago2 does not. This result may explain the distinct functions of *Drosophila* Ago1 and Ago2 in two separate (miRNA and siRNA) pathways.

A third mechanism for miRISC-mediated translation repression implies destabilization of target mRNA. In some cases, degradation rates of target mRNAs increase with miRNA-mediated gene repression, suggesting miRNAs may trigger target destabilization by a deadenylation mechanism (Bagga et al. 2005; Behm-Ansmant et al. 2006; Wu et al. 2006). These miRNA-regulated mRNAs may also translocate to P-bodies for accelerated decay.

P-bodies (also known as GW or DCP bodies) contain non-translating mRNAs as well as proteins involved in mRNA remodeling, translation repression, decapping, and 5′ to 3′ exonuclease activity (reviewed in Eulalio et al. 2007a). MiRISC and target mRNA have also been found in P-bodies, suggesting they serve as places for storage or decay of target mRNA in the miRNA-mediated gene-silencing pathway (Sen and Blau 2005; Liu et al. 2005b). While several P-body components, like GW182 and RCK/p54, function in translational repression by miRISC (Chu and Rana 2006; Jakymiw et al. 2005; Liu et al. 2005a; Rehwinkel et al. 2005), consequence of P-body structures for miRNA function is somewhat elusive.

The functions of siRISC and miRISC are not affected, although Ago2 is dispersed throughout the cytoplasm, after P-bodies are disrupted by depleting Lsm1, a P-body scaffold protein (Chu and Rana 2006). This evidence, together with similar results (Eulalio et al. 2007b) suggests, localization of miRISC and target mRNA in P-bodies is not a prerequisite or trigger for miRNA-mediated silencing; rather, it is a consequence of miRNA function. The functional significance of P-bodies in miRNA-mediated gene silencing is that they provide an environment with high concentrations of enzymes for RNA decay, like the CCR4/NOT deadenylase, DCP1/DCP2 decapping complex, and general translational repressors (eIF4ET, RCK/p54). Although mRNA starts being degraded before it enters P-bodies, its relocation to P-bodies may accelerate this decay in processes downstream of miRNA functions. Another function of P-bodies in translational repression is to sequester mRNA. Some miRNA-controlled mRNAs can be released from P-bodies in response to stress conditions (Bhattacharyya et al. 2006), suggesting repression by miRNA is a dynamic and reversible phenomenon in the cytoplasm and miRNA-repressed mRNAs are capable of resuming active translation.

In summary, experimental evidence to date supports at least three models for miRNA-mediated gene silencing. The complexity of miRISC structure and function in regulating cellular physiology, make these miRNA-mediated mechanisms act simultaneously or synergistically to repress translation. Given that relatively few examples of miRNAs have been studied, other cofactors or conditions may likely modulate miRNA function and remain to be uncovered in the future.

17.4 Piwi-Interacting RNAs

17.4.1 Germline-Specific Argonaute Proteins and Their Associated Small RNAs

The Argonaute protein family has many members that can be divided according to sequence homology into several subsets: the Argonaute subfamily, the Piwi subfamily, and the *C. elegans*-specific subfamily. As described earlier, the core effector for miRISC or siRISC belongs to the Argonaute subfamily -Ago1 and Ago2 in *Drosophila* and Ago1–4 in humans.

The second subset of Argonaute proteins, Piwi proteins, is expressed in certain animal tissues. Unlike the ubiquitously expressed subfamily of Argonaute proteins, Piwi proteins are found in germline cells of animals. These Piwi proteins in different organisms share some functional similarities. Genetic studies in flies (Cox et al. 1998), zebrafish (Houwing et al. 2007) and mice (Carmell et al. 2007; Deng and Lin 2002; Kuramochi-Miyagawa et al. 2004) have shown Piwi proteins are involved in germline development. In flies, the Piwi subfamily includes Ago3, Aubergine, and Piwi, whereas the mouse Piwi subfamily contains MIWI, MILI, and MIWI2.

Like other members of the Argonaute family, Piwi proteins also bind to small RNAs. Piwi proteins have recently been shown to associate with endogenous small 24–30 nt RNAs, called PIWI-interacting RNAs (piRNAs) (reviewed in Hartig et al.

2007). Sequence and structure analysis of cloned piRNAs from immuno-purified Piwi proteins indicate several interesting features of piRNA, distinct from those of miRNA. (1) the 5′ end of piRNAs contains a high percentage of uridine nucleotides. (2) genomic mapping of piRNAs in fly and mouse has shown piRNA genes are concentrated at several genomic loci; adjacent clusters of piRNA reveal the same orientation with a strand bias. (3) the 3′ end of piRNAs is 2′-*O*-methylated.

17.4.2 Biogenesis of piRNAs

Since the 24–30 nt piRNAs are longer than miRNAs, piRNA biosynthesis is believed to be Dicer-independent. This possibility is supported by genetic studies in flies (Vagin et al. 2006) and zebrafish (Houwing et al. 2007) suggesting Dicer processing is not involved in piRNA biogenesis. Insight into the mechanism by which Piwi proteins control piRNA production comes from large-scale sequencing of piRNAs isolated from purified Piwi protein. Genetic mapping of these sequenced piRNAs indicated that piRNA genes are located in several pericentromeric or telomeric heterochomatin loci containing retrotransposons. Strand bias is found in piRNAs from different Piwi proteins. For example, the fly Aub- and Piwi-associated piRNAs have sequences similar to anti-sense strand of retrotransposons, whereas Ago3-bound piRNAs have sequences similar to the sense strand (Brennecke et al. 2007; Gunawardane et al. 2007). Interestingly, the first ten nucleotides of sense-strand piRNAs associated with Ago3 complement the first ten nucleotides of Aub-interacting antisense-strand piRNAs, suggesting a unique mechanism is involved in producing two piRNA populations (see below). Additionally, the Aub- and Piwi-bound piRNAs contain high percentage of uridine at their 5′ end, whereas the Ago3-bound piRNAs contain a high percentage of adenine at the tenth nucleotide from the 5′ end (Brennecke et al. 2007; Gunawardane et al. 2007).

Given all Piwi proteins exhibit cleavage activities, a unique amplification loop has been proposed for piRNA biogenesis, in which each piRNA-mediated cleavage creates the 5′ end of a new piRNA (Brennecke et al. 2007; Fig. 17.3). Long transcripts of RNA are cleaved by Ago3 associated with sense-strand piRNA, forming the 5′ end of antisense piRNAs. This 5′-end product of Ago3 cleavage is subsequently bound to Aub or Piwi and cleaved at ~24–30 nt from the 5′ end to generate the antisense strand of piRNAs. The Aub or Piwi complexed with antisense-strand piRNA then functions as a slicer to bind and cleave sense-strand RNAs, resulting in the 5′ end of sense strand RNA coupled with Ago3 protein. This 5′-end processing is followed by processing at the 3′ end, thus generating more mature sense-strand piRNAs (Brennecke et al. 2007).

While this model explains the amplification of both piRNA strands, it is unclear how the 3′ end is cleaved. Mutation of two *Drosophila* genes, *zucchini* and *squash*, results in loss of piRNAs, suggesting the encoded putative nucleases, Zucchini and Squash, may be involved in processing the 3′ end during piRNA maturation (Pane et al. 2007).

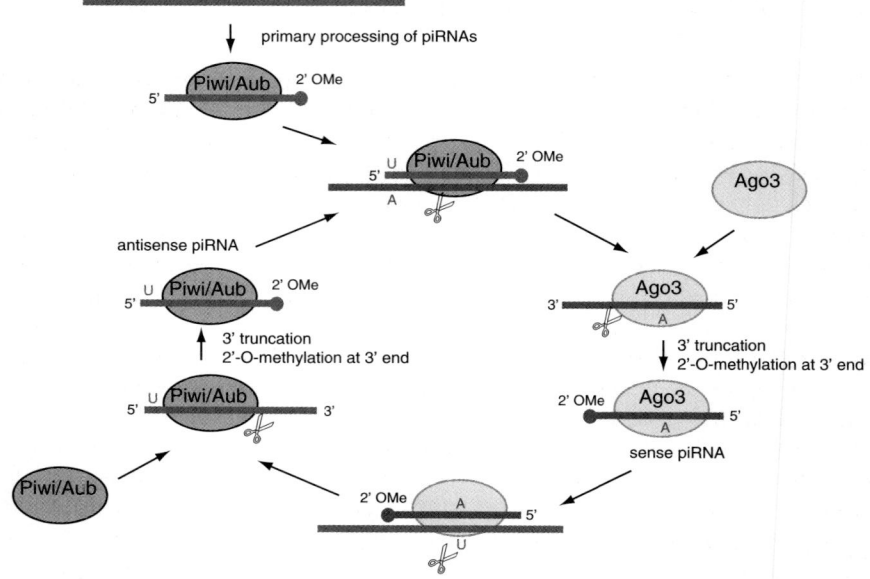

Fig. 17.3 Model for piRNA biogenesis. Primary piRNA transcripts, such as transposon regulatory regions of heterochromatin, are incorporated into Piwi and Aub and trigger the piRNA amplification loop. Piwi and Aub bind anti-sense strand piRNA and cleave target transcript to generate the 5′ end of sense strand piRNA; this is subsequently associated with Ago3. After 3′ end cleavage and 3′-end 2′-O-methylation, the Ago3-bound sense strand complex with piRNA functions as another slicer, which recognizes and cleaves the anti-sense strand precursor, to generate more Piwi/Aub-associated anti-sense strand piRNAs (See figure insert for color reproduction)

17.4.3   3′-End Modification of piRNAs

As mentioned earlier, one feature of piRNA in the mouse, fly and zebrafish is that its 3′ end is 2′-O-methylated (Kirino and Mourelatos 2007; Ohara et al. 2007). This 2′-O-methylation of 3′ termini is also found in plant miRNAs and *Drosophila* siR-NAs. The 3′ termini of fly piRNAs and siRNAs are methylated by a single-strand-specific methyltransferase, Pimet/DmHen1 (Horwich et al. 2007; Saito et al. 2007). Pimet/DmHen1 has been shown to interact with Aub, Piwi, and Ago3, but not with Ago1, suggesting its role in piRNA maturation. Mutating Pimet/DmHen1 results in slightly shortened piRNAs, indicating that 2′-O-methylation can protect piRNAs and siRNAs from attack by other exonucleases and interfere with the interaction between piRNAs and other proteins. Nonetheless, genetic evidence that *pimet/hen1*-mutant flies are fertile and viable (Saito et al. 2007) indicates the 3′-end modification may not be crucial for piRNA function.

17.4.4 Biological Functions of piRNAs

Function of piRNAs in germline development is supported by evidence from genetic studies in mice and flies. Mouse Piwi proteins, including MIWI, MILI and MIWI2, are expressed specifically in male testis at different stages of spermatogenesis (Carmell et al. 2007; Kuramochi-Miyagawa et al. 2001). Knockout mutations of MIWI, MILI, and MIWI2 result in the arrest of spermatogenesis at different stages. In MILI-deficient mice, spermatogenesis is arrested at the early pachytene stage of spermatocytes, while in MIWI-knockout, mice spermatogenesis is stopped at the round spermatid stage. Loss of germ cell phenotype is observed in all Piwi-deficient mice and is accompanied by derepression of transposable elements, suggesting the MILI-, MIWI-, and MIWI2-associated piRNAs are involved in maintaining germline stem cells as well as transposon control (Aravin et al. 2007; Carmell et al. 2007).

Biological function of piRNAs has also been elucidated in the *Drosophila* system. piRNAs were first identified in the fruit fly as repeat-associated RNAs (rasiRNAs) which silence genomic tandem repeats and transposons (Aravin et al. 2001) as depleting all proteins involved in piRNA production induced a significant overexpression of retrotransposons (Aravin et al. 2001; Vagin et al. 2006). Both the biogenesis of piRNAs and piRNA mechanisms has been implicated in repression of selfish genetic elements such as transposons (Brennecke et al. 2007). One discrete genomic locus of piRNA clusters, *flamenco*, has been shown to control activity of *gypsy*, *Idefix*, and *ZAM* retrotransposons. Mutating the *flamenco* locus leads to decreased levels of mature piRNA (Brennecke et al. 2007), and mutating *piwi* results in increased *gypsy* RNA levels (Sarot et al. 2004), suggesting Piwi proteins and piRNAs are required to silence retrotransposons and that this process is coupled with the biogenesis of piRNAs. However, the detailed mechanisms of piRNAs are still unknown.

In addition to repressing genetic elements, Piwi and piRNAs have recently been implicated in playing roles in *Drosophila* activation of heterochromatin (Yin and Lin 2007). Localization studies showed that Piwi, unlike other Piwi family proteins, is mainly distributed in the nuclei of fly oocytes, suggesting the nucleus is the functional compartment for piRNAs bound to Piwi (Brennecke et al. 2007; Saito et al. 2006). Piwi protein has been shown to promote modifications in euchromatic histones and enhance piRNA transcription in the heterochromatin of subtelomeric chromosome regions (Yin and Lin 2007). Depleting Piwi results in the loss of euchromatic histone modification and accumulation of heterochromatin, indicating Piwi can modify heterochromatin to become euchromatic and increase its transcription (Yin and Lin 2007). This finding, which contradicts the known role of piRNAs in gene silencing, indicates small RNAs may achieve epigenetic control through multiple mechanisms, either silencing or activating target genes, at the transcriptional or post-transcription level of gene expression.

17.5   Concluding Remarks

Though RNAi phenomenon was discovered only 10 years ago, studies have provided a wealth of clues to uncover the mystery of RNAi and revealed salient roles of small RNAs in regulating gene expression. Structural and functional analyses of RISC components and small RNAs have highlighted the complexity of this cellular regulatory network. However, several questions remain to be answered regarding the details of the silencing mechanism and the therapeutic application of RNAi. How does miRNA mediate gene silencing within the three existing models for translation repression? Do other factors facilitate RISC function? Can some small RNAs activate gene expression rather than silence it? How can the potency and specificity of in vivo RNAi be optimized for application to RNAi-based therapies? Future genetic and biochemical work on these topics will help understand diverse mechanisms of gene regulation by small RNAs. With recent fascinating and rapid progress in studies of small RNA biology, researchers can look forward to applying RNAi as a powerful research tool and therapeutic strategy.

References

Amarzguioui M, Prydz H (2004) An algorithm for selection of functional siRNA sequences. Biochem Biophys Res Commun 316:1050–1058

Amarzguioui M, Holen T, Babaie E et al. (2003) Tolerance for mutations and chemical modifications in a siRNA. Nucleic Acids Res 31:589–595

Ameres SL, Martinez J, Schroeder R (2007) Molecular basis for target RNA recognition and cleavage by human RISC. Cell 130:101–112

Aravin AA, Naumova NM, Tulin AV et al. (2001) Double-stranded RNA-mediated silencing of genomic tandem repeats and transposable elements in the D. melanogaster germline. Curr Biol 11:1017–1027

Aravin AA, Sachidanandam R, Girard A et al. (2007) Developmentally regulated piRNA clusters implicate MILI in transposon control. Science 316:744–747

Bagga S, Bracht J, Hunter S et al. (2005) Regulation by let-7 and lin-4 miRNAs results in target mRNA degradation. Cell 122:553–563

Bartel DP (2004) MicroRNAs: genomics, biogenesis, mechanism, and function. Cell 116:281–297

Behm-Ansmant I, Rehwinkel J, Doerks T et al. (2006) mRNA degradation by miRNAs and GW182 requires both CCR4:NOT deadenylase and DCP1:DCP2 decapping complexes. Genes Dev 20:1885–1898

Bhattacharyya SN, Habermacher R, Martine U et al. (2006) Relief of microRNA-mediated translational repression in human cells subjected to stress. Cell 125:1111–1124

Birmingham A, Anderson E, Sullivan K et al. (2007) A protocol for designing siRNAs with high functionality and specificity. Nat Protoc 2:2068–2078

Bohnsack MT, Regener K, Schwappach B et al. (2002) Exp5 exports eEF1A via tRNA from nuclei and synergizes with other transport pathways to confine translation to the cytoplasm. EMBO J 21:6205–6215

Bohnsack MT, Czaplinski K, Gorlich D (2004) Exportin 5 is a RanGTP-dependent dsRNA-binding protein that mediates nuclear export of pre-miRNAs. RNA 10:185–191

Braasch DA, Jensen S, Liu Y et al. (2003) RNA interference in mammalian cells by chemically-modified RNA. Biochemistry 42:7967–7975

Brennecke J, Stark A, Russell RB et al. (2005) Principles of microRNA-target recognition. PLoS Biol 3:e85

Brennecke J, Aravin AA, Stark A et al. (2007) Discrete small RNA-generating loci as master regulators of transposon activity in Drosophila. Cell 128:1089–1103

Brown KM, Chu CY, Rana TM (2005) Target accessibility dictates the potency of human RISC. Nat Struct Mol Biol 12:469–470

Carmell MA, Hannon GJ (2004) RNase III enzymes and the initiation of gene silencing. Nat Struct Mol Biol 11:214–218

Carmell MA, Girard A, van de Kant HJ et al. (2007) MIWI2 is essential for spermatogenesis and repression of transposons in the mouse male germline. Dev Cell 12:503–514

Chendrimada TP, Gregory RI, Kumaraswamy E et al. (2005) TRBP recruits the Dicer complex to Ago2 for microRNA processing and gene silencing. Nature 436:740–744

Chendrimada TP, Finn KJ, Ji X et al. (2007) MicroRNA silencing through RISC recruitment of eIF6. Nature 447(7146):823–828

Chiu YL, Rana TM (2002) RNAi in human cells: basic structural and functional features of small interfering RNA. Mol Cell 10:549–561

Chiu YL, Rana TM (2003) siRNA function in RNAi: a chemical modification analysis. RNA 9:1034–1048

Chu CY, Rana TM (2006) Translation repression in human cells by microRNA-induced gene silencing requires RCK/p54. PLoS Biol 4:e210

Cox DN, Chao A, Baker J et al. (1998) A novel class of evolutionarily conserved genes defined by piwi are essential for stem cell self-renewal. Genes Dev 12:3715–3727

Czauderna F, Fechtner M, Dames S et al. (2003) Structural variations and stabilising modifications of synthetic siRNAs in mammalian cells. Nucleic Acids Res 31:2705–2716

Deng W, Lin H (2002) miwi, a murine homolog of piwi, encodes a cytoplasmic protein essential for spermatogenesis. Dev Cell 2:819–830

Doench JG, Sharp PA (2004) Specificity of microRNA target selection in translational repression. Genes Dev 18:504–511

Elbashir SM, Lendeckel W, Tuschl T (2001) RNA interference is mediated by 21- and 22-nucleotide RNAs. Genes Dev 15:188–200

Enright AJ, John B, Gaul U et al. (2003) MicroRNA targets in Drosophila. Genome Biol 5:R1

Eulalio A, Behm-Ansmant I, Izaurralde E (2007a) P bodies: at the crossroads of post-transcriptional pathways. Nat Rev Mol Cell Biol 8:9–22

Eulalio A, Behm-Ansmant I, Schweizer D et al. (2007b) P-body formation is a consequence, not the cause, of RNA-mediated gene silencing. Mol Cell Biol 27:3970–3981

Fire A, Xu S, Montgomery MK et al. (1998) Potent and specific genetic interference by double-stranded RNA in *Caenorhabditis elegans*. Nature 391:806–811

Förstemann K, Horwich MD, Wee L et al. (2007) Drosophila microRNAs are sorted into functionally distinct argonaute complexes after production by Dicer-1. Cell 130:287–297

Forstemann K, Tomari Y, Du T et al. (2005) Normal microRNA maturation and germ-line stem cell maintenance requires Loquacious, a double-stranded RNA-binding domain protein. PLoS Biol 3:e236

Grimson A, Farh KK, Johnston WK et al. (2007) MicroRNA targeting specificity in mammals: determinants beyond seed pairing. Mol Cell 27:91–105

Grun D, Wang YL, Langenberger D et al. (2005) microRNA target predictions across seven Drosophila species and comparison to mammalian targets. PLoS Comput Biol 1:e13

Gunawardane LS, Saito K, Nishida KM et al. (2007) A slicer-mediated mechanism for repeat-associated siRNA 5′ end formation in Drosophila. Science 315:1587–1590

Hartig JV, Tomari Y, Forstemann K (2007) piRNAs – the ancient hunters of genome invaders. Genes Dev 21:1707–1713

Horwich MD, Li C, Matranga C et al. (2007) The Drosophila RNA methyltransferase, DmHen1, modifies germline piRNAs and single-stranded siRNAs in RISC. Curr Biol 17:1265–1272

Houwing S, Kamminga LM, Berezikov E et al. (2007) A role for Piwi and piRNAs in germ cell maintenance and transposon silencing in Zebrafish. Cell 129:69–82

Jackson AL, Burchard J, Leake D et al. (2006) Position-specific chemical modification of siRNAs reduces "off-target" transcript silencing. RNA 12:1197–1205

Jakymiw A, Lian S, Eystathioy T et al. (2005) Disruption of GW bodies impairs mammalian RNA interference. Nat Cell Biol 7:1267–1274

John B, Enright AJ, Aravin A et al. (2004) Human microRNA targets. PLoS Biol 2:e363

Khvorova A, Reynolds A, Jayasena SD (2003) Functional siRNAs and miRNAs exhibit strand bias. Cell 115:209–216

Kim VN (2005) MicroRNA biogenesis: coordinated cropping and dicing. Nat Rev Mol Cell Biol 6:376–385

Kim DH, Behlke MA, Rose SD et al. (2005) Synthetic dsRNA Dicer substrates enhance RNAi potency and efficacy. Nat Biotechnol 23:222–226

Kiriakidou M, Nelson PT, Kouranov A et al. (2004) A combined computational-experimental approach predicts human microRNA targets. Genes Dev 18:1165–1178

Kiriakidou M, Tan GS, Lamprinaki S et al. (2007) An mRNA m(7)G Cap Binding-like Motif within Human Ago2 Represses Translation. Cell 129:1141–1151

Kirino Y, Mourelatos Z (2007) Mouse Piwi-interacting RNAs are 2′-O-methylated at their 3′ termini. Nat Struct Mol Biol 14:347–348

Krek A, Grun D, Poy MN et al. (2005) Combinatorial microRNA target predictions. Nat Genet 37:495–500

Kuramochi-Miyagawa S, Kimura T, Yomogida K et al. (2001) Two mouse piwi-related genes: miwi and mili. Mech Dev 108:121–133

Kuramochi-Miyagawa S, Kimura T, Ijiri TW et al. (2004) Mili, a mammalian member of piwi family gene, is essential for spermatogenesis. Development 131:839–849

Lall S, Grun D, Krek A et al. (2006) A genome-wide map of conserved microRNA targets in C. elegans. Curr Biol 16:460–471

Lee RC, Feinbaum RL, Ambros V (1993) The C. elegans heterochronic gene lin-4 encodes small RNAs with antisense complementarity to lin-14. Cell 75:843–854

Lee YS, Nakahara K, Pham JW et al. (2004) Distinct roles for Drosophila Dicer-1 and Dicer-2 in the siRNA/miRNA silencing pathways. Cell 117:69–81

Lee Y, Hur I, Park SY et al. (2006) The role of PACT in the RNA silencing pathway. EMBO J 25:522–532

Lewis BP, Shih IH, Jones-Rhoades MW et al. (2003) Prediction of mammalian microRNA targets. Cell 115:787–798

Lewis BP, Burge CB, Bartel DP (2005) Conserved seed pairing, often flanked by adenosines, indicates that thousands of human genes are microRNA targets. Cell 120:15–20

Liu J, Carmell MA, Rivas FV et al. (2004) Argonaute2 is the catalytic engine of mammalian RNAi. Science 305:1437–1441

Liu J, Rivas FV, Wohlschlegel J et al. (2005a) A role for the P-body component GW182 in microRNA function. Nat Cell Biol 7:1261–1266

Liu J, Valencia-Sanchez MA, Hannon GJ et al. (2005b) MicroRNA-dependent localization of targeted mRNAs to mammalian P-bodies. Nat Cell Biol 7:719–723

Lund E, Guttinger S, Calado A et al. (2004) Nuclear export of microRNA precursors. Science 303:95–98

Ma JB, Yuan YR, Meister G et al. (2005) Structural basis for 5′-end-specific recognition of guide RNA by the A. fulgidus Piwi protein. Nature 434:666–670

Maroney PA, Yu Y, Fisher J et al. (2006) Evidence that microRNAs are associated with translating messenger RNAs in human cells. Nat Struct Mol Biol 13:1102–1107

Martinez J, Tuschl T (2004) RISC is a 5′ phosphomonoester-producing RNA endonuclease. Genes Dev 18:975–980

Mathonnet G, Fabian MR, Svitkin YV et al. (2007) MicroRNA inhibition of translation initiation in vitro by targeting the cap-binding complex eIF4F. Science 317(5845):1764–1767

Matranga C, Tomari Y, Shin C et al. (2005) Passenger-strand cleavage facilitates assembly of siRNA into Ago2-containing RNAi enzyme complexes. Cell 123:607–620

Meister G, Landthaler M, Patkaniowska A et al. (2004) Human argonaute2 mediates RNA cleavage targeted by miRNAs and siRNAs. Mol Cell 15:185–197

Mello CC, Conte D (2004) Revealing the world of RNA interference. Nature 431:338–342

Mochizuki K, Fine NA, Fujisawa T et al. (2002) Analysis of a piwi-related gene implicates small RNAs in genome rearrangement in tetrahymena. Cell 110:689–699

Morrissey DV, Lockridge JA, Shaw L et al. (2005) Potent and persistent in vivo anti-HBV activity of chemically modified siRNAs. Nat Biotechnol 23:1002–1007

Nottrott S, Simard MJ, Richter JD (2006) Human let-7a miRNA blocks protein production on actively translating polyribosomes. Nat Struct Mol Biol 13:1108–1114

O'Donnell KA, Boeke JD (2007) Mighty Piwis defend the germline against genome intruders. Cell 129:37–44

Ohara T, Sakaguchi Y, Suzuki T et al. (2007) The 3′ termini of mouse Piwi-interacting RNAs are 2′-O-methylated. Nat Struct Mol Biol 14:349–350

Okamura K, Ishizuka A, Siomi H et al. (2004) Distinct roles for Argonaute proteins in small RNA-directed RNA cleavage pathways. Genes Dev 18:1655–1666

Okamura K, Hagen JW, Duan H et al. (2007) The Mirtron pathway generates microRNA-class regulatory RNAs in Drosophila. Cell 130:89–100

Olsen PH. Ambros V (1999) The lin-4 regulatory RNA controls developmental timing in *Caenorhabditis elegans* by blocking LIN-14 protein synthesis after the initiation of translation. Dev Biol 216:671–680

Pane A, Wehr K, Schupbach T (2007) zucchini and squash encode two putative nucleases required for rasiRNA production in the Drosophila germline. Dev Cell 12:851–862

Peragine A, Yoshikawa M, Wu G et al. (2004) SGS3 and SGS2/SDE1/RDR6 are required for juvenile development and the production of trans-acting siRNAs in Arabidopsis. Genes Dev 18:2368–2379

Petersen CP, Bordeleau ME, Pelletier J et al. (2006) Short RNAs repress translation after initiation in mammalian cells. Mol Cell 21:533–542

Pillai RS, Bhattacharyya SN, Artus CG et al. (2005) Inhibition of translational initiation by Let-7 MicroRNA in human cells. Science 309:1573–1576

Prakash TP, Allerson CR, Dande P et al. (2005) Positional effect of chemical modifications on short interference RNA activity in mammalian cells. J Med Chem 48:4247–4253

Rana TM (2007) Illuminating the silence: understanding the structure and function of small RNAs. Nat Rev Mol Cell Biol 8:23–36

Rand TA, Petersen S, Du F et al. (2005) Argonaute2 cleaves the anti-guide strand of siRNA during RISC activation. Cell 123:621–629

Rehmsmeier M, Steffen P, Hochsmann M et al. (2004) Fast and effective prediction of microRNA/target duplexes. RNA 10:1507–1517

Rehwinkel J, Behm-Ansmant I, Gatfield D et al. (2005) A crucial role for GW182 and the DCP1: DCP2 decapping complex in miRNA-mediated gene silencing. RNA 11:1640–1647

Reynolds A, Leake D, Boese Q et al. (2004) Rational siRNA design for RNA interference. Nat Biotechnol 22:326–330

Rivas FV, Tolia NH, Song JJ et al. (2005) Purified Argonaute2 and an siRNA form recombinant human RISC. Nat Struct Mol Biol 12:340–349

Robb GB, Rana TM (2007) RNA helicase A interacts with RISC in human cells and functions in RISC loading. Mol Cell 26:523–537

Ruby JG, Jan CH, Bartel DP (2007) Intronic microRNA precursors that bypass Drosha processing. Nature 448:83–86

Rusinov V, Baev V, Minkov IN et al. (2005) MicroInspector: a web tool for detection of miRNA binding sites in an RNA sequence. Nucleic Acids Res 33:W696–700

Saetrom V, Snove O, Jr., Saetrom P (2005) Weighted sequence motifs as an improved seeding step in microRNA target prediction algorithms. RNA 11:995–1003

Saito K, Ishizuka A, Siomi H et al. (2005) Processing of pre-microRNAs by the Dicer-1-Loquacious complex in Drosophila cells. PLoS Biol 3:e235

Saito K, Nishida KM, Mori T et al. (2006) Specific association of Piwi with rasiRNAs derived from retrotransposon and heterochromatic regions in the Drosophila genome. Genes Dev 20:2214–2222

Saito K, Sakaguchi Y, Suzuki T et al. (2007) Pimet, the Drosophila homolog of HEN1, mediates 2′-O-methylation of Piwi- interacting RNAs at their 3′ ends. Genes Dev 21:1603–1608

Sarot E, Payen-Groschene G, Bucheton A et al. (2004) Evidence for a piwi-dependent RNA silencing of the gypsy endogenous retrovirus by the *Drosophila melanogaster* flamenco gene. Genetics 166:1313–1321

Schwarz DS, Hutvágner G, Du T et al. (2003) Asymmetry in the assembly of the RNAi enzyme complex. Cell 115:199–208

Schwarz DS, Hutvagner G, Haley B et al. (2002) Evidence that siRNAs function as guides, not primers, in the Drosophila and human RNAi pathways. Mol Cell 10:537–548

Sen GL, Blau HM (2005) Argonaute 2/RISC resides in sites of mammalian mRNA decay known as cytoplasmic bodies. Nat Cell Biol 7:633–636

Siolas D, Lerner C, Burchard J et al. (2005) Synthetic shRNAs as potent RNAi triggers. Nat Biotechnol 23:227–231

Taverna SD, Coyne RS, Allis CD (2002) Methylation of histone h3 at lysine 9 targets programmed DNA elimination in tetrahymena. Cell 110:701–711

Thermann R, Hentze MW (2007) Drosophila miR2 induces pseudo-polysomes and inhibits translation initiation. Nature 447:875–878

Tolia NH, Joshua-Tor L (2007) Slicer and the argonautes. Nat Chem Biol 3:36–43

Tomari Y, Du T, Zamore PD (2007) Sorting of Drosophila small silencing RNAs. Cell 130:299–308

Ui-Tei K, Naito Y, Takahashi F et al. (2004) Guidelines for the selection of highly effective siRNA sequences for mammalian and chick RNA interference. Nucleic Acids Res 32:936–948

Vagin VV, Sigova A, Li C et al. (2006) A distinct small RNA pathway silences selfish genetic elements in the germline. Science 313:320–324

Wakiyama M, Takimoto K, Ohara O et al. (2007) Let-7 microRNA-mediated mRNA deadenylation and translational repression in a mammalian cell-free system. Genes Dev 21:1857–1862

Wang B, Love TM, Call ME et al. (2006) Recapitulation of short RNA-directed translational gene silencing in vitro. Mol Cell 22:553–560

Wu L, Fan J, Belasco JG (2006) MicroRNAs direct rapid deadenylation of mRNA. Proc Natl Acad Sci U S A 103:4034–4039

Yi R, Qin Y, Macara IG et al. (2003) Exportin-5 mediates the nuclear export of pre-microRNAs and short hairpin RNAs. Genes Dev 17:3011–3016

Yin H, Lin H (2007) An epigenetic activation role of Piwi and a Piwi-associated piRNA in *Drosophila melanogaster*. Nature 450:304–308

Zeng Y, Cullen BR (2004) Structural requirements for pre-microRNA binding and nuclear export by Exportin 5. Nucleic Acids Res 32:4776–4785

Index

Color Plates

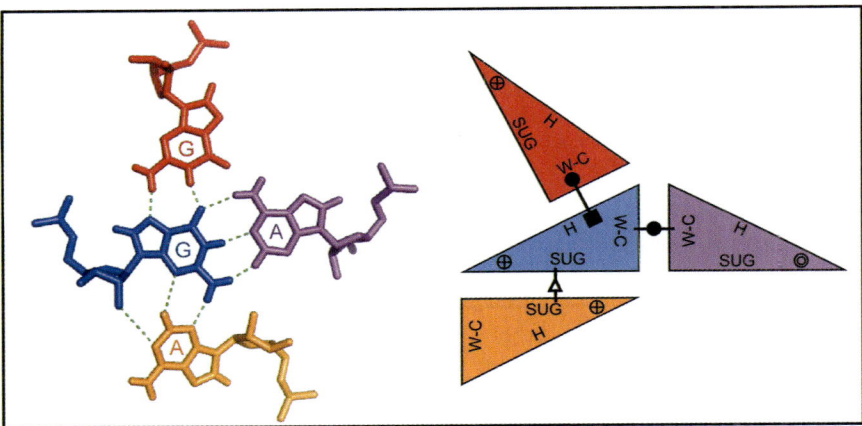

Fig. 1.3 Triangle abstraction for RNA bases. As implied by the triangle abstraction, RNA bases can interact with three different bases using their three edges, Watson–Crick (WC), Hoogsteen (H) and Sugar (Sug), forming "saturated" base quadruples. (*Left*) Example of a base quadruple of this type from *T. thermophilis* 16S rRNA (PDB file 1j5e) in which G68 (blue) forms a cWW pair with A101 (magenta), a cHW pair with G64 (red) and a tsS pair with A152 (yellow). The green dotted lines indicate Hydrogen-bonds. (*Right*) Schematic representation showing each base as a triangle with edges labeled. The base-pairing type is given using the symbols from Fig. 1.2

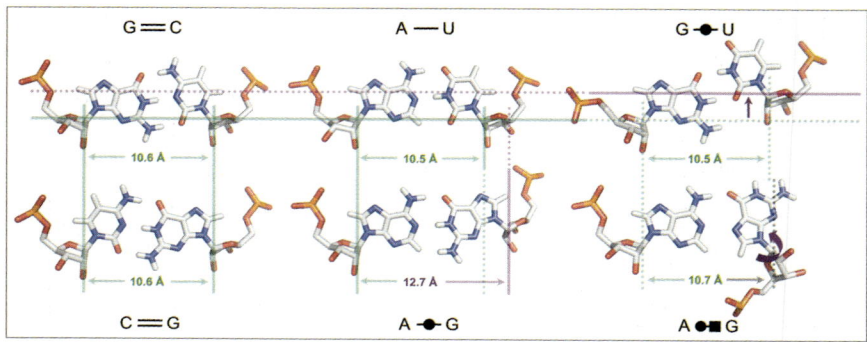

Fig. 1.4 Isosteric relationships between base pairs. Two base pairs are isosteric when they meet three criteria: (1) The C1′–C1′ distances are the same; (2) the paired bases are related by same rotations in 3D space; and (3) H-bonds are formed between equivalent base positions. The cWW GC, CG, and AU base pairs (upper and lower left and upper center) meet all three criteria and are isosteric to each other, as shown. The cWW AG pair (*lower center*) and GU pair (*upper right*) belong to the same geometric family and so the paired bases are related by the same 3D rotation. However, the cWW AG pair has a significantly longer C1′–C1′ distance (12.7 Å) and so is not isosteric to the other pairs, even though it meets the other two criteria. The C1′–C1′ distance in the cWW GU (wobble) pair is about the same, but the U is shifted toward the major groove, so H-bonding does not occur between the same positions as in the other cWW pairs. This change is more subtle and so GU is considered near isosteric to the canonical cWW pairs AU, UA, GC, and CG, consistent with its ability to substitute in Watson–Crick helices for these pairs. The last example, cWH AG (*lower right*), has about the same C1′–C1′ distance as the canonical cWW pairs, but belongs to a different geometric family. The bases are related by a very different 3D rotation so it is not isosteric or near isosteric to any of the cWW base pairs

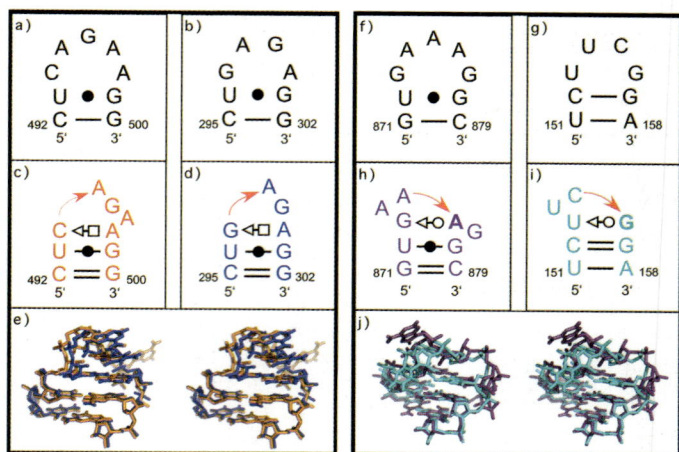

Fig. 1.5 Structurally similar hairpin loop motifs. **a–e** Comparison of sequence and structure annotations of two geometrically similar hairpin loop motifs, only one of which is a tetraloop and conforms to the consensus sequence "GNRA." **f–j** Comparison of sequence and structure annotations of geometrical hairpin loop motifs, where only one is a tetraloop and conforms to the consensus sequence "UNCG." **e** Stereo superpositions of motifs in **c** and **d**. **j** Stereo superpositions of motifs in **h** and **i**

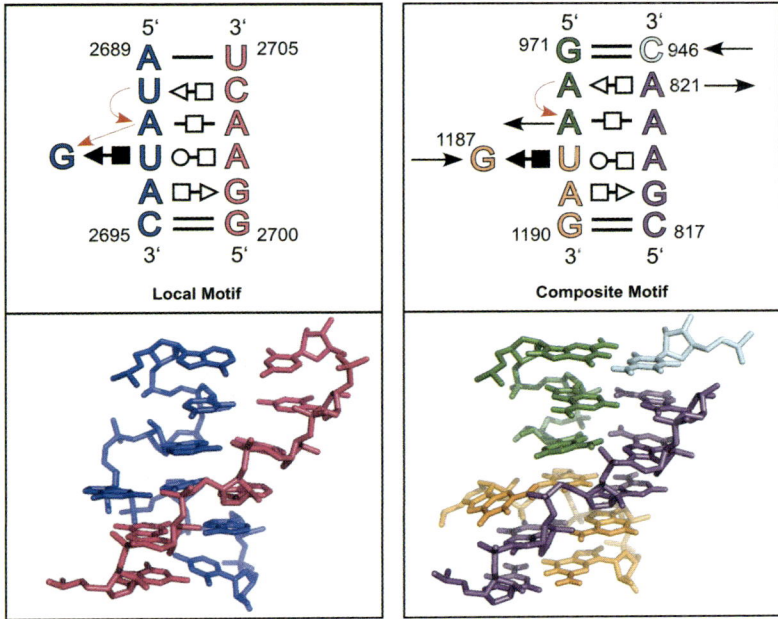

Fig. 1.8 Local vs. Composite motifs. *Left*: Local (internal loop) sarcin/ricin motif from *H. maris-mortui* 23S rRNA comprising two strand segments. *Right*: Composite sarcin/ricin motif from *E. coli* 23S rRNA comprising four different strand segments. The 3D structure of each motif is shown below each annotated diagram (PDB files 1s72 and 2aw4)

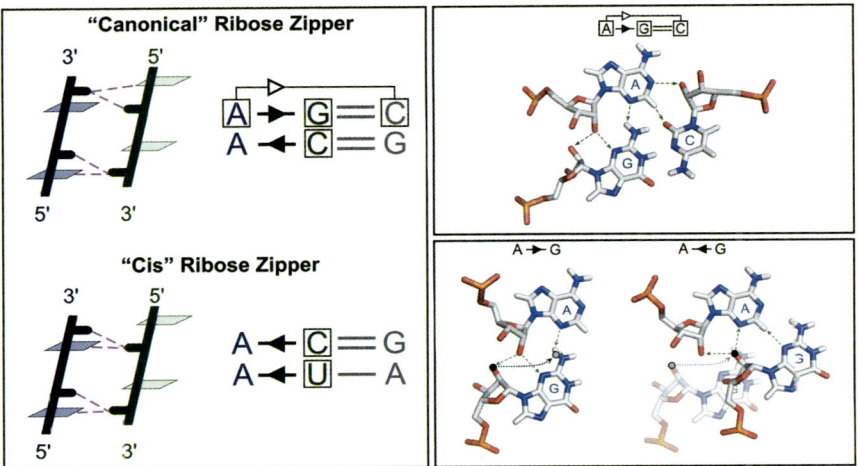

Fig. 1.9 Ribose zippers are tertiary interaction motifs composed of two Sugar-edge base pairs. *Left*: Schematic representation adapted from (Tamura and Holbrook 2002) and base-pair annotation (Leontis and Westhof 2001) of "canonical" and "cis" Ribose Zipper (RZ) tertiary motifs. *Upper Right*: Base triple composed of GC cWW, AG cSs and AC tSs base pairs. The A forms two

Fig. 1.9 (*continued*) pairs with its Sugar edge and is assigned higher priority in each interaction, as explained in the text. *Lower Right*: Comparison of A/G cSs (*left*) and A/G csS (*right*) base pairs. The dotted black arrow indicates the lateral shift that transforms one type into the other. In the A/G cSs, the A is the dominant base so the arrow points from the A to the G. The roles are reversed in the A/G csS pair. The dashed green arrows indicate hydrogen-bonds

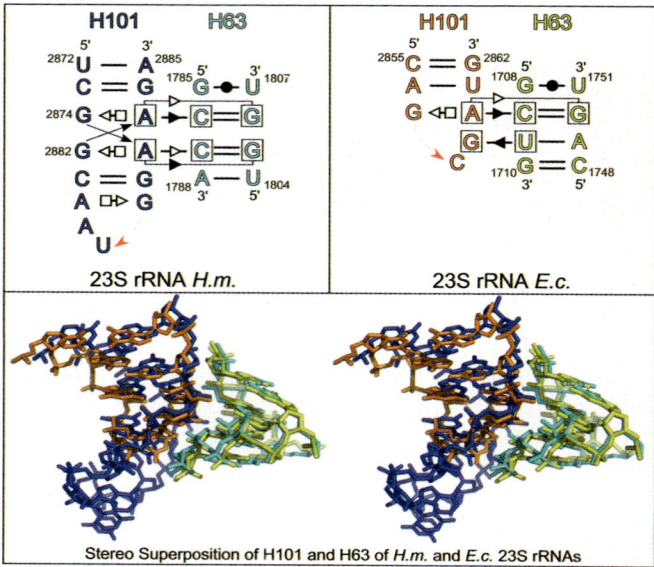

Fig. 1.10 Conserved tertiary interaction in 23S rRNA mediated different motifs. *Upper panels*: Annotated secondary structures of conserved interaction between Helices 101 (H101) and 63 in 23S rRNA of *H. marismortui* (*left*) and *E. coli* (*right*). In 23S of *H. marismortui*, the interaction is mediated by an internal loop in H101 (nucleotides 2,874; 2,875; 2,882; and 2,883), whereas in the *E. coli* structure it is mediated by a GNRA hairpin loop at the equivalent position of H101 (nucleotides 2,857–2,860). *Lower panel*: Stereo superposition of the 3D structures of Helices 101 and 63 from 23S rRNA of *H. marismortui* and *E. coli*. (PDB files 1s72 and 2aw4.) Color coding: *H. marismortui* Helix 101 (blue), Helix 63 (cyan), *E. coli* Helix 101 (orange), Helix 63 (yellow)

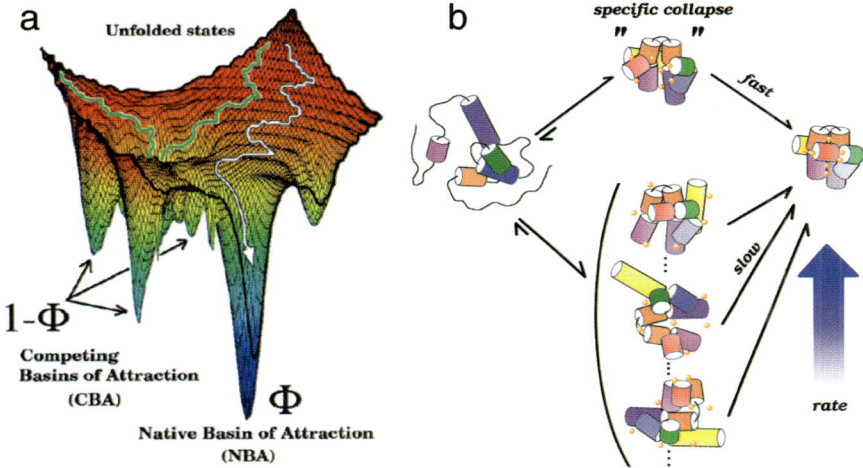

Fig. 2.2 (**a**) Schematic sketch of the rugged folding landscape of RNA. Conformational entropy and electrostatic repulsion between the phosphate groups favor the high free energy unfolded structures at low ionic strength. Under folding conditions a fraction of molecules (Φ) reach the NBA directly. A sketch of a trajectory for a fast track molecule that starts in a region of the energy landscape and which connects directly to the NBA is given in white. Trajectories (shown in green) that begin in other regions of the energy landscape can be kinetically trapped in the CBAs with probability ($1-\Phi$). The low dimensional representation of the complex energy landscape suggests that the initial conditions, which can be changed by counterions, stretching force, or denaturants, can alter the folding pathways. (**b**) Representation of RNA folding by KPM. Based on theory it is suggested that the fast track molecules specifically collapse into near native-like structures that rearrange to the native state without being trapped in the CBA. In contrast, the slow track molecules collapse to one of the manifold of misfolded structures. The collapse time scale, that depends on the nature of ions, for fast and slow track molecules, is similar. A spectrum of rates determine the transition from the CBAs to NBA

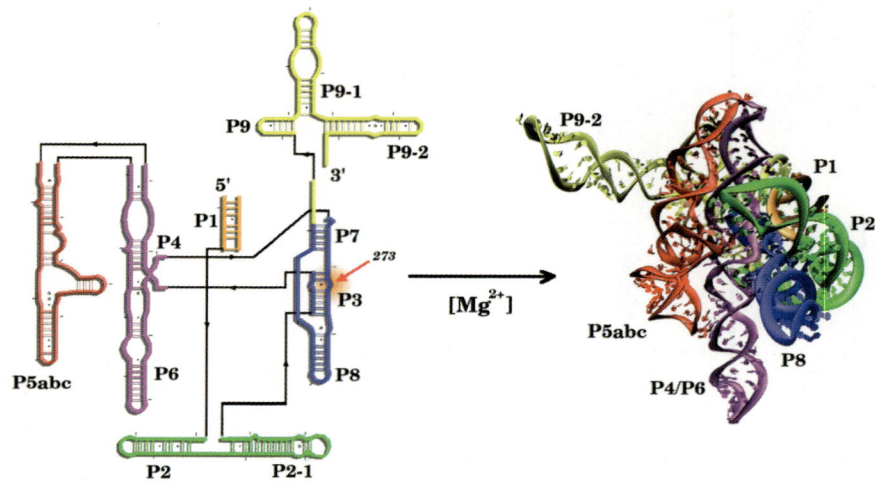

Fig. 2.3 Secondary structure of the most extensively studied group I intron from Tetrahymena. The secondary structure has a number of paired helices indicated by P1 through P9. Upon addition of excess Mg transition to compact tertiary structure, occurs (shown on the right) that is stabilized by the catalytic core formed by an interface involving the P5–P4–P6 and P3–P7–P8 helices. The structure of the independently folding P4–P6 domain is known in atomic detail (Cate et al. 1996). The structure on the right is a model proposed by Westhof and Michel (Lehnert et al. 1996)

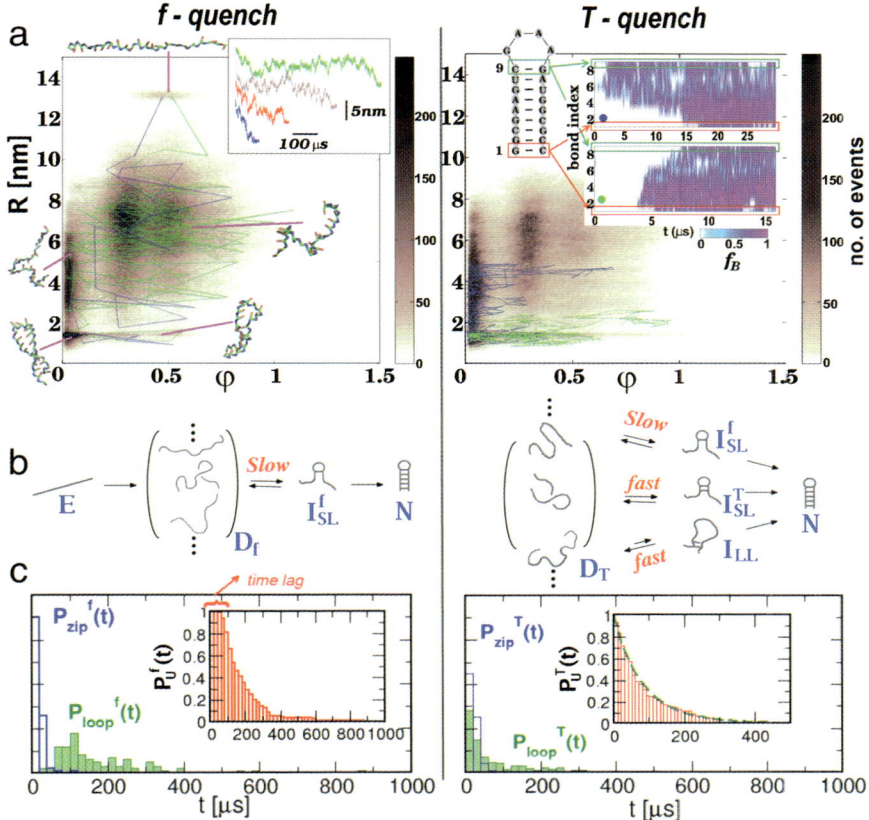

Fig. 2.4 Kinetic analysis of the refolding trajectories upon f-quench and T-quench. (**a**) Conformational space navigated by the refolding trajectories projected onto the (R, φ) plane. The trajectories of individual molecules are overlapped onto the (R, φ) plane. The corresponding trajectories monitored using a single parameter are shown in the insets. (**b**) Summary of the pathways to the NBA inferred from the dynamics depicted in (**a**). (**c**) Statistical analysis of refolding kinetics. The refolding time for each molecule is decomposed into looping and zipping time as $\tau_{FP} = \tau_{loop} + \tau_{zip}$. The fraction of unfolded molecules ($P_u(t)=1-\int_o^t d\tau P_{FP}(\tau)$) where $P_{FP}(\tau)$ is the refolding or first passage time distribution) is plotted in the inset. The probability of the hairpin remaining unfolded upon f-quench $P_u^f(t)$ shows a lag phase (left hand side of C) suggesting the presence of an intermediate, while $P_u^T(t)$ is well fit using $P_u^T(t)=0.4\exp(t/62\,\mu s) + 0.6\exp(-t/100\,\mu s)$

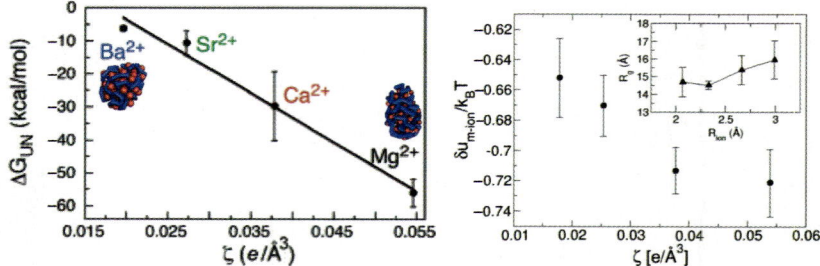

Fig. 2.5 (*Left*) Stability of *Tetrahymena* L-21Sca ribozyme (ΔG_{UF}) vs. cation charge density (ζ). (*Right*) The Brownian Dynamics simulations of polyelectrolyte collapse for the average stabilization energy between monomer of polyelectrolyte and ion as a function of ζ. The remarkable linear dependence on the left is captured by ion-induced collapse of flexible polyelectrolytes. The radius of gyration of the collapsed polyelectrolyte ($N = 120$) is plotted as a function of group II metal ion size in the inset

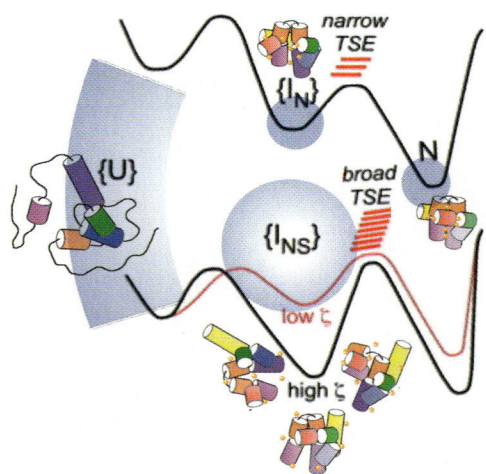

Fig. 2.6 Coupling of diversity of the folding pathways and heterogeneity of the transition state structures of *Tetrahymena* ribozyme to the charge density of counterions (ζ). The majority of the ribozyme folds through intermediates in which the core P3 helix is replaced by a non-native helix alt-P3. The pathway diversity increases as ζ decreases (lower part of the figure). For the fraction $((1-\Phi))$ slow track molecules, the transition state ensemble (TSE) along the $\{U\}\rightarrow\{I_{NS}\}\rightarrow N$ pathway becomes broader and less structured as the ζ decreases. Thus, the system is more dynamic in polyamines (low ζ) than Mg^{2+} (high ζ). The fast track molecules that fold via $\{U\}\rightarrow\{I_N\}\rightarrow N$ (Φ) (upper part of the figure) first form specifically collapsed compact structure that becomes increasingly native-like as the folding reaction proceeds. For $\{U\}\rightarrow\{I_N\}\rightarrow N$ we suggest that the TSE is narrow with little structural heterogeneity. The pathway diversity is expected to increase as ζ decreases

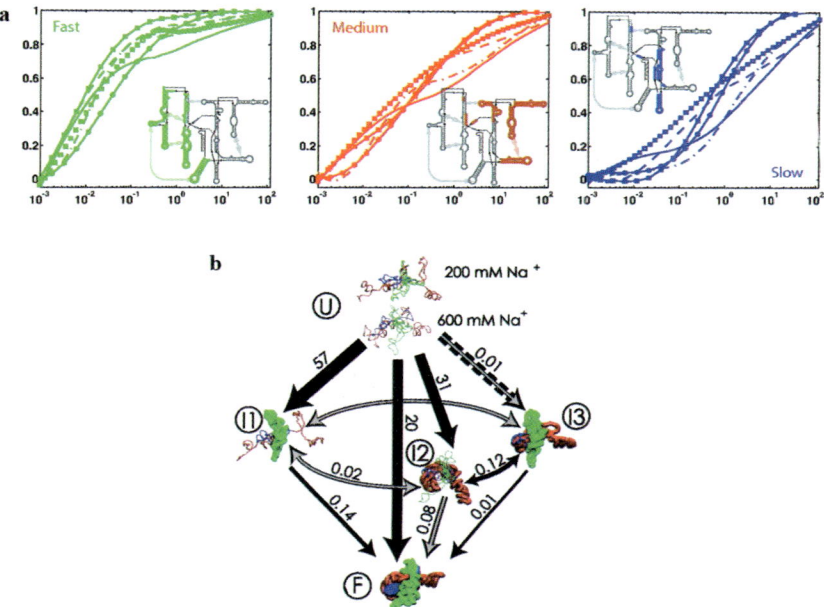

Fig. 3.3 Hydroxyl footprinting shows different local folding of L-21 group I intron ribozyme from Tetrahymena thermophila. (**a**) Fast, medium and slow folding of domains. Each curve represents one ionic condition. (**b**) A multi-pathway kinetic folding model involves three intermediates (I1, I2 and I3) from unfolded (U) to folded (F) state. Adapted from (Laederach et al. 2007)

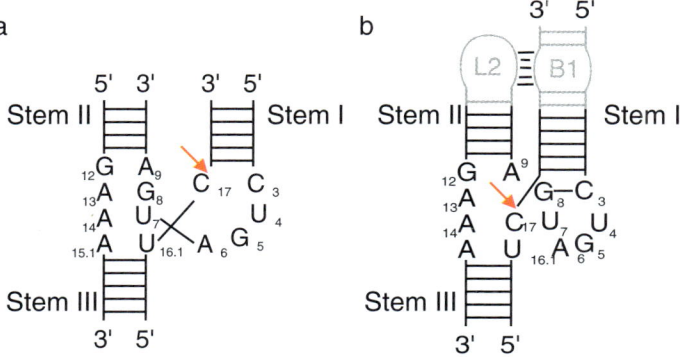

Fig. 4.7 The minimal and full-length hammerhead ribozyme secondary structures. Figure A shows the minimal hammerhead ribozyme sequence, the focus of study between about 1987 and 2003, when it became apparent that an additional tertiary contact having very limited sequence conservation and no readily discernible co- variance is also present in natural hammerhead ribozyme sequence. The presence of the contact can enhance catalysis up to 1,000-fold. The crystal structure of the full-length hammerhead, when compared to that of the minimal hammerhead, as shown in Fig. 4.8, reveals that the presence of the tertiary contact stabilizes an active site conformation, not observed in the minimal hammerhead structures, in which several of the invariant residues shown explicitly in the above figure are arranged to orient the substrate for in-line attack and to position it for acid-base catalysis, as shown in Fig. 4.9

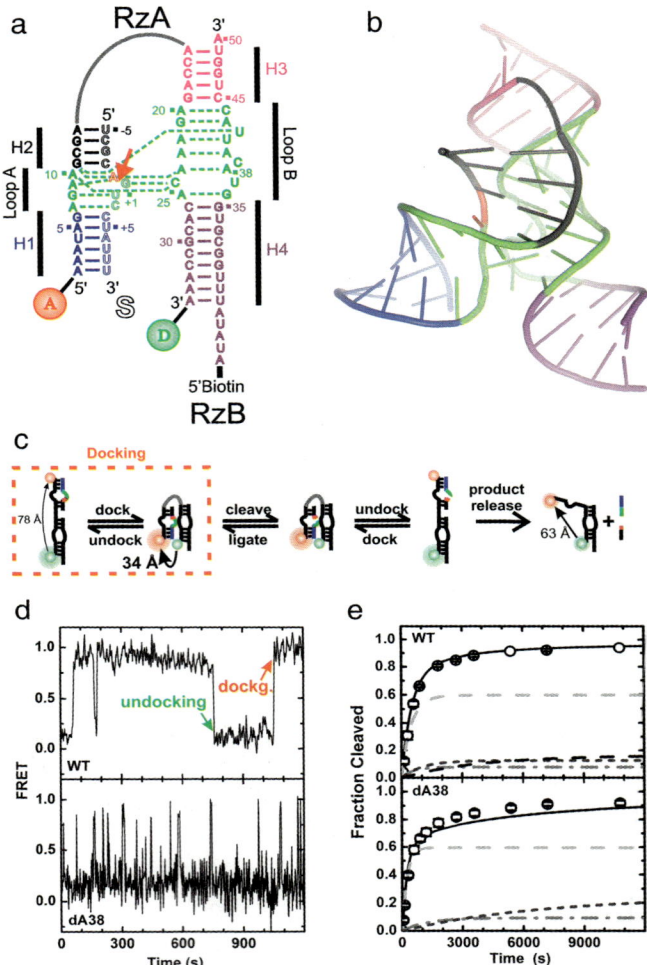

Fig. 5.3 FRET studies of the hairpin ribozyme. (**a**) Secondary structure of the hairpin ribozyme from the tobacco ringspot satellite RNA utilized in these studies. Helices are color coded, the catalytic core is highlighted in green, and the cleavage site in the substrate (outlined, S) is marked by a closed red arrow. The RzA strand is labeled with donor (D) and acceptor (A) fluorophores, as indicated, which undergo FRET. (**b**) Cartoon representation of the crystal structure of a junction-less, all-RNA hairpin ribozyme (PDB ID: 2OUE) (Salter et al. 2006), color-coded as in panel A; a gray line was added to indicate the connectivity of the two domains. (**c**) Reaction pathway of the hairpin ribozyme as revealed by ensemble and single molecule FRET. Fluorophore distances measured by time-resolved FRET are indicated. (**d**) Representative single molecule FRET time trace of wild-type (WT) and dA38 mutant ribozymes after binding of non-cleavable (2′-OMe-A$_{-1}$) substrate analog under standard conditions (50 mM Tris-HCl, pH 7.5, 12 mM Mg^{2+}, 25°C). One undocking and one docking event are indicated. (**e**) Experimental (symbols) and simulated (lines) cleavage time courses of the WT and dA38 mutant ribozymes under standard conditions. The predicted contributions from four and three experimentally detected molecule sub-populations, respectively, are indicated by gray and dashed lines. In part modified with permission from (Rueda et al. 2004)

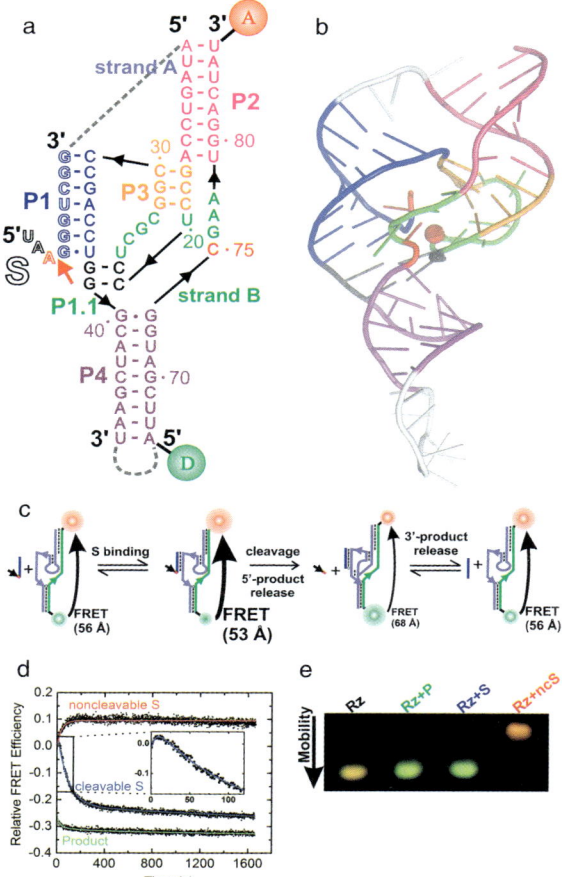

Fig. 5.4 FRET studies of the HDV ribozyme. (**a**) Secondary structure of the HDV ribozyme D1 utilized in these studies. Helices are color coded, the catalytic core is highlighted in green, and the cleavage site in the substrate (outlined, S) is marked by a closed red arrow. Strand B of the ribozyme is labeled with donor (D) and acceptor (A) fluorophores, as indicated, which undergo FRET. Dashed gray lines indicate connections that are present in the naturally occurring, cis-acting, genomic and antigenomic HDV ribozymes. (**b**) Cartoon representation of the crystal structure of the genomic HDV ribozyme (PDB ID: 1SJ3) (Ke et al. 2004), color-coded as in panel A with connections removed for the trans-acting D1 construct shown in silver. Red sphere, presumably catalytically involved Mg^{2+} ion. (**c**) Reaction pathway of the HDV ribozyme as revealed by ensemble steady-state FRET. Fluorophore distances measured by time-resolved FRET are indicated. (**d**) Relative FRET efficiency (calculated as ratio of the acceptor:donor fluorescence signals) over time of the doubly labeled D1 construct upon addition of a fivefold excess of non-cleavable substrate analogue (ncS3), cleavable substrate (S3; which binds, leading to an increase [inset], followed by a decrease upon cleavage), or 3′ product (3′P), as indicated, under standard conditions (40 mM Tris-HCl, pH 7.5, 11 mM Mg^{2+}, 25 mM DTT, 25°C). (**e**) Non-denaturing gel electrophoresis of the doubly labeled ribozyme alone (Rz) and in complex with 3′-product (Rz + P), cleavable substrate (Rz + S; this complex undergoes catalysis and is indistinguishable from the 3′-product complex), and non-cleavable substrate (Rz + ncS), as indicated. Detection is by FRET in the gel. In part modified with permission from (Pereira et al. 2002)

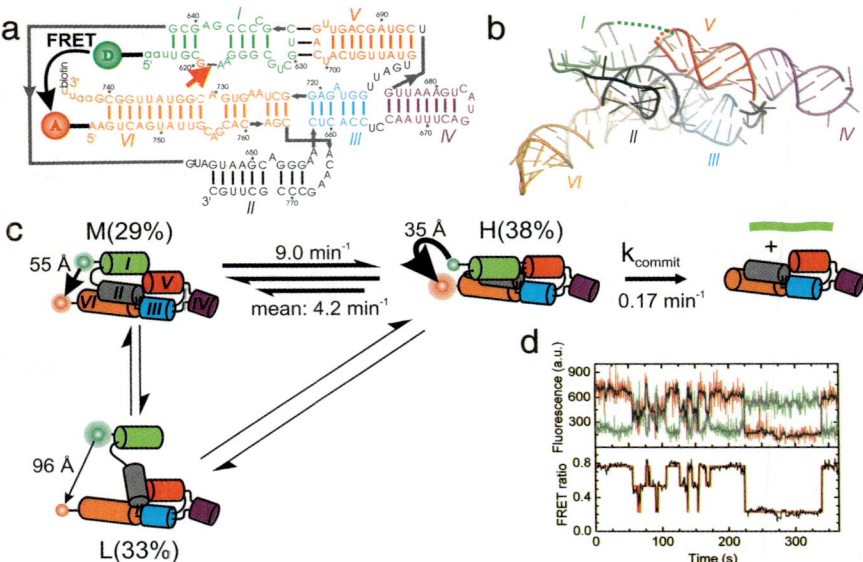

Fig. 5.5 FRET studies of the VS ribozyme. (**a**) Secondary structure of the VS ribozyme G11 uti-lized in these studies. Helices are color coded, the cleavage site is marked with a closed red arrow, and the FRET donor (D) and acceptor (A) and biotin labeling sites are indicated. (**b**) Tertiary struc-ture model of the VS ribozyme, using the same color code as in panel A. Some strand connectivi-ties are incomplete in the model; the ones in the I–V kissing loop are indicated by dashed lines. (**c**) Structural and kinetic model of the reaction pathway of the WT VS ribozyme as derived from single molecule FRET. The helix colors match those in panel A. (**d**) Single molecule FRET time trace of a VS ribozyme (Hiley and Collins 2001). The raw Cy3 donor and Cy5 acceptor fluores-cence signals are green and red, respectively (upper panel). Superimposed in gray are the data after applying a non-linear filter (Haran 2004). The FRET ratio (lower panel, black trace) is calculated from the filtered data as acceptor/(donor + acceptor) and reveals three distinct states by Hidden Markov modeling (red line). In part modified with permission from (Pereira et al. 2008)

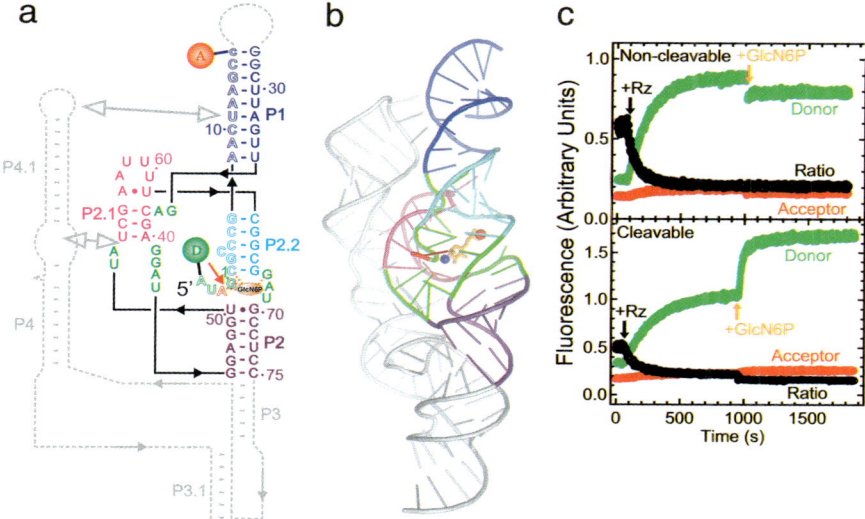

Fig. 5.6 FRET studies of the *glmS* ribozyme. (**a**) Secondary structure of the trans-acting *glmS* ribozyme utilized in these studies. Helices are color coded, the catalytic core is highlighted in green, the cleavage site in the substrate (outlined, S) is marked by a closed red arrow, and the cofactor GlcN6P is shown as an orange oval interacting with several core residues (dashed orange lines). The substrate strand is labeled with donor (D) and acceptor (A) fluorophores, as indicated, which undergo FRET. Dashed gray lines indicate connections and a downstream pseudoknot that are present in the naturally occurring, cis-acting ribozyme. (**b**) Cartoon representation of the crystal structure of the *Thermoanaerobacter tengcongensis glmS* ribozyme (PDB ID: 2H0Z) (Klein and Ferre-D'Amare 2006), color-coded as in panel A with removed parts shown in silver. Red sphere, ligand-chelating Mg^{2+} ion; blue sphere, amino group of the ligand presumably involved in catalysis. (**c**) Changes over time in donor, acceptor, and acceptor:donor ratio signals upon addition of a tenfold excess of *glmS* ribozyme (Rz) and subsequently saturating (10 mM) GlcN6P ligand to either the non-cleavable or cleavable substrate under standard conditions (50 mM HEPES-KOH, pH 7.5, 200 mM K^+, 10 mM Mg^{2+}, 25 mM DTT, 25°C), as indicated. In part modified with permission from (Tinsley et al. 2007)

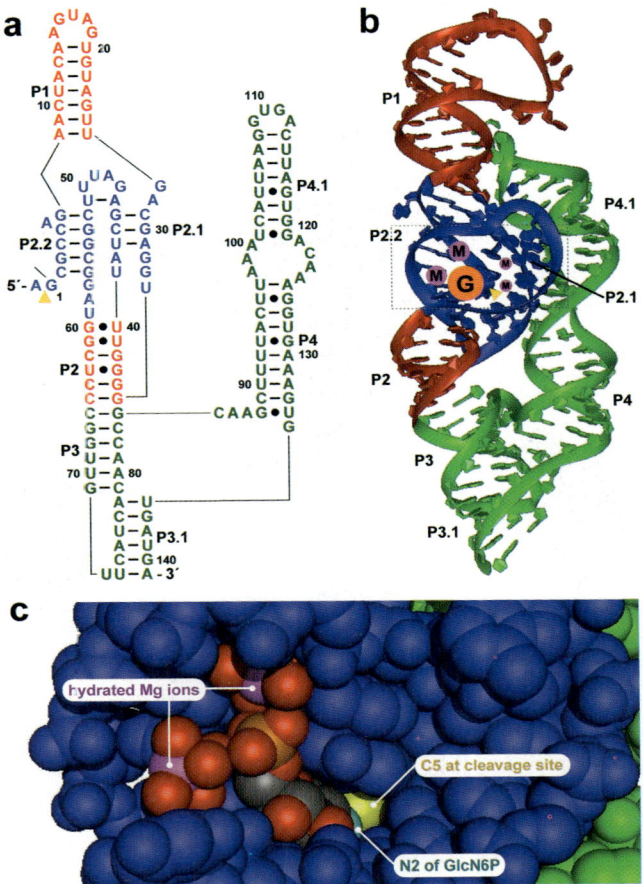

Fig. 6.1 Overview of *glmS* ribozyme structure. (**a**) Secondary structure. Depicted is the secondary structure of the *B. cereus* ribozyme based upon phylogenetic comparison (Winkler et al. 2004) and crystal structures of the *T. tengcongensis* (Klein and Ferré-D'Amaré 2006) and *B. anthracis* ribozymes (Cochrane et al. 2007). The highly conserved catalytic core (blue) comprises nucleotides adjacent to and including paired regions P2.1 and P2.2; the minimal portion that supports GlcN6P-dependent self-cleavage additionally includes P1 and P2 (red). The remaining portion including P3–P4 (green) enhances ribozyme activity, but is not required. Nucleotides are numbered relative to the site of cleavage indicated by the yellow arrowhead. (**b**) Tertiary structure. The ribbon model depicts the structure of the *B. anthracis* ribozyme determined by X-ray crystallography (Cochrane et al. 2007) in complex with U1A protein (not shown) bound to a site inserted into the loop of P1. Importantly, nucleotide identities and numbering within the catalytic core (blue) are identical to that of *B. cereus* ribozyme. The scissile phosphodiester bond is denoted by the yellow arrowhead, and the approximate site of GlcN6P binding (orange encircled G) and four sites of divalent metal ion binding (each magenta encircled M) are indicated. (**c**) GlcN6P binding site. The space-filling model shows the approximate region indicated in Fig. 6.1b (dashed box) with two hydrated metal ions and GlcN6P bound adjacent to the scissile phosphate. Ligands are colored by atom (Mg, magenta; O, red; P, orange; C, gray; and N, cyan), whereas C5′ forming the scissile phosphate ester is colored yellow

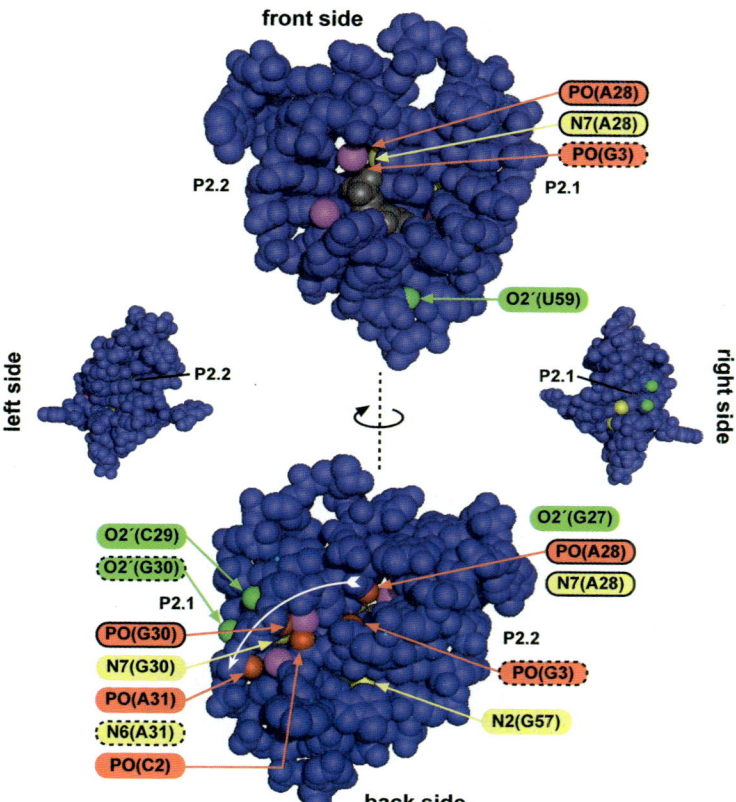

Fig. 6.4 Functional group requirements for metal ion binding and GlcN6P recognition determined by NAIM and NAIS. Depicted is the catalytic core only of the *glmS* ribozyme (blue) shown from each side relative to the "front" or GlcN6P (gray)-bound side (Cochrane et al. 2007). Functional groups within the catalytic core for which chemical mutagenesis results in interference (Jansen et al. 2006) are colored indicating requisite R_p phosphate oxygens (red), 2′ hydroxyls (green), or nucleobase functional groups (yellow). Each group is labeled in detail to provide nucleotide identity and number. Sites that cannot be observed in the space-filling model lack arrows, and sites at which interference suppression are observed with GlcN (Jansen et al. 2006) are outlined with solid or dashed lines to illustrate full or partial suppression, respectively. Magnesium ions (magenta) are shown as unhydrated for simplicity, and the white arrow denotes the backbone from A28 to A31 along which interference suppressions with GlcN are largely observed

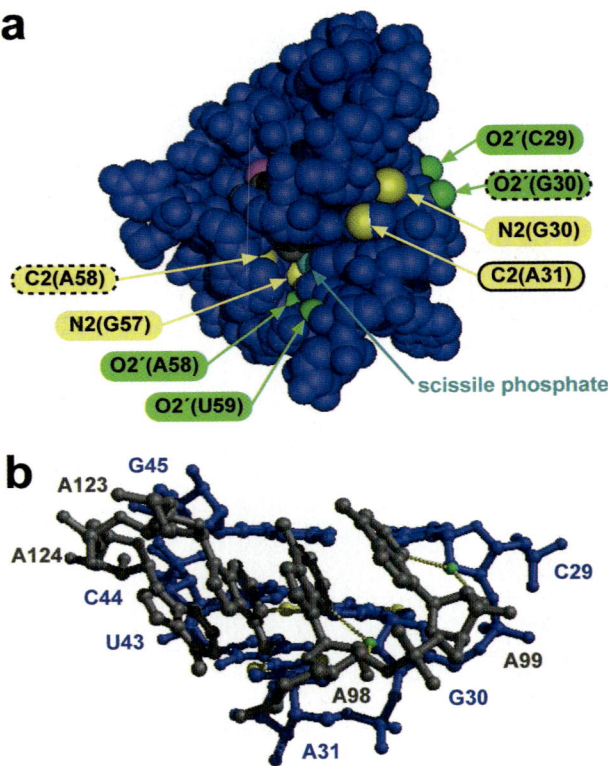

Fig. 6.5 Functional group requirements for P4 domain interaction and active site catalysis determined by NAIM and NAIS. (**a**) Functional group interferences and suppressions within the active site (*left*) and P2.1 (*right*). Depicted is the catalytic core only of the *glmS* ribozyme as described in the legend to Fig. 6.4 and shown from a lower right side angle. The scissile phosphate is colored cyan. (**b**) Interaction of P2.1 with P4 domain adenosines. Shown are four adenosines from the internal loop of the P4 domain (gray) that lie in the minor groove of P2.1 (blue). Nucleotide identities and numbering are depicted. Interferences support hydrogen-bonding interactions (dotted yellow lines) between A123 and G30 and formation of a ribose zipper between C29–G30 and A98–A99

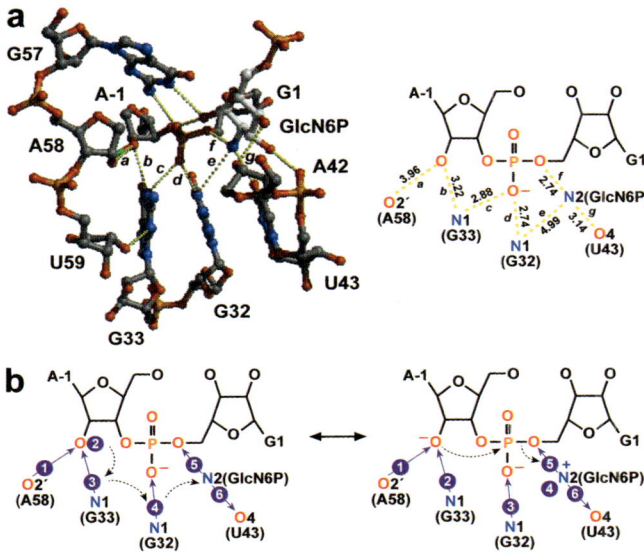

Fig. 6.7 *glmS* ribozyme active site and proposed comprehensive mechanism of action. (**a**) Active site composition. Depicted is a partial structure (*left*) of the catalytic core surrounding the scissile phosphate and including GlcN6P (Cochrane et al. 2007). Nucleotide identities and positions are denoted, where nucleobases for A-1, G1, A42, A58, and U59 are omitted for clarity. Atoms are colored by type with carbon (gray in RNA and white in GlcN6P), oxygen (red), nitrogen (blue), and phosphorus (orange). The methyl group inactivating the 2′ oxygen nucleophile at A-1 is colored green. The diagram at right shows active site functional groups with interatomic distances (dotted yellow lines) given in angstroms. (**b**) Proposed mechanism of action through coordinated proton transfer. The schematic diagram depicts the transfer of active site protons (numbered purple circles) and their hydrogen bonding interactions (purple arrows) between active site functional groups before and after general base catalysis initiated by GlcN6P. In this manner, G33 may be activated to serve as the ultimate general base for deprotonation of the 2′ hydroxyl at A-1 while GlcN6P is concomitantly activated to serve as the general acid catalyst for protonation of the 5′ oxygen at G1

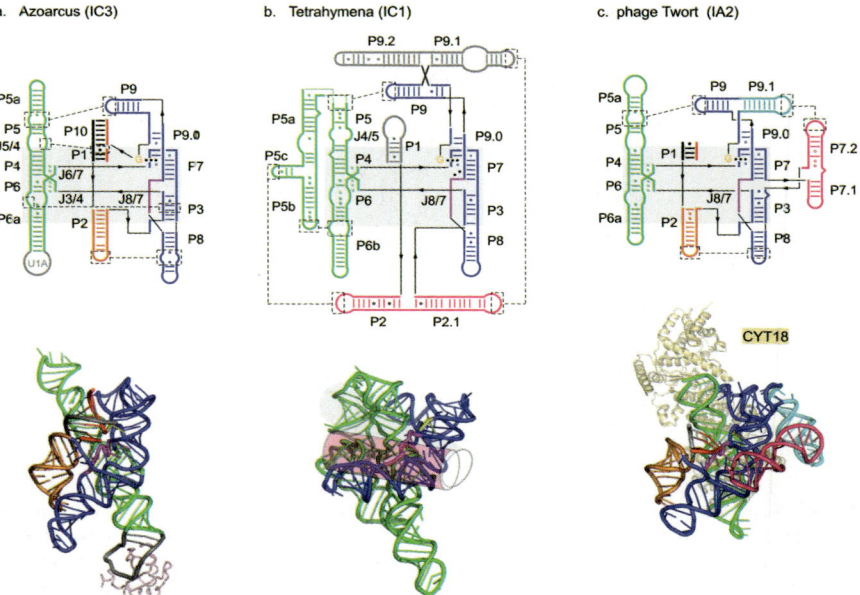

Fig. 7.2 Structures of group I ribozymes. 2D schematics and 3D ribbons are colored by domain as indicated. Red5′ and 3′ exons. (**a**) *Azoarcus* pre-tRNA^{ile} ribozyme (*1u6b*; Adams et al, 2004). The gray ribbon is U1A protein, which was used to aid crystallization. (**b**) *Tetrahymena thermophila* LSU rRNA (*1x8w*; Guo et al. 2004). Pink and grey cylinders indicate the predicted position of the P2/P2.1 and P9.1/P9.2 helices (Lehnert et al. 1996), which were not present in the crystal structure. (**c**) Phage Twort orf142 ribozyme complexed with the C-terminal fragment of *N. crassa* CYT18 (*2rkj*; Paukstelis et al. 2008). Figure adapted from Woodson (2005a)

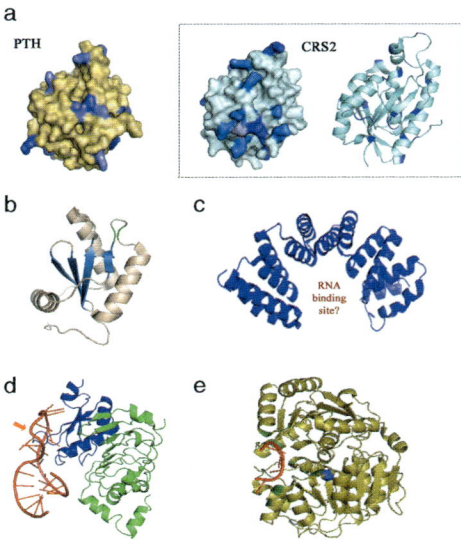

Fig. 8.2 Structures of selected proteins associated with RNA processing. (**a**) Electrostatic surface representations of peptidyl-tRNA hydrolases (PTH) and CRS2 (also in ribbon form, right). Basic residues (blue) are sites for tRNA and prospective group II intron binding, respectively (Ostersetzer et al. 2005; Schmitt et al. 1997). (**b**) Crystal structure of an ancient RNA domain (YhbY) homologous with CRS1. The β sheet face (blue) contains conserved basic residues, and the GxxG motif (green), are both implicated with nucleic acid recognition (Ostheimer et al. 2002). (**c**) Model of PPR motifs adapted from the crystal structure of a closely related TPR protein (Kim et al. 2006; Tavares-Carreon et al. 2008). (**d**) Spliceosomal RRM protein U2B' (blue) interacting with the hairpin region of U2 snRNA (red); shown with arrow. This association occurs only when U2B' is interacting with the U2A', a leucine-rich protein (green) (Maris et al. 2005; Price et al. 1998). (**e**) Crystal structure of the DEAD-box protein, Vasa, showing the DEAD-box site (blue) and RNA binding sites (green) in the helicase domain (Sengoku et al. 2006). In (**d**) and (**e**), RNA strands are shown in red. Figures were adapted from the references indicated and generated using PyMol from protein data bank accession numbers: 2PTH, 1RYB, 1LN4, 2FI7 and IA9N, respectively

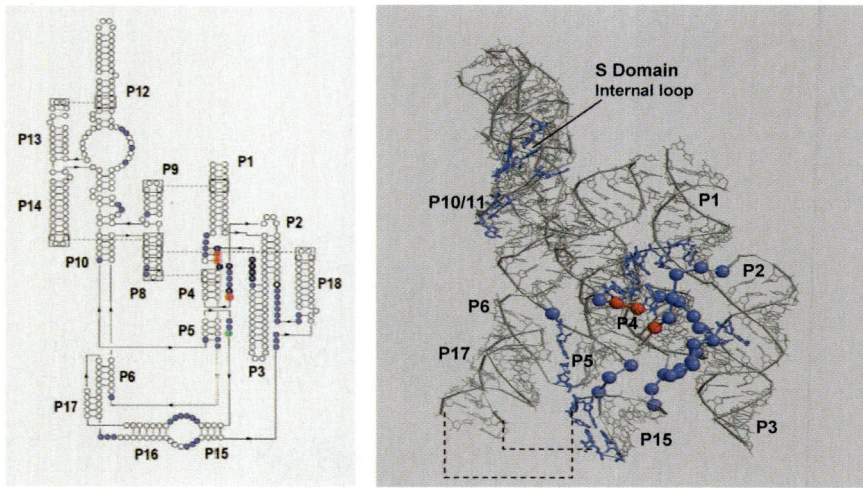

Fig. 9.4 Sites of metal ion dependent cleavage and phosphorothioate interference indicate residues involved in Mg²⁺ interactions. The secondary structure of *E. coli* P RNA is shown on the left and the three-dimensional structure of *T. thermophilus*, both are Type A P RNAs. The sites of Tb³⁺ cleavage of *E. coli* P RNA are shown on the secondary structure diagram as blue circles. Blue nucleotides in the three-dimensional structure are the homologous residues in the *T. thermophilus* structure. Similarly, the sites of strong phosphorothioate interference observed in both *E. coli* and *B. subtillis* P RNAs are indicated by red spheres

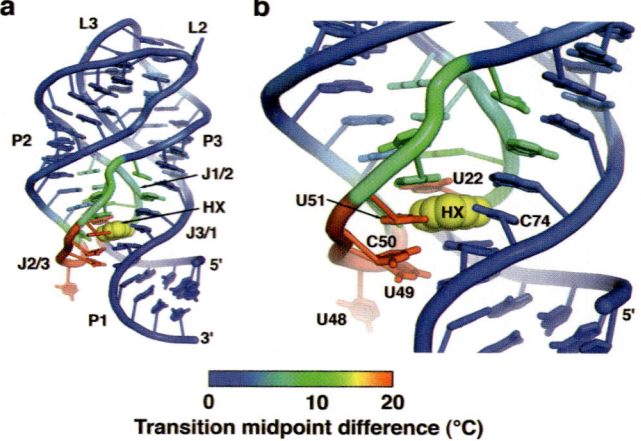

Fig. 10.2 NMIA probing of the purine riboswitch (adapted from (Stoddard et al. 2008)). (**a**) Global structure of the *xpt-pbuX* aptamer domain RNA colored to reflect the difference in the melting temperature (ΔT_m) for each nucleotide between the hypoxanthine bound and unbound states. Blue colors reflect nucleotide positions whose T_m is unaffected by ligand binding whereas red reflects significant changes in the T_m between the two states (see color bar). (**b**) Close-up view of the three-way junction emphasizing that most of the nucleotides affected by the presence of the ligand are centered about the turn in J2/3 (U48–U51, red) whereas J3/1 appears to be static (blue)

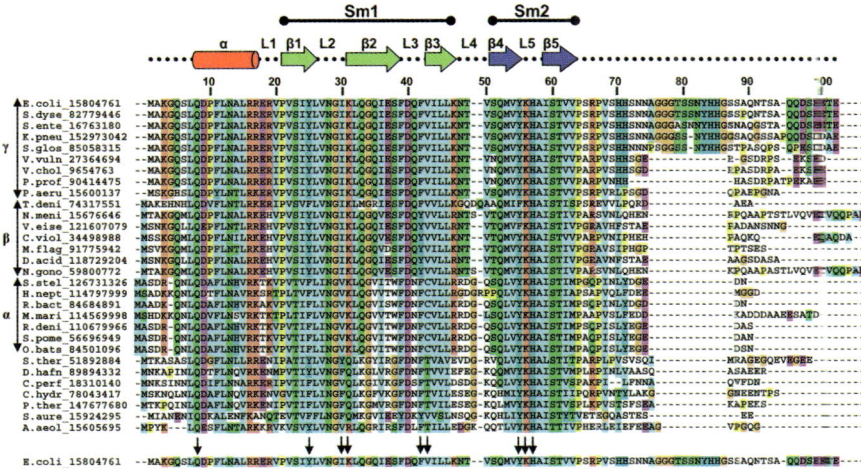

Fig. 12.1 Multiple sequence alignment of Hfq proteins from 30 bacteria. Organisms corresponding to sequences are listed with their accession numbers. Letters α, β, and γ on the left of the organism list indicate α-, β-, and γ-proteobacteria, respectively. Other organisms belong to firmicutes and aquificae phyla. The numbering at the top corresponds to *E. coli* sequence. Secondary structure of Hfq is schematized based on the *E. coli* Hfq crystal structure. Sm1 and Sm2 motifs are colored in green and blue, respectively. *E. coli* Hfq sequence is presented once more at the bottom to indicate those residues that are specifically mentioned in the text (*arrows*). Species (from top to bottom): γ-proteobacteria: *Escherichia coli, Shigella dysenteriae, Salmonella enterica, Klebsiella pneumoniae, Sodalis glossinidius, Vibrio vulnificus, Vibrio cholerae, Photobacterium profundum, Pseudomonas aeruginosa.* β-proteobacteria: *Thiobacillus denitrificans, Neisseria meningitides, Verminephrobacter eiseniae, Chromobacterium violaceum, Methylobacillus flagellatus, Delftia acidovorans, Neisseria gonorrhoeae.* α-proteobacteria: *Sagittula stellata, Hyphomonas neptunium, Rhodobacterales bacterium, Maricaulis maris, Roseobacter denitrificans, Silicibacter pomeroyi, Oceanicola batsensis.* Firmicutes: *Symbiobacterium thermophilum, Desulfitobacterium hafniense, Clostridium perfringens, Carboxydothermus hydrogenoformans, Pelotomaculum thermopropionicum, Staphylococcus aureus.* Aquificae: *Aquifex aeolicus*

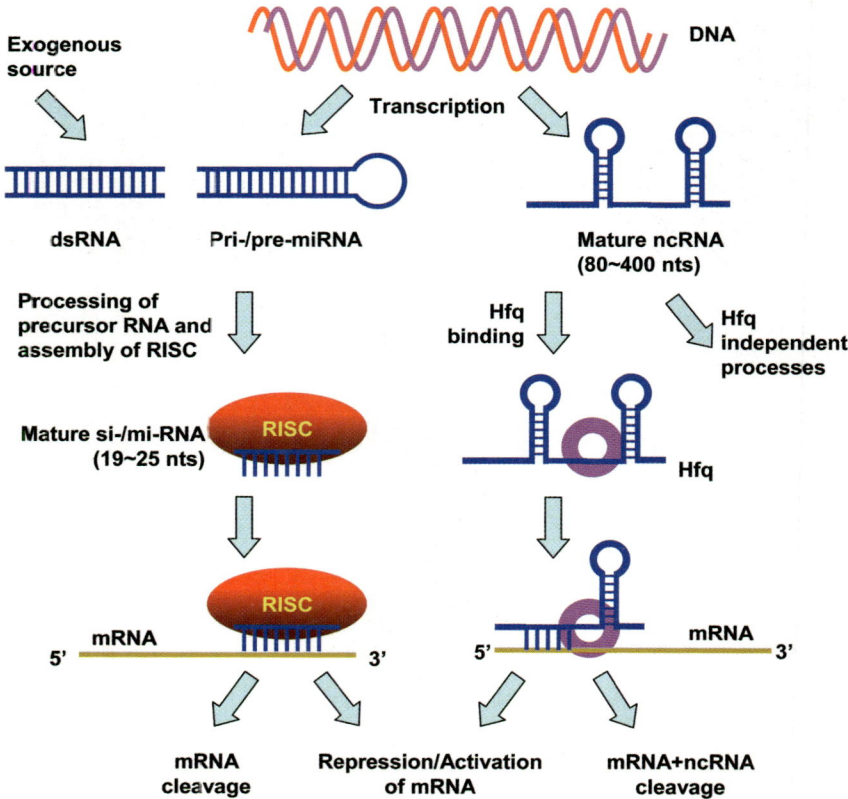

Fig. 12.2 Comparison of RNA interference to bacterial post-transcriptional gene regulation. Biogenesis of mi-/siRNA in eukaryotes is programmed genetically (miRNA) or can be triggered by double-stranded RNA from an exogenous source (siRNA). The initial precursors have to be cleaved to 19–25 nt double strand RNAs by Dicer, a multidomain enzyme of the RNase III family, and they are assembled into an RNA induced silencing complex (RISC) mediated by RISC loading complex (RLC). When RISC is loaded with RNA, only one strand of RNA is chosen. In contrast, ncRNAs are transcribed into mature single stranded form from DNA in response to stresses. The length of ncRNAs varies from 80 to 400 nts. Some of ncRNAs bind to Hfq to regulate the gene expression through base-pairing with their target mRNAs. ncRNAs usually anneal to the 5′ UTR in the proximity of ribosomal binding site (RBS) to repress/activate the expression of mRNA or to lead to the degradation of mRNA. mi-/siRISC, however, bind to the 3′ UTR to induce cleavage or to repress mRNA translation

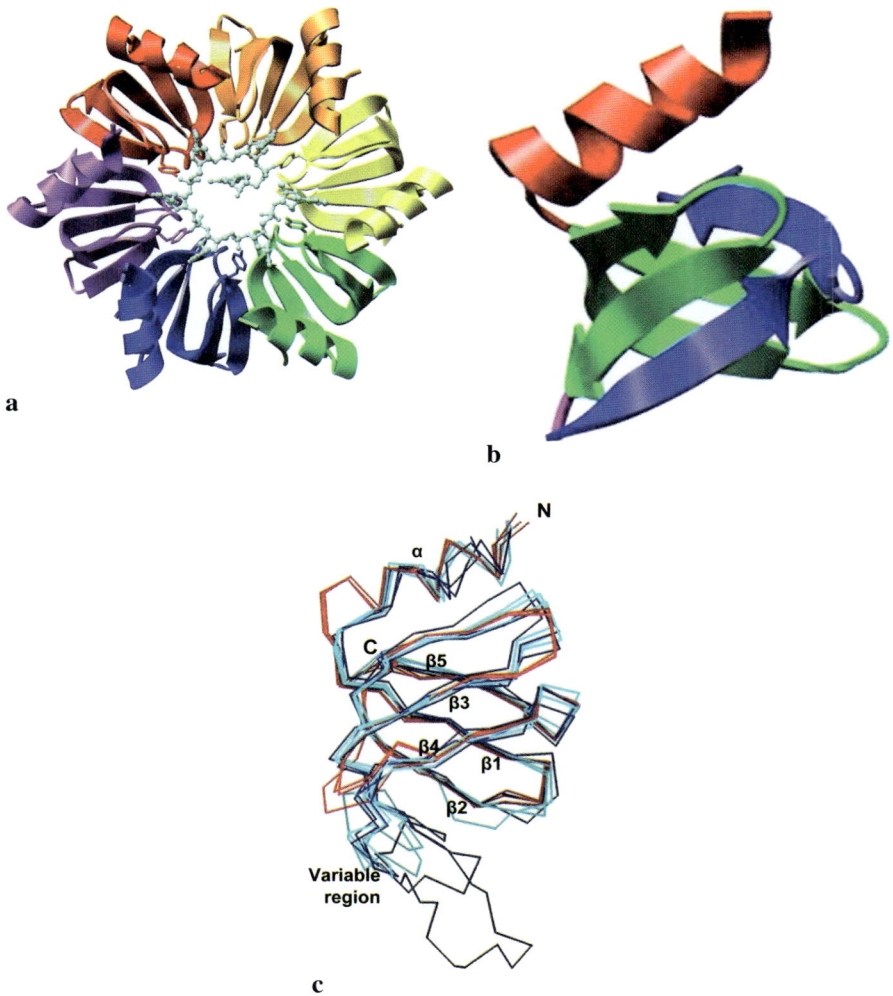

Fig. 12.5 Hfq structures. **a** Hexameric form of Hfq from *S. aureus* is displayed with AU₅G bound in the central cavity. For clarity each subunit is colored differently. Y55 residues are shown from each subunit to illustrate the stacking interactions with RNA bases. **b** Isolated monomeric unit form the crystal structure of *E. coli* Hfq shows the Sm fold containing an N-terminal α-helix followed by five β-strands forming a barrel-shaped structure. Sm1 and Sm2 motifs are colored in green and blue, respectively. Loop 4, which lies between the Sm1 and Sm2 motifs has variable length in Sm/Lsm proteins, is colored in magenta. **c** Superimposition of the monomeric units from bacterial Hfq, archaeal Lsm and eukaryotic Sm proteins clearly shows the structural similarity especially within Sm1 and Sm2 motifs of these proteins. Hfq monomers from *E. coli, Staphylococcus aureus,* and *Pseudomonas aeruginosa* are colored red. Lsm proteins from *Archaeoglobus fulgidus* (AF-Sm2), *Methanobacterium thermoautotrophicum,* and *Pyrobaculum aerophilum* are cyan, and Sm proteins from human (SmB and SmD3) and yeast (SmF) are navy blue

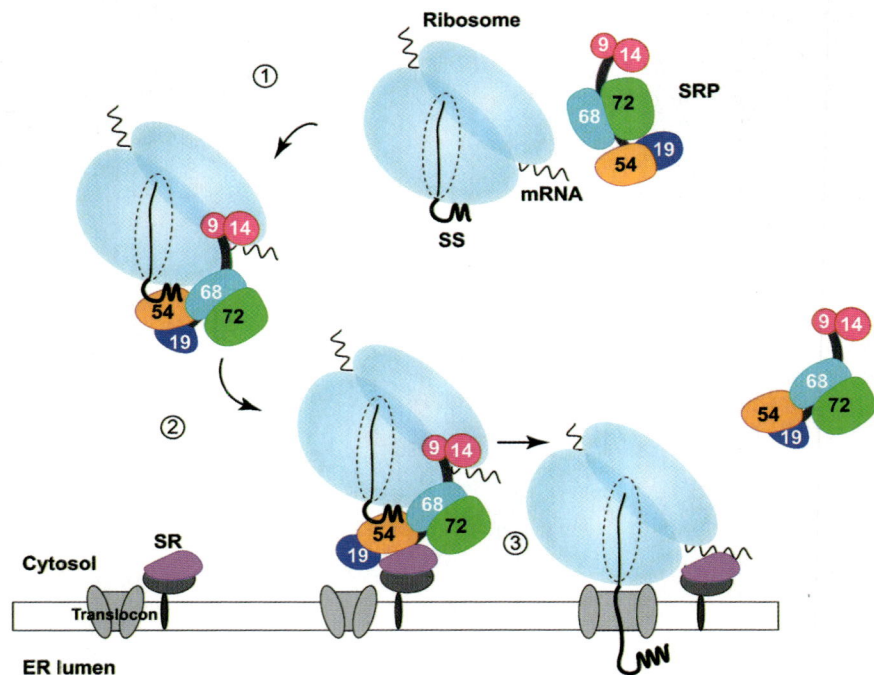

Fig. 13.1 Functional cycle of mammalian SRP. (**1**) Mammalian SRP is composed of a single RNA molecule, the 7SL RNA (brown), and six proteins (SRP72, SRP68, SRP54, SRP19, SRP14 and SRP9, shown as coloured ovals). SRP binds to the signal sequence (SS) of nascent secretory and membrane proteins as it emerges from the ribosome and halts the elongation. (**2**) The ribosome-nascent-chain (RNC)-SRP complex is targeted to the ER membrane where SRP interacts with the SRP receptor (SR, pink and dark brown) in a GTP dependent manner. (**3**) The ribosome is transferred to the translocon (light brown), SRP is released and protein synthesis resumes (figure adapted from Schwartz and Blobel 2003)

Fig. 13.2 Schematic representation of mammalian SRP. Thedomain comprises the first 100 nucleotides of 7SL RNA and approximately 50 nucleotides of the 3′-end, in addition to the SRP9/14 heterodimer. The S domain is composed of the remaining four proteins and the central region of the 7SL RNA (nucleotides G101–C251, thick line), which form helices 5 (nucleotides 101–128 and 222–251), 6, 7 and 8. Symmetric loop (sl) and asymmetric loop (al) in helix 8 are indicated

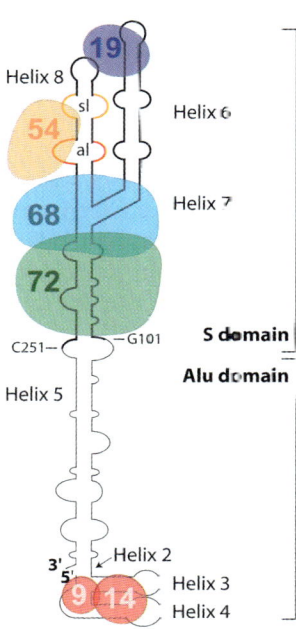

Fig. 13.3 Structural changes of 7SL RNA in two assembly intermediates revealed by crystallography (**a**) Binary complex with SRP19; (**b**) Ternary complex with SRP19 and the M domain of SRP54 (Kuglstatter et al 2002; Oubridge et al. 2002)

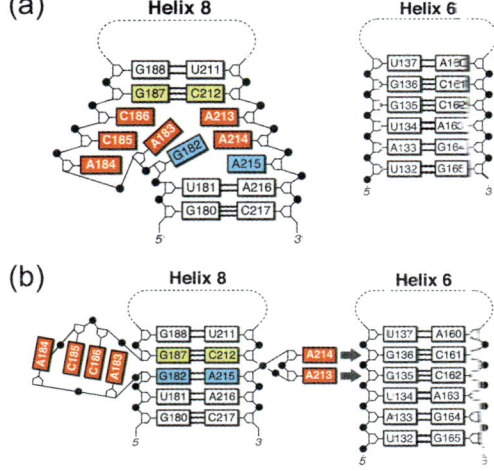

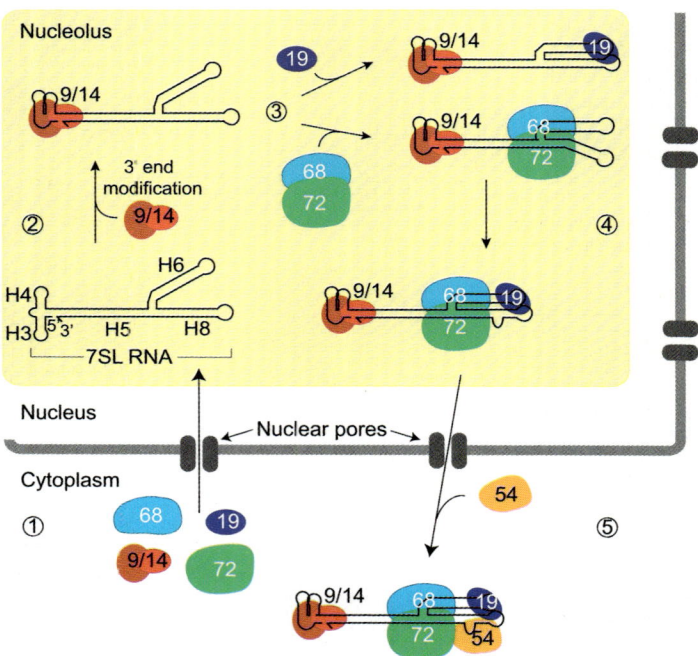

Fig. 13.4 Proposed mechanism for SRP biogenesis and assembly. (**1**) All SRP proteins, except SRP54, are imported into the nucleolus, where assembly of SRP initiates. (**2**) The binding of SRP9/14 to the 5′-end of 7SL RNA is thought to occur co-transcriptionally. (**3**) SRP68/72 binds near the three-way junction, SRP19 binds to the tips of helices 6 and 8. (**4**) Simultaneous binding of SRP68/72 and SRP19 pre-organizes the binding site of SRP54. (**5**) SRP becomes active after the pre-SRP is exported to the cytoplasm and bound by SRP54

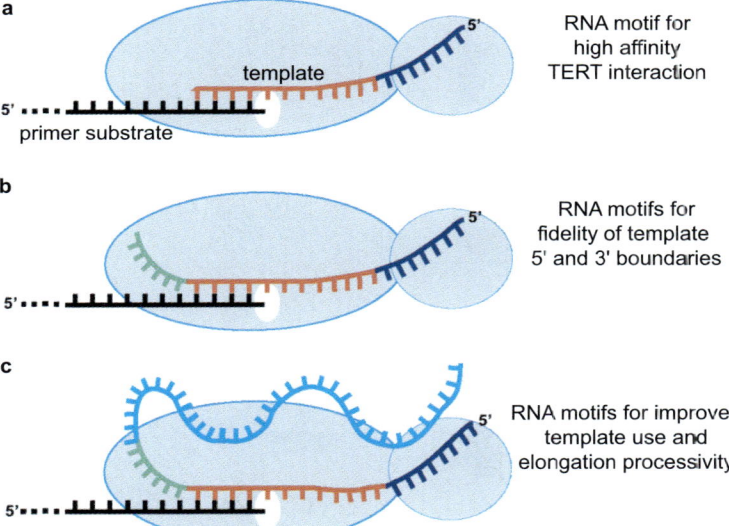

a — RNA motif for high affinity TERT interaction
template
5′ primer substrate
5′

b — RNA motifs for fidelity of template 5′ and 3′ boundaries
5′
5′

c — RNA motifs for improved template use and elongation processivity
5′
5′

Fig. 14.4 A speculative evolutionary route for gains of TER non-template motif functions. See text for explanation

a

b

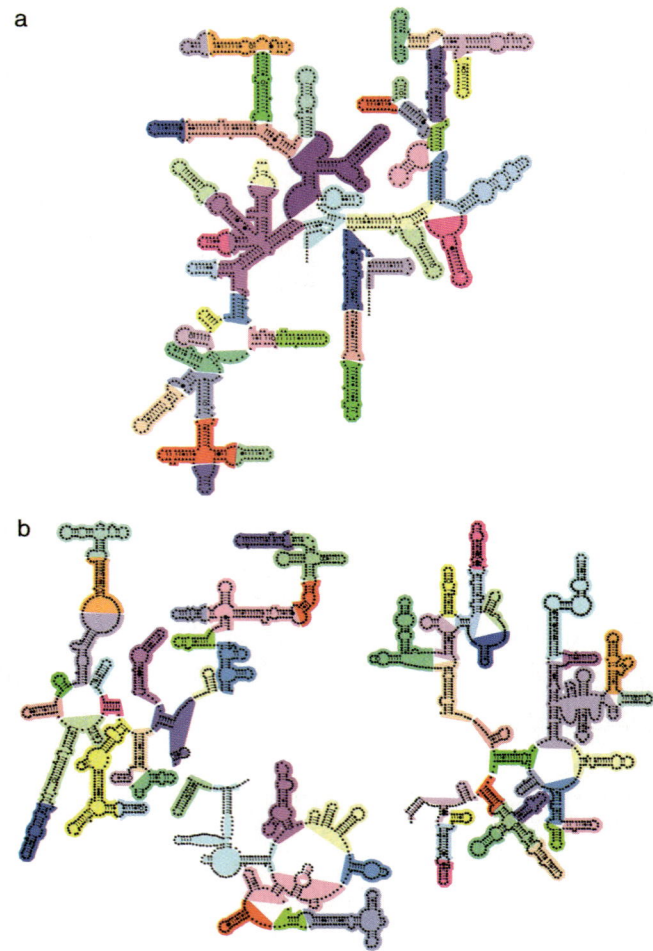

Fig. 15.2 Rigid-body assignment of *E. coli* 70S ribosomal RNAs used in the real-space refinement. (**a**) 16S rRNA (43 pieces). (**b**) 23S rRNA (62 pieces)

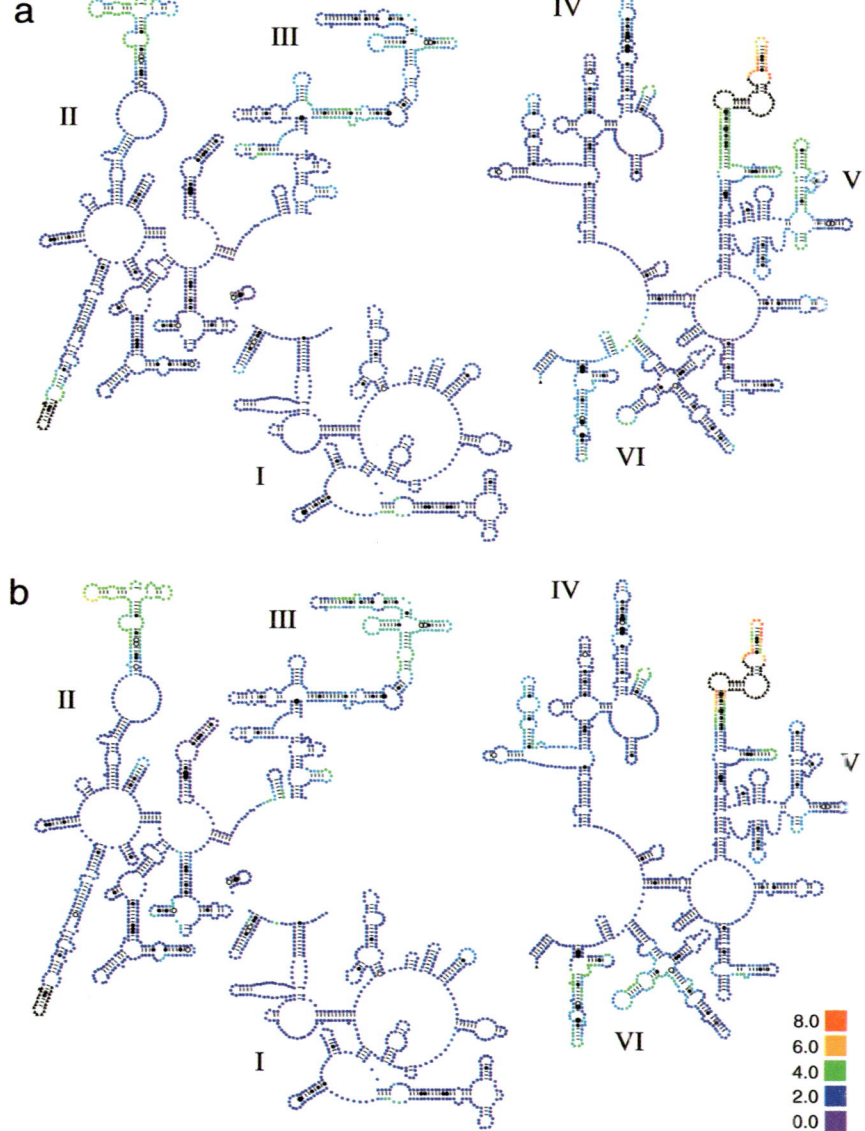

Fig. 15.5 Extent of factor-related ratchet motions depicted on the secondary structure of 23S rRNA. (**a** and **c**) EF-G-related ratchet motion with color bar on an expanded scale. (**b** and **d**) RF₃-related ratchet motion with color bar on an expanded scale

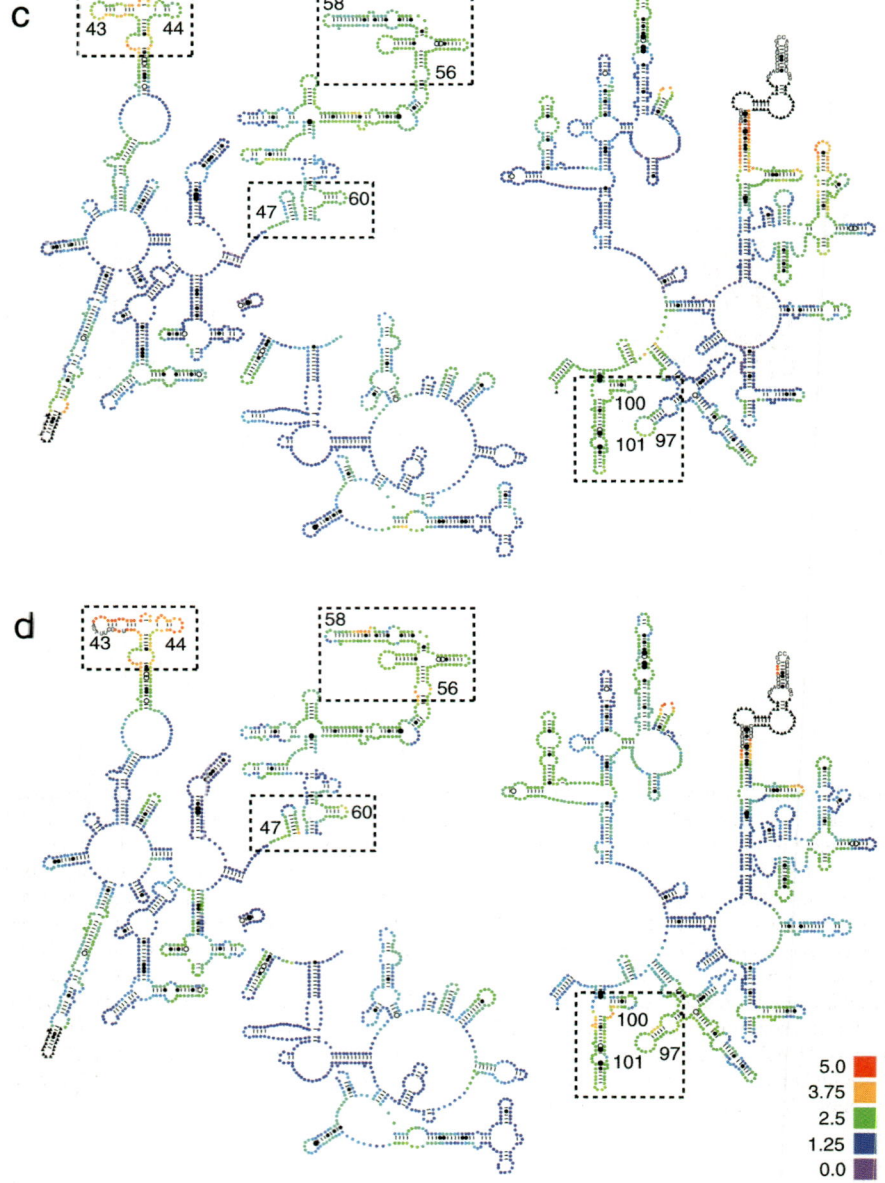

Fig. 15.5 (Continued)

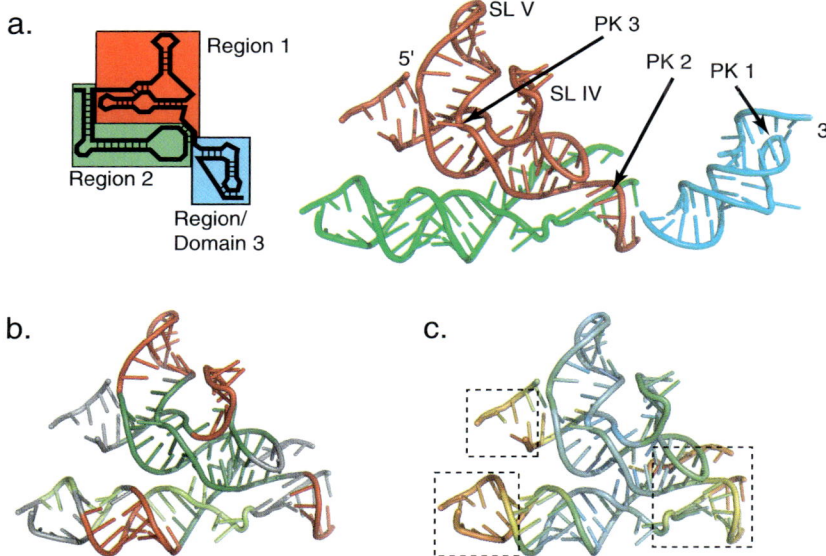

Fig. 16.6 Structural features of the IGR IRES from crystal structures. (**a**) Diagram of secondary structure of an IGR IRES RNA is shown at left, with the three regions colored. Region 1 and 2 form the ribosome binding domain. At right are ribbon representations of the two crystal structures of the two domains, colored to match the diagram. Important structural features mentioned in text are noted. (**b**) Hydroxyl radical probing data overlaid on a ribbons representation of ribosome binding domain of the PSIV IGR IRES. Red indicates enhanced cleavage upon folding and solvent accessibility. Green indicates reduced cleavage upon folding and protection from solvent. Lighter green areas were protected, but to a lesser extent. Pattern agrees well with the crystal structure, linking the folded free form of the RNA to the crystal structure. (**c**) Ribbon representation of the ribosome binding domain of the PSIV IGR IRES, colored to reflect crystallographic B-factors. "Cool" colors (blue/green) denote low B-factors and "warm" colors (red/orange/yellow) have higher B-factors. These B-factors reflect disorder in the crystal structure; higher B-factors are more disordered and these portions are boxed

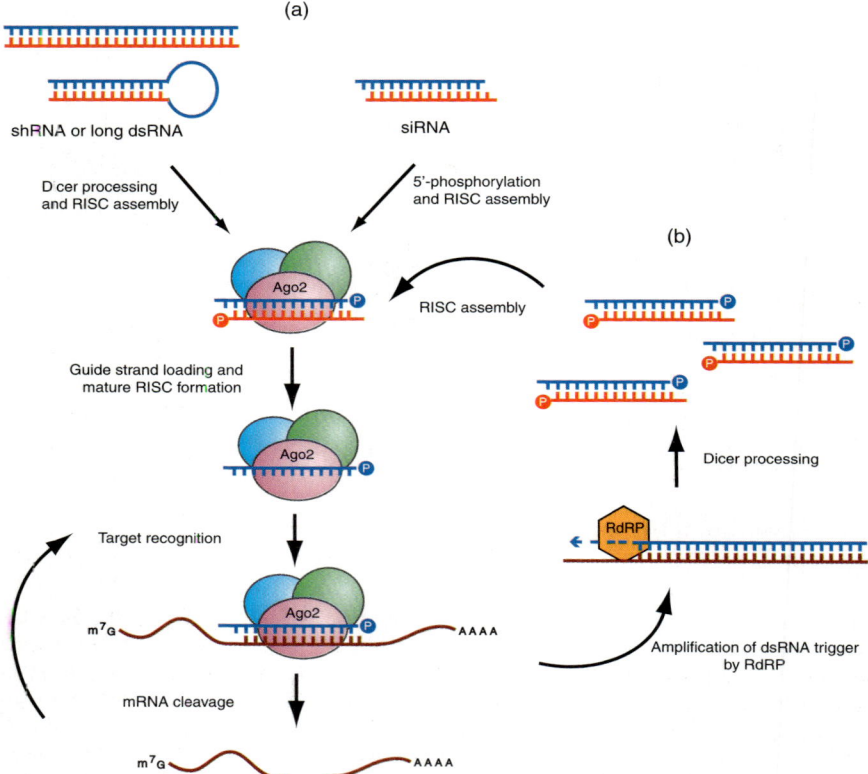

Fig. 17.1 Short-interfering RNA (siRNA)-mediated gene silencing. (**a**) In vertebrates, 21–22 nt siRNA duplexes are processed from long double-stranded RNAs (dsRNAs) or short hairpin RNA (shRNA) by the Dicer/TRBP complex and incorporated into RISC. Similar mechanism is used to introduce synthetic 21 nt siRNAs into cells and load them onto RISC. Active RISC containing guide strand of siRNA recognizes and cleaves complementary target mRNA. After target cleavage, RISC is recycled for multiple rounds of catalysis. (**b**) In plants and worms, the RNAi response is amplified by RNA-dependent RNA polymerases (RdRPs). RdRPs coupled with RISC generate dsRNA using target mRNA as the template. This long dsRNA is subsequently cleaved by Dicer to form secondary siRNAs that can target more mRNA and amplify the RNAi effect

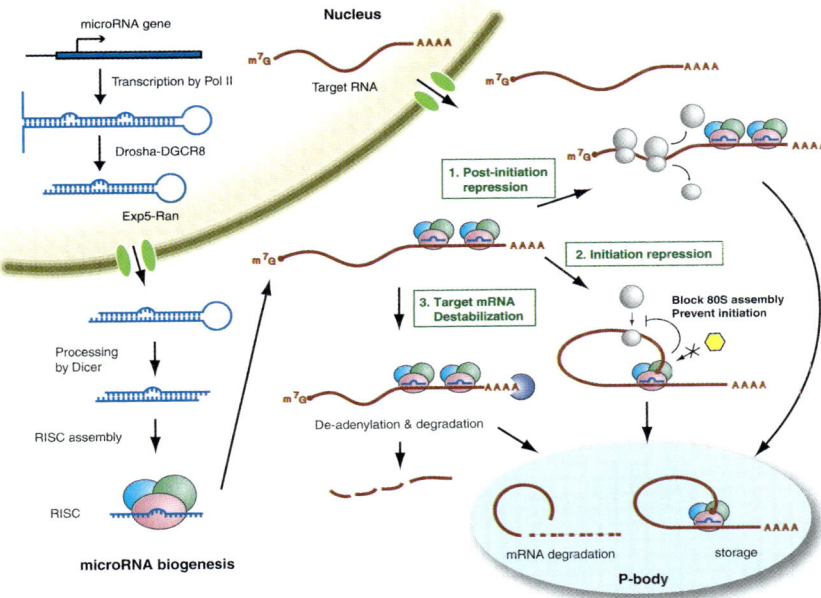

Fig. 17.2 microRNA-mediated gene silencing. miRNA genes are transcribed by RNA Pol II into primary miRNAs (pri-miRNAs), which are processed in the nucleus by Drosha/DGCRB into pre-miRNAs and exported into the cytoplasm by Exp5-Ran. The hairpin miRNA precursors (pre-miR-NAs) are subsequently cleaved by Dicer/TRBP (transactivation-response element RNA-binding protein) to ~22 nt miRNA/miRNA* duplexes; these are then loaded into RISC. Active RISC then recognizes the 3′ UTR of target mRNAs with imperfect base-pairing and induces translation repression through three possible mechanisms: post-initiation repression, initiation repression, or target mRNA destabilization. In post-initiation repression, RISC would inhibit protein translation after the initiation step and prevent protein synthesis. To repress initiation, RISC would bind to the cap structure of target mRNA and block the initiation of protein translation by preventing the translation initiation factor, eIF4E, from association with the mRNA 5′ cap. In this case, RISC would also block assembly of the 80S ribosome. To destabilize the target mRNA, its recognition by RISC would induce deadenylation and decay of the target mRNA. Destabilization and increased turnover of target mRNA would result in reduced protein synthesis. In all three mechanisms, RISC-targeted mRNAs would be sequestered in P-bodies for storage or decay, although mRNA might also be degraded outside P-bodies

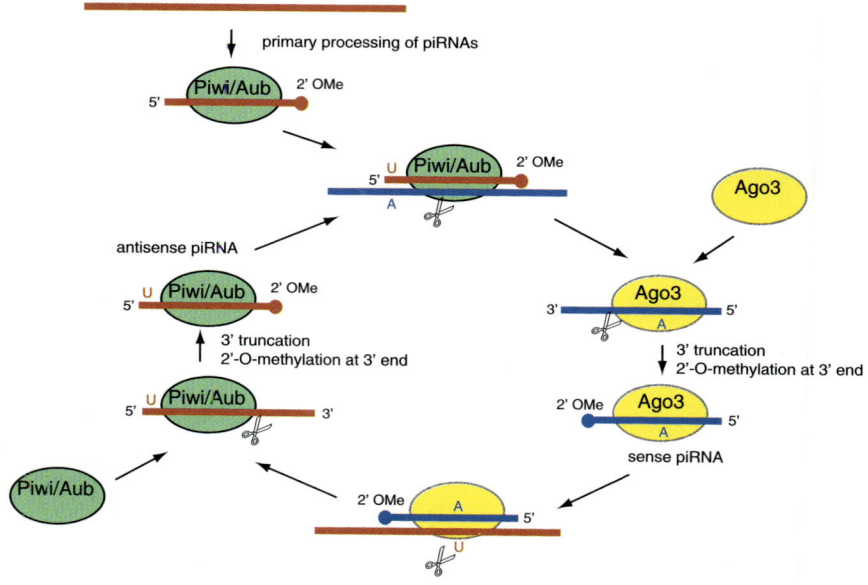

Fig. 17.3 Model for piRNA biogenesis. Primary piRNA transcripts, such as transposon regulatory regions of heterochromatin, are incorporated into Piwi and Aub and trigger the piRNA amplification loop. Piwi and Aub bind anti-sense strand piRNA and cleave target transcript to generate the 5′ end of sense strand piRNA; this is subsequently associated with Ago3. After 3′ end cleavage and 3′-end 2′-*O*-methylation, the Ago3-bound sense strand complex with piRNA functions as another slicer, which recognizes and cleaves the anti-sense strand precursor, to generate more Piwi/Aub-associated anti-sense strand piRNAs

Printing: Krips bv, Meppel, The Netherlands
Binding: Stürtz, Würzburg, Germany